# Safety and Security of Cyber-Physical Systems

Frank J. Furrer

# Safety and Security of Cyber-Physical Systems

Engineering dependable Software using
Principle-based Development

 Springer Vieweg

Frank J. Furrer
Computer Science Faculty
Technical University of Dresden
Dresden, Germany

ISBN 978-3-658-37184-5      ISBN 978-3-658-37182-1   (eBook)
https://doi.org/10.1007/978-3-658-37182-1

Responsible Editor: Leonardo Milla

This Springer Vieweg imprint is published by the registered company Springer Fachmedien Wiesbaden GmbH, part of Springer Nature.
The registered company address is: Abraham-Lincoln-Str. 46, 65189 Wiesbaden, Germany

# Foreword

This book is about *cyber-physical systems*. A Google search for this term results in 405'000'000 hits in 0.64 seconds (25.01.2022). Together with the strongly linked use of the term and the concept with other related popular terms such as *Industry 4.0* and *(IoT)* Internet of Things, this fact clearly shows the importance of cyber-physical systems in the present and the upcoming world!

The NIST characterization: *"Cyber-Physical Systems (CPS) comprise interacting digital, analog, physical, and human components engineered for function through integrated physics and logic. These systems will provide the foundation of our critical infrastructure, form the basis of emerging and future smart services, and improve our quality of life in many areas. Cyber-physical systems will bring advances in personalized health care, emergency response, traffic flow management, and many more"* (https://www.nist. gov/el/cyber-physical-systems, [Last accessed: 16.06.2021]).

Suffices by itself to underline the importance of the field.

This characterization does not explicitly mention the correct operation and continuous availability of cyber-physical systems and is apparently taken for granted. However, we need and somehow expect such fundamental properties to become a "commodity" for the functioning of our modern CPS-based society.

Unfortunately, reaching this goal will require a lot of effort, research, and practice. In fact, all existing cyber-physical systems are and *will ever be* under the continuous influence of cyber-attacks, faults, and failures hitting their software or underlying hardware, possibly causing unavailability or improper behavior of their operating environment, human errors, or being impacted by malicious activities.

This book precisely addresses those challenges that are at the basis of proper functioning of cyber-physical systems and the infrastructures they compose: It deals with the focused and fundamental issue of *engineering principles* for safety and security.

The book is divided into two main parts, each providing significant contributions toward managing safety and security.

The first addresses all the relevant concepts and explains their relations and links to the reader. Here the book offers many definitions and provides the conceptual framework for cyber-physical systems. All the facets of CPSs and the possible safety and security implications are linked to the fundamental notion of *risk*. Finally, all the impairments to proper functioning at all possible locations are explored and exemplified so that the reader can have a comprehensive view of the complexity that needs to be managed appropriately.

While the first part provides and structures the essential knowledge on safety, security, and risk, the second part elaborates on proper responses regarding a few paramount questions that must be asked: 1) How are good safety and security defined? 2) How is good safety and security formalized? 3) How is good safety and security taught? 4) How is good safety and security enforced?

This book provides enlightening solutions by defining, formalizing, strictly applying, and enforcing safety and security principles. In fact, good safety and security principles provide reliable knowledge for the successful architecting of trustworthy cyber-physical systems.

The second part of the book introduces and then details the concept of *principle-based engineering* to explain and justify a number of safety and security principles. This approach to architecting safe and secure systems is complemented and enriched by many instructive examples and practical observations, which add value to the book and allow the message to arrive clearly and sound.

I also want to remark the high value of the book as an instrument for teaching and education. Given its style and the richness of the examples made, the book can be used both as a textbook for graduate students but also as a guideline for the practitioner and the engineer as a help for a correct application of the *safety and security principles* when architecting and implementing cyber-physical systems of any of the types and variations described in the book.

The reader will appreciate how the book grew out of the long-running activities of the author and shows the admirable maturity he has achieved, both in the vastity of the material included and the presentation style. All this makes the book a source of inspiration.

I wish the readers to enjoy reading this rewarding book as I had!

June 2021

Prof. Dr. Andrea Bondavalli
Head of the Resilient Computing Lab
Department of Mathematics and Informatics
University of Florence, Italy
I-50134 - Firenze, Italy

# Preface

Today, *Cyber-Physical Systems* (CPS) are at the core of our industrial society. We rely heavily on cyber-physical systems in all areas of work and life.

Cyber-physical systems combine a *cyber-part*, i.e., a software-controlled computer system, with a *physical part*, i.e., a real system. The *software* operates the physical system. Examples of cyber-physical systems are cars, airplanes, trains, communications and energy infrastructures, heart pacemakers, medical diagnosis equipment, water treatment plants, and many more.

Controlling a physical system by software has tremendous benefits: It allows nearly unlimited functionality, high flexibility for extensions and changes, and cost-effective implementations. Because of these benefits, a large, capable, and profitable software industry has evolved.

However, cyber-physical systems introduce a new *danger*: Because they directly affect the real, physical world, they have the potential to cause harm, injury, or loss to the users of the CPS. A cyber-physical system's failure, malfunction, or unavailability may have grave consequences, such as accidents, loss of life, damage to property or the environment, or severe legal repercussions. A CPS's trustworthiness is the most crucial factor for its safe use and acceptance by its users. It is, therefore, the critical responsibility of the CPS industry to build, maintain, and evolve *trustworthy cyber-physical systems* (TwCPS).

Unfortunately, all cyber-physical systems today cannot operate undisturbed: They are under the continuous influence of negative factors, such as cyber-attacks, faults, and failures in their software or underlying hardware, unavailability or dysfunction of their operating environment, human errors, or malicious activities. Therefore, the task of creating a trustworthy CPS is a considerable one!

Because of a CPS's possibility to generate harm, the concept of *risk* and *risk management* becomes central. All sources of negative impact on and within the CPS must be methodically, ultimately, and competently identified, assessed, mitigated, and reduced to an acceptable residual risk. Therefore, the science of "dealing with risks" is the foundation of a trustworthy CPS.

Trustworthiness is a CPS property: It is an umbrella term that heads a taxonomy of constituent properties. The most important are *safety* and *security*, embedded in a *legal framework*. This monograph focuses exclusively on safety and security.

Safety and security are properties of the *complete* CPS, i.e., including the hardware, networks, systems and communications software, application software, the operating environment, and the users. This monograph's focus is additionally narrowed to the *application software*, i.e., the software providing the *control functionality* of the CPS.

This monograph is *not* a "cookbook" for creating and evolving safe and secure CPS's. The objective is to provide fundamental *engineering principles* which are proven and powerful for building trustworthy cyber-physical systems.

> **Quote**
> *"I would advise students to pay more attention to the fundamental ideas rather than the latest technology. The technology will be out-of-date before they graduate. Fundamental ideas never get out-of-date"*
> David L. Parnas, 2001 (in [Hoffman01])

The literature contains thousands of principles and patterns applicable to safe and secure applications software for CPSs—here, the most general and influential are chosen. Wherever necessary, the monograph points specifically to valuable literature that delves deeper into a specific subject matter—this is the reason for the extensive list of references.

A note on terminology: Unfortunately, the formidable knowledge available in the references uses a lot of unprecise and even conflicting terminology. This monograph, therefore, precisely defines all terms used—as part of the conceptual integrity.

This "guidebook" addresses safety and security teams in the CPS industry, as well as software engineers, both in practice and during university education. It is also well suited to a one-semester, graduate-level lecture at Technical universities.

This monograph relies on and has some overlap with the author's previous book: "Future-Proof Software-Systems—A Sustainable Evolution Strategy", Springer Vieweg-Verlag, 2019, ISBN 978-3-658-19937-1 ([Furrer19]). Some parts are reused here.

A final observation: This book contains hundreds of principles on more than 600 pages. Which do we apply to a specific project or product? The answer is: *risk analysis*! The risk analysis identifies the threats, the possible failure modes, uncovers the vulnerabilities, specifies the mitigation measures, and estimates the acceptable residual risks. **The required risk mitigation measures to reach acceptable residual risk dictate the applicable principles**. No more, no less!

Stein am Rhein (Switzerland)                                                         Frank J. Furrer
in June 2022

# Acknowledgments

A book is never the achievement of one person. This book grew out of my lectures as Honorary Professor at the faculty of Computer Science at the Technical University of Dresden (Germany), which started in the winter term of 2013/2014. Special thanks go to the chair of the software technology, Prof. Dr. Uwe Aßmann, for his continued support. In addition, my numerous students contributed to the quality of the content and the didactic flow by their active participation during the lectures and their valuable engagement in my seminars.

I repeat my sincere thanks to Stephan Murer and Bruno Bonati, who started me on this journey ([Murer11], [Furrer19]).

A book is never the achievement of one author: He stands on the shoulders of giants—of which the extensive reference section for each chapter of the book is proof.

Authoring an English-language book for a German-native speaker is not a simple task. I would like to acknowledge the invaluable help from the language checker "Grammarly" (https://www.grammarly.com).

Finally, I wish to express my gratitude to Springer-Verlag for the extensive support during the creation of this monograph and especially to my editors Sybille Thelen, David Imgrund, Leonardo Milla, Heike Jung, and the copy-editor Roopashree Polepalli for their highly valuable assistance. They all made the publication process a successful pleasure.

Finally, thanks go to you—my reader—for investing your valuable time reading this book.

Frank J. Furrer
frank.j.furrer@bluewin.ch

# Contents

# About the Author

**Frank J. Furrer** graduated in July 1970 as Diplomelektro-Ingenieur at the Eidgenössische Technische Hochschule in Zürich, Switzerland, and earned his Ph.D. (Doktor der Technischen Wissenschaften) in July 1974 from the same institution.

From 1975 to 2009, he was active in the industry, both as an entrepreneur and as a management consultant for Information Technology and IT Systems Architecture.

In March 2013, he was invited to start teaching by the Technische Hochschule Dresden, Germany (Faculty for Computer Science). In July 2015, he was appointed as an honorary professor by the Technische Hochschule Dresden.

This book results from the accumulated knowledge and experience of Prof. Dr. Frank J. Furrer in a 48-year professional lifetime in systems and software engineering, both in industry and as a teacher. This book can be seen as the last of three volumes:

I.   Stephan Murer, Bruno Bonati, Frank J. Furrer: **Managed Evolution—A Strategy for Very Large Information Systems**. Springer Verlag, Berlin, Germany, 2011. ISBN 978-3-642-01632-5.

II.  Frank J. Furrer: **Future-Proof Software-Systems—A Sustainable Evolution Strategy**. Springer Vieweg Verlag, Wiesbaden, Germany, 2019. ISBN 978-3-658-19937-1.

III. Frank J. Furrer: **Engineering Principles for Safety and Security of Cyber-Physical Systems**. Springer Vieweg Verlag, Wiesbaden, Germany, 2022 [This monograph].

# List of Figures

# List of Tables

# List of Examples

# List of Definitions

# List of Principles

# Introduction

*Cyber-physical systems are the foundation of our modern world and are sources of wealth, comfort, and a functioning society. They permeate all areas of work and life. Therefore, engineering cyber-physical systems has become an essential activity of our modern industrial world. Unfortunately, apart from the tremendous benefits that cyber-physical systems provide, they are also sources of danger. Because they directly impact the real world, they can cause accidents, loss of life or property, or damage to society or the environment. Therefore, a fundamental requirement is to build the CPSs with sufficient safety and security.*

## 1.1 Cyber-Physical Systems

The term "cyber-physical systems (CPS)" was introduced around 2006 by Helen Gill at the National Science Foundation in the USA ([Lee15]). Cyber-physical systems combine a *cyber-part*, i.e., a software-controlled computer system, with a *physical part*, i.e., a real system. The *software* operates the physical system, providing the control functionality. Figure 1.1 shows the essential elements of a CPS: The physical part is represented by the globe. Physical measurements, such as temperatures, velocities, flows, and pressures, are read from the physical part via *sensors* and fed to the cyber-part, where they are converted into digital values for processing by the control software. The control software executes the control algorithms and outputs digital values to steer the physical part. The digital values are converted to analog values and use *actuators* to control physical quantities, such as speed, altitude, braking, or temperatures.

The critical element is the applications or *control software*: It provides all the functionality of the CPS, supervises its correct operation, and interacts with the users.

F. J. Furrer, *Safety and Security of Cyber-Physical Systems*, https://doi.org/10.1007/978-3-658-37182-1_1

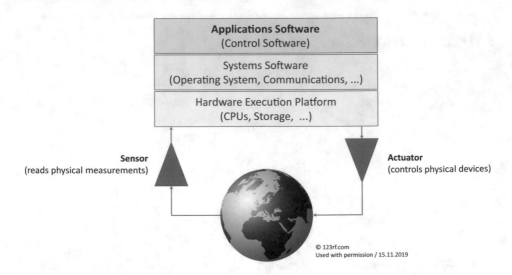

**Fig. 1.1** Cyber-physical system

Controlling a physical system by software has tremendous benefits: It allows nearly unlimited functionality, high flexibility for extensions and changes, and cost-effective systems development and implementation. Because of these benefits, a large, capable, and profitable software industry has evolved around CPSs.

This monograph's topic is the applications (control) software for CPSs, especially its safety and security.

> **Quote**
> *"As an intellectual challenge, CPS is about the intersection, not the union, of the physical and the cyber"*
>     Edward A. Lee, 2015

Cyber-physical systems are abundant—we encounter them in nearly any activity we do. A Google search for "cyber-physical systems examples" returns 181,000,000 results (25.01.2022). Some examples are as follows:

**Embedded systems:**

- Cardiac pacemaker;
- Anti-skid brake control of a car;
- The automatic pilot of a plane;
- etc.

**Devices:**

- Robots (for all sorts of applications);
- Autonomous cars;
- Self-piloted trains;
- Unmanned aerial vehicles (UAV);
- Smart buildings;
- Medical diagnosis equipment;
- etc.

**Infrastructure:**

- Telecommunications (voice, data, and video);
- Electricity grid;
- Water treatment and distribution;
- Manufacturing lines;
- Smart cities;
- etc.

## 1.2   Risk in Cyber-Physical Systems

The cyber-physical systems are the foundation of our modern world and are the source of wealth, comfort, and a functioning society. Unfortunately, cyber-physical systems are also a severe source of *danger*: Because they directly affect the real, physical world, they have the potential to cause harm or injury to the users of the CPS. A cyber-physical system's failure, malfunction, or unavailability may have grave impacts, such as accidents, loss of life, damage to property or the environment, reputation damage, liability, or severe legal repercussions.

Unfortunately, all cyber-physical systems today cannot operate undisturbed: They are under the continuous influence of *threats* or *failures* (Fig. 1.2), such as cyber-attacks, faults, and failures in their software or underlying hardware, unavailability or dysfunction of their operating environment, human errors, or malicious activities. Building CPSs with a sufficient degree of trustworthiness—especially *safety* and *security*—is paramount.

> **Quote**
> *"Trustworthiness of a software, application, service or infrastructure is a key success factor for its use and acceptance by end-users"*
>    Nazila Gol Mohammadi, 2019

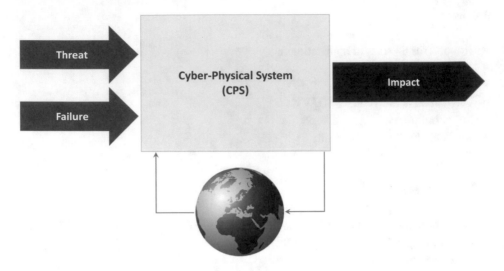

**Fig. 1.2**  Danger potential of a cyber-physical system

Because of the possibility of a CPS to generate real harm, the concept of *risk* and *risk management* becomes central. All sources of negative impact on and within the CPS must be methodically, ultimately, and competently identified, assessed, mitigated, and reduced to an acceptable residual risk. Therefore, the science of "dealing with risks" is the foundation of a trustworthy CPS.

Trustworthiness is a property of a CPS: It is an umbrella term that heads a taxonomy of constituent quality properties. The most important are *safety* and *security*, embedded in a dependable *legal framework*. Safety and security are properties of the complete CPS, i.e., including the hardware, networks, systems and communications software, application software, and even the operating environment (Fig. 1.1). This monograph's focus is additionally narrowed to the application software, i.e., the software providing the *control functionality* of the CPS.

Risk is unavoidable. Sound engineering means identifying, recognizing, assessing, and mitigating all risks as well as possible to reduce the risks to low, acceptable, residual risks (Fig. 1.3).

> **Quote**
> *"The consequences of failure to adequately manage risk can be disastrous"*
>    *Paul Hopkin, 2018*

Therefore, an effective risk management process is the central focus of safety and security engineering for cyber-physical systems.

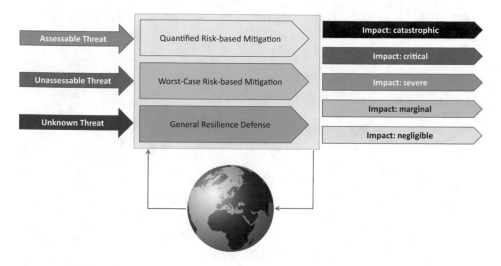

**Fig. 1.3**  Risk in a cyber-physical system

## References

[Lee15]   Edward A. Lee: *The Past, Present, and Future of Cyber-Physical Systems - A Focus on Models* Sensors (Basel), March 2015. Published online, February 2015, https://doi.org/10.3390/s150304837. Downloadable from: https://www.mdpi.com/1424-8220/15/3/4837 [last accessed: 29.3.2020]

# Cyber-Physical Systems

**2**

*Cyber-physical systems are the technical foundation of our modern world. They combine a cyber-part (software-based computer control) and a physical part (real system, such as a car, an airplane, and a cardiac pacemaker). These software-controlled systems allow nearly unlimited functionality, flexible functional changes, and cost-effective development and implementation. However, because they interact with the real world, they have the potential to cause harm to persons, property, society, or the environment. Therefore, safety and security are essential concerns when developing and operating cyber-physical systems.*

## 2.1 Cyber-Physical Systems[1]

*Cyber-physical systems* have a tremendous proliferation: They can be found in myriads of devices where they interact with and control a specific part of the real world. A definition is given in Definition 2.1.

▶ **Definition 2.1: Cyber-Physical System**
A cyber-physical system (CPS) consists of a collection of computing devices communicating with one another and interacting with the physical world, often in a feedback loop.
  Rajeev Alur, 2015.

---

[1] Part of this section has been reused from the authors previous book "Future-Proof Software-Systems", Springer Vieweg Verlag, Wiesbaden, Germany, 2019. ISBN 978-3-658-19937-1 ([Furrer19]).

F. J. Furrer, *Safety and Security of Cyber-Physical Systems*, https://doi.org/10.1007/978-3-658-37182-1_2

A comprehensive sample of a cyber-physical (embedded) system is given in Example 2.1 below.

---

**Example 2.1: Automotive Anti-Skid Braking Control (ABS)**

**Quote**
*"Your car used to be a mechanical device with some computers in it. Now, it is a 100+ computer distributed system with four wheels and an engine"*
   *Bruce Schneier, 2018*

A car must be stopped in critical situations in the shortest possible time, even if the road and weather conditions are adverse. For a short stopping distance, all modern vehicles have computer-controlled anti-skid braking control systems on-board ([Zaman15]).

The ABS-system is shown in Fig. 2.1.

The *physical part* of the system consists of the following:

- The four wheels of the car;
- The four wheel rotation sensors (measuring the rotation speed of each wheel);

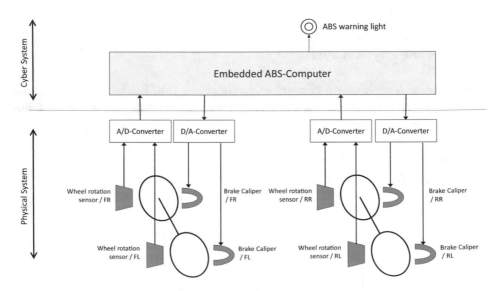

**Fig. 2.1**   Automotive anti-skid braking system (ABS)

- The four hydraulic braking calipers (braking each wheel individually);
- Four analog/digital converters (converting the analog rotation speed signals to a digital format);
- Four digital/analog converters (converting the digital control signal back to an analog signal);
- The ABS warning light indicating an ABS-system failure.

The *cyber-part* of the system is the embedded ABS-computer and its *software*.

The ABS-system is not a stand-alone system but is networked with many other electronic systems in the car, such as automatic distance control and electronic stability control computer.

The operating mode of the ABS-system is as follows:

- The four wheel rotation sensors (front right, …, rear left) measure the wheel rotation rate of each wheel (100 times/sec);
- The embedded ABS-computer identifies differences in the wheel rotation rates (e.g., indicating a skid), calculates the situation-adequate rotation rate for each wheel (the ABS-computer may receive additional information from other systems in the car, e.g., the gyroscope or the electronic stability system), and issues individual braking signals for each hydraulic caliper;
- The wheels are individually slowed down to stabilize the car.

As many cyber-physical systems are, this system is a closed control loop system governed by software ([Astrom11]).

Many modern cyber-physical systems have to work correctly in highly complex, fast-changing, and uncertain environments—such as driverless cars ([Maurer16]), unmanned aircraft ([Valavanis14], [Nichols19]), conductorless trains, intelligent traffic control systems ([Bestugin19]), medical robots ([Schweikard15]), rescue robots ([Sreejith12]), planetary explorer vehicles ([Gao16]), and many more.

In order to complete their mission successfully and safely, such systems need a high degree of *autonomy*, i.e., they must be able to adapt to unknown situations, learn from their environment, and take stand-alone decisions. Autonomy thus becomes an important capability (property) of such advanced systems.

Autonomy has been studied for a long time. In 2006, a reference architecture for autonomic computing "MAPE-K" ([IBM06], [Lalanda13] was presented, which has found many applications. Such challenging autonomic systems rely on artificial intelligence ([Tianfielda04], [Russell17]) and machine learning ([Alpaydin16]). Example 2.2 shows the MAPE-K reference architecture.

**Fig. 2.2** MAPE-K reference
architecture

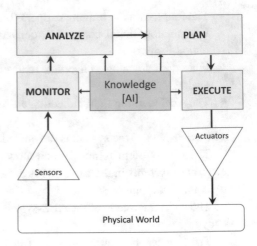

The Monitor-Analyze-Plan-Execute-Knowledge (MAPE-K) reference architecture
([IBM06]) is shown in Fig. 2.2: Basically, this is a software-governed feedback control loop. The physical world is sensed via sensors (measuring physical quantities
such as temperature and speed—but also including cameras, gyroscopes, etc.). The
input data from the physical world is read in, preprocessed, and then analyzed. The
analysis results are transferred to the planning functionality, which decides the actions
to be taken. Finally, the execute block generates the necessary signals to control the
physical world, which are then applied via actuators. All four functional blocks use
the central block's knowledge and artificial intelligence capabilities.

## 2.2   Cyber-Physical Systems-of-Systems[2]

Most of today's exciting applications are implemented as *systems-of-systems* (SoS), very
often as cyber-physical systems-of-systems (CPSoS). A CPSoS is defined in Definition 2.2.

▶ **Definition 2.2: System-of-Systems (SoS)** A system-of-systems (SoS) is a set or
arrangement of constituent systems that results when independent and useful systems are
integrated into a larger system that delivers unique capabilities.

---

[2] Part of this section has been reused from the authors previous book "Future-Proof Software-Systems",
Springer Vieweg Verlag, Wiesbaden, Germany, 2019. ISBN 978-3-658-19937-1 ([Furrer19]).

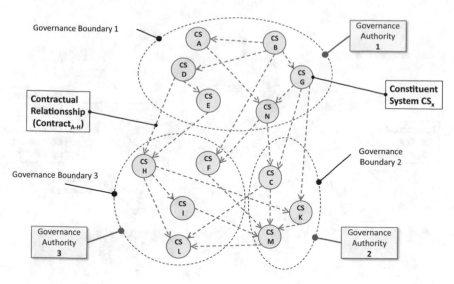

**Fig. 2.3**  Characteristics of a system-of-systems

The exciting part of the definition is that individual systems—the *constituent systems* (CS)—are combined to achieve a higher objective, which could not be accomplished by any single constituent system (Fig. 2.3).

Figure 2.3 shows the characteristics of a system-of-systems (the so-called Maier criteria [Maier98]:

- The system-of-systems is assembled from individual systems: The constituent systems **CSn**;
- Each of the constituent systems **CSn** has its own objective, which contributes to the overall objective of the system-of-systems;
- Some of the constituent systems **CSn** are under different *governance*, i.e., the authority to build, change, operate, and evolve the **CSn** lies with various organizations. This may complicate the forming and the operation of SoS difficult;
- The *cooperation* of the constituent systems **CSn** must be formally agreed by *contracts*.

Fortunately, *system-of-systems engineering* (SoSE) is a mature field of engineering, and many proven processes exist ([Jamshidi09a], [Jamshidi09b], [Luzeaux11], [MITRE14], [Romanovsky17], [Stevens11], [Eisner20], [SEI21]).

In *cyber-physical systems-of-systems*, one or several constituent systems interact with the real world, i.e., are cyber-physical systems (Fig. 2.4, Definition 2.3).

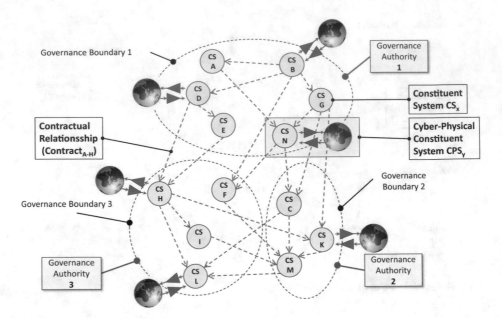

**Fig. 2.4** Cyber-physical system-of-systems

▶ **Definition 2.3: Cyber-Physical System-of-Systems (CPSoS)**
A cyber-physical system-of-systems (CPSoS) is a set or arrangement of constituent systems—of which at least one is a cyber-physical system—that results when independent and useful systems are integrated into a larger system that delivers unique capabilities.
   Adapted from: US DoD, 2004.

For the rest of this monograph, "cyber-physical system (CPS)" means both the single cyber-physical system and also a cyber-physical system-of-systems. It is assumed that all management and operational issues in the cyber-physical system-of-systems are resolved.

## 2.3   Emergence

The most intriguing property of a system-of-systems is *emergence* ([Bedau08], [Bondavalli16], [Mittal18], [Charbonneau17], [Sethna06], [Rainey18], [Mobus15/ Sect. 10.4], [Furrer19]) in the form of *emergent properties*, *emergent behavior* or *emergent information* (Definition 2.4, Fig. 2.5).

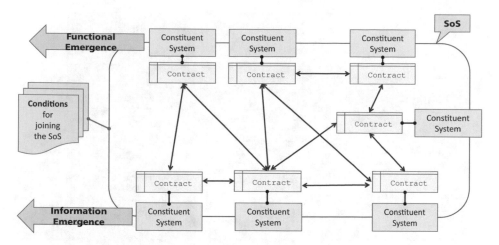

**Fig. 2.5** Functional and information emergence in a system-of-systems

▶ **Definition 2.4: Emergence**
An emergent property/behavior is a property, behavior, or aggregated information, which a collaboration of constituent systems has, but which the individual constituent systems do not have.

Unfortunately, emergence in a SoS or CPSoS is only predictable in a very limited way. SoSs very quickly become so complex and complicated that they are difficult to understand: Even when all the constituent systems are well understood and mastered, the behavior or the properties of the ensemble of the interacting systems, i.e., the SoS, can be a surprise.

Figure 2.6 shows the *emergence matrix*: It has four quadrants, $Q_1$ to $Q_4$. $Q_1$ is the expected, *positive emergence* and is the reason why the SoS is built and operated. The cooperating constituent systems together achieve functionality or information, which none of the individual systems could. $Q_2$ is an engineering challenge: An expected *negative emergence* is present. This quadrant must be handled by sound engineering, including *risk analysis* and mitigation. $Q_3$ is a rare surprise: An unexpected, beneficial, and positive emergence may appear. Finally, $Q_4$ is the dangerous quadrant: unexpected, harmful emergence, which may result in *safety accidents* or *security incidents*. A recent occurrence (2016) is shown in Example 2.3.

**Example 2.3: Schiaparelli Mars Lander Crash**

The European Space Agency (ESA), on November 23, said its Schiaparelli lander's crash landing on Mars on October 19, 2016, followed an unexplained saturation of its inertial measurement unit (IMU). The IMU delivered bad data to the lander's navigation computer and forced a premature release of its parachute (Fig. 2.7).

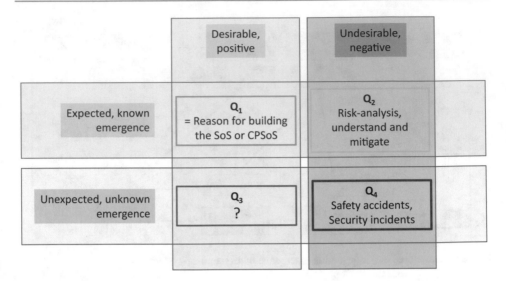

**Fig. 2.6**  Emergence matrix

**Fig. 2.7**  Schiaparelli Mars Lander

Polluted by the IMU data, the lander's computer apparently thought it had either already landed or was just about to land. The parachute system was released, the braking thrusters were fired only briefly, and the on-ground systems were activated.

Instead of being on the ground, Schiaparelli was still 3.7 km above the Mars surface. It crashed after a nearly free fall and was destroyed on impact.

The Schiaparelli crash is an example of $Q_4$ in Fig. 2.6: It started with an unforeseen situation, led to unexpected behavior of the IMU (= first constituent system), forced the navigation computer (= second constituent system) into maloperation, and resulted in the crash of the lander.

Source: https://spacenews.com [November 23, 2016].

## 2.4  Infrastructure

### 2.4.1  Introduction

One particularly important type of system-of-systems is *infrastructure* (Definition 2.5, [Ackermann17], [Johnson15]). Our modern world is based on many different infrastructures, such as:

- The power grid;
- The telecommunication networks;
- The traffic systems (road, air, rail);
- The postal delivery channels;
- The financial system;
- The water treatment and distribution facilities;
- The national and international government structures;
- The emergency services (medical, police, fire brigades, …);
- etc.

▶ **Definition 2.5: Infrastructure**
The basic systems and services, such as transport, communications, and energy distribution, that a country or organization uses in order to work effectively.
Adapted from: https://dictionary.cambridge.org/dictionary/english/infrastructure

All infrastructures are enormous, vulnerable, and target-rich structures. They are very difficult to defend, not at least because many of them are based on old, sometimes outdated, technology. Significant concerns exist that the critical infrastructures could become the targets of cyber-war activities ([Shakarian13], [Prunckun18], [HomelandSecurity17], [Saadawi13]). Several attacks have already been carried out and have shown tremendous damage potential (Example 2.4, e.g., https://www.nec.com/en/global/techrep/journal/g17/n02/170204.html).

> **Quote**
> *"Technology is power. Whoever controls the global digital infrastructure controls the world"*
>    The Economist, April 11, 2020

**Example 2.4: US Department of Energy "Aurora Vulnerability"**

DoE's experiment used a 2.25 MW diesel generator. The Aurora vulnerability allows an attacker to disconnect the generator from the grid just long enough to get slightly out of phase with the grid and then reconnect it. This desynchronization puts a sudden, severe strain on the rotor of the generator, which causes a pulse of mechanical energy to shake the generator, damaging the bearings and causing sudden increases in temperature. By disconnecting and reconnecting the generator's circuit to the grid, the Aurora vulnerability led to the generator's *destruction* in about three minutes.

[Source: https://blog.trendmicro.com/the-aurora-power-grid-vulnerability-and-the-blackenergy-trojan/].

## 2.4.2   ICS Architecture

The foundation of computerized infrastructure control is *industrial control systems* (ICS, Definition 2.6, Fig. 2.8, [Hanssen15], [Rabiee17], [Manoj19], [Stouffer15], [Ackerman21]).

▶ **Definition 2.6: Industrial Control Systems (ICS)**
"Industrial control system" is an all-encompassing term used for various automation systems and its devices, such as programmable logic controllers (PLC), human–machine interfaces (HMI), supervisory control and data acquisition (SCADA), distributed control systems (DCS), safety-instrumented systems (SIS), and many others—including various types of networks.
[Ackermann17].

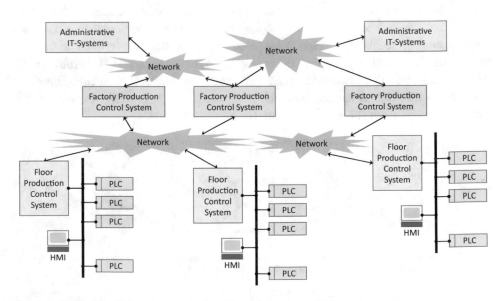

**Fig. 2.8**   Typical architecture on an industrial control system (ICS)

Unfortunately, the wide variety of computing, controlling, and communications devices and technologies used in an industrial control system (Fig. 2.8) results in a large *attack surface*. Because a successful attack, however, generates enormous damages, the incentive for potential attackers is tempting. Defending an industrial control system—i.e., building and keeping it *safe* and *secure*—is a colossal task (Example 2.5, e.g., [Ackerman21], [Knapp14], [Krutz16], [Macaulay11], [McCarthy21], [Colbert16], [ISA22], [IEC62443], [McCarthy22]).

---

**Quote**

*"From the traffic lights on your drive to work, or the collision avoidance system of the train or metro, to the delivery of electricity that powers the light you use to read this book, to the processing and packaging that went into creating the jug of milk in your fridge, to the coffee grinds for that cup of joe that fuels your day: What all these things have in common are the industrial control systems driving the measurements, decisions, corrections, and actions that result in the end products and services that we take for granted each day"*
Pascal Ackermann, 2015

---

**Example 2.5: Stuxnet Malware**

Around the year 2010, an amazing, very sophisticated *malware* was detected on many computers worldwide. Its purpose was entirely unclear, and researchers around the world started its analysis. The malware was labeled "Stuxnet" ([Zetter14], [Poroshyn14]). Its underlying architecture consisted of two parts: the delivery system for spreading the malicious payload and the payload, which performs the attack.

---

**Quote**

*"The story of Stuxnet is a story of its evolution from an ordinary malware to a cyber-weapon"*
Roman Poroshyn, 2014

---

However, the delivery system was quickly understood—it consisted of a mix of known vulnerabilities and an as yet unknown zero-day exploits. Soon it was detected that the payload target was the SIEMENS SCADA software (https://new.siemens.com/) and the SIEMENS programmable logic controllers (PLCs with their STEP-7 programming language). Only systems containing these two specific software products were attacked.

If SIEMENS WinCC was identified, a central piece of software—the interface DLL (Window dynamic-link library, https://support.microsoft.com/en-us/help/815065/

what-is-a-dll)—was exchanged against the rogue DLL introduced by Stuxnet. This DLL is the driver which transfers commands and data between the WinCC supervisory/control program and the programmable logic controllers.

Whenever an operator tried to send a command to one of the PLCs, the original command was replaced with a rogue, malicious command. The same with the reporting way: Whenever an error message or warning was received from the PLCs, it was superseded by an inconspicuous rogue message—so no warnings or errors appeared on the SCADA screens.

The purpose of Stuxnet became clearer during an inspection in 2010 by the International Atomic Energy Agency (IAEA) to a facility near Natanz in Iran. The officials noted that an unusual percentage of Natanz's uranium enrichment centrifuges broke down. The centrifuges are very delicate devices, spinning at nominal 1′064 revolutions/sec. Running them at higher speeds mechanically *destroys* the centrifuges.

This was precisely the malicious payload of Stuxnet: Interfering with commands between the SCADA-system and the PLCs controlling the operation of the centrifuges. Running up the speed of the centrifuges until they broke, and suppressing warnings coming back to the operator screens (Fig. 2.9). The IAEA reported that 1′148 centrifuges were taken out of operation at Natanz. It is estimated that Stuxnet delayed the Iranian Atomic Program for several years. A powerful piece of malware had become the first digital weapon to cause military harm!

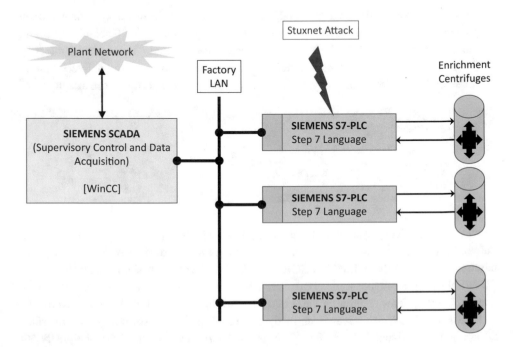

**Fig. 2.9** Stuxnet attack

In 2012, researchers became certain that at least three more cyber-weapons derived from Stuxnet have been created and are waiting to be launched ([Poroshyn14]).

## 2.5  Autonomous Cyber-Physical Systems

In Example 2.2, the architecture for *autonomic software* was introduced by IBM ([Kephart03]). This concept laid the foundation for *autonomous cyber-physical systems* (Definition 2.7, e.g., [Liu18]). Many of today's and future cyber-physical systems operate in a challenging, fast-changing environment with significant uncertainties. These CPSs, therefore, require various degrees of autonomy. The fascinating and somewhat frightening fact is that these systems take *autonomous decisions* (Fig. 2.10). Such autonomous decisions are primarily based on the application of *artificial intelligence* ([Stone19], [Hutter05]).

▶ Definition 2.7: Autonomous Cyber-Physical System
Autonomous cyber-physical systems refer to systems capable of operating in a real-world environment with little or without any form of external control for extended periods of time.

  https://www.sciencedirect.com/topics/computer-science/autonomous-system

The future will introduce more and more cyber-physical systems that are—at least partially—controlled by *artificial intelligence* (AI) and make decisions of their own. These types of systems introduce new, severe safety and security concerns (Definition 2.8, [Yampolskiy18], [Juliano16], [Zivic20], [SASWG19], [Lawless17], [Mittu16]).

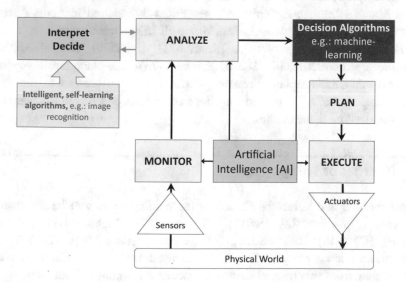

**Fig. 2.10**  Autonomous system software architecture

▶ **Definition 2.8: Artificial Intelligence Safety and Security**
**Artificial Intelligence Safety:** Property of a system, which is wholly or partially controlled by AI, to have a verifiable and acceptable residual risk for a safety accident.

**Artificial Intelligence Security:** Property of a system, which is wholly or partially controlled by AI, to have a verifiable and acceptable residual risk for a security incident.

*Risk management*—which is already challenging for deterministic systems—becomes even more difficult for autonomous systems. A sample of failed risk management is given in Example 2.6.

---

**Example 2.6: Boeing 737MAX Crash**

The Boeing 737 MAX is the fourth generation of the Boeing 737, a narrow-body airliner manufactured by Boeing Commercial Airplanes (https://www.boeing.com/commercial/737max/). Its first flight was on January 29, 2016.

The Boeing 737 MAX had to be fitted with new, more fuel-efficient engines. Because of their larger diameter, they were positioned slightly forward and higher up on the wing. This changed the plane's flight characteristics: When accelerating, the aircraft increased its angle-to-climb, which could lead to dangerous flight situations, eventually causing it to stall. To compensate for this risk, engineers at Boeing implemented an autonomous, cyber-physical system Maneuvering Characteristics Augmentation System (MCAS). When MCAS detects a dangerous angle of climb, it autonomously pushes the nose of the plane down to prevent stalling. MCAS is activated without the pilot's command, i.e., it is autonomous.

On October 29, 2018, Lion Air Flight 610 took off from Jakarta. Twelve minutes after takeoff, the plane crashed into the Java Sea, killing the crew and 189 passengers.

On March 10, 2019, Ethiopian Airlines Flight 302 started in Addis Abeba. Six minutes after takeoff, the plane crashed, killing all 157 people aboard.

The investigation reports (e.g., [Robison21], [HCTI20], [Daniels20], https://www.theverge.com/2019/3/22/18275736/boeing-737-max-plane-crashes-grounded-problems-info-details-explained-reasons) attribute the two crashes to the fact that the autonomous MCAS overruled the pilot's commands and forced the planes into a nose-dive, leading to the crashes.

---

## 2.6 Internet of Things

The most massive and influential CPSoS in the coming years will be the *Internet of Things* (IoT, Definition 2.9, [Raj17], [Greengard15], [Mahmood19], [Jeschke16], [Davies20], [Cirani18], [Lakhwani20], [Mohamed19], [Banafa18]). The IoT extends the well-known Internet of Computing & Communications by a colossal number of new devices, ranging from small, smart sensors and actuators, robots, and autonomous cars to smart cities. IoT enables an imposing number of new applications, such as smart homes, smart airports, or interesting healthcare solutions (e.g., [Raj17]).

▶ **Definition 2.9: Internet of Things (IoT)**
The Internet of Things (IoT) not only comprises millions of networked computing machines and software services, but also billions of personal and professional devices, diminutive sensors and actuators, robots, etc., and trillions of sentient, smart, and digitized objects.
  Adapted from [Raj17].

Because of the gigantic spread of the IoT—into all areas of life, work, and society—*safety* and *security* are a grave concern. In addition to the direct impact on the physical world, which generates sources of *danger*, many IoT devices are small and suffer from limited processing capacity and very restricted power sources. This poses a strong and dangerous constraint for the implementation of dependable safety and security controls ([Fagan20a], [Fagan20b], [Fagan20c], [Fagan20d], [Fagan20e], [Fagan20f]).

> **Quote**
> *"IoT is all about enabling extreme connectivity among various objects across the industry domains"*
>   Pethuru Raj, 2017

## 2.7    Cloud-Based Cyber-Physical Systems

### 2.7.1    Conceptual Architecture

The term *cloud computing* was introduced to describe platforms for distributed computing as early as 1993. In the first decade of 2000, the commercial business case for IT cloud computing (Definition 2.10, [Erl13], [Ruparelia16]) became strong: Instead of buying, installing, operating, and maintaining their own IT infrastructure, organizations outsourced parts or all of their IT operations. The advantages were cost effectiveness, flexibility, scalability, reliable disaster recovery, and dependable backup solutions. Today (2021), cloud computing is the prevalent paradigm in commercial data processing (https://www.globenewswire.com).

▶ **Definition 2.10: (IT-) Cloud**
Cloud computing is the delivery of different services through the Internet. These resources include tools and applications like data storage, servers, databases, networking, and software. Cloud computing is named as such because the functionality and information being accessed are found remotely in the cloud or in the virtual space.
  Adapted from: https://www.investopedia.com/terms/c/cloud-computing.asp

Unfortunately, cloud computing has one crucial *risk*: The responsibility for the quality attributes, such as dependability and trustworthiness, is transferred from the organization to a third party: to the *IT cloud service provider* (e.g., [Brender13], [Morrow18], [Alosaimi16]).

> **Quote**
> *"Using a cloud provider will extend the perimeter of the organization, creating a larger attack surface. The concentration of clients using cloud services could make a cloud provider a high value target"*
>    Richard Piggin, 2014

In the last decade, part of the functionality and data from some cyber-physical systems have been transferred to the cloud (e.g., [Colombo14], [Olding20], [Diez14]), and this tendency is accelerating. Such *cloud-based cyber-physical systems* (Definition 2.10, [Chaari16]) significantly increase the capabilities and widely expand the applications range of the CPS.

▶ **Definition 2.11: (IT-) Cloud-Based Cyber-Physical System** CPS in which intensive computations, large data amounts, and laborious analytical tasks are offloaded to the much more powerful IT infrastructure in the cloud.

The *conceptual architecture* of a cloud-based cyber-physical system is shown in Fig. 2.11, identifying three hierarchical layers:

a. The Cyber-Physical Systems Level (CPS level): Home to the individual CPSs and CPSoSs (e.g., [Marwedel18], [Wang18], [Cao20a], [Song16], [Colombo14], [Kravets20], [Kravets20], [Rawat16], [Arseniev20]);
b. The Communications Network: Data and control exchange technology (such as the 5G cellular network [Sauter21], WLANs [Perahia13], [Lee20a]);
c. The Cloud: Network-accessible computing and storage infrastructure, (Definition 2.10, [Erl13], [Gai20], [Ruparelia16], [Haque20]). The cloud's mission is twofold: a) provide communication, integration, and synchronization between the connected CPSs and b) supply computing power and data storage to the individual CPSs.

Moreover, the *secured channels* between the individual CPS and the cloud are drawn in Fig. 2.11.
    Examples of cyber-physical systems utilizing cloud resources are as follows:

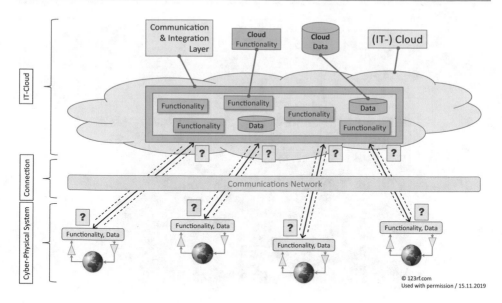

**Fig. 2.11** Cloud-based cyber-physical systems architecture

- Industrial control systems (ICS, e.g., [Colombo14], [Olding20], [Wang18], [Piggin14]);
- Cloud robotics (Example 2.7, [Stakem18], [Koubaa16], [Asama02], [Chalup19]);
- Smart energy grid ([Kabalci19], [Knapp13]);
- Transportation systems (Example 2.8, [Paul17], [Milani18], [Milani19]);
- Manufacturing ([Wang18], [Yáñez17]);
- Health industry ([Pezoulas20], [McCaffrey20]);
- … and many more.

---

**Example 2.7: RoboCup**

The RoboCup (Fig. 2.12, https://www.robocup.org) competition was established to develop a team of autonomous humanoid robot soccer players complying with the FIFA rules (www.fifa.com).

Autonomous soccer-playing robot teams playing against another team pose complex robotics problems. Although this application of cloud robotics is *non-mission critical*, it has all the technical challenges of autonomous, mobile robots and represents an ideal research ecosystem (e.g., [Chalup19]). The individual robots require massive sensor technology, massive real-time processing power, and low latency communication and synchronization—a typical cloud mission!

Both teams have their proprietary applications/data in the cloud, which control the movements of the mobile robots.

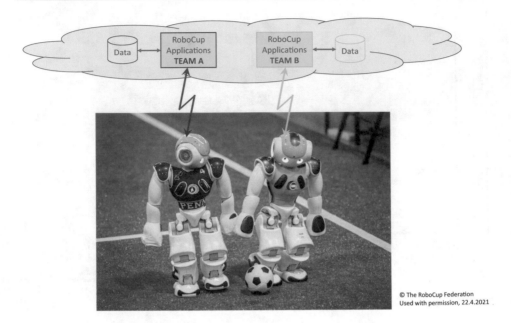

**Fig. 2.12**  RoboCup

---

**Example 2.8: Vehicular Cloud**

Individual mobility using road vehicles is essential for the current and future economy. The massive growth in traffic causes increasing problems, such as congestion, accidents, and environmental damage. The *vehicular cloud* (Fig. 2.13, e.g., [Paul17], [Milani18], [Milani19]) is a current solution: The cloud captures information from the vehicles, from the transportation infrastructure, from the police, and the road authorities, from the weather situation, and forecast, etc.

Equipped with this massive, dynamic information, many beneficial applications running in the vehicular cloud become possible, such as (Fig. 2.13)

- *Traffic management* (e.g., [Treiber13]): Steer the traffic flow to optimize efficiency, i.e., avoid traffic congestion and optimize throughput;
- *Accident avoidance* (e.g., [Ward19]): Warnings to the drivers or direct intervention to the vehicle in case of dangerous situations;
- *Rescue service optimization* (e.g., [Prosser19]): Efficiency enhancement of the incident response services;
- *Fleet management* (e.g., [Goel08]): Minimizing time, cost, and empty running of commercial deliveries;

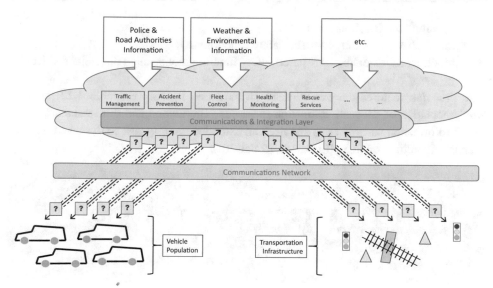

**Fig. 2.13** Vehicular cloud

- *Health monitoring* (e.g., [Rammler21]): Remotely supervise the health of the vehicle drivers and warn/intervene in case of suspicious symptoms;
- *Computer power sharing* (e.g., [Ghafoor13]): Numerous vehicles are unused most of the time. Their considerable on-board computing resources remain idle. With the consent of the owners, these resources could be used remotely by cloud applications;
- *Surveillance* (e.g., [Petersen12]): Modern vehicles are equipped with many sensors, such as cameras and radar. The density of this sensorium is high. Collecting and analyzing this information can be used for accident reconstruction, crime prevention, or anti-terrorism actions.

### 2.7.2   Cloud Safety, Security, and Real Time

The use of *cloud services* in the cyber-physical system introduces new *failure modes* (= possible sources of safety accidents) and increases the *attack surface* (= possible attack vectors for security incidents). Therefore, three areas need specific attention: 1) safety, 2) security, and 3) real-time behavior.

*Cloud-based CPS safety* To divide safety-critical functionality between the CPS resources and the cloud is a serious decision. Such a split opens at—least three—new risks:

1. New failure modes (Definition 2.5);
2. A direct attack on the cloud-based safety functionality;
3. An indirect attack on the CPS-based safety functionality through the cloud services.

The first risk requires the execution of an extended FMEA (Example 4.22) and the implementation of adequate mitigation measures. The second and third risks are "classical" information security risks and need an extended security risk analysis and the corresponding mitigation (e.g., [Loske15], [Rakas20]).

> **Quote**
> *"A successful break will not only affect the hacked IIoT device, but also provides loopholes that can be exploited to access an enterprise's entire cloud-based infrastructure"*
>    Mark Olding, 2020

**Cloud-based CPS security:** When CPS-related functionality—both mission critical or non-mission critical—is executed in the cloud, a new, perilous field of security risks is opened (valid for the public, private, community, and hybrid clouds).

One primary source is the *divided responsibility* for security between the cloud provider and the cloud service consumer (e.g., [Greer20]). Vague assignments and weak accountability of security responsibilities are the primary sources of security incidents. Therefore, the divided responsibility must be codified in the *shared cloud responsibility model* (Definition 2.12, [Greer20]).

▶ **Definition 2.12: Shared Cloud Security Responsibility Model** Legally binding, complete, detailed, continuously updated model of the security responsibilities of cloud provider and cloud consumer, covering all aspects of security, such as architecture, implementation, processes, legal and compliance obligations, incident handling, maintenance, and evolution—and (possibly) liability.

The shared cloud responsibility *model*—often supplied by the cloud service provider— is a normative document contractually agreed between cloud provider and cloud service consumer.

**Example 2.9: Weak Container Security**

A shared cloud responsibility model aims to clarify the boundaries of accountability for all elements and processes in the cloud. This can be tricky, e.g., if a cloud consumer installs containers ([Rice20], [Scholl19]) with poor security. Such a flawed application container may open the cloud to a successful attack.

> **Quote**
> *"Cloud data breaches are serious, costly, and consequential. They can be distress-*
> *ing for consumers and disastrous for companies. However, the stakes are differ-*
> *ent—and often higher—when hackers target industrial and critical infrastructure"*
>    Jeff Zindel (Honeywell Corporation), 2021

Before an organization migrates to the cloud, a cloud policy must be generated, especially a *cloud security policy* (e.g., [Pranic20], [TechTarget21]). The cloud security
policy is a formal, mandatory regulation to ensure an organization's safe and secure
operations in the cloud. The cloud policy must also describe a *cloud exit strategy*, e.g., in
case of leaving the cloud or changing the cloud service provider (e.g., [EBF20]).

Trustworthy cloud-based CPSs have some specific safety and security assurance
requirements ([McCarthy21], [Diez14], [Dotson19], [Jenner18], [Gai20]). Two specific
topics are *cloud security auditing* (Definition 4.29, [Majumdar20]) and *log file reviews.*
*Real-time behavior of cloud-based CPS* When exporting real-time sensitive tasks—or
parts of them (Definition 2.14, [Kopetz11], [Nath17])—to the cloud, two cloud obstacles
are met. First, the *cloud communications and execution platform's latency* ([Zhang20])
may make temporal commitments difficult to ensure under all operating conditions.
Second, the *worst-case execution time* (WCET, Definition 2.15, [Franke18]) determination in these mixed-platform tasks is very difficult. Therefore, cloud-based real-time
CPSs require deep, well-justified analysis, reproducible decisions, and careful design.

## 2.8   Token Economy

A nine-page white paper by an as yet (2021) unknown author, "Satoshi Nakamoto",
introduced the revolutionary concept of *bitcoin* and *blockchain* ([Nakamoto08],
[Nakamoto19], [Champagne14]). Following this breakthrough, many new ideas for cryp-
tocurrencies, distributed ledgers, and tokens developed: A great, promising, and crucial
new industry emerged ([Werbach18], [Mougayar16], [Xu19], [WEF20]). This development is often described as the *Internet of Value* (= Web3).

> **Quote**
> *"If we assume that the WWW has revolutionized information and the Web2 revo-*
> *lutionizes interactions, the Web3 has the potential to revolutionize agreements and*
> *value exchange"*
>    Shermin Voshmgir, 2019

One key concept of this new blockchain world is the *token*—or more precisely: the *cryptographic token* (Definition 2.13, [Voshmgir19], [Spencer19]). Such tokens are entities which are strongly protected by modern *cryptography* (e.g., [Paar10], [Shetty19]).

▶ **Definition 2.13: Cryptographic Token**
A cryptographic token represents a programmable, digital asset or access right, managed by a smart contract, and an underlying distributed ledger.
    Shermin Voshmgir, 2019.

Why is the token economy a cyber-physical system? The answer is given in Fig. 2.14: The cryptographic token represents a real-world value, such as money, securities, or real estate. Losing, damaging, or altering the token, therefore, results in real-world loss!

    The key innovation of the blockchain—and all following developments of various distributed ledgers—is the architecture of *trust* ([Werbach18]). Blockchains allow secure, non-disputable, transparent transactions between untrusted parties without a central trust authority. They rely on various consensus mechanisms to establish the validity of an entry into the blockchain. Once accepted into the blockchain, entries can never be modified or deleted: They are protected by *strong cryptography*.

    However, a first dark danger lurks at the horizon: *quantum computing*. Many of today's widely used cryptographic methods, such as RSA ([Coutinho99]) or elliptic curves ([Hankerson13]), will be broken by quantum algorithms. This danger becomes more real every year and is termed *Cryptography Apocalypse* ([Grimes19]).

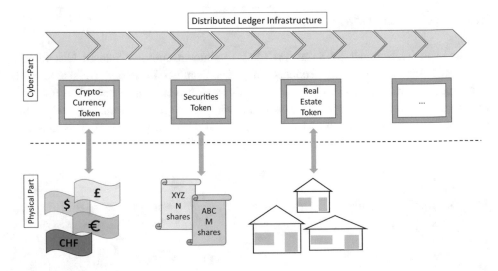

**Fig. 2.14**  Token infrastructure as cyber-physical system

The second danger is the *legal and regulatory framework*. Without a proven, stable, and accepted legal foundation, many applications of blockchains are not possible ([DeFilippi19], [Finck18]). Fortunately, many countries have started or already introduced legislation and regulation of these new digital technologies.

The third danger is careless or faulty implementations. While cryptography, especially post-quantum cryptography ([NIST16]), is mathematically sound and safe for a long time to come, many of the current implementations have faults, bugs, and defects that make them vulnerable (Example 2.10, [Werbach18]).

> **Quote**
> *"Trusted and trustworthy blockchains depend on messy efforts by communities of human beings, just like anything else of similar importance in society"*
>   Kevin Werbach, 2018

Attacks on blockchain based, weakly implemented cyber-physical systems are one of the grave menaces—today and even more in the future (e.g., [CipherTrace21]).

**Example 2.10: DAO Incident**

Decentralized autonomous organization (DAO) is a virtual venture capital fund governed by DAO investors and relies on blockchain technology. It went live in April 2016.

On June 16, 2016, an attacker maliciously retrieved approximately 3.6 million Ether ([Diedrich16]) from the DAO fund by abusing a loophole, known as a "recursive call exploit" in the DAO *blockchain implementation*. The theft represented one-third of the value of the fund at that time.

This incident resulted in a period of chaos in the Ethereum community. Many ways to reverse this illegal transaction were discussed. Finally, a hard fork was implemented, which reversed the unlawful operation. However, this action revealed that Ethereum transactions were not immune from centralized interference and weakened trust in blockchain implementations as a whole.

## 2.9 Cyber-crime and Cyber-war

The transformation of our world to a *digital* economy, society, life, democracy, and also the military is progressing at an amazing speed ([Hanna16], [Dolgin11], [Gupta18], [Moore18], [Scharre19]). More and more applications are computer-controlled (= software-controlled), networked, and decisions are taken by software, often assisted by artificial intelligence.

Besides the valuable, significant advantages of digitization, the potential dangers have risen: *Cyber-crime* and *cyber-war* have become an everyday reality—often with highly unpleasant consequences.

> **Quote**
> *"We live in a society exquisitely dependent on science and technology, in which hardly anyone knows anything about science and technology"*
>     Carl Sagan, 2006

### 2.9.1   Cyber-crime

*Cyber-crime* (Definition 2.14, [Kshetri10], [Clough15], [Lusthaus18], [Goodman15], [Benny13], [Knight20], [Zongo18], [Sangster20]) is a daily occurrence. Every day the news reports successful cyber-attacks, often with amazing damage.

Cyber-crime is executed by individuals, groups, and organized crime. It has become a vast and fast-growing, high-profit industry. Cyber-crime and cyber-crime defense is a continuous, relentless *battle* between criminals and cyber-crime fighters. Unfortunately, cyber-criminals seem to be a step ahead. One reason—probably the most grave—is the irresponsibility of the management of many companies, organizations, and government units. Successfully guarding against cyber-crime is a costly matter, both in terms of money and expert people, which is often not recognized and not acted upon: The result is vulnerable systems!

> **Quote**
> *"Cyber-risk is a business risk, not a technology problem"*
>     Philimon Zongo, 2018

▶ **Definition 2.14: Cyber-crime**
1. Crimes in which the computer or computer network is the target of the criminal activity, for example, hacking, malware, ransomware, pishing, and DoS attacks;
2. Existing offenses where the computer is a tool used to commit the crime, for example, child pornography, stalking, criminal copyright infringements, fraud, and industrial espionage.

Adapted from: https://www.justice.gov/sites/default/files/criminal-ccips/legacy/2015/01/14/ccmanual.pdf

Most cyber-crimes follow an *attack scheme*, as shown in Fig. 2.15. It has three phases: preparation, execution, and exploitation. A crucial factor for success is *cyber-crime tools* or hacking tools ([Kim15], [Diogenes19], [Erickson08], [Yaworski19]). Unfortunately, such powerful and effective tools are available in abundance, many of them free over the Internet ([Pauli13]) or Darknet ([Senker17]). Many tools have been developed by "white hat hackers", i.e., people with good intentions to find and report vulnerabilities, e.g., by penetration testing ([Baloch17], [Weidman14], [Chebbi18]). Example 2.11 describes a modern Internet search engine for Internet-connected devices.

---

**Example 2.11: Search Engine for Internet-connected Devices**

Shodan (https://www.shodan.io/) is a search engine for Internet-connected devices. It lets users find specific types of computers (webcams, routers, servers, etc.) connected to the Internet, guided by various filters.

Shodan currently collects data mostly on web servers using various Internet protocols ([Fall11]), such as (HTTP/HTTPS, multiple ports), as well as FTP, SSH, Telnet, SNMP, IMAP, SIP, real-time streaming protocol (RTSP). Shodan has since been used to find industrial systems, including control systems for water plants, power grids, traffic lights, construction machines, and a cyclotron (https://money.cnn.com/2013/04/08/technology/security/shodan/index.html, last accessed: 25.04.2020). A typical Shodan result screen is shown in Fig. 2.16.

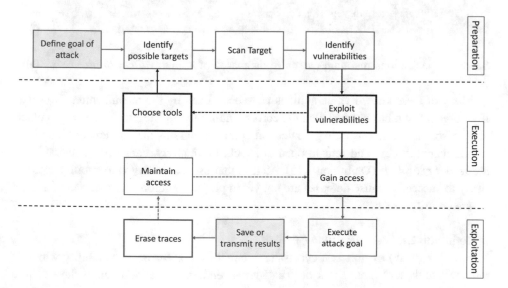

**Fig. 2.15**  Cyber-crime attack scheme

```
47.112.112.169
Hangzhou Alibaba Advertising Co.,Ltd.

Added on 2020-04-23 09:36:07 GMT
China

Technologies:
HTTP/1.0 200 OK
Content-Type: text/html; charset=utf-8
Content-Length: 42304
X-AspNetMvc-Version: 5.2
X-AspNet-Version: 4.0.30319
X-Powered-By: ASP.NET
X-Check: 3c12dc4d54f8e22d666785b733b0052100c53444
X-Language: english
X-Template: tpl_CleanPeppermintBlack_twoclick
X-Cache: miss
```

**Fig. 2.16**  Shodan result screen

Several organizations maintain extensive databases of known *vulnerabilities* (e.g., Table 2.1).

Many of these known vulnerabilities have been fixed by the manufacturer. However, if an organization has not installed the corresponding fix/update, it remains very vulnerable. Also, older versions of, e.g., industrial control programs may no longer be updated by the manufacturers and remain open to attack. Particularly, dangerous vulnerabilities are *zero-day exploits* (Definition 2.15). Attacks that use a zero-day vulnerability are very likely to succeed because defenses are not yet in place. Therefore, zero-day attacks are a severe security threat.

▶ **Definition 2.15: Zero-Day (0day) Exploit**
A zero-day (0day) exploit is a cyber-attack targeting a software vulnerability which is unknown to the software vendor or to anti-virus vendors. The attacker spots the software

**Table 2.1**   Cyber-vulnerability lists

| Institution | Access |
| --- | --- |
| MITRE Common Vulnerabilities and Exposures List | https://cve.mitre.org/ |
| CVE Vulnerability Database | https://www.cvedetails.com/ |
| NIST—US NATIONAL VULNERABILITY DATABASE | https://nvd.nist.gov/ |
| Vulnerability Data Base | https://vuldb.com/ |
| Software Engineering Institute: CERT Coordination Center Vulnerability Notes Database | https://www.kb.cert.org/vuls/ |
| Rapid7 Vulnerabilities and Exploits database | https://www.rapid7.com/de/db |

vulnerability before any parties are interested in mitigating it, quickly creates an exploit, and uses it for an attack.

https://www.imperva.com/learn/application-security/zero-day-exploit/

### 2.9.2   Cyber-war

Cyber-crime's goals are to gain economic advantages: steal money, blackmail for ransom money, obtain trade secrets by espionage, deal with illegal things (such as drugs, weapons, or child pornography), etc. Cyber-war tries to inflict as much damage and harm as possible to the adverse nation-state.

*Cyber-war* (Definition 2.16, [Clarke12], [Stiennon15], [Perlroth21]) is a reality: NATO recognized *cyberspace* as a new, decisive domain of military operations.

▶ **Definition 2.16: Cyber-war**

Cyber-war includes all actions by a nation-state to penetrate another nation's computers or networks for the purpose of causing damage or disruption, or for hindering or inhibiting the use of weapons force.

Adapted from [Clarke12].

> **Quote**
> *"NATO declared cyberspace as the 5th domain of operations—just like air, land, sea, and space at the Warsaw Summit in 2016"*
>     https://www.nato.int/nato_static_fl2014/assets/pdf/pdf_2019_02/20190208_1902-factsheet-cyber-defence-en.pdf

The probably first use of electronic means to defeat a conventional weapons system was the blinded Syrian air defense (Example 2.12, [Clarke12]).

---

**Example 2.12: Blinded Syrian Air Defense**

After the year 2000, Syria was suspected of secretly building a facility for mass destruction weapons. After midnight of September 6, 2007, a new kind of attack on the facility began, originating in cyberspace.

The Israeli air force attacked the facility with F-15 Eagles and F-16 Falcons and completely destroyed the facility. After the successful strike, all planes banked north and left without resistance from the heavily deployed Syrian air defense. Syria had spent billions of dollars on its Russian air defense system.

---

**Quote**
*"The formation of Eagles and Falcons should have lit up the Syrian radars like the Christmas tree illuminating New York's Rockefeller Plaza in December. But they didn't"*
     Richard A. Clarke, 2012

---

Israel had penetrated the Syrian radar systems and inhibited the radar from reporting the fighter jets' incoming formation. Consequently, the air defense system did not react.

Until today, it is not publicly known which electronic means were used to render the air defense system blind. However, it was very effective and avoided a substantial loss in the squadrons of the fighter planes, which returned unattacked.

Cyber-war is the fifth domain of military operations and is entirely different from the domains of land, sea, air, and space ([Richards14]). Without physical force or physical presence, the adversary damages, shuts down, or misuses the *infrastructure*, such as

- The supply of:
  - Electricity (Example 2.13);
  - Water;
  - Transportation (land, sea, air);
- The postal system;
- The financial transaction system (banks, insurances);
- The telecommunications (including satellite links);
- Manufacturing plants:

  – Chemical;
  – Automotive;
  – Food processing;
- Military communications and command systems;
- Government infrastructure;
- Autonomous vehicles;
- Air traffic control systems;
- etc.

---

**Example 2.13: Power System Cyber-sabotage**

In 2015, various places in Ukraine suffered a power failure two days before Christmas. It was night, and the temperature was almost below freezing at the time of the attack. Within a few hours, the engineers were able to restore the power manually. This activity was the first cyber-attack that succeeded in shutting down a power station. The attack also involved a concurrent DDoS attack that sabotaged customer service.

  https://www.rathenau.nl/en/digital-society/cyberspace-without-conflict/ cyber-attacks-cybersabotage

---

The most often used technique is *cyber-sabotage* (Definition 2.17, [Bochman21]). A computer connected to a network, especially to the Internet, is penetrated, malware or back doors are planted, and the attacker takes control over the computer's functionality or data.

▶ **Definition 2.17: Cyber-sabotage**
A cyber-attack for the purpose of sabotaging a digital system and causing damage. Various forms of sabotage are possible because many different applications are connected digitally.

  Adapted from: https://www.rathenau.nl/en/digital-society/cyberspace-without-conflict/ cyber-attacks-cybersabotage

In many cases, such cyber-attacks are not detected by the owner of the affected system because the malware is inactive—such malware is termed a *logic bomb* (Definition 2.18, [Hintzbergen15], [Lysne18], [Sanger19], [Perlroth21]). The malware will only execute when certain conditions are met, e.g.,

- At a specific date/time;
- When particular internal conditions are fulfilled (such as the start of a specific application, a rise in data traffic);
- When triggered by the adversary through the network or other means.

▶ **Definition 2.18: Logic Bomb**

A logic bomb is a piece of code inserted into an operating system or software application that executes a malicious function after a certain amount of time, when specific conditions are met, or when triggered over the network.

Adapted from: https://www.computerhope.com/jargon/l/logibomb.htm

Logic bombs are hard to detect and eliminate because they can be inserted from anywhere. A remote attacker can plant a logic bomb through numerous ways on multiple platforms, e.g., by hiding the malicious code in a script, virus, worm, or deploying it on a SQL server.

> **Quote**
> *"Of the three things about cyberspace that make cyberwar possible, the most important may be the flaws [= Vulnerabilities, Fig. 6.1] in the software and hardware [Note: The other two are: (2) Flaws in the design of the Internet, and (3) The move to put more and more critical systems online]"*
> Richard A. Clarke, 2012

One of the most rewarding targets for cyber-war is military *C4ISR systems* (Definition 2.19, [Shakarian13], [VanPutte17], [VanPutte19]). Today's *weapon systems* are highly dependent on networks, and the loss or compromise of these networks will render whole attack and defense weapons useless. Therefore, many nation-states prepare the *cyber-battlefield* today by efforts to compromise the potential adversary's C4ISR systems.

▶ **Definition 2.19: C4ISR**

C4ISR is an acronym used by the US Department of Defense, US intelligence agencies, and the defense community which stands for command, control, communications, computers, intelligence, surveillance, and reconnaissance.

https://www.baesystems.com/en-us/definition/what-is-c4isr

Although little is published about these activities, at least one detailed report is available: well-written, informative, and open ([VanPutte17], Example 2.14).

---

**Example 2.14: Terminal Fury**

"Terminal Fury" ([VanPutte17]) is a two-week, vast war-game with over 3'000 service members, government officials, and defense contractors connected by a web of electronic communications systems for planning large-scale military operations in the Pacific theater. Decisions of actors are fed into a giant computer game, simulating all situations, movements, and encounters.

In the year 2004 exercise, a group of government hackers (including the author of this example) was tasked to attack the highly protected network and war simulation computer. By the third day of the exercise, the hackers were deep inside the network and "owned" many computers. They noticed a US Navy carrier strike group, complete with an aircraft carrier and its supporting destroyers, submarines, and tenders steaming toward an adversary's shores. With a few presses on the keyboard, the hackers caused bright red foreign submarines to appear magically in front of the fleet in every command center across the Pacific theater. The white blips of the US carrier group slowly turned away from the ghost submarines. Typing on a keyboard had manipulated the Navy into changing the direction of a carrier strike group.

## 2.10  Diffuse Computer Crime

### 2.10.1  Supply Chain Dangers

Today's software-systems are highly complex constructs ([Furrer19]). Many people and organizations have to contribute to the development and operation of these systems. A whole *supply chain* is required to bring these software-systems to market to provide valuable products or services to their users. New dangers emerge through the supply chain members (Definition 2.20, e.g., [Essig13]). Any development step and each hardware and software brought into the software-system may introduce *some accidental* or *intentional vulnerabilities* (= supply chain dangers, Definition 2.20). Therefore, systems in operation may contain *dormant* logic bombs, back doors, or malware (Fig. 2.17, Example 2.15, Example 2.16).

▶ **Definition 2.20: Supply Chain Danger**
Property of a hardware or software third-party product which has the power to make it do things which are not in the interest of its owner or user (accidentally or intentionally).
　　Adapted from [Lysne18].

Identifying such latent, dormant vulnerabilities in today's complex systems is believed to be impossible in most cases ([Comparetti10], [Lysne18]).

> **Quote**
> *"Distrust in the design prompted several countries to ban equipment from companies such as Huawei and ZTE, based on the fear that it could have been deliberately designed to act against the interests of its owners"*
> 　　Olav Lysne, 2018

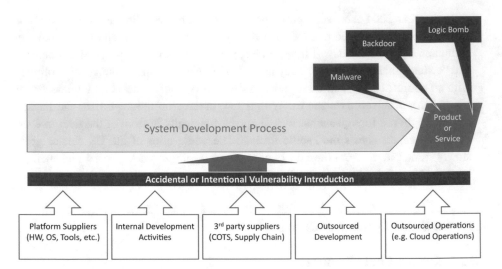

**Fig. 2.17**   Supply chain dangers

---

**Example 2.15: Supply Chain Dangers**

1. In September 2010, the US Missile Defense Agency found that the memory in a high-altitude missile's mission computer was counterfeit. Had the bomb launched, it most likely would have failed, the agency said.
2. Two years earlier, the FBI seized US$ 76 million of counterfeit CISCO routers that the Bureau said could have provided Chinese hackers a back door into US government networks. A number of government agencies bought the routers from an authorized Cisco vendor, but that legitimate vendor purchased the routers from a high-risk Chinese supplier.

   https://money.cnn.com/2012/11/08/technology/security/counterfeit-tech/index.html

---

**Quote**
*"Cyber criminals have begun to penetrate the supply chains for both computer hardware and software manufacturers to inject malicious code"*
   Richard A. Clarke, 2012

---

**Example 2.16: SolarWinds Orion Code Compromise**

December 13, 2020: Malicious actors are currently exploiting SolarWinds Orion products. The Orion platform is a suite of products to monitor the health of IT

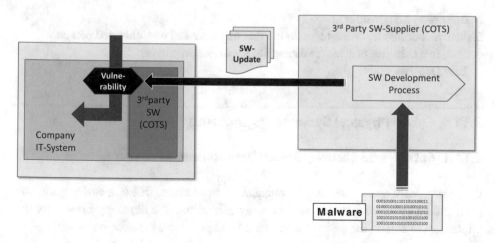

**Fig. 2.18**  Software distribution attack path

networks (https://www.solarwinds.com). SolarWinds acknowledged that hackers had inserted malware into its *software update distribution* mechanism. This security incident resulted in malicious code being pushed to more than 16'000 customers, many of them across the US Federal Government and Fortune 500 firms. The malware permits attackers to directly access the affected organizations' network management systems (Fig. 2.18)—therefore, attacking many resources.

The security incident is so extensive that the US Secretary of Homeland Security issued the *Emergency Directive 21–01* through its Cyber-security and Infrastructure Security Agency (13.12.2020: https://cyber.dhs.gov/ed/21-01/). Such an attack path unbolted by a third party is extremely difficult to control and constitutes a growing concern in the industry.

## 2.10.2 Insider Crime

Threats to the system come either:

- Over connected *networks* (such as the Internet, e.g., [Saxe18], [Selby17], [Diogenes18], or the 5G mobile network ([NTIA21]);
- From *insiders* (Intranet, e.g., [Mehan16], [Stolfo09]).

Defending against network attacks is difficult. However, protecting against *insider attacks* is even more challenging. Insiders with malicious intent often have valid credentials for access to their Intranet or can get them via a criminal act (Definition 2.21). Insider crime often has devastating consequences. Many examples are known (e.g., [CSI20]).

▶ **Definition 2.21: Insider Crime**
Insider risk or insider crime refers to the fact that a trusted or authorized person will participate in a behavior that causes damage to his or her employer.
Eric D. Shaw, 2009.

## 2.11   Cyber-Physical Systems Engineering

### 2.11.1  Safety- and Security-Aware Development Process

*Cyber-physical systems* are built by humans. In most cases, this is a challenging, complex, and complicated undertaking with many constraints. The development activity must implement or extend the desired functionality and guarantee the *acceptable residual risk* of the systems transferred to operation.

The foundation of safe and secure cyber-physical systems is the *safety- and security-aware development process* (Definition 2.22, [Leveson11], [Furrer19]).

▶ **Definition 2.22: Safety- and Security-Aware Development Process**
Software development process which embeds safety and security concerns and decisions in all phases, values safety and security higher than functionality, and is backed by a suitable governance.

> **Quote**
> *"One key to having a cost-effective safety effort is to embed it into a system engineering process from the very beginning and to design safety into the system as the design decisions are made"*
>     Nancy G. Leveson, 2011

The most important process is the *software development process*. A very large number of different software development processes exist (Examples: [Thayer12a], [Thayer12b], [Thayer12c], [Kuhrmann16], [Münch14], [Kruchten03], [Ahern08], [Oram10], [Shuja07], [Armour03], [Jeffries15], [Humble10], [Nygard16],). Also, a rich literature on "lessons learned" is available (e.g., [Eisner20]).

The essential elements of safety- and security-aware development processs are shown in a generic form in Fig. 2.19. The insights are as follows:

- *Safety requirements* and *security requirements* for a project or product must be explicitly formulated, documented, and justified, at least with the same precision and completeness as the functional requirements. The requirements must be finalized before any development phase starts;

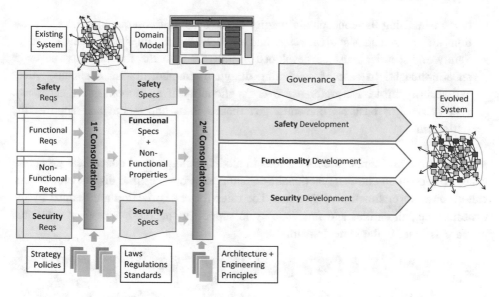

**Fig. 2.19**  Safety- and security-aware development process

> **Quote**
> *"Contrary to common belief, software rarely fails. More often than not, the software behaved exactly as it was required to, but it was the requirements that were flawed"*
>    M. Bialy, 2017 (in [Griffor16])

- The requirements must be technology and implementation independent. No safety or security means or technologies should be specified. These will be introduced by the risk management process, applicable policies, and industry standards at a later phase;
- All requirements must be carefully aligned and integrated into the *existing system*. The requirements must conform to all strategies and policies in the organization and must respect all applicable laws, regulations, and industry standards. An appropriate mechanism, such as reviews, must be in place to align, approve, or correct requirements;
- The requirements must be accurately transformed into the *specifications*. These deliver the information for system architecting and design ([Rupp15], [Wiegers13], [Robertson12]). Whenever possible, use formal requirements and specifications representation;
- The following alignment step for the specifications is the correct integration into the *existing system*. First, the *organization's domain model* (Definition 2.23, [Furrer19]) is used to assign the new functionality, i.e., the individual specifications, to the correct part of the existing system, thus avoiding functional and information redundancy and

thus maintaining the conceptual integrity of the system. Subsequently, the specifications are checked against *architecture* and *engineering* principles ([Furrer19]);

- Safety and security must be developed *in parallel* with the functionality, never as an afterthought! In case of conflicting design or implementation decisions, safety and security must have preference over functionality. If *compromises* are necessary, e.g., because of business demands, governance must understand and approve the settlement.

▶ **Definition 2.23: Domain Model** A domain model is a high-level conceptual grouping/categorization of all entities, functionalities, and their properties into disjunct containers, called (business or functional) domains. The rules for categorization are defined by the cohesion, i.e., all entities and functionality to support a specific business or functional area are assigned to the same domain.

> **Quote**
> *"As the systems we build and operate increase in size and complexity, the use of sophisticated systems engineering approaches becomes more critical. Important system-level (emergent) properties, such as safety or security, must be built into the design of these systems: They cannot be effectively added on or simply measured afterwards"*
>    Nancy G. Leveson, 2011

One crucial process step is the final step: the *explementation* of *decommissioned applications* (Definition 2.24).

▶ **Definition 2.24: Application Decommissioning** Application decommissioning (or application retirement) means the total removal of all code, information, data, relationships, and references, including the respective structures of a retired application from all parts of the IT system.

Decommissioning is an essential final step of the development process: A retired application must be eliminated entirely from the IT system. A specific legacy application is often replaced by a modern, technology-affine, new application. However, in some cases, the retiring application remains in the IT system—resulting in a significant risk! In most cases, the condemned application is still used by unknown users and is not maintained correctly and not updated anymore by the IT department. Therefore, existing vulnerabilities are not remedied, unwanted functionality is still available, new threats are not mitigated, and misuse of the code, information, or access rights is possible.

> **Quote**
>
> *"Forgotten, decayed code or data intentionally or accidentally left in an IT system constitutes a grave, imponderable risk for safety and security"*

Unfortunately, many legacy replacement projects do not get the time or the funding for the complete explementation of the superseded application. Such projects are cynically called «Plus-1 projects», because they leave the application landscape with one more application, i.e., the old and the new one!

Note that completely decommissioning an application is challenging, laborious, and even risky work: All dependencies—within and outside of the organization's IT system—must be identified and acceptably terminated (e.g., Example 2.14, [Seacord03], [Warren99], [SoftLanding19], [Wortmann13]). Principle 9.24 lists the key points of a dependable development process.

> **Quote**
>
> *"Retiring an application that has been part of an enterprise's technology landscape for years, perhaps even decades, is not a simple case of clicking «uninstall». Users, business processes, interdependencies between different applications, and many other factors need to be considered before decisions and actions can be taken"*
>
> Alex Wortmann, 2013

## 2.11.2 Governance

Any safety- and security-aware development process needs strong, competent, and consequent *governance* (Definition 2.25). During system development, many conflicts can emerge, such as

- Disputes between business unit interests and safety/security requirements;
- A clash between short time-to-market of the functionality and longer development time for safety or security measures;
- Disagreement over the allocation of the development money between functionality delivery and safety/security risk mitigation measures;
- Questioning or dissent about the assessment of specific risks and of the effort and technologies to mitigate them;
- Arguments about the budget to be allocated to enhance or correct safety/security in the existing system (corrective and preventive maintenance);
- … and some more.

▶ **Definition 2.25: Governance**

Corporate governance is the organization and the system of rules, practices, and processes by which a company is directed and controlled.

　　Adapted from: https://www.icsa.org.uk/about-us/policy/what-is-corporate-governance

Governance concerning *safety* and *security* means well-established processes to fairly and competently resolve any conflict during the development and operation of the cyber-physical systems ([Murer11], [Maleh19a], [Fitzgerald11], [Rost11], [Furrer19]).

## 2.11.3 Competence Center

Both *safety engineering* ([Leveson16], [Smith20], [Gullo18]), and *security engineering* ([Mead16], [Shrobe18], [Koç18], [Song17]) for cyber-physical systems are today highly specialized, knowledge-intensive, and fast-moving engineering disciplines. Successful safety engineers and security engineers are highly educated, experienced, and lifelong learning individuals.

　　Both fields—safety engineering and security engineering—have become broad fields of knowledge and are moving forward at high speed. No individual can master this complexity. Therefore, many organizations have built *competence centers* for safety and security (Definition 2.26). Such competence centers house the safety and security specialists, usually under the leadership of the chief architect ([Murer11], [Furrer19]). They act as consultants in all projects and assure the conceptual integrity of all projects.

▶ **Definition 2.26: Competence Center**

An organizational structure is used to coordinate IT skills—such as safety engineering and security engineering—within an enterprise. Competence centers provide expertise for project or program support, acting both as repositories of knowledge and resource pools for multiple business areas.

　　https://www.gartner.com/en/information-technology/glossary/competency-center

One of the most essential tasks of the competence center is to guide higher management's decision finding—defending and justifying safe and secure solutions! This process requires a high degree of trust between the business units and the information technology/development departments ([Roughton19], [Trim16]).

> **Quote**
> *"Decision makers often don't understand operating in cyberspace. This lack of understanding forces leaders to speak in colorful, yet vague figures of speech"*
> 　Michael A. Vanputte, 2017

## 2.11.4 Contract-Based Engineering

Cooperation *contracts* (formal or semi-formal) between parts of the software-system are a powerful instrument to manage both complexity of the system and contribute to cyber-physical systems' *safety* and *security*.

Contracts (Definition 2.27) as a specification for cooperation are used on all software hierarchy levels (Fig. 2.20). Software engineering based on contracts was introduced by Bertrand Meyer ([Meyer98]). He also introduced a software development methodology "design-by-contract", leading to significantly higher software quality.

▶ **Definition 2.27: Contract**

A contract is a precise, verifiable—preferably formal—agreement between two cooperating systems to assure interoperability, i.e., precisely specify the mechanisms to exchange and make use of data/information, to access functionality, and to transfer control between systems. Contracts include preconditions, post-conditions, and invariants.

Two types of contracts are used in the software-system as follows:

a. *Interface contracts*: Agreed on the level of modules and components ([Meyer09], [Benveniste12]),
b. *Service contracts*: Established on the higher levels (Fig. 2.20), i.e., between applications and constituent systems of CPSoS's ([Erl05], [Erl08], [Erl11], [Erl16], [Erl17]).

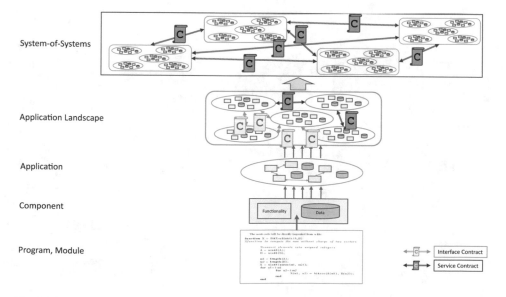

**Fig. 2.20**  Contracts in software engineering

### 2.11.4.1  Interface Contracts

**Quote**

*"The idea of design-by-contract considers the construction of software-systems as the implementation of a number of individual contractual relations between modules, each characterized by a precise specification of obligations and benefits"*
    Bertrand Meyer, 2009

An *interface contract* (Fig. 2.21) is a formalized agreement between two (or more) software-systems parts—mainly on the level of components and applications. The interface contract specifies the functionality between the collaborating partners.

A valuable contribution to contracts is *preconditions* and *post-conditions* (Fig. 2.21). A precondition is a set of constraints which must be fulfilled before the call can be executed. The set of post-conditions is the guaranteed result after the call has been executed. *Invariants* are properties that are assumed on entry and guaranteed on the exit of the call. Example 2.17 lists a small program segment with pre and post-conditions.

---

**Example 2.17: Preconditions and Post-conditions**

The program segment below adds a customer identification and a customer name to the customer information file (CIF)

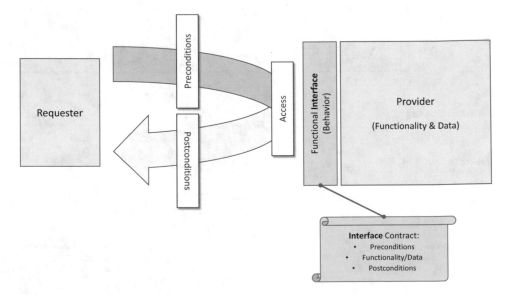

**Fig. 2.21**  Contracts, preconditions, and post-conditions

```
Precondition:
       1. ID not already active
Functionality: addCustomerID
Postconditions:
       1. ID now active
       2. #ofCustomers + 1

Precondition:
       1. ID active
Functionality: addCustomerName
Postcondition:
       1. Name registered
```

### 2.11.4.2 Service Contracts

The difference between an *interface contract* and a *service contract* is shown in Fig. 2.22: An interface contract is primarily a technical agreement between two cooperating parts, whereas a service contract includes additional properties of the exchange, such as the *operational* properties, the *commercial* parameters, and in many cases, some *guarantees*.

The guarantees are essential both for security and safety: In this section of the service contract, essential properties such as *fault containment* in safety-critical applications or *access rights* in security-critical applications are defined and enforced.

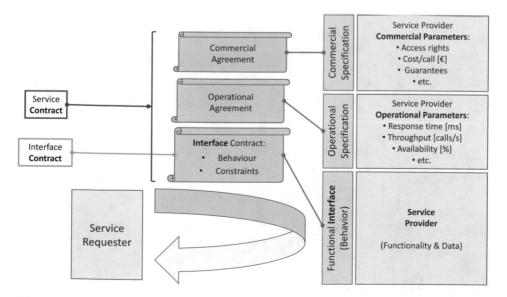

**Fig. 2.22**  Interface and service contracts

**Example 2.18: The Ariane Example**

*"Design by contract is the principle that the interfaces between modules of a software system—especially a mission-critical one—should be governed by precise specifications. The contracts cover mutual obligations (pre-conditions), benefits (post-conditions), and consistency constraints (invariants). Together, these properties are known as assertions and are directly supported in some design and programming languages. A recent $500 million software error provides a sobering reminder that this principle is not just a pleasant academic ideal. On June 4, 1996, the maiden flight of the European Ariane 5 launcher crashed about 40 seconds after takeoff. The rocket was uninsured. The French space agency, CNES (Centre National d'Etudes Spatiales), and the European Space Agency (ESA) immediately appointed an international inquiry board. The board makes several recommendations with respect to software process improvement. There is a simple lesson to be learned from this event: Reuse without a precise, rigorous specification mechanism is a risk of potentially disastrous proportions"*

Source: [Jézéquel97].

The definition of contracts can have several forms, such as *programming constructs* (for modules and components, [Faella20], [Mitchell01], [Pugh06], [Ploesch04], [Meyer98], [Meyer09], [Jong18], [Tockey19], [Czarnecki00], [Lecessi19], [Cambell16]), *templates* ([Bachmann02], [Wheatcraft10]), or *language-based* service contracts (for applications or SoS' constituent systems, e.g., using WSDL [Peterson11], [Yang18], [Benveniste12], [Erl08], [Alonso04], [Daigneau11], Fig. 2.23).

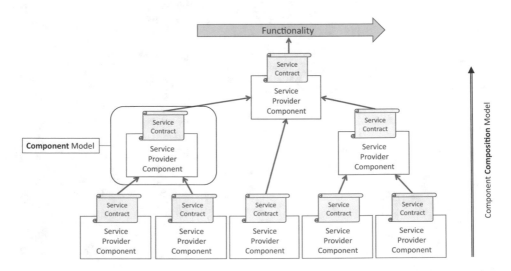

**Fig. 2.23**   Contract-based engineering

### 2.11.4.3 Contract-Based Engineering

The concept of interface/service contracts generates the field of *contract-based engineering* (Definition 2.28, Fig. 2.23, [Ploesch12], [Meyer09], [Witt09]).

▶ **Definition 2.28: Contract-Based Engineering (Design by Contract)**

Under the design by contract theory, a software-system is viewed as a set of communicating components whose interaction is based on precisely—whenever possible formally—defined specifications of the mutual rights and obligations (https://www.eiffel.com/values/design-by-contract/introduction/).

Contract-based engineering relies on the contract as carrier of functionality, operational parameters, commercial conditions, and guarantees.

### 2.11.5  Agile Methods in Safety and Security[3]

### 2.11.5.1  The Agile Manifesto

In 2001, 17 software engineers—many of them distinguished, respected scientists—published a new software development methodology: the *Agile software development* method (Definition 2.29, e.g., [Douglass15], [Meyer14], [Ashmore14], [Adkins10]). This proposal was a reaction to the increasing heaviness, slowness, and overhead of the current software development processes, such as, e.g., CMMI (capability maturity model integration, e.g., [Chrissis11]) or the ISO/IEC 25010:2011 (systems and software engineering/systems and software quality requirements and evaluation (SQuaRE)/system and software quality models [https://www.iso.org/obp/ui/#iso:std:iso-iec:25010:ed-1:v1:en]).

▶ **Definition 2.29: Agile Methods**

The Agile method is a software development methodology based on four key statements:

1. Individuals and interactions over processes and tools;
2. Working software over comprehensive documentation;
3. Customer collaboration over contract negotiation;
4. Responding to change ov following a plan.

   http://agilemanifesto.org

They published the fundamental statements of the Agile method in their Manifesto for *Agile Software Development* (http://agilemanifesto.org). The Manifesto was later supplemented by 12 principles, such as, e.g., principle 11: "The best architectures,

---

[3] Part of this section has been reused from the authors previous book "Future-Proof Software-Systems", Springer Vieweg Verlag, Wiesbaden, Germany, 2019. ISBN 978-3-658-19937-1 ([Furrer19]).

requirements, and designs emerge from self-organizing teams" [http://agilemanifesto.
org/principles.html].

The Agile method immediately triggered heated discussions between "Agilists"
and "Traditionalists". Agilists saw the dawn of a new, more productive software area,
whereas traditionalists identified an unforgivable step backward into the dark ages of
software engineering. The truth is somewhere in the middle: If Agile methods are adequately utilized in an adapted organization, they can indeed bring significant progress
concerning time-to-market and cost (e.g., [Larmann10], [Bloomberg13], [Douglass15],
[Merkow19], [Scheerer18], [Larman16]). However, suppose one has to deal with large,
complex, mission-critical, long-lived systems with stringent quality of service properties (such as security, safety, or business continuity): In that case, Agile methods may
become *dangerous* because of their curtailing of conceptual, modeling, architecture, and
design effort ([Meyer14], [Boehm04]).

> **Quote**
> *"In today's fast-paced, often "agile" software development, how can the secure
> design be implemented? In my experience, tossing requirements, architectures, and
> designs 'over the wall' and into the creative, dynamic pit of Agile development is a
> sure road to failure"*
>     Brook S.E. Schoenfield, 2015

### 2.11.5.2  Agile Application Spectrum

The embittered battle between agilists and traditionalists continues to this day. However,
there is no doubt that Agile methods, especially in conjunction with the SCRUM™
framework ([Rubin12], [Schwaber17]), have brought significant advantages under the
following conditions:

- A *small* team can implement the targetted requirements (~ 8 people) working face to
  face and including stakeholder representatives;
- The requirements have only a minimal, *strictly local* impact on the software-
  system. This is often the case if the system is based on a *microservice* architecture
  ([Amundsen16], [Newman15]);
- A first overarching, effective process is in place to ensure the adequate *overall architecture* and its evolution. The process must detect and prevent any architecture violation, architecture erosion, and technical debt accumulation;
- A second overarching, compulsive process is in place to enforce the global quality of
  service properties (e.g., security, safety, business continuity, conformance to laws and
  regulations).

### 2.11.5.3  Agile Methods and CPS Safety and Security

The focus of this monograph is the safety and security of cyber-physical systems.
Traditionally, such systems were associated with a *solid foundation* based on systems

architecture, (semi-) formal modeling, reliable development processes, and extensive documentation. Such a reliable foundation makes the construction and operation of today's—and much more of tomorrow's—trustworthy CPS manageable!

On the other hand, the commercial pressure to react in a significantly shorter time-to-market to new requirements, changes in the environment, and new legal and compliance regulations are rising every year. The promise of the Agile methods was seen as the light in the tunnel. However, the Agile method often traded damage to the foundation (= architecture erosion, technical debt accumulation) against the desired time-to-market. The Agile methods advocated severely reducing the amount of work invested in architecture, modeling, and documentation—especially the up-front work, i.e., the work done before the coding starts ([Meyer14], [Boehm04], [Furrer19]).

So, the question comes up: What can trustworthy cyber-physical systems safely take over from the Agile community? How far can speed against quality be traded? The inescapable answer is:

1. No compromise in architecture policies, principles, and standards (principle-based architecting);
2. No compromise in quality of service properties (security, safety, business continuity, …);
3. Using an adapted Agile, incremental *development process*, e.g., Scrum™, is possible with additional, accompanying, mandatory processes.

The possible gain in speed lies in an adapted development process. This, however, also requires an adapted organization. One possible solution is shown in Fig. 2.24 ([Leffingwell07], [Leffingwell10], [Bloomberg13], [Erder16], [Coplien10], [Rumpe19], [Beine18]). The Scrum™ cycle is at the center. However, there are now *two sets* of requirements driving the cycle:

- *Architecture requirements*, leading to architecture specifications backlog. These represent the architectural demands, such as the actions to maintain or improve the quality of service properties. They also include the reengineering or refactoring tasks that have to be done in the actual Sprint run;
- *Business (functional) requirements*, leading to the Sprint business requirements backlog, handled by the standard Scrum™ methods, roles, and tools.

Three elements cannot be transferred from the traditional to the adapted, now Agile development environment:

1. The *architecture* governance, organization, and process: The imperative power of architecture remains the foundation of trustworthy cyber-physical systems;
2. The development process must respect all architecture principles, standards, and patterns;

3. A new role is added to the Scrum™ method: The *architecture escort* (Definition 2.30, [Furrer19]). Experienced architects from all disciplines involved (application architects, security architects, business continuity experts, compliance specialists, etc.) are delegated from the central architecture organization and work as esteemed members of the Scrum™ team, although not coding, but consulting.

> **Quote**
> *"In some cases we are agile—and, therefore, we run faster into technical debt"*
>     Philippe Kruchten, 2020

Through the *architecture escort* (Definition 2.30, Fig. 2.24), conformity to adequate architecture is established, enforced, and assured. The architecture escort will ensure that all necessary safety and security concerns are correctly implemented. He/she will do that by continuous involvement with the project and the team, gently guiding safety and security. In (hopefully) rare cases when the architecture escort cannot convince the team to implement the necessary safety and security requirements, he/she will escalate. The team will then be forced by management to comply.

▶ **Definition 2.30: Architecture Escort**
Experienced, specialized architect (application architect, security architect, safety architect, …) delegated to be a member of a SCRUM™ team with the responsibility to:

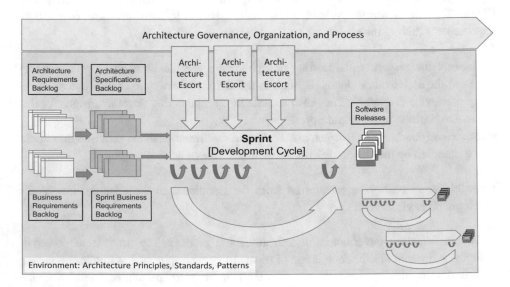

**Fig. 2.24**  Architecture escort for the Agile methods

- Consult and guide the team to conform to the architectures (horizontal and vertical architecture layers) defined by the central architecture team;
- (Gently) enforce the architecture policies, principles, and standards;
- Escalate (as a last resort) if the SCRUM™ team violates any of the architecture policies, principles, and standards.

### 2.11.5.4 Agility Against Architecture?

Agile methods entail a loss of *discipline*. The reduced front-end work, less documentation, and omitted reviews are some causes. Less discipline endangers architecture and the quality properties of the cyber-physical system. Therefore, a recipe is required to ensure the long-lived architecture, the required quality of service properties, and the governance of the trade-off between agility and architecture.

Two solutions are possible (e.g., [Furrer19]):

1. Embedding of the development process into an environment, which ensures the adherence to the intentional architecture;
2. Enriching the Agile development process through specific activities and artifacts (e.g., by SafeScrum™, Example 2.19).

In an organization, which produces trustworthy cyber-physical systems, both solutions will have to be implemented well at the same time. If the structure and the management processes of the organization are not adapted to Agile, the effect of introducing Agile may well be negative overall ([Moran14], [Moran15], [Crowder16]).

> **Quote**
> *"I have seen security and Agile work seamlessly together as a whole, more than once. This process requires that security expertise be seen as part of the team"*
> Brook S.E. Schoenfield, 2015

**Example 2.19: SafeScrum™**

SafeScrum™ is a typical example of an enriched Agile development process ([Hanssen18], [Myklebust18]). SafeScrum™ was motivated by requirements from the *railway signaling domain*—a highly safety-critical application of cyber-physical systems. The development of safety-critical systems is guided by document-centric standards and by heavy processes. The enriched Agile process (Fig. 2.25) brings flexibility in Agile methods while still complying with the governing safety standards EN 50128:2011 and IEC 61508-3.

This is achieved by adding the artifacts and activities shown in Fig. 2.25. An important part is adding a second backlog: the *safety* and *security requirements backlog*. This allows separating the frequently changing functional requirements from the more stable safety and security requirements. ◀

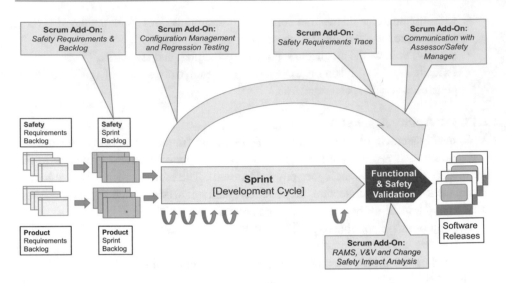

**Fig. 2.25**   SafeScrum™ model

# References

| | |
|---|---|
| [Ackerman21] | Pascal Ackerman: **Industrial Cybersecurity** - *Efficiently monitor the Cybersecurity Posture of your ICS Environment* Packt Publishing, Birmingham, UK, 2nd edition, 2021. ISBN 978-1-800-20209-2 |
| [Ackermann17] | Pascal Ackerman: **Industrial Cybersecurity** - *Efficiently secure critical Infrastructure Systems*. Packt Publishing, Birmingham, UK, 2017. ISBN 978-1-788-39515-1 |
| [Adkins10] | Lyssa Adkins: **Coaching Agile Teams** - *A Companion for ScrumMasters, Agile Coaches, and Project Managers in Transition* Addison-Wesley, Upper Saddle River, NJ, USA, 2010. ISBN 978-0-321-63770-3 |
| [Ahern08] | Dennis M. Ahern, Aaron Clouse, Richard Turner: **CMMI Distilled** - *A Practical Introduction to Integrated Process Improvement*. Addison-Wesley Professional, Upper Saddle River, NJ, USA, 3rd edition, 2008. ISBN 978-0-321-46108-7 |
| [Alonso04] | Gustavo Alonso, Fabio Casati, Harumi Kuno, Vijay Machiraju: **Web Services** - *Concepts, Architectures and Applications*. Springer Verlag, Berlin, Germany, 2004. ISBN 978-3-540-44008-6 |
| [Alosaimi16] | Rana Alosaimi, Mohammad Alnuem: *Risk Management frameworks for Cloud Computing - A critical Review* International Journal of Computer Science & Information Technology (IJCSIT), Vol 8, No 4, August 2016 https://www.researchgate.net/publication/307899076_Risk_Management_Framework_for_Cloud_Computing_A_Critical_Review |
| [Alpaydin16] | Ethem Alpaydin: **Machine Learning** - *The New AI* The MIT Press, Cambridge, MA, USA, 2016 ISBN 978-0-262-52951-8 |

[Amundsen16]     Mike Amundsen, Matt Mclarty: **Microservice Architecture - *Aligning***
                 ***Principles, Practices, and Culture*** O'Reilly Media, Inc., Sebastopol,
                 CA, USA, 2016. ISBN 978-1-491-95625-0

[Armour03]       Philip G. Armour: **Laws of Software Process - *A New Model for***
                 ***the Production and Management of Software***. Auerbach Publishers
                 Inc. (Taylor & Francis), Boca Raton, FL, USA, 2003. ISBN
                 978-0-849-31489-6

[Arseniev20]     Dmitry G. Arseniev, Ludger Overmeyer, Heikki Kälviäinen, Branco
                 Katalinic (Editors): **Cyber-Physical Systems and Control** Springer
                 Nature Switzerland, Cham, Switzerland, 2020. ISBN 978-3-030-34982-0

[Asama02]        Haijme Asama, Tamio Aray, Toshio Fukuda, Tsutomu Hasegawa
                 (Editors): **Distributed Autonomous Robotic Systems** Springer Verlag,
                 Tokyo, Japan, 2002. ISBN 978-4-431-65943-3

[Ashmore14]      Sondra Ashmore, Kristin Runyan: **Introduction to Agile Methods**
                 Addison-Wesley, Upper Saddle River, NJ, USA, 2014. ISBN
                 978-0-321-92956-3

[Astrom11]       Karl Johan Astrom, Bjorn Wittenmark: **Computer-Controlled Systems**
                 **- *Theory and Design*** Dover Books on Electrical Engineering, 3rd edi-
                 tion, 2011. ISBN 978-0-486-48613-0

[Bachmann02]     Felix Bachmann, Len Bass, Paul Clements, David Garlan, James
                 Ivers, Reed Little, Robert Nord, Judith Stafford: **Documenting**
                 **Software Architecture: *Documenting Interfaces*** Carnegie Mellon
                 University, Pittsburg, USA, June 2002. TECHNICAL NOTE CMU/
                 SEI-2002-TN-015. Downloadable from: https://resources.sei.cmu.edu/
                 asset_files/TechnicalNote/2002_004_001_13973.pdf  [Last accessed:
                 27.1.2021]

[Baloch17]       Rafay Baloch: **Ethical Hacking and Penetration Testing Guide CRC**
                 **Press** (Taylor & Francis Ltd.), Boca Raton, FL, USA, 2017. ISBN
                 978-1-138-43682-4

[Banafa18]       Ahmed Banafa: **Secure and Smart Internet of Things (IoT) - *Using***
                 ***Blockchain and Artificial Intelligence*** River Publishers, Delft, NL,
                 2018. ISBN 978-87-7022-030-9

[Bedau08]        Mark A. Bedau, Paul Humphreys (Editors): **Emergence - *Contemporary***
                 ***Readings in Philosophy and Science*** Massachusetts Insitute of
                 Technology (MIT) Press, USA, 2008. ISBN 978-0-262-02621-5

[Beine18]        Gerritt Beine: **Technical Debts - *Economizing Agile Software***
                 ***Architecture*** De Gruyter Oldenbourg, Oldenbourg, Germany, 2018.
                 ISBN 978-3-1104-6299-9

[Benny13]        Daniel J. Benny: **Industrial Espionage** CRC Press (Taylor & Francis),
                 Boca Raton, FL, USA, 2013. ISBN 978-1-466-56814-3

[Benveniste12]   Albert Benveniste, Benoît Caillaud, Dejan Nickovic, Roberto Passerone,
                 Jean-Baptiste Raclet, Philipp Reinkemeier, Alberto Sangiovanni-
                 Vincentelli, Werner Damm, Tom Henzinger, Kim Larsen: *Contracts for*
                 *Systems Design* INRIA RESEARCH REPORT, N° 8147, November
                 2012 ISSN 0249-6399 Downloadable from: http://hal.inria.fr/
                 docs/00/75/85/14/PDF/RR-8147.pdf [last accessed: 23.9.2017]

[Bestugin19]     A.R. Bestugin, A.A. Eshenko, A.D. Filin, A.P. Plyasovskikh, A.Y.
                 Shatrakov, Y.G. Shatrakov: **Air Traffic Control Automated Systems**
                 Springer Nature, Singapore, Singapore, ISBN 978-9-811-39385-3

[Bloomberg13]        Jason Bloomberg: **The Agile Architecture Revolution** - *How Cloud Computing, REST-Based SOA, and Mobile Computing Are Changing Enterprise IT* John Wiley & Sons, Inc., New York, NY, USA, 2013. ISBN 978-1-118-40977-0

[Bochman21]          Andrew A. Bochman, Sarah Freeman: **Countering Cyber Sabotage** - *Introducing Consequence-Driven, Cyber-Informed Engineering (CCE)* CRC Press (Taylor & Francis), Boca Raton, FL, USA, 2021. ISBN 978-0-367-67371-0

[Boehm04]            Barry Boehm, Richard Turner: **Balancing Agility and Discipline** - *A Guide for the Perplexed* Pearson Education (Addison-Wesley), Upper Saddle River, NJ, USA, 2004. ISBN 978-0-321-18612-6

[Bondavalli16]       Andrea Bondavalli, Sara Bouchenak, Hermann Kopetz (Editors): **Cyber-Physical Systems of Systems: Foundations - A Conceptual Model and Some Derivations: The AMADEOS Legacy** Springer Lecture Notes in Computer Science, Heidelberg, Germany, 2016. ISBN 978-3-319-47589-9

[Brender13]          Nathalie Brender, Iliya Markov: **Risk Perception and Risk Management in Cloud Computing - Results from a Case Study of Swiss Companies** International Journal of Information Management, Vol. 33, Nr. 5, pp. 726–733. Downloadable from: https://doc.rero.ch/record/208726/files/Brender_Markov_2013_risk_perception.pdf    [Last accessed: 18.4.2021]

[Cambell16]          Edward Cambell: **Microservices Architecture** - *Make the Architecture of a Software as simple as possible* CreateSpace Independent Publishing Platform, North Charleston, S.C., USA, 2016. ISBN 978-1-5300-0053-1

[Cao20a]             Yijia Cao, Yong Li, Xuan Liu, Christian Rehtanz: **Cyber-Physical Energy and Power Systems** - *Modeling, Analysis, and Application* Springer Nature Singapore, Singapore, 2020. ISBN 978-9-811-50064-0

[Cao20b]             Zongyu Cao, Wanyou Lv, Yanhong Huang, Jianqi Shi, Qin Li: *Formal Analysis and Verification of Airborne Software based on DO-333* Electronics 2020, Vol. 9, Nr. 2, (Design and Applications of Software Architectures), February 2020). Downloadable from: https://doi.org/10.3390/electronics9020327 [Last accessed: 23.1.2021]

[Chaari16]           Rihab Chaari, Fatma Ellouze, Anis Koubaa, Basit Qureshi, Nuno Pereira, Habib Youssef, Eduardo Tovar: *Cyber-Physical Systems Clouds: A Survey* Computer Networks, The International Journal of Computer and Telecommunications Networking, Elsevier, Amsterdam, September 2016. Downloadable from: https://www.researchgate.net/publication/307996575_Cyber-Physical_Systems_Clouds_A_Survey [Last accessed: 20.4.2021]

[Chalup19]           Stephan Chalup, Tim Niemueller, Jackrit Suthakorn, Mary-Anne Williams (Editors): **RoboCup 2019** - *Robot World Cup XXIII* Springer Nature Switzerland, Cham, Switzerland, 2019 (Lecture Notes in Computer Science, Vol. 11531). ISBN 978-303-035698-9

[Champagne14]        Phil Champagne: **The Book Of Satoshi** - *The Collected Writings of Bitcoin Creator Satoshi Nakamoto* e53 Publishing LLC, 2014. ISBN 978-0-9960-6130-8

[Charbonneau17]      Paul Charbonneau: **Natural Complexity** - *A Modeling Handbook* Princeton University Press, Princeton, USA, 2017. ISBN 978-0-691-17035-0

[Chebbi18]        Chiheb Chebbi: **Advanced Infrastructure Penetration Testing -** *Defend your Systems from methodized and proficient Attackers* Packt Publishing, Birmingham, UK, 2018. ISBN 978-1-78862-448-0

[Chrissis11]      Mary Beth Chrissis, Mike Konrad, Sandy Shrum: **CMMI for Development -** *Guidelines for Process Integration and Product Improvement* Addison Wesley Publishing Inc., USA (The SEI Series in Software Engineering), 3rd revised edition, 2011. ISBN 978-0-321-71150-2

[CipherTrace21]   CipherTrace: *Cryptocurrency Crime and Anti-Money Laundering Report 2021* CipherTrace Report-CAML-20210128, February 2021. CipherTrace Inc., Los Gatos, CA, USA, 2021. Downloadable from: https://ciphertrace.com/2020-year-end-cryptocurrency-crime-and-anti-money-laundering-report/ [Last accessed: 31.1.2021]

[Cirani18]        Simone Cirani, Gianluigi Ferrari, Marco Picone, Luca Veltri: **Internet of Things -** *Architectures, Protocols and Standards* John Wiley & Sons, Inc., Hoboken, NJ, USA, 2018. ISBN 978-1-119-35967-8

[Clarke12]        Richard A. Clarke, Robert Knake: **Cyber War -** *The Next Threat to National Security and What to Do About It* Ecco (Harper Collins Publishers), New York, N.Y., USA, Reprint edition 2012. ISBN 978-0-061-96224-0

[Clough15]        Jonathan Clough: **Principles of Cybercrime** Cambridge University Press, Cambridge, UK, 2nd edition, 2015. ISBN 978-1-107-69816-1

[Colbert16]       Edward J. M. Colbert, Alexander Kott (Editors): **Cyber-Security of SCADA and Other Industrial Control Systems** Springer International Publishing, Cham, Switzerland, 2016. ISBN 978-3-319-32123-3

[Colombo14]       Armando W. Colombo, Thomas Bangemann, Stamatis Karnouskos, Jerker Delsing, Petr Stlutka, Robert Harrison, François Jammes, Jose L. Martinez Lastra (Editors): **Industrial Cloud-Based Cyber-Physical Systems -** *The IMC-AESOP Approach* Springer International Publishing, Cham, Switzerland, 2014 (2016 Softcover reprint of the original 1st edition 2014). ISBN 978-3-319-38265-4

[Comparetti10]    Paolo Milani Comparetti, Guido Salvaneschi, Engin Kirda, Clemens Kolbitsch, Christopher Kruegel, Stefano Zanero: *Identifying Dormant Functionality in Malware Programs* 2010 IEEE Symposium on Security and Privacy, 16–19 May 2010, Berkeley/Oakland, CA, USA. DOI: https://doi.org/10.1109/SP.2010.12. Downloadable from: http://citeseerx.ist.psu.edu/viewdoc/download?doi=10.1.1.642.2105&rep=rep1&type=pdf [last accessed: 30.4.2020]

[Coplien10]       James O. Coplien, Gertrud Bjornvig: **Lean Architecture for Agile Software Development** John Wiley & Sons Inc., New York, NY, USA, 2010. ISBN 978-0-470-68420-7

[Coutinho99]      S. C. Coutinho: **The Mathematics of Ciphers -** *Number Theory and RSA Cryptography* A K Peters Publishers, Natick, MA, USA, 1999. ISBN 978-1-5688-1082-9

[Crowder16]       James A. Crowder, Shelli Friess: **Agile Project Management -** *Managing for Success* Springer-Verlag, Heidelberg, Germany, 2016. ISBN 978-3-319-34922-0

[CSI20]           Cybersecurity Insiders: *2020 Insider Threat Report* Cybersecurity Insiders, USA, 2020. Available at: https://www.cybersecurity-insiders.com/portfolio/2020-insider-threat-report/ [last accessed: 2.5.2020]

[Czarnecki00]        Krzysztof Czarnecki, Ulrich Eisenecker: **Generative Programming -** *Methods, Tools, and Applications* Addison Wesley, Upper Saddle River, NJ, USA, 2000. ISBN 978-0-201-30977-5

[Daigneau11]         Robert Daigneau: **Service Design Patterns -** *Fundamental Design Solutions for SOAP/WSDL and RESTful Web Services* Addison-Wesley, Indianapolis, IL, USA, 2011. ISBN 978-0-321-54420-9

[Daniels20]          Dewi Daniels: *The Boeing 737 MAX Accidents* In: Mike Parsons, Mark Nicholson (Editors): **Assuring Safe Autonomy** Proceedings of the 28th Safety-Critical Systems Symposium (SSS'20), York, UK, 11–13 February 2020. ISBN 978-1-713305-66-8

[Davies20]           John Davies, Carolina Fortuna (Editors): **The Internet of Things - From Data to Insight** John Wiley & Sons, Inc., Hoboken, NJ, USA, 2020. ISBN 978-1-119-54526-2

[DeFilippi19]        Primavera De Filippi: **Blockchain and the Law -** *The Rule of Code* Harvard University Press, Cambridge, MA, USA, 2019. ISBN 978-0-674-24159-6

[Diedrich16]         Henning Diedrich: **Ethereum -** *Blockchains, Digital Assets, Smart Contracts, Decentralized Autonomous Organizations* CreateSpace Independent Publishing Platform, North Charleston, S.C., USA, 2016. ISBN 978-1-5239-3047-0

[Diez14]             Oscar Diez: **Resilience of Cloud Computer in Critical Systems** CreateSpace Independent Publishing Platform, 2014. ISBN 978-1-4929-1205-7

[Diogenes18]         Yuri Diogenes, Erdal Ozkaya:  **Cybersecurity –** *Attack and Defense Strategies - Counter modern Threats and employ state-of-the-art Tools and Techniques to protect your Organization against Cybercriminals*. Packt Publishing, Birminghm, UK, updated 2nd edition, 2019. ISBN 978-1-838-827793 01-9

[Diogenes19]         Yuri Diogenes, Erdal Ozkaya: **Cybersecurity - Attack and Defense Strategies:** *Counter modern threats and employ state-of-the-art tools and techniques to protect your organization against cyber-criminals* Packt Publishing, Birmingham, UK, 2nd edition, 2019. ISBN 978-1-83882-779-3

[Dolgin11]           Alexander Dolgin: **Manifesto of the New Economy -** *Institutions and Business Models of the Digital Society* Springer-Verlag, Heidelberg, Germany, 2012. ISBN 978-3-642-21276-5

[Dotson19]           Chris Dotson: **Practical Cloud Security -** *A Guide for Secure Design and Deployment* O'Reilly Media, Sebastopol, CA, USA, 2019. ISBN 978-1-492-03751-4

[Douglass15]         Bruce Powel Douglass: **Agile Systems Engineering** Morgan Kaufmann (Elsevier), Waltham, MA, USA, 2015. ISBN 978-0-128-02120-0f

[EBF20]              Cloud Banking Forum (EBF): **Cloud Exit Strategy** Technical Paper, Cloud Banking Forum, Brussels, Belgium, 2020. Downloadable from: https://www.ebf.eu/wp-content/uploads/2020/09/Cloud-exit-strategy-Testing-of-exit-plans.pdf [Last accessed: 5.5.2021]

[Eisner20]           Howard Eisner: **Systems Engineering -** *Fifty Lessons Learned* CRC Press (Taylor & Francis), Boca Raton, FL, USA, 2020. ISBN 978-0-367-53430-1

[Erder16]            Murat Erder, Pierre Pureur: **Continuous Architecture -** *Sustainable Architecture in an Agile and Cloud-Centric World* Morgan Kaufmann (Elsevier), Waltham, MA, USA, 2016. ISBN 978-0-12-803284-8

[Erickson08]     Jon Erickson: **Hacking: The Art of Exploitation** No Starch Press, San Francisco, CA, USA, 2nd edition, 2008. ISBN 978-1-5932-7144-2

[Erl05]     Thomas Erl: **Service-oriented Architecture - *Concepts, Technology, and Design*** Prentice-Hall, Upper Saddle River, N.J., USA, 2005, 2nd edition, 2016. ISBN 978-0-133-85858-7

[Erl08]     Thomas Erl: **Web Service Contract Design and Versioning for SOA** Prentice-Hall, Upper Saddle River, N.J., USA, 2008. ISBN 978-0-136-13517-3

[Erl11]     Thomas Erl et. al.: **SOA Governance - *Governing Shared Services On-Premise and in the Cloud*** Pearson Education (Prentice Hall Service-Oriented Computing Series), Upper Saddle River, N.J., USA, 2011. ISBN 978-0-138-15675-6

[Erl13]     Thomas Erl, Zaigham Mahmood, Ricardo Puttini: **Cloud Computing - *Concepts, Technology & Architecture*** Prentice-Hall, Upper Saddle River, N.J., USA, 2013. ISBN 978-0-133-38752-0

[Erl16]     Thomas Erl: **Service-Oriented Architecture: *Analysis and Design for Services and Microservices*** Prentice-Hall, Upper Saddle River, N.J., USA, 2nd edition, 2016. ISBN 978-0-133-85858-7

[Erl17]     Thoms Erl: **Service Infrastructure- *On-Premise and in the Cloud*** Pearson Education (Prentice Hall Service Technology Series), Upper Saddle River, N.J., USA, 2017. ISBN 978-0-133-85872-3

[Essig13]     Michael Essig, Michael Hülsmann, Eva-Maria Kern, Stephan Klein-Schmeink (Editors): **Supply Chain Safety Management - *Security and Robustness in Logistics*** Springer Verlag, Berlin, Germany, 2013. ISBN 978-3-642-32020-0

[Faella20]     Marco Faella: **Seriously Good Software - *Code that works, survives, and wins*** Manning Publications, Shelter Island, NY, USA, 2020. ISBN 978-1-617-29629-1

[Fagan20a]     Michael Fagan, Jeffrey Marron, Kevin G. Brady, Barbara B. Cuthill, Katerina N. Megas, Rebecca Herold: **IoT Device Cybersecurity Guidance for the Federal Government - *Establishing IoT Device Cybersecurity Requirements*** Draft NIST Special Publication 800–213, NIST, Washington, USA, December 2020, Downloadable from: https://nvlpubs.nist.gov/nistpubs/SpecialPublications/NIST.SP.800-213-draft.pdf [Last accessed: 16.12.2020]

[Fagan20b]     Michael Fagan, Katerina N. Megas, Karen Scarfone, Matthew Smith: **IoT Device Cybersecurity Capability Core Baseline** NISTIR 8259A, NIST, Washington, USA, May 2020, Downloadable from: https://nvl-pubs.nist.gov/nistpubs/ir/2020/NIST.IR.8259A.pdf [Last accessed: 16.12.2020]

[Fagan20c]     Michael Fagan, Jeffrey Marron, Kevin G. Brady, Barbara B. Cuthill, Katerina N. Megas, Rebecca Herold: **IoT Non-Technical Supporting Capability Core Baseline** Draft NISTIR 8259B, NIST, Washington, USA, December 2020, Downloadable from: https://nvlpubs.nist.gov/nistpubs/ir/2020/NIST.IR.8259b-draft.pdf [Last accessed: 16.12.2020]

[Fagan20d]     Michael Fagan, Jeffrey Marron, Kevin G. Brady, Barbara B. Cuthill, Katerina N. Megas, Rebecca Herold: **Creating a Profile Using the IoT Core Baseline and Non-Technical Baseline Draft** NISTIR 8259C, NIST, Washington, USA, December 2020, Downloadable from: https://nvlpubs.nist.gov/nistpubs/ir/2020/NIST.IR.8259c-draft.pdf [Last accessed: 16.12.2020]

[Fagan20e]       Michael Fagan, Jeffrey Marron, Kevin G. Brady, Barbara B. Cuthill, Katerina N. Megas, Rebecca Herold: **Profile Using the IoT Core Baseline and Non-Technical Baseline for the Federal Government** Draft NISTIR 8259D, NIST, Washington, USA, December 2020, Downloadable from: https://nvlpubs.nist.gov/nistpubs/ir/2020/NIST. IR.8259D-draft.pdf [Last accessed: 16.12.2020]

[Fagan20f]       Michael Fagan, Katerina N. Megas, Karen Scarfone, Matthew Smith: **Foundational Cybersecurity Activities for IoT Device Manufacturers** NISTIR 8259, NIST, Washington, USA, May 2020, Downloadable from: https://nvlpubs.nist.gov/nistpubs/ir/2020/NIST.IR.8259.pdf [Last accessed: 16.12.2020]

[Fall11]         Kevin R. Fall, W. Richard Stevens: **TCP/IP Illustrated, Volume 1 - *The Protocols*** Pearson Education (Addison-Wesley), Upper Saddle River, NJ, USA, 2nd edition, 2011. ISBN 978-0-321-33631-6

[Finck18]        Michèle Finck: **Blockchain Regulation and Governance in Europe** Cambridge University Press, Cambridge, UK, 2018. ISBN 978-1-108-46545-8

[Fitzgerald11]   Todd Fitzgerald: **Information Security Governance Simplified - *From the Boardroom to the Keyboard*** CRC Press (Taylor & Francis), Boca Raton, FL, USA, 2012. ISBN 978-1-439-81163-4

[Franke18]       Björn Franke: **Embedded Systems.** *Lecture 11: Worst-Case Execution Time* University of Edinburgh, Edinburgh, Scotland, 2018. Downloadable from: http://www.inf.ed.ac.uk/teaching/courses/es/PDFs/ lecture_11.pdf [Last accessed: 31.1.20121]

[Furrer19]       Frank J. Furrer: **Future-Proof Software-Systems - *A Sustainable Evolution Strategy*** Springer Vieweg Verlag, Wiesbaden, Germany, 2019. ISBN 978-3-658-19937-1

[Gai20]          Silvano Gai: **Building a Future-Proof Cloud Infrastructure - *A Unified Architecture for Network, Security, and Storage Services*** Addison-Wesley Educational Publishers Inc., Upper Saddle River, NJ, USA, 2020. ISBN 978-0-136-62409-7

[Gao16]          Yang Gao (Editor): **Contemporary Planetary Robotics - *An Approach Toward Autonomous Systems*** Wiley-VCH Verlag GmbH & Co. KGaA, Weinheim, Germany, 2016. ISBN-13: 978-3-527-41325-6

[Ghafoor13]      Kayhan Zrar Ghafoor, Kamalrulnizam Abu Bakar, Marwan Aziz Mohammed, Jaime Lloret: *Vehicular Cloud Computing: Trends and Challenges* White Paper, 2013. Downloadable from: https://www. researchgate.net/publication/262773787_Vehicular_Cloud_Computing_ Trends_and_Challenges/link/0f317538d851178338000000/download [Last accessed: 01.05.2021]

[Goel08]         Asvin Goel: **Fleet Telematics - *Real-time Management and Planning of Commercial Vehicle Operations*** Springer Science+Business Media LLC, New York, NY, USA, 2008. ISBN 978-1-441-94524-2

[Goodman15]      Marc Goodman: **Future Crimes - *Inside The Digital Underground and the Battle For Our Connected World*** Transworld Publishers, London, UK, 2016. ISBN 978-0-552-17080-2

[Greengard15]    Samuel Greengard: **The Internet of Things** MIT Press, Cambridge, MA, USA, 2015. ISBN 978-0-262-52773-6

[Greer20]        Melvin B. Greer, Kevin L. Jackson: **Practical Cloud Security - *A Cross-Industry View*** CRC Press (Taylor & Francis Group), Boca Raton, FL, USA, 2020. ISBN 978-0-367-65842-7

[Griffor16]       Edward Griffor: **Handbook of System Safety and Security - *Cyber Risk and Risk Management, Cyber Security, Threat Analysis, Functional Safety, Software Systems, and Cyber-Physical Systems*** Syngress (Elsevier), Cambridge, MA, USA, 2016. ISBN 978-0-128-03773-7

[Grimes19]        Roger A. Grimes: **Cryptography Apocalypse - *Preparing for the Day When Quantum Computing Breaks Today's Crypto*** John Wiley & Sons, Inc., Hoboken, NJ, USA, 2019. ISBN 978-1-119-61819-5

[Gullo18]         Louis J. Gullo, Jack Dixon: **Design for Safety** John Wiley & Sons, Inc., Hoboken, NJ, USA, 2018. ISBN 978-1-118-97429-2

[Gupta18]         Sunil Gupta: **Driving Digital Strategy** Harvard Business Review Press, Boston, MA, USA, 2018. ISBN 978-1-633-69268-8

[Hankerson13]     Darrel Hankerson, Alfred Menezes, Scott Vanstone: **Guide to Elliptic Curve Cryptography** Springer Science+Business Media, New York, N.Y., USA, 2013 (Softcover reprint of the original 1st edition 2004. ISBN 978-1-441-92929-7

[Hanna16]         Nagy K. Hanna: **Mastering Digital Transformation - *Towards a Smarter Society, Economy, City, and Nation*** Emerald Group Publishing, Bingley, UK, 2016. ISBN 978-1-7856-0465-2

[Hanssen15]       Dag H. Hanssen: **Programmable Logic Controllers** John Wiley & Sons, Inc., Chichester, UK, 2015. ISBN 978-1-118-94924-5

[Hanssen18]       Geir Kjetil Hanssen, Tor Stålhane, Thor Myklebust: **SafeScrum - *Agile Development of Safety-Critical Software*** Springer-Verlag, Heidelberg, Germany, 2018. ISBN 978-3-319-99333-1

[Haque20]         Enamul Haque: **The Ultimate Modern Guide to Cloud Computing - *Everything from Cloud Adoption to Business Value*** ENEL Publications, London, UK, 2020. ISBN 979-8-666-05063-7

[HCTI20]          House Committee on Transportation and Infrastructure (HCTI): The Boeing 737 MAX Aircraft: **Costs, Consequences, and Lessons from its Design, Development, and Certification - *Preliminary Investigative Findings*** March 2020. Downloadable from: https://transportation. house.gov/imo/media/doc/TI%20Preliminary%20Investigative%20 Findings%20Boeing%20737%20MAX%20March%202020.pdf   (last accessed: 27.4.2020)

[Hintzbergen15]   Jule Hintzbergen, Kees Hintzbergen, André Smulders, Hans Baars: **Foundations Of Information Security Based on ISO27001 And ISO27002** Van Haren Publishing, Zaltbommel, Nederlands, 3rd edition, 2015. ISBN 978-94-018-0012-9

[HomelandSecurity17]  U.S. Homeland Security: **Emerging Cyber Threats to the United States** Security Technologies of the Committee on Homeland Security House of Representatives Subcommittee on Cybersecurity (Infrastructure Protection), CreateSpace Independent Publishing Platform, North Charleston, S.C., USA,2017. ISBN 978-1-5464-8512-4

[Humble10]        Jez Humble, David Farley: **Continuous Delivery - *Reliable Software Releases Through Build, Test, and Deployment Automation*** Addison Wesley, Upper Saddle River, NJ, USA, 2010. ISBN 978-0-321-60191-9

[Hutter05]        Marcus Hutter: **Universal Artificial Intelligence - *Sequential Decisions Based on Algorithmic Probability*** Springer-Verlag, Berlin, Germany, 2005. ISBN 978-3-540-22139-5

[IBM06]             IBM Business Consulting Services: **An Architectural Blueprint for Autonomic Computing** IBM Autonomic Computing, 4th edition, June 2006. Downloadable from: http://www-01.ibm.com/software/tivoli/autonomic/

[IEC62443]          IEC 62443: **Security for Industrial Automation and Control Systems (14 Parts)** IEC (International Electrotechnical Commission), Geneva, Switzerland, 2020. Available from: https://webstore.iec.ch/home [Last accessed: 26.01.2022]

[ISA22]             ISA: **Quick Start Guide - An Overview of the ISA/IEC 62443 Standards** International Society of Automation (ISA), Global Cybersecurity Alliance, Durham, NC, USA, 2022. Downloadable from: https://gca.isa.org/blog/download-the-new-guide-to-the-isa/iec-62443-cybersecurity-standards [Last accessed: 26.01.2022]

[Jamshidi09a]       Mo Jamshidi (Editor): **Systems of Systems Engineering -** *Principles and Applications* CRC Press, Taylor & Francis Group, Boca Raton, USA, 2009. ISBN 978-1-4200-6588-6

[Jamshidi09b]       Mo Jamshidi (Editor): **Systems of Systems Engineering -** *Innovations for the 21st Century* John Wiley & Sons Inc., Hoboken, New Jersey, USA, 2009. ISBN 978-0-470-19590-1

[Jeffries15]        Ron Jeffries: **The Nature of Software Development -** *Keep It Simple, Make It Valuable, Build It Piece by Piece*. The Pragmatic Bookshelf, Raleigh, NC, USA, 2015. ISBN 978-1-941-22237-9

[Jenner18]          Nate Jenner: **Cloud Security -** *Introduction to Cloud Security and Data Protection* CreateSpace Independent Publishing Platform, North Charleston, S.C., USA, 2018. ISBN 978-1-7170-1843-4

[Jeschke16]         Sabina Jeschke, Christian Brecher, Houbing Song, Danda B. Rawat (Editors): **Industrial Internet of Things -** *Cybermanufacturing Systems* Springer International, Cham, Switzerland, 2017. ISBN 978-3-319-42558-0

[Jézéquel97]        Jean-Marc Jézéquel, Bertrand Meyer: *Design by Contract - The Lessons of Ariane*. White Paper, EiffelSoft, Goleta, CA, USA, 1997. Downloadable from: http://se.ethz.ch/~meyer/publications/computer/ariane.pdf [Last accessed: 5.7.2022]

[Johnson15]         Thomas A. Johnson (Editor): **Cybersecurity -** *Protecting Critical Infrastructures from Cyber Attack and Cyber Warfare* CRC Press, Taylor & Francis Ltd., Boca Raton, FL, USA, 2015. ISBN 978-1-482-23922-5

[Jong18]            Jos Jong: **Vertically Integrated Architectures -** *Versioned Data Models, Implicit Services, and Persistence-Aware Programming* Apress Media LLC, New York, N.Y., USA, 2018. ISBN 978-1-4842-4251-3

[Juliano16]         Dustin Juliano: **AI Security** CreateSpace Independent Publishing Platform, North Charleston, S.C., USA, 2016. ISBN 978-1-5351-1900-9 Undine Press, Fort Myers, FL, USA, 2016, ISBN 978-1-5351-1900-9 Read online: www. http://aisecurity.org/ [Last accessed: 16.3.2021]

[Kabalci19]         Ersan Kabalci, Yasin Kabalci (Editors): **Smart Grids and Their Communication Systems** Springer Nature Singapore Pte Ltd., Singapore, Singapore, 2019. ISBN 978-9-811-34680-4

[Kephart03]         Jeffrey O. Kephart, David M. Chess: **The Vision of Autonomic Computing** IEEE Computer Society, New York, N.Y., USA, 2003, pp. 41–50. Downloadable from: http://130.18.208.80/~ramkumar/acvi-sion.pdf (last accessed: 26.4.2020)

[Kim15]  Peter Kim: **The Hacker Playbook 2** - *Practical Guide To Penetration Testing* CreateSpace Independent Publishing Platform, North Charleston, S.C., USA, 2015. ISBN 978-1-51221-456-7

[Knapp13]  Eric D. Knapp, Raj Samani: **Applied Cyber Security and the Smart Grid** - *Implementing Security Controls into the Modern Power Infrastructure* Syngress (Elsevier), Amsterdam, Netherlands, 2013. ISBN 978-1-597-49998-9

[Knapp14]  Eric D. Knapp, Joel Thomas Langill: **Industrial Network Security** - *Securing Critical Infrastructure Networks for Smart Grid, SCADA, and Other Industrial Control Systems* Syngress (Elsevier), Waltham, MA, USA, 2nd edition 2014. ISBN 978-0-124-20114-9

[Knight20]  Alissa Knight: **Hacking Connected Cars** - *Tactics, Techniques, and Procedures* John Wiley & Sons, Inc., Hoboken, New Jersey, USA, 2020. ISBN 978-1-119-49180-4

[Kopetz11]  Hermann Kopetz: **Real-Time Systems** - *Design Principles for Distributed Embedded Applications* Springer Science & Business Media, New York, N.Y., USA, 2nd edition, 2011. ISBN 978-1-461-42866-4

[Koç18]  Çetin Kaya Koç (Editor): **Cyber-Physical Systems Security**. Springer Nature Switzerland, Cham, Switzerland, 2018. ISBN 978-3-319-98934-1

[Koubaa16]  Anis Koubaa, Elhadi Shakshuki (Editors): **Robots and Sensor Clouds** Springer International Publishing, Cham, Switzerland, 2016. ISBN 978-3-319-35863-5

[Kravets20]  Alla G. Kravets, Alexander A. Bolshakov, Maxim V. Shcherbakov (Editors): **Cyber-Physical Systems** - *Industry 4.0 Challenges* Springer Nature Switzerland, Cham, Switzerland, 2020. ISBN 978-3-030-32650-0

[Kruchten03]  Philippe Kruchten: **The Rational Unified Process** - *An Introduction*. Addison-Wesley Professional, Upper Saddle River, NJ, USA, 3rd edition, 2003. ISBN 978-0-321-19770-2

[Krutz16]  Ronald L. Krutz: **Industrial Automation and Control System Security** International Society of Automation (ISA), Research Triangle Park, N.C., USA, 2nd edition 2016. ISBN 978-1-9415-4682-6

[Kshetri10]  Nir Kshetri: **The Global Cybercrime Industry** - *Economic, Institutional and Strategic Perspectives* Springer-Verlag, Berlin, Germany, 2010. ISBN 978-3-642-11521-9

[Kuhrmann16]  Marco Kuhrmann, Jürgen Münch, Ita Richardson, Andreas Rausch, He Zhang (Editors): **Managing Software Process Evolution:** *Traditional, Agile and Beyond – How to Handle Process Change*. Springer International Publishing, Cham, Switzerland, 2016. ISBN 978-319-31543-0

[Lakhwani20]  Kamlesh Lakhwani, Hemant Kumar Gianey, Joseph Kofi Wireko, Kamal Kant Hiran: **Internet of Things (IoT)** - *Principles, Paradigms, and Applications of IoT* BPB Publications, New Delhi, India, 2020. ISBN 978-9-389-42336-5

[Lalanda13]  Philippe Lalanda, Julie A. McCann, Ada Diaconescu: **Autonomic Computing** - *Principles, Design, and Implementation* Springer-Verlag, London, UK, 2013. ISBN 978-1-4471-5006-0

[Larman16]          Craig Larman: **Large-Scale Scrum** - *More with Less* Addison-Wesley Professional, Upper Saddle River, NJ, USA, 2016. ISBN 978-0-321-98571-2

[Larmann10]         Craig Larmann, Bas Vodde: **Practices for Scaling Lean & Agile Development** - *Large, Multisite, and Offshore Product Development with Large-Scale Scrum* Pearson Education (Addison-Wesley), Upper Saddle River, NJ, USA, 2010. ISBN 978-0-321-63640-9

[Lawless17]         W.F. Lawless, Ranjeev Mittu, Donald Sofge, Stephen Russell (Editors): **Autonomy and Artificial Intelligence** - *A Threat or Savior?* Springer International Publishing, Cham, Switzerland, 2017. ISBN 978-3-319-59718-8

[Lecessi19]         Ralph Lecessi: **Functional Interfaces in Java** - *Fundamentals and Examples* Apress Media LLC, New York, N.Y., USA, 2019. ISBN 978-1-484-24277-3

[Lee20a]            Kang B. Lee, Richard Candell, Hans-Peter Bernhard, Dave Cavalcanti, Zhibo Pang, Inaki Val: *Reliable, High-Performance Wireless Systems for Factory Automation* NIST White Paper, NIST, Gaithersburg, MD, USA, 2020. Downloadable from: https://www.nist.gov/publications/reliable-high-performance-wireless-systems-factory-automation   [Last accessed: 4.5.2021]

[Lee20b]            Robert M. Lee *2020 SANS Cyber Threat Intelligence (CTI) Survey* SANS Insitute Report, SANS, bethesda, MA, U.S.A., February 2020. Downloadable   from:   https://www.domaintools.com/content/SANS_CTI_Survey_2020.pdf [Last accessed: 23.12.2020]

[Leffingwell07]     Dean Leffingwell: **Scaling Software Agility** - *Best Practices for Large Enterprises* Addison-Wesley Professional, Upper Saddle River, NJ, USA, 2007. ISBN 978-0-321-45819-3

[Leffingwell10]     Dean Leffingwell: **Agile Software Requirements** - *Lean Requirements Practices for Teams, Programs, and the Enterprise* Addison Wesley, Upper Saddle River, NJ, USA, 2010. ISBN 978-0-321-63584-6

[Leveson11]         Nancy G. Leveson: **Engineering a Safer World** - *Systems Thinking applied to Safety* MIT Press, Cambridge MA, USA, 2011. ISBN 978-0-262-01662-9

[Leveson16]         Nancy G. Leveson: **Engineering a Safer World** - *Systems Thinking Applied to Safety* MIT Press Ltd., Massachusetts, MA, USA, 2016. ISBN 978-0-262-53369-0

[Liu18]             Shaoshan Liu, Liyun Li, Jie Tang: **Creating Autonomous Vehicle Systems** Morgan & Claypool Publishers, San Rafael, CA, USA, 2018. ISBN 978-1-681-73007-3

[Loske15]           André Loske: **IT Security Risk Management in the Context of Cloud Computing** - *Towards an Understanding of the Key Role of Providers' IT Security Risk Perceptions* Springer Fachmedien, Wiesbaden, Germany, 2015. ISBN 978-3-658-11339-1

[Lusthaus18]        Jonathan Lusthaus: **Industry of Anonymity** - *Inside the Business of Cybercrime* Harvard University Press, Cambridge, MA, USA, 2018. ISBN 978-0-674-97941-3

[Luzeaux11]         Dominique Luzeaux, Jean-René Ruault, Jean-Luc Wipplere (Editors): **Complex Systems and Systems of Systems Engineering** iSTE Publishing Ltd., London UK, 2011. Distributed by John Wiley & Sons Inc., NY, USA. ISBN 978-1-848-21253-4

[Lysne18]      Olav Lysne: **The Huawei and Snowden Questions** - *Can Electronic Equipment from Untrusted Vendors be Verified? Can an Untrusted Vendor Build Trust into Electronic Equipment?* Springer International Publishing AG, Cham, Switzerland. ISBN-13: 978-3-319-74949-5

[Macaulay11]   Tyson Macaulay, Bryan L. Singer: **Cybersecurity for Industrial Control Systems** - *SCADA, DCS, PLC, HMI, and SIS* CRC Press (Taylor & Francis), Boca Raton, FL, USA, 2011. ISBN 978-1-439-80196-3

[Mahmood19]    Zaigham Mahmood (Editor): **The Internet of Things in the Industrial Sector** - *Security and Device Connectivity, Smart Environments, and Industry 4.0* Springer International, Cham, Switzerland, 2019. ISBN 978-3-030-24891-8

[Maier98]      Mark W. Maier: **Architecting Principles for Systems-of-Systems** Systems Engineering (Journal), Volume1, Issue 4, 1998, Pages 267–284. https://onlinelibrary.wiley.com/doi/10.1002/(SICI)1520-6858(1998)1:4%3C267::AID-SYS3%3E3.0.CO;2-D

[Majumdar20]   Suryadipta Majumdar, Taous Madi, Yushun Wang, Azadeh Tabiban, Momem Oqaily, Amir Alimohammdadifar, Yosr Jarraya, Makan Pourzandi, Lingyu Wang, Mourad Debbabi: **Cloud Security Auditing** Springer Nature Switzerland, Cham, Switzerland, 2019. ISBN 978-3-030-23130-9

[Maleh19a]     Yassine Maleh, Abdelkebir Sahid, Mustapha Belaissaoui: **Strategic IT Governance and Performance Frameworks in Large Organizations** IGI Global Publishing (Business Science Reference), Hershey, PA, USA, 2019. ISBN 978-1-5225-7826-0

[Manoj19]      K. S. Manoj: **Industrial Automation with SCADA** - *Concepts, Communications, and Security* Notion Press, Chennai, India, 2019. ISBN 978-1-684-66828-1

[Marwedel18]   Peter Marwedel: **Embedded System Design** - *Embedded Systems Foundations of Cyber-Physical Systems, and the Internet of Things* Springer International Publishing, Cham, Switzerland, 3rd edition, 2018. ISBN 978-3-319-56043-4

[Maurer16]     Markus Maurer, J. Christian Gerdes, Barbara Lenz, Hermann Winnder (Editors): **Autonomous Driving** - *Technical, Legal and Social Aspects* Springer Verlag, Germany, 2016. ISBN 978-3-662-48845-4

[McCaffrey20]  Peter McCaffrey: **An Introduction to Healthcare Informatics** - *Building Data-Driven Tools* Academic Press (Elsevier), London, UK, 2020. ISBN 978-0-128-14915-7

[McCarthy21]   Jim McCarthy, Don Faatz, Nik Urlaub, John Wiltberger, Tsion Yimer: **Securing the Industrial Internet of Things** -*Cybersecurity for Distributed Energy Resources* Volume B: Approach, Architecture, and Security Characteristic NIST Special Publication 1800–32B, Draft April 2021. Downloadable from: https://www.nccoe.nist.gov/sites/default/files/library/sp1800/energy-iiot-sp1800-32b.pdf This publication is available free of charge from https://www.nccoe.nist.gov/iiot

[McCarthy22]   Jim McCarthy, Eileen Division, Don Faatz, Nik Urlaub, John Wiltberger, Tsion Yimer: **Securing Distributed Energy Resources** - *An Example of Industrial Internet of Things Cybersecurity* NIST Special Publication 1800–32, February 2022, US National Institute of Standards and Technology, Washington, USA. Downloadable from: https://www. nccoe.nist.gov/energy/securing-distributed-energy-resources          [Last accessed: 03.02.2022

[Mead16]          Nancy R. Mead, Carol C. Woody: **Cyber Security Engineering** - *A Practical Approach for Systems and Software Assurance* Addison-Wesley Professional, Boston, USA, 2016. ISBN 978-0-134-18980-2

[Mehan16]         Julie E. Mehan: **Insider Threat** - *A Guide to Understanding, Detecting, and Defending Against the Enemy from Within*. ITGP (IT Governance Publishing), Ely, UK, 2016. ISBN 978-1-8492-8839-2

[Merkow19]        Mark S. Merkow: **Secure, Resilient, and Agile Software Development** CRC Press (Taylor & Francis Ltd.), Boca Raton, FL, USA, 2019. ISBN 978-0-367-33259-4

[Meyer09]         Bertrand Meyer: **A Touch of Class** - *Learning to Program Well with Objects and Contracts* Springer-Verlag, Berlin, 2009. ISBN 978-3-540-92144-5

[Meyer14]         Bertrand Meyer: **Agile! - The Good, the Hype, and the Ugly** Springer International Publishing, Cham, Switzerland, 2014. ISBN 978-3-319-05154-3

[Meyer98]         Bertrand Meyer: **Object-Oriented Software Construction** Prentice-Hall, Upper Saddle River, N.J., USA, 2nd edition, 1998. ISBN 978-0-136-29155-8

[Milani18]        Farzaneh Milani, C. Beidl: *Cloud-based Vehicle Functions - Motivation, Use-Cases, and Classification* IEEE Vehicular Networking Conference (VNC), 2018, December 5–7, 2018, Taipei, Taiwan.

[Milani19]        Farzaneh Milani, Mike Foell, Christian Beidl: *Suitability Analysis for Cloud-based Vehicle Functions* 8th International Symposium on Development Methodology, Wiesbaden, Germany, November 2019. Downloadable from: https://www.researchgate.net/publication/337286049_Suitability_Analysis_for_Cloud-based_Vehicle_Functions [Last accessed: 17.4.2021]

[Mitchell01]      Richard Mitchell, Jim McKim: **Design by Contract, by Example** Addison-Wesley, Indianapolis, IL, USA, 2001. ISBN 978-0-201-634600

[MITRE14]         The MITRE Corporation: **MITRE Systems Engineering Guide -** *Collected wisdom from MITRE's systems engineering experts* The MITRE Corporation, Bedford, MA, USA, 2014. Downloadable from: http://www.mitre.org/sites/default/files/publications/se-guide-book-interactive.pdf (last accessed: 17.6.2017)

[Mittal18]        Saurabh Mittal, Saikou Diallo, Andreas Tolk (Editors): **Emergent Behaviour in Complex Systems** - *A Modeling and Simulation Approach* John Wiley & Sons, Inc., Hoboken, NJ, USA, 2018. ISBN 978-1-119-37886-0

[Mittu16]         Ranjeev Mittu, Donald Sofge, Alan Wagner, W.F. Lawless (Editors): **Robust Intelligence and Trust in Autonomous Systems** Springer Science & Business Media, New York, N.Y., USA, 2016. ISBN 978-1-489-97666-6

[Mobus15]         George E. Mobus, Michael C. Kalton: **Principles of Systems Science -** *Understanding Complex Systems* Springer Verlag, New York, NY, USA, 2015. ISBN 978-1-493-91919-2

[Mohamed19]       Khaled Salah Mohamed: **The Era of Internet of Things -** *Towards a Smart World* Springer Nature, Cham, Switzerland, 2019. ISBN 978-3-030-18132-1

[Münch14]    Tobias Münch: **System Architecture Design and Platform Development Strategies** - *An Introduction to Electronic Systems Development in the Age of AI, Agile Development, and Organizational Change*. Springer Nature Switzerland AG, Cham, Switzerland, 2022. ISBN 978-3-030-97694-1

[Moore18]    Martin Moore **Democracy Hacked** Oneworld Publications, London, UK, 2018. ISBN 978-1-78607-408-9

[Moran14]    Alan Moran: **Agile Risk Management** Springer-Verlag, Heidelberg, Germany, 2014. ISBN 978-3-319-05007-2

[Moran15]    Alan Moran: **Managing Agile** - *Strategy, Implementation, Organisation, and People* Springer-Verlag, Heidelberg, Germany, 2015. ISBN 978-3-319-16261-4

[Morrow18]    Timothy Morrow: *12 Risks, Threats, & Vulnerabilities in Moving to the Cloud* SEI Blog, March 5, 2018. Carnegie Mellon University, Software Engineering Institute, Pittsburgh, PA, USA. Downloadable from: https://insights.sei.cmu.edu/blog/12-risks-threats-vulnerabilities-in-moving-to-the-cloud/ [Last accessed: 18.4.2021]

[Mougayar16]    William Mougayar: **The Business Blockchain** - *Promise, Practice, and Application of the Next Internet Technology* John Wiley & Sons, Inc., Hoboken, NJ, USA, 2016. ISBN 978-1-119-30031-1

[Murer11]    Stephan Murer, Bruno Bonati, Frank J. Furrer: **Managed Evolution** - *A Strategy for Very Large Information Systems* Springer Verlag, Berlin, Germany, 2011. ISBN 978-3-642-01632-5

[Myklebust18]    Thor Myklebust, Tor Stålhane: **The Agile Safety Case** Springer-Verlag, Heidelberg, Germany, 2018. ISBN 978-3-319-70264-3

[Nakamoto08]    Satoshi Nakamoto: **Bitcoin** - *A Peer-to-Peer Electronic Cash System* BitCoin Organization, 2008. Downloadable from: https://bitcoin.org/bitcoin.pdf []last accessed: 13.4.2020]

[Nakamoto19]    Satoshi Nakamoto, Jaya Klara Brekke: **The White Paper - Original White Paper and a Guide** Ignota Books (www.ignota.org), 2019. ISBN 978-1-9996-7592-9

[Nath17]    Shyam Nath, Robert Stackowiak, Carla Romano: **Architecting the Industrial Internet** - *The Architect's Guide to designing Industrial Internet Solutions* Packt Publishing, Birmingham, UK, 2017. ISBN 978-1-7872-8275-9

[Newman15]    Sam Newman: **Building Microservices** - *Designing Fine-Grained Systems* O'Reilly and Associates, Sebastopol, CA, USA, 2015. ISBN 978-1-491-95035-7 2nd edition 2021, ISBN 978-1-492-03402-5

[Nichols19]    R.K. Nichols, JJCH Ryan, H.C. Mumm, W.D. Lonstein, C. Carter, J.P. Hood: **Unmanned Aircraft Systems in the Cyber-Domain** - *Protecting USA's Advanced Air Assets* New Prairie Press, Kansas State University, Kansas, USA, 2nd edition, 2019. ISBN 978-1-9445-4823-0

[NIST16]    Lidong Chen, Stephen P. Jordan, Yi-Kai Liu, Dustin Moody, Rene C. Peralta, Ray A. Perlner, Daniel C. Smith-Tone: **Report on Post-Quantum Cryptography** US National Institute of Standards and Technology, Gaithersburg, MD, USA, 2016. Downloadable from: https://nvlpubs.nist.gov/nistpubs/ir/2016/NIST.IR.8105.pdf [last accessed: 18.4.2020]

[NTIA21]    NTIA: *Potential Threat Vectors to 5G Infrastructure* United States National Telecommunications and Information Administration (NTIA),

Washington, USA, 2021. Downloadable from: https://media.defense. gov/2021/May/10/2002637751/-1/-1/0/POTENTIAL%20THREAT%20 VECTORS%20TO%205G%20INFRASTRUCTURE.PDF          [Last accessed: 25.01.2022]

[Nygard16]     Michael Nygard: **Release It!** - *Design and Deploy Production-Ready Software*. The Pragmatic Bookshelf, Raleigh, NC, USA, O'Reilly UK Ltd., 2nd edition, 2018. ISBN 978-1-680-50239-8

[Olding20]     Mark Olding: **Future Visions - Industrial Cloud Computing** Independently published, 2020. ISBN 979-8-58181-456-7

[Oram10]       Andy Oram, Greg Wilson: **Making Software - *What Really Works, and Why We Believe It***. O'Reilly Media, Sebastopol, CA, USA, 2010. ISBN 978-0-596-80832-7

[Paar10]       Christof Paar, Jan Pelzl: **Understanding Cryptography** - *A Textbook for Students and Practitioners* Springer-Verlag, Berlin, Germany, 2010. ISBN 978-3-642-04100-6

[Paul17]       Anand  Paul,  Naveen  Chilamkurti,  Alfred  Daniel,  Seungmin Rho: **Intelligent Vehicular Networks and Communications -** *Fundamentals, Architectures, and Solutions* Elsevier, Amsterdam, Netherlands, 2017. ISBN 978-0-128-09266-8

[Pauli13]      Josh Pauli: **The Basics of Web Hacking** - *Tools and Techniques to Attack the Web* Syngress (Elsevier), Waltham, MA, USA, 2013. ISBN 978-0-124-16600-4

[Perahia13]    Eldad Perahia, Robert Stacey: **Next Generation Wireless LANs -** *802.11n and 802.11ac* Cambridge University Press, Cambridge, UK, 2nd edition, 2013. ISBN 978-1-107-01676-7

[Perlroth21]   Nicole Perlroth: **This Is How They Tell Me the World Ends - The Cyber Weapons Arms Race** Bloomsbury Publishing, London, UK, 2021. ISBN 978-1-635-57605-4

[Petersen12]   J. K. Petersen: **Handbook of Surveillance Technologies** CRC Press (Taylor & Francis), Boca Raton, FL, USA, 3rd edition, 2012. ISBN 978-1-439-87315-1

[Peterson11]   Larry L. Peterson, Bruce S. Davie: **Computer Networks -** *A Systems Approach* Morgan Kaufmann (Elsevier), Waltham, MA, USA, 5th edition, 2011. ISBN 978-0-123-85059-1

[Pezoulas20]   Vasileios Pezoulas, Themis Exarchos, Dimitrios I Fotiadis: **Medical Data Sharing, Harmonization and Analytics** Academic Press (Elsevier), London, UK, 2020. ISBN 978-0-128-16507-2

[Piggin14]     Richard Piggin: *Are Industrial Control Systems ready for the Cloud?* International Journal of Critical Infrastructure Protection, Elsevier, Amsterdam, Netherlands, December 2014, Nr. 9. Downloadable from: https://www.researchgate.net/publication/269820923_Are_industrial_ control_systems_ready_for_the_cloud [Last accessed: 17.4.2021]

[Ploesch04]    Reinhold Ploesch: **Contracts, Scenarios, and Prototypes -** *An Integrated Approach to High-Quality Software* Springer Verlag, Berlin, Germany, 2004. ISBN 978-3-540-43486-3

[Ploesch12]    Reinhold Ploesch: **Contracts, Scenarios and Prototypes -** *An Integrated Approach to High Quality Software*. Springer Verlag, Berlin, Germany (Softcover reprint of the original 1st edition   2004), 2012. ISBN 978-3-642-62160-4

| | |
|---|---|
| [Poroshyn14] | Roman Poroshyn: **Stuxnet - *The True Story of Hunt and Evolution*** CreateSpace Independent Publishing Platform, North Charleston, S.C., USA, 2014. ISBN 978-1-4997-0922-3 |
| [Pranic20] | Robert Pranic:**Cloud Security Policy** Independently published, 2020. ISBN 979-8-6243-1208-1 |
| [Prosser19] | Tony Prosser, Mark Taylor: **Fire and Rescue Incident Command - *A Practical Guide to Incident Ground Management*** Pavilion Publishing and Media Ltd, Shoreham-by-Sea, UK, 2019. ISBN 978-1-9127-5509-7 |
| [Prunckun18] | Henry Prunckun (Editor): **Cyber Weaponry - *Issues and Implications of Digital Arms*** Springer International Publishing, Cham, Switzerland, 2018. ISBN 978-3-319-74106-2 |
| [Pugh06] | Ken Pugh: **Interface-Oriented Design** The Pragmatic Bookshelf, Raleigh, NC, USA, 2006. ISBN 978-0-9766-9405-2 |
| [Rabiee17] | Max Rabiee: **Programmable Logic Controllers Hardware and Programming** Goodheart-Willcox Co, Inc., Tinley Park, IL, USA, 4th edition, 2017. ISBN 978-1-631-26932-5 |
| [Rainey18] | Larry B. Rainey, Mo Jamshidi (Editors): **Engineering Emergence - *A Modeling and Simulation Approach*** CRC Press (Taylor & Francis), Boca Raton, FL, USA, 2019. ISBN 978-1-138-04616-0 |
| [Raj17] | Pethuru Raj, Anupama C. Raman: **The Internet of Things - *Enabling Technologies, Platforms, and Use Cases*** CRC Press (Taylor & Francis Inc.), Boca Raton, FL, USA, 2017. ISBN 978-1-498-76128-4 |
| [Rakas20] | Slavica V. BoStjancic Rakas, Mirjana D. Stojanovic: **Cyber Security of Industrial Control Systems in the Future Internet Environment** Information Science Reference (IGI Global), Hershey, PA, USA, 2020. ISBN 978-1-7998-2911-9 |
| [Rammler21] | Roman Rammler: **AI-Automated Telemedicine - *A New Approach to Remote Patient Monitoring*** Independently published, 2021. ISBN 979-8-5996-8950-8 |
| [Rawat16] | Danda B. Rawat, Joel J .P. C. Rodriques, Ivan Stojmenovic (Editors): **Cyber-Physical Systems - *From Theory to Practice*** CRC Press (Taylor & Francis), Boca Raton, FL, USA, 2016. ISBN 978-1-482-26332-9 |
| [Rice20] | Liz Rice: **Container Security - *Fundamental Technology Concepts that Protect Containerized Applications*** O'Reilly Media Inc., Sebastopol, CA, USA, 2020. ISBN 978-1-492-05670-6 |
| [Richards14] | Julian Richards: **Cyber-War - *The Anatomy of the Global Security Threat*** Palgrave McMillan, Basingstoke, UK, 2014. ISBN 978-1-137-39961-8 |
| [Robertson12] | Suzanne Robertson, James Robertson: **Mastering the Requirements Process - *Getting Requirements Right*** Pearson Education (Addison Wesley), Upper Saddle River, NJ, USA, 3rd edition, 2012. ISBN 978-0-321-81574-3 |
| [Robison21] | Peter Robison: **Flying Blind -*The 737 MAX Tragedy and the Fall of Boeing*** Doubleday (Penguin), New York, NY, USA, 2021. ISBN 978-0-385-54649-2 |
| [Romanovsky17] | Alexander Romanosky, Fuyuki Ishikawa (Editors): **Trustworthy Cyber-Physical Systems Engineering** CRC Press, Boca Raton FL, USA, 2017. ISBN 978-1-4978-4245-0 |
| [Rost11] | Johann Rost, Robert L. Glass: **The Dark Side of Software Engineering - *Evil on Computing Projects*** John Wiley & Sons, Hoboken, N.J., USA, 2011 (IEEE Computer Society), 2011. ISBN 978-0-470-59717-0 |

[Roughton19] James Roughton, Nathan Crutchfield, Michael Waite: **Safety Culture - *An Innovative Leadership Approach*** Butterworth-Heinemann (Elsevier), Kidlington, UK, 2nd edition, 2019. ISBN 978-0-128-14663-7

[Rubin12] Kenneth S. Rubin: **Essential Scrum - *A Practical Guide to the Most Popular Agile Process*** Addison Wesley, Upper Saddle River, NJ, USA, 2012. ISBN 978-0-137-04329-3

[Rumpe19] Bernhard Rumpe: **Agile Modeling with UML - *Code Generation, Testing, Refactoring*** Springer International Publishing, Cham, Switzerland, 2019. ISBN 978-3-319-86494-5

[Ruparelia16] Nayan B. Ruparelia: **Cloud Computing** (The MIT Press Essential Knowledge Series) MIT Press, Cambridge, MA, USA, 2016. ISBN 978-0-262-52909-9

[Rupp15] Chris Rupp, Klaus Pohl: **Requirements Engineering Fundamentals - *A Study Guide for the Certified Professional for Requirements Engineering Exam*** Rocky Nook Inc., Santa Barbara, CA, USA, 2nd edition, 2015. ISBN 978-1-9375-3877-4

[Stolfo09] Salvatore J. Stolfo, Steven M. Bellovin Shlomo Hershkop, Angelos Keromytis, Sara Sinclair, Sean W. Smith (Editors): **Insider Attack and Cyber Security - *Beyond the Hacker*.** Springer Science&Business Media, New York, NY, USA, 2009. ISBN 978-1-441-94589-1

[Russell17] Stuart J. Russell, Peter Norvig: **Artificial Intelligence** Prentice-Hall International, Upper Saddle River, N.J., USA, 3rd revised edition, 2017. ISBN 978-1-292-15396-4

[Saadawi13] Tarek Saadawi, Louis Jordan: **Cyber Infrastructure Protection** www.lulu.com Publishing, 2013. ISBN 978-1-3041-1120-3

[Sanger19] David E. Sanger: **The Perfect Weapon - *War, Sabotage, and Fear in the Cyber Age*** Broadway Books, New York, N.Y., USA, 2019. ISBN 978-0-451-49790-1

[Sangster20] Mark Sangster: **No Safe Harbor - *The Inside Truth About Cybercrime and How To Protect Your Business*** Page To Books, Vancouver, Canada, 2020. ISBN 978-1-989603-42-0

[SASWG19] Safety of Autonomous Systems Working Group (SASWG): **Safety Assurance Objectives for Autonomous Systems** Independently published, SASWG, 2019. ISBN 978-1-7904-2122-0

[Sauter21] Martin Sauter: **From GSM to LTE-advanced Pro and 5G - *An Introduction to Mobile Networks and Mobile Broadband*** John Wiley & Sons, Inc., Hoboken, NJ, USA, 4th edition 2021. ISBN 978-1-119-71467-5

[Saxe18] Joshua Saxe, Hillary Sanders: **Malware Data Science - *Attack Detection and Attribution*** No Starch Press Inc., San Francisco, USA, 2018. ISBN 978-1-5932-7859-5

[Scharre19] Paul Scharre: **Army of None - *Autonomous Weapons and the Future of War*** WW Norton & Company, New York, N.Y., USA, 2019. ISBN 978-0-393-35658-8

[Scheerer18] Alexander Scheerer: **Coordination in Large-Scale Agile Software Development - *Integrating Conditions and Configurations in Multiteam Systems*** Springer International Publishing, Cham, Switzerland, 2018. ISBN 978-3-319-85629-2

[Scholl19]        Boris Scholl, Trent Swanson, Peter Jausovec: **Cloud Native - Using Containers, Functions, and Data to build next-generation Applications** O'Reilly Media Inc., Sebastopol, CA, USA, 2019. ISBN 978-1-492-05382-8

[Schwaber17]      Ken Schwaber, Jeff Sutherland: **The Scrum Guide**™ - *The Definitive Guide to Scrum: The Rules of the Game* www.scrum.org, November 2017. Downloadable from: https://www.scrum.org/resources/what-is-scrum [last accessed 14.6.2021]

[Schweikard15]    Achim Schweikard, Floris Ernst: **Medical Robotics** Springer Verlag, Heidelberg, Germany, 2015. ISBN 978-3-319-22890-7

[Seacord03]       Robert C. Seacord, Daniel Plakosh, Grace A. Lewis: **Modernizing Legacy Systems** - *Software Technologies, Engineering Processes, and Business Practices* Addison-Wesley Professional, Upper Saddle River, NJ, USA, 2003. ISBN 978-0-321-11884-4

[Selby17]         Nick Selby, Heather Vescent: **Cyber Attack Survival Manual - ** *From Identity Theft to The Digital Apocalypse and Everything in Between*. Weldon Owen Publishing, Richmond, CA, USA, 2017. ISBN 978-1-681-88175-1

[SEI21]           SEI: **Architecting the Future of Software Engineering - ** *An US Agenda for Software Engineering and Development* SEI Study, Carnegie Mellon University, Software Engineering Institute (SEI), Pittsburgh, PA, USA, November 2021. Downloadable from: https://resources.sei.cmu.edu/asset_files/Book/2021_014_001_741195.pdf [Last accessed: 13.11.2021]

[Senker17]        Cath Senker: **Cybercrime and the Darknet** Arcturus Publishing Ltd., London, UK, 2017. ISBN 978-1-7842-8868-6

[Sethna06]        James P. Sethna: **Entropy, Order Parameters, and Complexity** Oxford University Press, Oxford, UK, 2006. ISBN 978-0-19-856677-9

[Shakarian13]     Paulo Shakarian, Jana Shakarian, Andrew Ruef: **Introduction to Cyber-Warfare** - *A Multidisciplinary Approach* Syngress (Elsevier), Waltham, MA, USA, 2013. ISBN 978-0-124-07814-7

[Shetty19]        Sachin S. Shetty, Charles A. Kamhoua, Laurent L. Njilla (Editors): **Blockchain for Distributed Systems Security** IEEE Press, John Wiley & Sons Inc., Hoboken, NJ, USA, 2019. ISBN 978-1-119-51960-7

[Shrobe18]        Howard Shrobe, David L. Shrier, Alex Pentland (Editors): **New Solutions for Cybersecurity** MIT Press Ltd., Cambridge, MA, USA, 2018. ISBN 978-0-262-53537-3

[Shuja07]         Ahmad K. Shuja, Jochen Krebs: **IBM Rational Unified Process Reference and Certification Guide** - *Solution Designer*. IBM Press, Indianapolis, IN, USA, 2007. ISBN-13 978-0-131-56292-9

[Smith20]         David J. Smith, Kenneth G. L. Simpson: **The Safety-Critical Systems Handbook** - *A Straightforward Guide to Functional Safety: IEC 61508 (2010 Edition), IEC 61511 (2015 Edition) and Related Guidance* Butterworth-Heinemann (Elsevier), Kidlington, UK, 5th edition, 2020. ISBN 978-0-128-20700-0

[SoftLanding19]   SoftLanding Systems: **Application Decommissioning - The practical Alternative to legacy Life Support** (*A Guide to Successful Application Decommissioning and how to choose the right Approach for your Business*) White paper, 2019, UNICOM® Systems, Mission

Hills, CA, USA. Downloadable from: https://www.softlanding.com/files/1815/5965/0240/softlanding-legacy-application-decommissioning-onto-ibm-1i-whitepaper.pdf [Last accessed: 27.6.2021]

[Song16]        Houbing Song, Danda B. Rawat, Sabina Jeschke, Christian Brecher (Editors): **Cyber-Physical Systems - *Foundations, Principles and Applications*** Academic Press (Elsevier), London, UK, 2016. ISBN 978-0-128-03801-7

[Song17]        Houbing Song, Glenn A. Fink, Sabina Jeschke (Editors): **Security and Privacy in Cyber-Physical Systems - *Foundations, Principles, and Applications*** John Wiley & Sons, Inc., Hoboken, NJ, USA, 2017. ISBN 978-1-119-22604-8

[Spencer19]     Bryan Spencer: **Crypto Teacher on the new Economy - *The Greatest Experiment in over 400 years*** Independently published (by author), 2019. ISBN 978-1-67585-403-7

[Sreejith12]    K.G. Sreejith, K.G. Ajith: **Emergency Responder:** *A Smart Rescue Robot - A Surveillance Robot for Disaster Sites* LAP LAMBERT Academic Publishing, Riga, Latvia, 2012. ISBN 978-3-6591-7568-8

[Stakem18]      Patrick Stakem: **Mobile Cloud Robotics** Independently published, 2018. ISBN 978-1-9804-8808-8

[Stevens11]     Renee Stevens: **Engineering Mega-Systems - *The Challenge of Systems Engineering in the Information Age*** Auerbach Publications, Taylor & Francis Group, Boca Raton, USA, 2011. ISBN 978-1-4200-7666-0

[Stiennon15]    Richard Stiennon: **There Will Be Cyberwar - *How The Move To Network-Centric War Fighting Has Set The Stage For Cyberwar*** IT-Harvest Press, Birmingham, MI, USA, 2015. ISBN 978-0-9854-6078-5

[Stone19]       James V. Stone: **Artificial Intelligence Engines - *A Tutorial Introduction to the Mathematics of Deep Learning*** Sebtel Press, Sheffield, UK, 2019. ISBN 978-0-9563-7281-9

[Stouffer15]    Keith Stouffer, Victoria Pillitteri, Suzanne Lightman, Marshall Abrams, Adam Hahn: **Guide to Industrial Control Systems (ICS) Security - *Supervisory Control and Data Acquisition (SCADA) Systems, Distributed Control Systems (DCS), and Other Control System Configurations such as Programmable Logic Controllers (PLC)*** NIST Special Publication 800–82, Revision 2, May 2015. National Institute of Standards and Technology, Gaithersburg, MD, USA, 2015. Downloadable from: https://nvlpubs.nist.gov/nistpubs/SpecialPublications/NIST.SP.800-82r2.pdf [Last accessed: 19.11.2021]

[TechTarget21]  TechTarget: **Cloud Security Policy Template** TechTarget Inc., Boston, MA, USA, 2021. Downloadable from: https://searchcloudsecurity.techtarget.com/tip/How-to-create-a-cloud-security-policy-step-by-step#:~:text=In%20short%2C%20a%20cloud%20security,be%20used%20in%20multiple%20arrangements [Last accessed: 5.5.2021]

[Thayer12a]     Richard Hall Thayer, Merlin Dorfman (Editors): **SOFTWARE ENGINEERING ESSENTIALS, *Volume I: The Development Process*.** Software Management Training Press, Carmichael, CA, USA, 4th edition, 2012. ISBN 978-0-9852-7070-4

[Thayer12b]     Richard Hall Thayer, Merlin Dorfman: **SOFTWARE ENGINEERING ESSENTIALS, *Volume II: The Supporting Processes*.** Software Management Training Press, Carmichael, CA, USA, 4th edition, 2012. ISBN 978-0-9852-7071-1

| | |
|---|---|
| [Thayer12c] | Richard Hall Thayer, Merlin Dorfman (Editors): **SOFTWARE ENGINEERING ESSENTIALS,** *Volume III: The Engineering Fundamentals.* Software Management Training Press, Carmichael, CA, USA, 4th edition, 2012. ISBN 978-0-9852-7072-8 |
| [Tianfielda04] | Huaglory Tianfielda, Rainer Unland: *Towards autonomic computing systems* Engineering Applications of Artificial Intelligence 17 (2004), 689–699 Downloadable from: https://www.researchgate.net/profile/Rainer_Unland3/publication/222433987_Towards_autonomic_computing_systems/links/00b7d51d039fb794b1000000.pdf [last accessed 5.4.2016] |
| [Tockey19] | Steve Tockey: **How to Engineer Software -** *A Model-Based Approach* IEEE Computer Society & John Wiley & Sons, Inc., Hoboken, NJ, USA, 2019. ISBN 978-1-119-54662-7 |
| [Treiber13] | Martin Treiber, Arne Kesting: **Traffic Flow Dynamics -** *Data, Models and Simulation* Springer Verlag, Heidelberg, Germany, 2013. ISBN 978-3-642-32459-8 |
| [Trim16] | Peter Trim, David Upton: **Cyber Security Culture -** *Counteracting Cyber Threats Through Organizational Learning and Training* Routledge Publishing (Taylor & Francis), Abingdon, UK, 2016. ISBN 978-1-409-45694-0 |
| [Valavanis14] | Kimon P. Kimon P. Valavanis (Editor): **Unmanned Aerial Vehicles** Springer Verlag, Heidelberg, Germany, 2014. ISBN 978-9-401-78174-9 |
| [VanPutte17] | Michael A. VanPutte: **Walking Wounded -** *Inside the US Cyberwar Machine* CreateSpace Independent Publishing Platform, North Charleston, S.C., USA, Updated version 2017. ISBN 978-1-5399-4561-1 |
| [VanPutte19] | Michael A. VanPutte, Thomas P. Sammel: **The Art of Cyberwar -** *The Principles of Conflict in Cyberspace* Independently published, 2019. ISBN 978-1-08110-757-4 |
| [Voshmgir19] | Shermin Voshmgir: **Token Economy -** *How Blockchains and Smart Contracts Revolutionize the Economy* BlockchainHub, Berlin, Germany, Germany, 2019. ISBN 978-3-9821-0382-2 |
| [Wang18] | Lihui Wang, Xi Vincent Wang: **Cloud-Based Cyber-Physical Systems in Manufacturing** Springer International Publishing, Cham, Switzerland, 2018. ISBN 978-3-319-88467-7 |
| [Ward19] | Nicholas John Ward, Barry Watson, Katie Fleming-Vogl (Editors): **Traffic Safety Culture -** *Definition, Foundation, and Application* Emerald Publishing Limited, Bingley, UK, 2019. ISBN 978-1-7871-4618-1 |
| [Warren99] | Ian Warren: **The Renaissance of Legacy Systems -** *Method Support for Software-System Evolution* Springer Verlag London, London, UK, 1999. ISBN 978-1-852-33060-6 |
| [WEF20] | World Economic Forum: **Inclusive Deployment of Blockchain for Supply Chains:** *Part 6 - A Framework for Blockchain Interoperability* World Economic Forum, Cologny-Geneva, Switzerland, April 2020. Downloadable from: http://www3.weforum.org/docs/WEF_A_Framework_for_Blockchain_Interoperability_2020.pdf [Last accessed: 16.4.2020] |
| [Weidman14] | Georgia Weidman: **Penetration Testing -** *A Hands-On Introduction to Hacking* No Starch Press, San Francisco, CA, USA, 2014. ISBN 978-1-5932-7564-8 |

[Werbach18]          Kevin Werbach: **The Blockchain and the New Architecture of Trust**
                     MIT Press Ltd., Cambridge, MA, USA, 2018. ISBN 978-0-262-03893-5

[Wheatcraft10]       Louis S. Wheatcraft: *Everything you wanted to know about Interfaces,*
                     *but were afraid to ask* Requirements Experts Inc., Austin, TX, USA,
                     2010.   Downloadable   from:   https://reqexperts.com/wp-content/
                     uploads/2016/04/Wheatcraft-Interfaces-061511.pdf   [Last   accessed:
                     27.1.2021]

[Wiegers13]          Karl Wiegers, Joy Beatty: **Software Requirements** Microsoft
                     Press, Redmond, Washington, USA, 3$^{rd}$ edition, 2013. ISBN
                     978-0-7356-7966-5

[Witt09]             Matthias Witt: **Zustandsdiagramme und Design by Contract**
                     **- UML-Zustandsdiagramme von Java-Klassen auf der Basis des**
                     **Vertragsmodells**. VDM Verlag, Saarbrücken, Germany, 2009. ISBN
                     978-3-63918439-6

[Wortmann13]         Alex Wortmann, Saurabh Laddha: *Application Retirement Methodology*
                     *-From Intent to Implementation (Introducing a practical, proven*
                     *Methodology)* White Paper, CapGemini Research Insitute, Paris, France,
                     2013. Downloadable from: https://www.capgemini.com/wp-content/
                     uploads/2017/07/2013-04-10_application_retirement_methodology_
                     whitepaper_web.pdf [Last accessed: 27.6.2021]

[Xu19]               Xiwei Xu, Ingo Weber, Mark Staples: **Architecture for Blockchain**
                     **Applications** Springer International, Cham, Switzerland, 2019. ISBN
                     978-3-030-03034-6

[Yampolskiy18]       Roman V. Yampolskiy: **Artificial Intelligence Safety and Security**
                     CRC Press (Taylor & Francis Inc.) Boca Raton, FL, USA, 2018. ISBN
                     978-0-815-36982-0

[Yáñez17]            Fran Yáñez: **The 20 Key Technologies of Industry 4.0 and**
                     **Smart Factories -** *The Road to the Digital Factory of the Future*
                     Independently published, 2017. ISBN 978-1-9734-0210-7

[Yang18]             Herong Yang:**WSDL Tutorials -** *Herong's Tutorial Examples*
                     Independently published, 2018. ISBN 978-1-72396-590-6

[Yaworski19]         Peter Yaworski: **Real-World Bug Hunting -** *A Field Guide to Web*
                     *Hacking* No Starch Press, San Francisco, CA, USA, 2019. ISBN
                     978-1-5932-7861-8

[Zaman15]            Najamuz Zaman: **Automotive Electronics Design Fundamentals**
                     Springer Verlag, Germany, 2015. ISBN 978-3-319-17583-6

[Zetter14]           Kim Zetter: **Countdown to Zero Day -** *Stuxnet and the Launch of the*
                     *World's First Digital Weapon* Broadway Books, New York, USA, 2014.
                     ISBN 978-0-7704-3619-3

[Zhang20]            Lei Zhang, Guodong Zhao, Muhammad Ali Imran (Editors): **Internet**
                     **of Things and Sensors Networks in 5G Wireless Communications**
                     MDPI AG, Basel, Switzerland, 2020. ISBN 978-3039-28148-0

[Zivic20]            Natasa Zivic, Obaid Ur-Rehman: **Security in Autonomous Driving** De
                     Gruyter Oldenbourg, Oldenbourg, Germany, ISBN 978-3-110-62707-7

[Zongo18]            Phillimon Zongo: **The Five Anchors of Cyber-Resilience -** *Why some*
                     *Enterprises are hacked into Bankruptcy while others easily bounce*
                     *back* Broadcast Books, Australia, 2018. ISBN 978-0-6480078-4-5

# Three Devils of Safety and Security

*Vulnerabilities of the systems are at the core of safety accidents and security incidents. Threats and failures utilize such vulnerabilities to damage the systems. Therefore, vulnerabilities, threats, and failures can be considered the "three devils of safety and security". These three devils accompany both the developers and the operators of the cyber-physical systems at all times. Guarding against the three devils is the foremost task while developing, evolving, and operating the CPSs.*

During the engineering process for safety and security, three perilous *devils* are encountered:

1. **The First Devil**: *Vulnerabilities* (Fig. 3.1). Vulnerabilities are weaknesses of the system which allow a failure to cause damage or which generate an entry point for malicious activities,
2. **The Second Devil**: *Threats* (Fig. 3.2). A threat is an intentional or unintentional *danger* aimed at the CPS in order to generate operational disruptions, such as security breaches, malfunctions, or unavailability,
3. **The Third Devil**: *Failures* (Fig. 3.3). Failure causes a part of a system (hardware, software, communications, or infrastructure) to become inoperable or unavailable to the services depending on this part.

> **Quote**
>
> *"The combination of physical systems with complex computer systems opens up new concerns and threats that are more than the sum of traditional safety engineering and computer security"*
>
> Marilyn Wolf and Dimitrios Serpanos, 2020

© The Author(s), under exclusive license to Springer Fachmedien Wiesbaden GmbH, part of Springer Nature 2022
F. J. Furrer, *Safety and Security of Cyber-Physical Systems*,
https://doi.org/10.1007/978-3-658-37182-1_3

**Fig. 3.1** First devil of safety
and security—vulnerabilities

**Fig. 3.2** Second devil of
safety and security—threats

## 3.1    Vulnerabilities

*Vulnerabilities* (Definition 3.1) in a CPS are the root of all safety accidents and security
incidents. A (hypothetical) system with zero vulnerabilities would be immune against
all threats and failures and thus be intrinsically safe and secure. Therefore, avoiding

**Fig. 3.3**  Third devil of safety
and security—failures

© http://clipart-library.com
Used with permission / 01.01.2020

and eliminating vulnerabilities is the *first priority* of engineering for safety and security. A whole industry has developed which discovers security vulnerabilities in operating systems, browsers, application programs, CPUs, etc. ([Ludwig19], [Lusthaus18], [Willems19], [White20]) and sells these vulnerabilities to any buyer. Incredibly valuable for malicious actors are "zero-day vulnerabilities", i.e., new vulnerabilities which are not yet known to the manufacturer or the community ([Saxe18]).

▶ **Definition 3.1: Vulnerability**

A vulnerability is a weakness in an element of the CPS which can be exploited by a threat actor, such as an attacker, to perform unauthorized actions within the computer system **or** a flaw, which permits a failure of an element to cause faults, malfunctions, or unavailability of parts of the CPS.

Adapted from: https://en.wikipedia.org/wiki/Vulnerability_(computing).

Vulnerabilities represent safety and security's first devil (Fig. 3.1). Hence, the primary responsibility of engineering and operating CPSs is to avoid, eliminate, and mitigate the possible consequences of vulnerabilities.

In complex systems—which is the case for all useful systems today—both failures of elements and threats are unavoidable. The operators of the CPSs can feel sure that both failures and threats will occur. The CPSs are, therefore, endangered at any time!

> **Quote**
> *"It is not a question* if *you will be attacked or* if *a component will fail – It is just a question of* when*"*
>    Anonymous

## 3.2    Threats

*Threats* (Definition 3.2)—the second devil of safety and security engineering (Fig. 3.2)—are menaces to the CPS and the organizations involved ([Vélez15], [Shostack14]). A threat is an intentional or unintentional danger aimed at the CPS in order to generate operational disruptions, such as security breaches, safety accidents, malfunctions, or unavailability. Note that a *CPS vulnerability must be present* to allow a threat to become successful.

▶ **Definition 3.2: Threat**
A potential negative event whose source might cause either a tangible or intangible impact to the system, the business/organization and its operations, such as loss of revenue, monetary loss, legal lawsuits and fines, or indictments.
    Tony Uceda Velez, Marco M. Morana, 2015.

Threats exploit *security vulnerabilities* in the system. Today's operating systems, browsers, file systems, drivers, etc., are full of known and unknown vulnerabilities. The known vulnerabilities are published by several institutions (Example 3.1), e.g.,

- MITRE Common Vulnerabilities and Exposures (CVE: https://cve.mitre.org/);
- NIST Information Technology Laboratory: National Vulnerability Database (NVD: https://nvd.nist.gov/)
- Software Engineering Institute, CERT Coordination Center (https://www.kb.cert.org/vuls/)
- Security Focus Vulnerability Database (https://www.securityfocus.com/vulnerabilities)

---

**Example 3.1: CVE-2020–3207 Detail [Published 2.6.2020]**

(see: https://nvd.nist.gov/vuln/detail/CVE-2020-3207#vulnDescriptionTitle).

A vulnerability in the processing of boot options of specific Cisco IOS XE Software switches could allow an authenticated, local attacker with root shell access to the underlying operating system (OS) to conduct a command injection attack during device boot. This vulnerability is due to insufficient input validation checks while processing boot options. An attacker could exploit this vulnerability by modifying device boot options to execute attacker-provided code. A successful exploit may allow an attacker to bypass the Secure Boot process and execute *malicious code* on an affected device with root-level privileges.

## 3.3    Failures

A *failure* (Definition 3.3) causes a part of a system (hardware, software, communications, or infrastructure) to become inoperable or unavailable to the services depending on the failed part. Failures are unavoidable in today's modern, complex, interconnected systems. Failures can occur in any part of the system and have different effects—from negligible to catastrophic. Failures can have single sources, such as a *single point of failure* in a system, or multiple sources, generating chains of failures. Failures can also be permanent or intermittent ([Gullo18], [Raheja12], [Leveson16]).

▶ **Definition 3.3: Failure**
A failure is a basic abnormal occurrence in a component of the cyber-physical system. Failure is the inability of a function to meet its requirement specification. Failures can be permanent or intermittent.
   Adapted from: Louis J. Gullo, Jack Dixon, 2018.

Therefore, the *third devil* of safety and security is *failures* (Fig. 3.3).
   An especially dangerous failure mode is *single points of failure* (SPOF, Example 3.2).

---

**Example 3.2: Single Point of Failure**

Whenever a function is executed in a cyber-physical system, the *flow of control* traverses many components or subsystems—both hardware and software. A significant number of control paths exist, with many decision points in the flow. In this example, the flow of control for the function **braking** of a passenger car is considered (Fig. 3.4): The control flow depends on the trigger: (a) standard braking, (b) emergency braking, and (c) braking by the distance control unit. Three control paths are programmed, one for each trigger. The control paths use some common components. In Fig. 3.4, all flow paths have to pass the same single component: If this component fails, the functionality **braking** is completely lost! This unique component (either an HW or SW component) constitutes a *single point of failure*.

Single points of failure are a severe danger to safety: They must be identified and eliminated, usually by introducing appropriate *fault tolerance*, such as redundancy ([Dubrova13], [Butler09], [Gullo18, p. 155]).

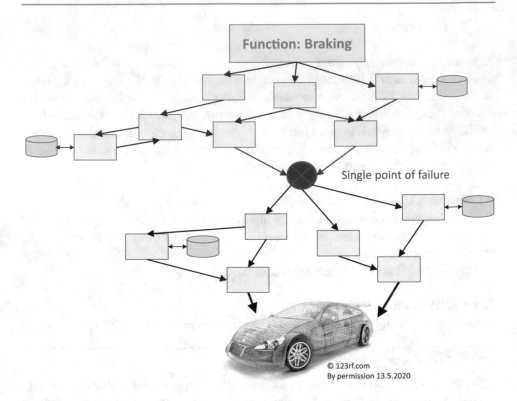

**Fig. 3.4** Single point of failure

> **Quote**
> *"Single point failure modes relate to one specific independent failure or hazard event that leads to a mission-critical or safety–critical effect, which involves catastrophic personal injuries or equipment damage"*
>     Louis J. Gullo, 2018

## 3.4 Risk Introduction

*Risk* (Definition 3.4) is a constant companion of cyber-physical systems. There is no way to eliminate all risks from CPSs entirely. The task of CPS engineering is to identify, recognize, assess, and mitigate as many risks as possible and reduce their impact to an acceptable residual risk (= risk management process). Failure to identify a risk, neglect or underestimate its impact, or insufficiently mitigate it with controls, is a sure recipe for *safety accidents* and *security incidents*.

▶ **Definition 3.4: Risk**

A probability or threat of damage, injury, liability, loss, or any other negative occurrence that is caused by external or internal vulnerabilities, and that may be avoided or mitigated through preemptive action.

   Adapted from: http://www.businessdictionary.com/definition/risk.html

*Risks* originate from a *vulnerability* in any of the elements of the CPCs (Fig. 3.5). The vulnerability is then exploited by either a threat or a failure and generates risk. The successful exploitation of vulnerabilities leads to consequences in the CPC, such as safety accidents or security incidents of various severity.

   Note the *legal framework* in Fig. 3.5: The legal framework—usually different in different nations—is the basis for enforceable preventive, corrective, and punitive actions concerning the developers, manufacturers, operators, and users of the CPS ([Delerue20], [Wong18], [Iglezakis16], [DeFilippi19], [Schreider20], [Buchan18], [Krauss19], [Geistfeld20], [Owen14]).

> **Quote**
>
> *"Companies are undergoing a digital transformation. An information-centric world requires a seismic shift in the way projects are delivered, businesses are run, and risks are managed"*
>
>    *Christopher J. Hodson, 2019*

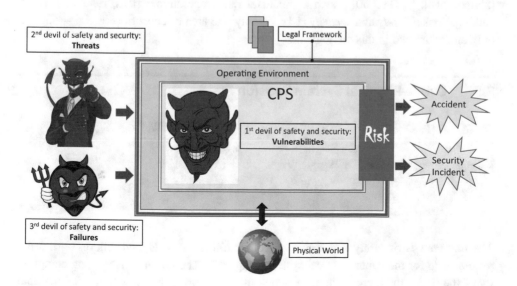

**Fig. 3.5** Risk in a cyber-physical system

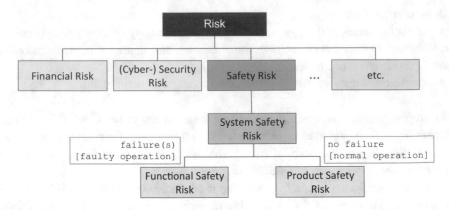

**Fig. 3.6**  Risk pyramid

Fortunately, a mature and responsible risk culture (e.g., [Banks12], [Hodson17]) exists in many of today's CPS development communities. The relevant development processes integrate an effective *risk management process*. In addition, many CPSs have to be validated or even *certified* by authorities, such as the USA Federal Aviation Authority FAA (https://www.faa.gov/), the European Union Agency for Railways ERA (https://www.era. europa.eu/), or the German Bundesamt für Sicherheit in der Informationstechnik (https:// www.bsi.bund.de).

A company or organization faces *multiple risks* (see, e.g., [Lam17], [Olson20], [Hutchins18], [ISO31000]), such as financial risks, operational risks, reputation risks, and legal risks. *Safety* and *security risks* are only one area of concern (Fig. 3.6)—and the only one addressed in this monograph.

## 3.5    Cyber-Physical System Tension Field

> **Quote**
> *"The connecting of everything into a single, complex, hyper-connected system quickly makes the risks catastrophic"*
>   *Bruce Schneier, 2018*

The *three devils* of safety and security—vulnerabilities, threats, and failures—form the *tension field* for the cyber-physical systems (Fig. 3.7). The continuously rising complexity of the CPSs, the increase in dependency from networks, the growth in the functional richness of the infrastructures, the perils of the supply chain, and the solid connection for third-party products open up many new possibilities for *vulnerabilities*.

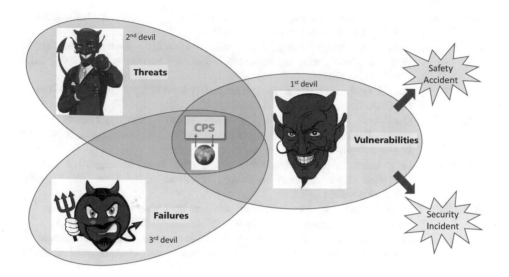

**Fig. 3.7** Tension field for cyber-physical systems

Simultaneously, the *threat landscape* evolves daily: More threats are developed, they become more sophisticated and dangerous, and their frequency rises. Cyber-crime, cyber-espionage, cyber-war, cyber-sabotage, etc., have become large, viable, illegal, and highly profitable industries.

Also, the increasing complexity—in the number of parts and connections—of the CPS generates new and more frequent opportunities for *failures*. Failures are becoming not the exception but the norm in CPSs.

… the three devils will persist as our precarious companions in CPS's!

# References

[Banks12]      Erik Banks: **Risk Culture - *A Practical Guide to Building and Strengthening the Fabric of Risk Management*** Palgrave Macmillan, London, UK, 2012. ISBN 978-1-137-26371-1

[Buchan18]     Russell Buchan: **Cyber Espionage and International Law** Hart Publishing, Oxford, UK, 2018. ISBN 978-1-7822-5734-9

[Butler09]     Michael Butler, Cliff B. Jones, Alexander Romanovsky, Elena Troubitsyna: **Methods, Models, and Tools for Fault Tolerance** Springer Verlag, Berlin, Germany, 2009 (LNCS 5454). ISBN 978-3-642-00866-5

[DeFilippi19]  Primavera Filippi De 2019 Blockchain and the Law - The Rule of Code Harvard University Press MA, USA Cambridge 978-0-674-24159-6

[Delerue20]    François Delerue: **Cyber Operations and International Law** Cambridge University Press, Cambridge, UK, 2020. ISBN 978–1–108–49027–6

[Dubrova13]      Elena Dubrova 2013 Fault-Tolerant Design Springer Science & Business Media
                 N.Y., USA New York 978-1-461-42112-2
[Geistfeld20]    Mark A. Geistfeld: **Principles of Products Liability - *Concepts and Insights***
                 West Academic Press, St. Paul, MN, USA, 3rd revised edition, 2020. ISBN
                 978–1–6424–2582–6
[Gullo18]        Louis J. Gullo, Jack Dixon: **Design for Safety** John Wiley & Sons, Inc., Hoboken,
                 NJ, USA, 2018. ISBN 978–1–118–97429–2
[Hodson17]       Christopher J Hodson: **Cyber Risk Management - *Prioritize Threats, Identify***
                 ***Vulnerabilities and Apply Controls*** Kogan Page, New Dehli, India, 2019. ISBN
                 978–0–749–48412–5
[Hutchins18]     Greg Hutchins: **ISO 31000:2018 Enterprise Risk Management** Certified
                 Enterprise Risk Manager Academy, Portland, OR, USA, 2018. ISBN
                 978–0–9654–6651–6
[Iglezakis16]    Ioannis Iglezakis: **The Legal Regulation of Cyber Attacks** Kluwer Law
                 International, Aalphen an den Rijn, NL, 2016. ISBN 978–9–0411–6901–3
[ISO31000]       ISO 31000:2018 (ISO/TC 262 Risk Management) **Risk Management —**
                 **Guidelines** Downloadable from: https://www.iso.org/standard/65694.html [last
                 accessed: 7.6.2020]
[Krauss19]       I Michael 2019 Krauss: Principles of Products Liability West Academic
                 Publishing 3 St. Paul MN, USA 978-1-6402-0128-6
[Lam17]          James Lam: **Implementing Enterprise Risk Management - *From Methods***
                 ***to Applications*** John Wiley & Sons, Inc., Hoboken, N.J., USA, 2017. ISBN
                 978–0–471–74519–8
[Leveson16]      Nancy G. Leveson: **Engineering a Safer World - *Systems Thinking***
                 ***Applied to Safety*** MIT Press Ltd., Massachusetts, MA, USA, 2016. ISBN
                 978–0–262–53369–0
[Ludwig19]       Mark Ludwig 2019 The Giant Black Book of Computer Viruses American Eagle
                 Books 2 San Francisco USA 978-1-6435-4313-0
[Lusthaus18]     Jonathan Lusthaus 2018 Industry of Anonymity - Inside the Business of
                 Cybercrime Harvard University Press MA, USA Cambridge 978-0-674-97941-3
[Olson20]        L David 2020 Olson, Desheng Wu: Enterprise Risk Management Models Springer
                 Verlag 3 Berlin Germany 978-3-662-60607-0
[Owen14]         G David 2014 Owen: Products Liability in a Nutshell West Academic Press 9 St.
                 Paul MN, USA 978-0-3142-6840-2
[Raheja12]       Dev G. Raheja, Louis J. Gullo (Editors): **Design for Reliability** John Wiley &
                 Sons, Inc., Hoboken, NJ, USA, 2012. ISBN 978–0–470–48675–7
[Saxe18]         Joshua Saxe, Hillary Sanders: **Malware Data Science - *Attack Detection***
                 ***and Attribution*** No Starch Press Inc., San Francisco, USA, 2018. ISBN
                 978–1–5932–7859–5
[Schreider20]    Tari Schreider 2020 Cybersecurity Law 2 Standards and Regulations Rothstein
                 Publishing Brookfield, USA 978-1-9444-8056-1
[Schreider20]    Tari Schreider 2020 Cybersecurity Law 2 Standards and Regulations Rothstein
                 Publishing Brookfield, Connecticut, USA 978-1-9444-8056-1
[Shostack14]     Adam Shostack: **Threat Modeling - *Designing for Security*** John Wiley & Sons,
                 Inc., Indianapolis, IN, USA, 2014. ISBN 978-1-118-80999-0
[Vélez15]        Tony Uceda Vélez, Marco M. Morana: **Risk Centric Threat Modeling - *Process***
                 ***for Attack Simulation and Threat Analysis*** John Wiley & Sons Inc., Hoboken,
                 NJ, USA, 2015. ISBN 978-0-470-50096-5

[White20]    Geoff White 2020 Crime Dot Com - From Viruses to Vote Rigging How Hacking Went Global Reaktion Books London, UK 978-1-7891-4285-3

[Willems19]    Eddy Willems: **Cyberdanger - *Understanding and Guarding Against Cybercrime*** Springer Nature, Cham, Switzerland, 2019. ISBN 978-3-030-04530-2

[Wong18]    Helen Wong: **Cyber Security - *Law and Guidance*** Bloomsbury Professional, London, UK, 2018. ISBN 978-1-526-50586-6

# Safety, Security, and Risk

**4**

*The main properties of trustworthy cyber-physical systems are safety and security. Both safety and security are the results of careful, responsible, risk-guided engineering. The real danger of cyber-physical systems is vulnerabilities in the protection mechanisms of the assets. Such vulnerabilities allow threats and failures to attack or blackout the systems—leading to safety accidents or cyber incidents. A careful, accountable risk management is, therefore, at the core of trustworthy cyber-physical systems. Risk management leads to protection measures, which should reduce the accident and incident risks to an acceptable, quantified residual risk. Management of risk is necessary from the requirements phase, all through the operations phase, mainly when accidents or incidents occur. Therefore, risk management must define and enforce adequate protection/mitigation measures in all system evolution and operation stages.*

## 4.1 Context

The fundamental concepts of this chapter are *safety*, *security*, and *risk*. The English language has a clear distinction between *safety* and *security*. German only knows "Sicherheit", French only lists "sécurité", encompassing both terms. In the context of this monograph, the difference between safety (Definition 4.2) and security (Definition 4.14, Definition 4.15) is of great importance!

> **Quote**
> *"Solving the safety and security deficiencies in next generation cyber-physical systems will require contributions from every branch of engineering, from mechanical*

*and power engineering to computer science and mathematics. Partners from university research labs, governments, and industry must come together. It is necessary that we establish an engaged, multidisciplinary, cyber-physical security community committed to developing unified foundations, principles, and technologies".*

Çetin Kaya Koç, 2018

Safety and security are strongly required properties of a cyber-physical system, especially of *safety-critical* ([Hobbs20], [Rierson13], [Bozzano10]) or *mission-critical* ([Ackerman17], [Krutz16], [Johnson15]) cyber-physical systems. The context for this chapter is shown in Fig. 4.1. At the center of attention are the *protection assets* (Fig. 4.1). The protection assets may be *cyber-assets*, such as customer information, bank accounts, credit card data, marketing plans, product engineering documentation, etc., or *physical assets*, such as a safe car, a dependable heart pacemaker, the energy infrastructure, a medical diagnostic device, an airplane autopilot, and many more.

A shell of protection measures protects the *assets*. Protection measures try to preclude attacks and absorb hardware and software failures. Unfortunately, as reality shows daily,

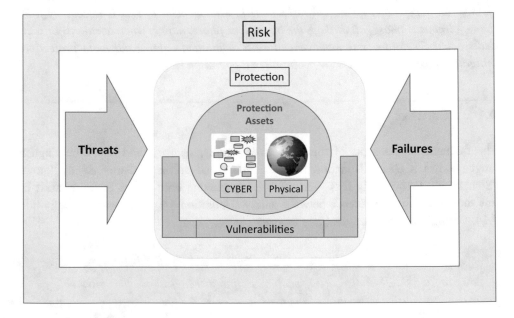

**Fig. 4.1**   Risk context

the systems and their protection measures have *vulnerabilities*. Vulnerabilities—acciden-tal or intentional—are the entry points for *attacks* or *failures*. Vulnerabilities are exploited, e.g., for cyberattacks, cybercrime, and cyberespionage. Vulnerabilities are also the reason for faults, errors, unavailability, and malfunctions of cyber-physical systems.

The cyber-physical system is continuously exposed to *threats* (Definition 3.2) and *failures* (Definition 3.3).

When exploited by threats or hit by failures, vulnerabilities generate *risks* (Definition 3.4). In Fig. 4.1, the risk is, therefore, the all-encompassing concept to build and evolve safe and secure cyber-physical systems.

> **Quote**
> *"Because of the many critical applications where CPS's are employed, either kind of attack or failure can result in dire real-world consequences"*.
> Glen A. Fink in [Song17], 2018

## 4.2 General Resilience

*Failures* in a cyber-physical system can be either *general* (such as a power failure, a nat-ural disaster, or an unexpected failure propagation) or *specific* (such as the failure of an integrated circuit in an electronic control unit with known consequences, discovered dur-ing the risk management process). General failures require the *resilience* (Definition 4.1) of the CPS. Resilience can be created by applying *resilience principles* during the archi-tecting and development of the CPS.

▶ **Definition 4.1: System Resilience**
System resilience is the system's ability to withstand a major disruption within accept-able degradation parameters and recover within an acceptable time.
www.igi-global.com

Important resilience principles are listed in Table 4.1 and explained in Part 2 of this mon-ograph. Many of them apply both to safety and security.

There are two categories of safety and security principles (Fig. 4.2):

a) *General* resilience principles: Defense against *unspecific* or *unknown* failures or threats;

b) *Specific* principles targeting *specific* safety and security quality properties, i.e., spe-cific mitigation measures against safety or security vulnerabilities identified and assessed during risk management.

**Table 4.1**   Resilience principles

| # | Principle | Summary |
|---|-----------|---------|
| G6_1 | Software Integrity | All software artifacts controlling the runtime CPS must be provably complete, consistent, authentic, and integer |
| G6_2 | Timing Integrity | All timing constraints in the runtime CPS must be met under all operating conditions |
| G6_3 | Fault Containment Regions | Prevent a failure in a part of the system to propagate and cause follow-up failures, leading to a chain of failures |
| G6_4 | Single Points of Failure | Avoid single points of failure by implementing adequate redundancy |
| G6_5 | Multiple Lines of Defense | Whenever possible, implement cascaded mitigation measures to absorb a deficient operation of a mitigation measure |
| G6_6 | Fail-Safe States | Define, implement, and enforce the transition of the CPS into a safe state following failures |
| G6_7 | Graceful Degradation | Design the system in such a way that it can work with reduced functionality in case of failures |
| G6_8 | Fault Tolerance | Use systematic redundancy and diversity to neutralize faults and malfunctions |
| G6_9 | Dependable Foundation (Infrastructure) | Provide a safe, reliable, available execution infrastructure for the software |
| G6_10 | Error & Failure Management | Ensure that all possible errors and failures are recognized, diagnosed, and correctly handled |
| G6_11 | Monitoring | Monitor the runtime behavior of the CPS, detect deviations from the expected behavior, and take real-time corrective actions |

A grave consequence of the neglect of the *software integrity principle* is given in Example 4.1.

---

**Example 4.1: Software Integrity Failure leads to A400M Crash**

Benjamin Zhang: *A Software Problem caused a brand-new Airbus* A400M *Military Plane to crash* ([Business Insider, June 3, 2015]):

On May 9, 2015, an Airbus A400M military plane on a test flight crashed after take-off less than 5 km from Seville Airport, killing four of the six crew.

In an interview with Handelsblatt, Airbus Chief Strategy Officer Marwan Lahoud blamed the crash on engine control software that was *incorrectly installed* during the final assembly.

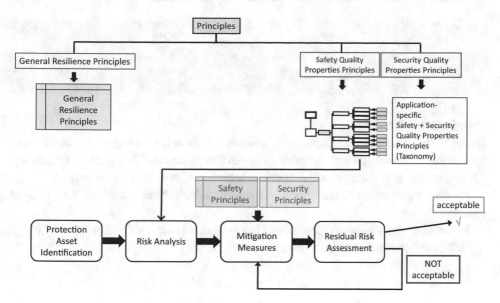

**Fig. 4.2**  Resilience principles and specific safety principles

"Engines 1, 2, and 3 experienced power frozen [to high power] after lift-off and did not respond to the crew's attempts to control the power setting in the normal way. When the power levers were set to 'flight idle' in an attempt to reduce power, the power reduced but then remained at 'flight idle' on the three affected engines for the remainder of the flight despite attempts by the crew to regain power", Airbus said in a statement.

https://www.businessinsider.com/a-software-problem-caused-an-airbus-a400m-to-crash-2015-6?r=US&IR=T [Last accessed: 2.10.2020].

A later inquiry revealed that the three engines' *configuration data* had not been installed. The missing configuration data had not been detected during the software's power-up: A fatal violation of the resilience principle G6_1 (Table 4.1)!

## 4.3  Safety

### 4.3.1  Introduction

*Safety* and *security* in a cyber-physical system are two very different concepts. Both are required to a sufficient degree to make a trustworthy cyber-physical system. *Safety* (Definition 4.2) targets the avoidance of *accidents* (e.g., [Hollnagel14], [Roughton19], [FAA19]).

**Quote**
*"Safety is the sum of all accidents which didn't happen".*
   Anonymous

▶ **Definition 4.2: Safety**
Safety is the state of being protected against faults, errors, failures, or any other event that could be considered non-desirable to achieve an acceptable level of risk concerning the loss of property, damage to life, health or society, or harm to the environment.

   *Product safety* refers to the operational safety under normal conditions, i.e., without failures.

   *Functional safety* refers to the safety of the system when it malfunctions.
   ISO 26262 [https://www.iso.org/standard/68383.html].

System safety—or just safety—has two facets (e.g., [Miller20]):

- *Product safety*: This term refers to the safety of a system *without* any failure or malfunction, i.e., the safety of the intended functions (SOTIF). An example is the correct operation of the automated emergency braking system of a road vehicle under all environmental conditions;
- *Functional safety:* This term has a limited scope, including only safety when there is a failure or malfunction in the system. An example is the automated distance control in a road vehicle: The loss of this intended function due to a component failure generates a visible and audible warning for the driver but does not stop the car ([Dearden18]).

Safety is a *global property* of a product, i.e., a cyber-physical system. Today's products are not monolithic systems but are composed of many parts (subsystems, constituent systems). Such systems are called *composite systems* and are assembled from their parts and connections. Very often, some parts are safety-critical, and others are not safety-critical. The safety-critical parts determine the safety of the composite system. Such systems are *mixed-criticality systems*.

## 4.3.2    Composite Systems

Most practical cyber-physical systems are *composite systems* (Definition 4.3) and *mixed-criticality systems* (Definition 4.4).

▶ **Definition 4.3: Composite System**  A cyber-physical system or product which is not monolithic, but composed from many interacting parts or subsystems.

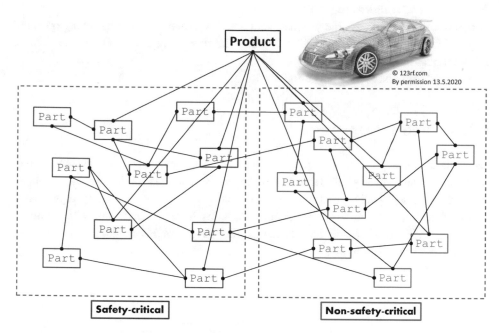

**Fig. 4.3** Mixed-criticality system

A typical composite, mixed-criticality cyber-physical system is shown in Fig. 4.3. The system or product—e.g., a car—is assembled from a large number of parts or subsystems. Some of these are safety-critical—such as the braking subsystem, and others are not—such as the car audio system.

▶ **Definition 4.4: Mixed-Criticality System** A composite system which consists of safety-critical parts or subsystems and non-safety-critical parts or subsystems, interacting to contribute to the functionality of the composite system.

Safety is a *global* property of the cyber-physical system. The safety is determined by the individual safety properties of the safety-critical parts or subsystems *and* their interactions. Unfortunately, even if all safety-critical parts possess the required safety properties, the composed system may be unsafe! The reason is the *emergent properties* (Definition 2.4): The assembly through interactions may generate new, unforeseen *failure modes* (Definition 4.5), which can create unexpected negative consequences. Therefore, the entire cyber-physical system must undergo complete risk analysis and mitigation, not only the individual parts!

▶ **Definition 4.5: Failure Mode**
A failure mode is one possible way a system can fail. When a system has many potential
ways of failing, it has multiple failure modes. The more complex a system is, the more
failure modes there are.

Adapted from: https://support.minitab.com

### 4.3.3   Safety Taxonomy

*Safety* is a global property—or macro-property—of the cyber-physical system. Safety
comprises many quality of service sub-properties, such as fault tolerance, robustness, graceful degradation, and many more. A sample *taxonomy* for safety is presented
in Fig. 4.4. Note that different application domains have different safety-taxonomies,
adapted to their specific safety and certification requirements. Figure 4.4 is only a sample
instance of a safety taxonomy.

One important sub-property of safety is *single points of failure* (Example 3.2).

**Fig. 4.4**   Safety taxonomy example

### 4.3.4  Safety Metrics

Many of the quality of service properties in the safety taxonomy of Fig. 4.4 can be *measured*, i.e., expressed in a quantitative form. A simple instance is shown in Example 4.2: The metric is (a) the *number* of single points of failure in the safety-critical part of Fig. 4.3—which should, of course, be zero!—and (b) the *number* of single points of failure in the non-safety-critical part of Fig. 4.3—which may tolerate some single points of failure leading to degraded operation.

The safety quality of service metrics of Fig. 4.4 are termed product metrics ([Jayasri17], [DeLong05], [Cais14], [Cais15], [Fowler10b]).

> **Quote**
> *"If you measure it, it is a fact. If you don't measure it, it is an opinion"*.
>   Anonymous

---

**Example 4.2: Single Point of Failure Metric**

The single point of failure metric is the *number* of single points of failure in a system, as in Fig. 4.3: Ncritical for the safety-critical part and Nstandard for the non-safety-critical part. Note that in the safety-critical part, *no* single point of failure is tolerated (Ncritical = 0). In the non-safety-critical part, single points of failure leading to degraded operation may be acceptable.

The measurement starts with the individual subsystem: Carefully analyze each of the subsystems/parts in Fig. 4.3, locate all single points of failure, and eliminate them by introducing proper *redundancy*. However, the system is not yet free of single points of failure: The subsystems' interconnections must also be examined. If two or more subsystems are connected by only one interface or link, this constitutes a single point of failure. This newly formed single point of failure is generated by *emergence* (Definition 2.4), i.e., by the system's assembly. Of course, all single points of failure in the safety-critical subsystems' interconnection must also be eliminated by adequate *redundancy* (= multiple links/interfaces).

The selection and consequent application of safety metrics is an important factor for manufacturers of safety-critical systems and applications.

Example 4.3 introduces an often applied *high-level safety metric* for the automotive domain: The *Automotive Safety Integrity Level* (ASIL, [ISO26262-9:2018], [Dimitrakopoulos20], [Miller20], [Ross16], [Winner18], [Gebhardt13]).

**Example 4.3: Automotive Safety Integrity Level (ASIL)**

A modern road vehicle contains many functions, some *safety-critical* (braking, steering, driver-assistance systems, airbags, etc.) and some *non-safety-critical* (multimedia-system, airconditioning, connectivity, etc.). To ensure the safety of the vehicle, all functions must be classified according to an *automotive safety integrity level* ASIL, prescribed in the ISO 26262 (Part 9) standard ([ISO26262-9:2018]). ASIL is a risk-based rating. The required ASIL is determined before a new function is developed, and the entire development activity is organized to reach and prove the required ASIL (in conformance with the ISO 26262 standard).

Four ASIL grades exist: ASIL A for low-risk functions to ASIL D for high-risk functions (Table 4.2). They are evaluated based on three criteria: (1) Severity, (2) Exposure, and (3) Controllability (Table 4.3).

**Table 4.2**   Automotive safety integrity levels

| Severity S | Exposure E | Controllability C | | |
|---|---|---|---|---|
| | | **C1**<br>Simply controllable | **C2**<br>Normally controllable | **C3**<br>Difficult to control or uncontrollable |
| **S0** | | QM (not safety-critical) | QM (not safety-critical) | QM (not safety-critical) |
| **S1** | **E1** | QM (not safety-critical) | QM (not safety-critical) | QM (not safety-critical) |
| | **E2** | QM (not safety-critical) | QM (not safety-critical) | QM (not safety-critical) |
| | **E3** | QM (not safety-critical) | QM (not safety-critical) | **ASIL A** |
| | **E4** | QM (not safety-critical) | **ASIL A** | **ASIL B** |
| **S2** | **E1** | QM (not safety-critical) | QM (not safety-critical) | QM (not safety-critical) |
| | **E2** | QM (not safety-critical) | QM (not safety-critical) | **ASIL A** |
| | **E3** | QM (not safety-critical) | **ASIL A** | **ASIL B** |
| | **E4** | **ASIL A** | **ASIL B** | **ASIL C** |
| **S3** | **E1** | QM (not safety-critical) | QM (not safety-critical) | **ASIL A** |
| | **E2** | QM (not safety-critical) | **ASIL A** | **ASIL B** |
| | **E3** | **ASIL A** | **ASIL B** | **ASIL C** |
| | **E4** | **ASIL B** | **ASIL C** | **ASIL D** |

**Table 4.3**   ASIL assessment criteria

| Severity Classifications (S) | Exposure Classifications (E) | Controllability Classifications (C) |
|---|---|---|
| **S0:** No injuries | **E0:** Unlikely | **C0:** Controllable in general |
| **S1:** Light to moderate injuries | **E1:** Very low probability (injury could happen only in rare operating conditions) | **C1:** Simply controllable |
| **S2:** Severe to life-threatening (survival probable) injuries | **E2:** Low probability | **C2:** Normally controllable |
| **S3:** Life-threatening (survival uncertain) to fatal injuries | **E3:** Medium probability | **C3:** Difficult to control or uncontrollable |
|  | **E4:** High probability (injury could happen under most operating conditions) |  |

The assurance that the required ASIL for the functionality at hand is demonstrated by the corresponding *safety case*.

### 4.3.5   Elements of Safety

Safety is much more than just technology. Creating, maintaining, and operating *safe cyber-physical systems* relies on a consistent set of elements (Fig. 4.5).

> **Quote**
> *"There is, however, only one properly legitimate agenda: The reduction of (safety) risk to be «As Low as Reasonably Practicable»".*
> *Harvey T. Dearden, 2018*

### 4.3.6   Safety Culture

A pervasive *safety culture* (Definition 4.6, [Gilbert18], [Miller20], [Antonsen17], [Roughton19]) is the indispensable foundation for an organization creating, maintaining, and operating safety-critical cyber-physical systems.

▶ **Definition 4.6: Safety Culture**
Safety culture is a risk culture (Definition 8.1). A safety culture are the enduring values, attitudes, motivations, and knowledge of an organization in which safety is prioritized over competing goals in decisions and behaviors.
   ISO 26262 [www.iso.org/standard/68383.html].

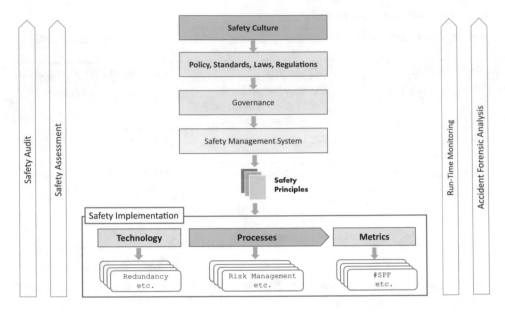

**Fig. 4.5** Elements of safety

> **Quote**
> *"Safety culture is a highly successful idea. Whatever your understanding of this idea, and wether you like it or not, you cannot ignore it".*
>   Hervé Laroche, 2018 (in [Gilbert18])

The safety culture must be deeply ingrained and lived on all organizational levels—from board members to top management, from sectional managers to project leaders, from developers to testers—even the users!

### 4.3.7   Safety Standards and Policies

#### 4.3.7.1 Safety Standards

For many centuries, products were manufactured by a few persons in the same room. When industrialization started, this changed: Different groups of people or several companies contributed, each building a part of the product. This cooperative manufacturing required agreements to ensure that the individually manufactured parts fit each other.

When the number of manufacturers grew, a new method of forcing cooperation appeared: The *standard* (Definition 4.7 ). One of the first industry standards was the "Deutsche Industrie Norm [DIN] 1", published on March 1, 1918 (https://www.din.de/en).

Today, indispensable standards exist in all areas of technology and processes—fortunately also for safety and security (e.g., Example 4.4 for the automotive industry).

▶ **Definition 4.7: Standard**
A standard is:
- A formally established norm for (technical) systems;
- A document which establishes uniform (engineering or technical) criteria, principles, methods, processes, and practices.

In many industries—such as aerospace, railway, road vehicles—compliance with a number of relevant standards is mandatory. Many standards are also the foundation of *certification* (e.g., [DeFlorio16]).

**Quote**
*"Compliance-based safety is formed around the idea, that accidents can be prevented when the system and its operation are designed in compliance with standards".*
Clifton A. Ericson, 2015

**Example 4.4: ISO/IEC 26,262—Automotive Functional Safety**

For *functional safety,* a number of very evolved and useful domain-specific *standards* exist. An interesting representative is the ISO/IEC 26,262 standard for functional safety of *road vehicles* ([ISO26262:18], [Miller20], [Ross16], [Pimentel19], [Gebhardt13], [ISO26262:18]). It was introduced in 2011 and revised in 2018. ISO/IEC 26,262 is a specialization of the more general "Functional Safety standard ISO/IEC 61,508 for Automotive Electric/Electronic Systems ([IEC61508])". An evolution of the ISO/IEC 26,262 for autonomous vehicles is the ISO/IEC 21,448 standard ([ISO21448:19]).

ISO/IEC 26,262 defines a risk-based approach covering all phases of systems development ([ISO26262:18], [Debouk18], [Gebhardt13], [Pimentel19]). The goal is to assess and account for only *acceptable residual risks* in the system—the *vehicle*—and all its relevant subsystems. The standard has 12 parts (Table 4.4).

ISO 26262 is not only an impressive normative work but, at the same time, also a serious and well-readable opus on automotive safety. It is worthwhile to read this official standard.

### 4.3.7.2 Safety Policies

Organizations involved with *safety-critical cyber-physical systems* are constrained by many standards, laws, and regulations (e.g., [Fowler10b]). Any organization needs to

**Table 4.4** Parts of the ISO/IEC 26,262 Standard

| Part | Topic | Overview |
|------|-------|----------|
| Part 1 | Vocabulary | This part introduces and precisely defines the terms used in the ISO 26262 parts of the standard |
| Part 2 | Management of Functional Safety | The functional *safety management system* covers planning, coordinating, and documenting all activities related to functional safety. This part also establishes the construction of the *safety case* |
| Part 3 | Concept Phase | The concept phase includes:<br>• The process for conceptualization<br>• The hazard analysis and risk assessment |
| Part 4 | Product Development at the System Level | Part 4 specifies the requirements for product development at the *system level*:<br>• The initiation of product development at the system level;<br>• The technical safety requirements;<br>• The technical safety concept;<br>• The system design;<br>• The integration and testing;<br>• The safety validation;<br>• The (final) functional safety assessment;<br>• The product release procedure |
| Part 5 | Product Development at the Hardware Level | Part 5 specifies the requirements for product development at the *hardware level*:<br>• General topics for product development at the hardware level;<br>• Specification of hardware safety requirements;<br>• Hardware design;<br>• Evaluation of the hardware architectural metrics;<br>• Evaluation of safety goal violations due to random hardware failures;<br>• Hardware integration and verification |
| Part 6 | Product Development at the Software Level | Part 6 refers specifically to the development of *software*:<br>• Initiation of product development at the software level;<br>• Derivation of software safety requirements from the system level;<br>• Verification of software safety requirements;<br>• Software architectural design;<br>• Software unit design and implementation;<br>• Software unit testing;<br>• Software integration and testing |
| Part 7 | Production and Operation | The objectives of part 7 are:<br>• To develop and maintain a *production process* for safety-related elements or items to be installed in road vehicles;<br>• To develop the necessary information concerning operation, service (maintenance and repair), and decommissioning, ensuring that functional safety is achieved throughout the vehicle's life cycle |

(continued)

**Table 4.4** (continued)

| Part | Topic | Overview |
|------|-------|----------|
| Part 8 | Supporting Processes | Part 8 is normative for the *supporting processes*:<br>• Interfaces within distributed developments;<br>• Overall management of safety requirements;<br>• Configuration management;<br>• Change management;<br>• Verification;<br>• Documentation management;<br>• Confidence in the use of software tools;<br>• Qualification of software components;<br>• Evaluation of hardware elements;<br>• Proven in use argument;<br>• Interfacing an application that is out of the scope of ISO 26262;<br>• Integration of safety-related systems not developed according to ISO 26262 |
| Part 9 | ASIL-oriented and Safety-Oriented Analysis | Requires ASIL (Automotive Safety Integrity Level) decomposition ([Frigerio18]), with the objectives:<br>• To ensure that a safety requirement is decomposed into redundant safety requirements at the next level of detail and that these are allocated to sufficiently independent design elements;<br>• To apply ASIL decomposition according to permitted ASIL decomposition schemas |
| Part 10 | Guideline on the Safety Standard ISO/IEC 26,262 (Informative) | Part 10 provides an *overview* of ISO 26262 and gives additional explanations, and is intended to enhance the understanding of the other parts of ISO 26262. It has an informative character only and describes the general concepts of ISO 26262 to facilitate comprehension<br>expands from general concepts to specific contents |

decide which standards, laws, and regulations apply to their business areas and fields of activities. These decisions are business-critical: Violation of a law or regulation may have severe consequences, e.g., product liability claims.

The medium "policy" is used to condense all applicable laws, regulations, and additional company-specific constraints and requirements (Definition 4.8). Policies are carefully authored documents covering specific areas of activity of the organization. To ensure a complete, consistent, and comprehensive set of policies in the organization, a *policy architecture* is required.

▶ **Definition 4.8: Policy**

The set of fundamental principles and associated guidelines, formulated and enforced by the governing body of an organization, to direct and limit its actions in pursuit of long-term goals.

http://www.businessdictionary.com/definition/policy.html

**Quote**

*"The definition of a policy architecture is one of the most important items that an information security team can do to assist in protecting the enterprise's assets".*
    Sandy Bacik, 2019

A specific set of policies is required for safety: The *safety policies* (e.g.: [Ericson11], [Ericson15], [Miller20], [Stolzer15] and Example 4.5).

---

**Example 4.5: European Union Agency for Railways Safety Policy (Extract)**

The structure and content of the European Union Agency for Railways Safety Policy [EUAR18, Sect. 2.2] are (only extract shown):

*A document describing the organization's safety policy is established by the top management and is:*

1. Appropriate to the organization's type, character, and extent of railway operations;
2. Approved by the organization's chief executive (or a representative(s) of the top management);
3. Actively implemented, communicated, and made available to all staff.

*The safety policy shall:*

1. Include a commitment to conform with all legal and other requirements related to safety;
2. Provide a framework for setting safety objectives and evaluating the organization's safety performance against these objectives;
3. Include a commitment to control safety risks that arise both from its own activities and those caused by others;
4. Include a commitment to continual improvement of the safety management system;
5. Be maintained in accordance with the business strategy and the evaluation of the safety performance of the organization.

## 4.3.8   Governance

Corporate *governance* is the system of rules, practices, and processes by which a company is directed and controlled. Therefore, the decisions and actions on all levels of the organization substantially impact safety. Unfortunately, a perpetual conflict of interests between *business interests* (profit, market share, time-to-market, e.g., [Rodin99]) and *quality requirements* (safety, security, etc., e.g., [Merkow10]) exists.

> **Quote**
> *"When greedy, ignorant executives who worry about losing a deal or getting fired themselves dictate an impossible deadline and tremendous scope, you must refuse it".*
>    Eben Hewitt, 2019

This conflict of interest manifests itself in every project ([Yourdon03]). Resolving this conflict requires a working trust between the business units (Commercial interests) and the architectural department (Software quality interests). This trust must be backed by corporate governance, which enforces *justifiable tradeoffs* ([Furrer19], [Ford21]).

## 4.3.9   Safety Management System

Dependable safety is, to a large extent, a question of "the right information at the right place at the right time". This is not a simple task with possibly thousands of employees working on safety-critical projects in larger organizations. A precious instrument is a *safety management system* (SMS, Definition 4.9).

▶ **Definition 4.9: Safety Management System (Safety-MS, SMS)**
Safety management systems provide organizations with a powerful framework of safety philosophy, tools, processes, and methodologies that improve their ability to understand, construct, and manage safety-critical systems.
   [Stolzer15].

Safety management systems are predominantly used in *aviation* ([Stolzer15], [ICAO13]), where they are mandatory. An aviation safety management system (SMS) is an instrument to support an aircraft's safe operation, both in the air and on the ground. SMS's are designed to continuously enhance safety by identifying and recognizing hazards, collecting and analyzing data, and assessing & proactively mitigating safety risks. Therefore, SMS's main objective is to eliminate or mitigate risks before they cause aviation accidents.

> **Quote**
> *"Remember, our Safety Management System demands that we look forward—that we hunt down and eliminate problems before they occur".*
> Alan J. Stolzer, 2015

Several aviation authorities have specified or developed safety management systems, e.g., the International Civil Aviation Organization (ICAO, Example 4.6).

---

**Example 4.6: ICAO Safety Management System**

Extract from the International Civil Aviation Organization Safety (ICAO) Management Manual (SMM, [ICAO13]):

*Chapter 5. Safety Management System (SMS)*
    5.1 Introduction
    5.2 Scope
    5.3 SMS framework
*SMS Component 1. Safety policy and objectives*:

- SMS Element 1.1 Management commitment and responsibility
- SMS Element 1.2 Safety accountabilities
- SMS Element 1.3 Appointment of key safety personnel
- SMS Element 1.4 Coordination of emergency response planning
- SMS Element 1.5 SMS documentation

*SMS Component 2. Safety Risk Management:*

- SMS Element 2.1 Hazard identification
- SMS Element 2.2 Safety risk assessment and mitigation

*SMS Component 3. Safety Assurance:*

- SMS Element 3.1 Safety performance monitoring and measurement
- SMS Element 3.2 The management of change
- SMS Element 3.3 Continuous improvement of SMS

*SMS Component 4. Safety Promotion:*

- SMS Element 4.1 Training and education
- SMS Element 4.2 Safety communication

*5.4 SMS Implementation Planning:*

- System description
- Integration of management systems
- Gap analysis
- SMS implementation plan
- Safety performance indicators

*5.5 Phased Implementation Approach:*

- 5.5.1 General
- 5.5.2 Phase 1: Integration & accountability
- 5.5.3 Phase 2: Essential safety management processes
- 5.5.4 Phase 3: Safety risk management processes
- 5.5.5 Phase 4: Periodic monitoring, feedback, and continuous corrective action.

## 4.3.10  Safety Principles

Safety *principles* (Definition 4.10) and security *principles* are the essences of dependable cyber-physical systems (and of this monograph). Strictly complying with the applicable, proven, and justified "eternal truths" leads the development process to construct trustworthy cyber-physical systems.

▶ **Definition 4.10: Safety Principle**
Safety principles are rules for the construction, evolution, and operation of safe systems. They are precisely formulated, well-justified, teachable, actionable, and enforceable.
    [Furrer19].

Figure 4.2 distinguishes between general resilience principles and specific safety principles. Specific safety principles target particular safety quality properties of a system and define mitigation measures.

> **Quote**
> *"Software engineering principles should capture the nature and behavior of software systems and guide their development. Such principles would help in restricting degrees of freedom in software development and achieving degrees of predictability and repeatability similar to those of classical engineering disciplines".*
>     Hadar Ziv, 1997

### 4.3.11  Safety Implementation

Mission-critical cyber-physical systems are often highly complex constructs. There is a colossal body of knowledge on building safe and secure systems (Note: Consult the extensive lists of references in this monograph!). Finally, however, the systems are implemented by teams of people. Safety and security manifest themselves in the implementation's architecture, design decisions, chosen technologies, programming quality, etc.

The implementation consists of (Fig. 4.5):

- Technology;
- Processes;
- Metrics.

Any organization needs to define, institutionalize, supervise, and enforce its own set of technologies, processes, and metrics. No universal advice can be given—with one exception: Aim for maximum *simplicity*, i.e., reduce *complexity* (Definition 4.11) as much as possible ([Mitchell11], [Bak09], [Jackson07], [Sessions08], [Sessions09], [Kopetz11], [Kopetz19], [Li20b], [Kanat-Alexander17], [Kanat-Alexander12]). A powerful method includes explicit *simplification steps* into the development process (Example 4.7).

▶ **Definition 4.11: Complexity**
Complexity is that property of an IT-system which makes it difficult to formulate its overall behavior, even when given complete information about its parts and their relationships.
   1. *Essential* complexity: This is caused by the problem to be solved. Nothing can remove it. It represents the inherent difficulty of the requirements to be implemented.
   2. *Accidental* complexity: This is caused by solutions that we create on our own or by impacts from our environment. It can be managed and minimized

---

**Example 4.7: Simplification Step in Development Process**

Figure 4.6 introduces two *simplification steps* into a schematic development process ([Furrer19]). The first formal step reduces the complexity of the requirements and the integration into the existing system. The second formal step reviews the system architecture and design and reduces this complexity.

The formal simplification steps are mandatory in the respective development process. A team of experienced experts executes them, possibly assisted by independent, external consultants. Any simplification recommendations are fed back into the corresponding development step and executed there. The power of such simplification reviews is potent and long-lasting.

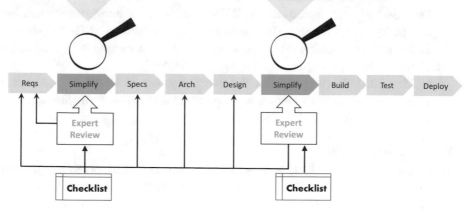

**Fig. 4.6** Simplification steps in the development process

**Quote**

*"One key to achieving dependability at reasonable cost is a serious and sustained commitment to simplicity, including simplicity of critical functions and simplicity in system interactions".*
  Daniel Jackson, 2007

## 4.3.12 Safety Assessment and Safety Audit

### 4.3.12.1 Safety Assessment

The safety task is not completed when an organization has effectively implemented safety policies, processes, metrics, etc. Safety is not a static issue: Unforeseen failure modes, changes in laws and regulations, adaptations of standards and certification, adjustments to environmental conditions, and many more causes require continuous realignment.

Therefore, a *safety assessment* is required during the product development cycle and must be repeated following any trigger condition. An interesting and comprehensive safety assessment report—with a view into the future—is given in Example 4.8.

---

**Example 4.8: BMW SAE Level 3 Automated Driving System**

*Automated driving systems* are offering more and more autonomy to their users. Because they are both safety-critical and under continuous development, the requirements for their safety assessment are high.

The German car manufacturer BMW (https://www.bmwgroup.com/de.html) is developing and marketing its automated driving system iNEXT, starting at SAE level 0 and aiming at SAE level 5 (https://www.researchgate.net/figure/The-SAE-levels-of-driving-automation_fig4_338655541).

In May 2020, BMW published the comprehensive "SAE Level 3 Automated Driving System Safety Assessment Report" ([BMW20])—a highly instructive document!

### 4.3.12.2  Safety Audit

A *safety audit* (Definition 4.12) is an essential constituent of an organization's safety management responsibilities. The safety auditor subjects the safety-related activities of the organization to systematic and critical evaluation. The aim is to discover weaknesses and rectify them. Three types of audit are possible:

i.   *Internal Safety Audit* (Self-Auditing): This audit is executed by an internal team, often under the compliance officer's leadership. The audit results are not communicated to external parties but serve to improve the organization's safety management instruments;

ii.  *Third-Party Audits* (External Auditing): The organization mandates an external, independent specialist who audits according to its own rules and procedures. The audit results are not communicated to external parties but serve to improve the organization's safety management instruments;

iii. *Regulatory Safety Audits* (External Auditing by an official or accredited Organization): Many industries and some certification agencies require safety audits by impartial, certified organizations. The audit results—especially safety enhancements—are binding for the organization under audit. Severe safety deficiencies uncovered during the audit may have consequences.

▶ **Definition 4.12: Safety Audit**

Safety audits of an organization are conducted to assess the degree of compliance with the applicable safety regulatory requirements, the mandatory safety standards, and the safety management system's procedural provisions. They are intended to provide assurance of the safety management functions, including staffing, compliance with applicable policies and regulations, competency levels, and training.

Adapted from: https://www.skybrary.aero/index.php/Safety_Audits

### 4.3.13 Safety Runtime Monitoring

Today's software for cyber-physical systems is a large, complex, and complicated achievement. Even sophisticated development processes cannot ultimately ensure freedom from faults and errors—and such faults and errors can result in malfunctions and *accidents* during runtime.

As a "last defense", *runtime monitoring/runtime verification* should be implemented (Definition 4.13, [Bartocci19]). A runtime monitor captures data from various software/ hardware system sources during actual operations. The verification part analyzes, examines, and interprets the data in real-time (Fig. 4.7).

▶ **Definition 4.13: Runtime Monitoring (Runtime Verification)**
Runtime monitoring is a dynamic verification technique that involves observing the internal operations of a software system and/or its interactions with other external entities to determine whether the system satisfies or violates a correctness specification or exhibits irregularities or anomalies in behavior. The aim is to timely prevent malfunctions of the software/hardware system.

Adapted from: Ian Cassar, 2017.

A runtime monitoring/runtime verification system includes the following functionality (Fig. 4.7):

1. The *functional software-system*, including the execution platform (Operating system, communications drivers, etc.);
2. The interfaces to the *external world*, including the sensor/actuators to the physical part of the cyber-physical system (Fig. 1.1);
3. The *monitoring system* (often also called instrumentation), which captures runtime behavior information from the hardware/software-system in real-time;
4. The *verification algorithms*, which analyze, examine, and interpret the data received from the monitoring and trigger actions to prevent, absorb, or rectify failures during execution;
5. The *rules*, *contracts*, and *policies* (often formally) define the correct behavior of the hardware/software-system. The verification checks the system's running actions against this set of verification norms, i.e., the model for correct behavior;

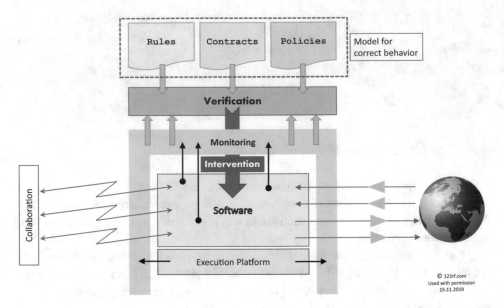

**Fig. 4.7** Runtime monitoring/runtime verification

6. A set of corrective *intervention actions* in case of violation of the rules, contracts, or policies or in case of other anomalies.

> **Quote**
> *"To improve the dependability of software-systems, the software running in the field can be equipped with solutions to prevent, detect, and react to failures".*
> Yliès Falcone, 2018

Many techniques to implement various levels of runtime monitoring/runtime verification exist ([Kane15], [Cassar17], [Drusinsky06], [Finkbeiner19]), and much active research is currently being done. A simple implementation of runtime monitoring/runtime verification is illustrated in Example 4.9.

**Example 4.9: Runtime System Watchdog**

Figure 4.8 shows a straightforward runtime monitoring/runtime verification system: A *runtime system watchdog*. The functional software is constructed in such a way that it issues an impulse—a call to "trigger watchdog"—in intervals of less than $\tau$ milliseconds ($\tau$=e.g., 5 ms). The watchdog is implemented in *hardware*: If any trigger impulse does *not* arrive within the watchdog interval $\tau$ the watchdog assumes a failure of the software ("dead").

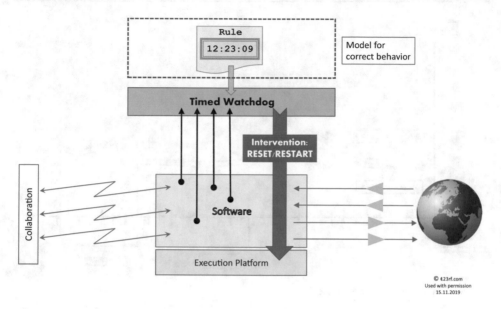

**Fig. 4.8** Runtime system watchdog

In this case, the watchdog intervenes:

a) If non-maskable *interrupts* (https://www.sciencedirect.com/topics/computer-science/interrupt-mechanism) are available in the processor, the watchdog activates one dedicated non-maskable interrupt. The associated, specific interrupt handling routine then executes code to restart the system and extract & archive forensic information (= *Soft restart*);

b) If *no* non-maskable interrupts are available in the processor, the watchdog executes a full hardware reset and thus restarts the system (= *Hard restart*).

The second application of runtime monitoring is shown in Example 4.10: Prediction of wheel rotation *sensor failure*. This example is based on the runtime evaluation of sensor data streams.

Note that runtime monitoring introduces some execution time overhead into the system, which must be considered when calculating the software's real-time behavior.

---

**Example 4.10: Wheel Rotation Sensor Failure Prediction**

For the *Anti-Skid Braking System* (ABS, Example 2.1), a vehicle needs wheel rotation sensors on all 4 wheels. The wheel rotation sensors transmit the current rotation rate of each wheel to the electronic control unit. Many commercially available wheel sensors have *two* redundant measurement channels 1 & 2 (e.g., https://www.infineon.com). By comparing and averaging the two channels and the values from all other

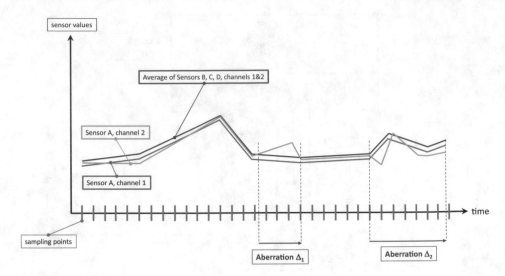

**Fig. 4.9**  Wheel rotation sensor aberrations

wheels, the ABS software can detect intermittent or permanent aberrations, as shown in Fig. 4.9: The data delivered by each sensor is continuously checked against the data provided by all other sensors. Thus, the software will detect two *aberrations* $\Delta_1$ and $\Delta_2$ of sensor A, channel 2. Not only will these measurements be corrected, but an early warning of forthcoming sensor failure will be generated. The sensor can then be replaced before a sensor failure occurs (*Predictive maintenance*).

### 4.3.14  How Much Safety is Enough?

Engineering, implementing, and maintaining *safety* is expensive. Therefore, the question "how much safety is enough?" must be answered during the engineering process. The *safety risk management process* provides the answer: A responsible balance between cost and residual risk must be achieved (Fig. 4.10). The safety cost curve has an exponential shape—100% safety means infinite cost—and is not realizable in any technical system!

Figure 4.10 shows the critical tradeoff point: It is determined by the *product's* or the *service's commercial viability* and the *acceptable residual safety risk*. Note that this tradeoff must be carefully evaluated, justified, and documented for each *failure mode*.

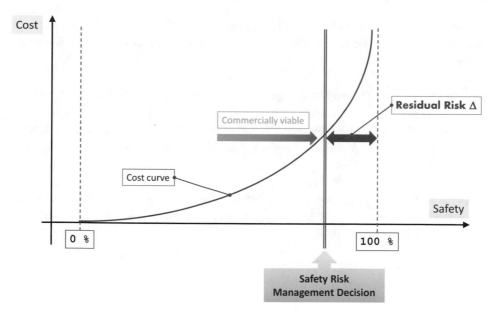

**Fig. 4.10** Safety cost curve

## 4.4   Security

### 4.4.1   Introduction

*Safety* and *security* in a cyber-physical system are two very different concepts. Both are required to a sufficient degree to ensure a trustworthy cyber-physical system. *Security* has the objective to defend against threats to the system's electronic assets.

> **Quote**
> *"The dramatic rise of cyberspace has presented unparalleled opportunities to improve our way of life. The dark side of this proliferation is an even more dramatic rise in the malevolent use of cyberspace".*
>    Michael A. Vanputte, 2017

Security has two areas of protection assets:

- *Functional security* (Definition 4.14): Protect the programs from modifications, malicious additions, or suppression—both during development and execution;
- *Information security* (Definition 4.15): Guard the information assets (Data, documents, plans, financial statements, business plans, etc.) against unauthorized access.

▶ **Definition 4.14: Functional Security** Functional security protects the software-system from malicious, infiltrated code, both from the outside and from the inside of the organization.

Security is a *global property* of the cyber-physical system: The CPS's weakest component can compromise the complete system!

An attacker will try many latent *attack paths* (Fig. 4.11) to reach a protection asset. In some cases, a first attack is used to find a vulnerability, and a second attack is attempted to gain access to the protection asset.

▶ **Definition 4.15: Information Security** Information Security protects the confidentiality, integrity, and availability (CIA) of computer system data and information from unauthorized and malicious accesses.

### 4.4.2   Security Taxonomy

The security of a CPS consists of many quality of service properties, i.e., security is a *macro-property*. Which quality of service properties constitutes a particular application's security depends on the risk potential. Therefore, for each application area, a specific *security taxonomy* must be generated (see, e.g., Fig. 4.12).

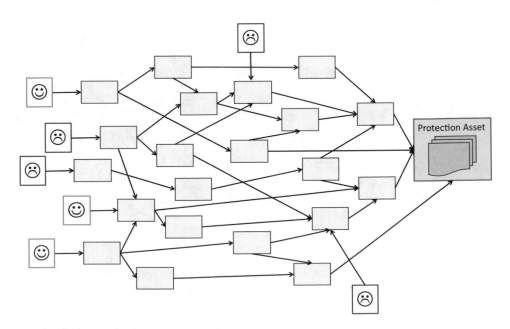

**Fig. 4.11**  Attack paths

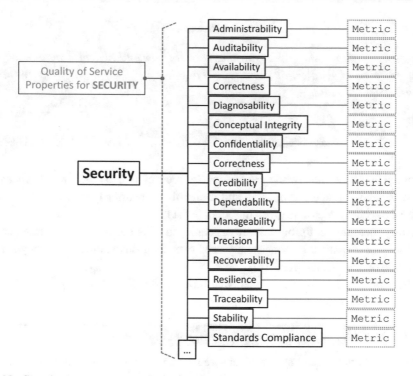

**Fig. 4.12** Security taxonomy example

### 4.4.3 Security Metrics

Many efforts have been made to make security measurable. Many *security metrics* (Definition 4.16) have been proposed and are in use ([Jaquith07], [Hubbard16], [Freund15], [Hayden10]).

▶ **Definition 4.16: Security Metric** Periodic, quantitative measurement value of a security-relevant property of the organization, the processes, the products, or the services for audit purposes, organizational improvements, management decisions, or investment assessments.

Security metrics are necessary for:

- Verify the compliance with laws and regulations;
- Measure the progress of security in the organization;
- Demonstrate the organization's adherence to policies and standards;
- Prove the security-relevant effect of the processes (development, deployment, operations, etc.);
- Provide evidence for the security of the products and services offered to the customers.

> **Quote**
> *"Security is hard to measure. With limited resources and budgets, and an endless list of problems to solve, as security practitioners we need the visibility of security metrics to enable better decision-making and drive change throughout our organizations".*
>   Caroline Wong, 2012

The foundation of a worthwhile set of security metrics is a *security measurement program* with a comprehensive mission and a significant charter (see, e.g., [Wong11]). A sample of a security metric is given in Example 4.11.

Security measurement programs are expensive to develop, maintain, operate, and use. The chosen metrics must be consistently used over long times; otherwise, they are of no value to the organization ([Mead16]).

Avoid *vanity metrics*—i.e., numbers that look good and seemingly provide value, but in reality, do not provide any answers ([Ries11]).

---

**Example 4.11: Network Security Controls Assessment**

Security in a networked system is ensured by *security controls* (Technical or procedural mitigation- or counter-measures to avoid, detect, counteract, or minimize security risks to the system). The security controls are defined during the *risk management process*. However, as the systems and operating environments continuously change, the implemented controls need regular reviews. A metric and measurement checklist for the review of the controls is presented in Table 4.5.

### 4.4.4   Elements of Security

Information security is a vast world of its own: To successfully secure a modern cyber-physical system in today's aggressive environment, many ingredients are necessary (Fig. 4.13). Security *implementation* is only the last step in a repeated chain of responsibilities.

> **Quote**
> *"While IT (Information Technology) itself is fairly mature, IT security is not. A single, agreed-upon methodology for securing IT-systems simply doesn't exist".*
>   Nancy R. Mead, 2017

**Table 4.5**  Review of security controls metric

| Critical Element | **Have the security controls of the system and the interconnected systems been reviewed?** |
|---|---|
| Subordinate Question | Are tests and examinations of crucial controls routinely made, i.e., network scans, analyses of router and switch settings, penetration testing? |
| **Metric** | **Percentage** of total systems for which security controls have been tested and evaluated in the past year |
| Purpose | To measure the level of **compliance** with requirements for system security control testing |
| Implementation Evidence | 1. Does your agency maintain a current system inventory?<br>☐ Yes      ☐ No<br>2. If yes, how many systems are there in your agency (or agency component, as applicable)? _____<br>3. For how many systems were system security controls tested in the past year? _____<br>4. How many systems have used the following testing methods in the past year to evaluate security controls:<br>Automated tools (e.g., password cracking and war dialing) _____<br>Penetration testing _____<br>Security test and evaluation (ST&E) _____<br>System audits _____<br>Risk assessments _____<br>Other (specify) _____<br>5. How many systems were tested using any of the methods in Question 4 in the following time frames? (Choose the nearest time frame for each system; do not count the same system in more than one time frame.)<br>Within the past 3 months _____<br>Within the past 6 months _____<br>Within the past 12 months _____<br>6. Are all testing instances and results recorded?<br>☐ Yes      ☐ No |
| Frequency | Annually |
| Formula | Number of systems with controls tested (Question 3 above) / Total number of systems in the inventory (Question 2 above) |
| Data Source | OMB Exhibits 53 and 300; budget office; audits; C&A database; automated tool reports; system testing logs/records |
| Indicators | The percentage trend should increase and approach or equal 100 percent. Overall, security controls must be tested once they are in place to ensure they are working as proposed. As changes occur within the security environment, the necessary controls also may change. To keep a control current and appropriate for the system, regular control testing and evaluation should be conducted |

Source: https://nvlpubs.nist.gov/nistpubs/Legacy/SP/nistspecialpublication800-55.pdf

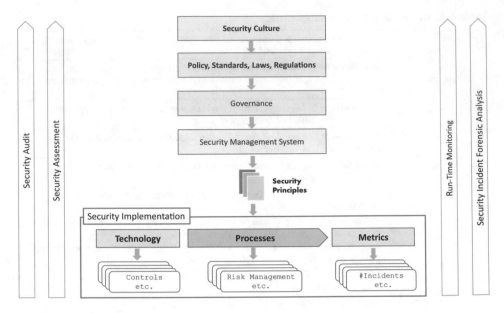

**Fig. 4.13**   Elements of security

### 4.4.5   Security Culture

The groundwork of secure IT-based systems is a sound and dependable *security culture* (Definition 4.17, [e.g.: Trim16], [Hayden16], [Roer15]).

▶ **Definition 4.17: Security Culture**
Security culture is a risk culture in which security is an overriding priority.

(Note: Overriding = Security requirements take priority over system requirements, such as functionality).

Adapted from [Gilbert18].

---

**Quote**
*"After having worked in information security for over 20 years, I have come to a simple conclusion: Unless we move beyond technology alone and start addressing the human element, we are in a no-win situation".*
Lance Hayden, 2016

---

The security culture co-exists with other company cultures, such as the safety culture, the human resource culture, or the legal & compliance culture. The clarity and consistency of all subcultures are of the highest importance: Inconsistencies or especially contradictions form a significant *risk for security* (Example 4.12).

---

**Example 4.12: Conflicting Cultural Values**

The following situation reveals a dangerous conflict between cultural values in a fictitious company (Source: [Hayden16]):

*"On-time, on-budget product delivery is part of the company's DNA. Clara (a long-standing programmer) knows this. Project stats tracked by the company are almost all about completion. How many milestones were completed on schedule? How often were they completed early? No one wants to talk about being late for a deadline. Missing project deadlines by even a couple of days throws everything in disarray and causes management to freak out. The managers' bonuses and hopes for promotion, just like Clara's and everyone else's on her team, are tied to project performance metrics. If you blow a deadline, management is suddenly in your face, and even your friends on the team begin to look at you funny, like an anchor that is dragging them down.*

*... during a particularly aggressive update project for the company's flagship product, Clara realizes early that meeting the product deadline is going to be extremely demanding.*

*... Clara realizes ten days before the project completion that finishing up her security reviews and testing is going to take two full weeks. In the end, the pressure of the deadline overwhelms the influence of the SDLC security review policies. So, she crosses her fingers and completes as much of the security review as she possibly can in ten days. No one notices that she's cut a few corners on the security reviews.*

*... Then the day comes when a vulnerability in one of the company's products spawns an exploit that is used to attack a customer, resulting in a very public security breach."*

The cultural conflict between "On-time, on-budget" versus "security imperatives" straightforwardly lead to this foreseeable disaster!

**Quote**
*"If an organization's security culture is weak and buggy, if it constantly competes or conflicts with other routines and processes running things, that organization is going to have problems".*
Lance Hayden, 2016

The building, maintaining, evolving, and enforcing the security culture is similar to other managed processes in the organization: A planned, continuous, controlled, and audited activity. Because of the security culture's fundamental significance for the security of the cyber-physical systems, it must have a high priority, sufficient funding, and the necessary attention throughout the organization.

### 4.4.6   Security Standards and Policies

#### 4.4.6.1  Security Standards

Security is not only a vast, complicated topic in today's cyber-physical systems—but it involves any number of people, organizations, and technology. Therefore, any help to structure and organize security efforts is valuable. In security, this role is played by *security standards* (Definition 4.18, e.g.: [Bishop18], [Scarfone10], [Landoll16]).

▶ **Definition 4.18: Security Standard**
A cyber-security standard defines both functional and assurance requirements within a product, system, process, or technology environment in order to optimize security.
   Adapted from: [Scarfone10].

A large number of security standards have been developed over time. Most of them are in continuous evolution, and new versions appear regularly.

   Security standards can be organized hierarchically, using the architecture framework of Fig. 5.5, as shown in Fig. 4.14. On top are the business standards, such as the US Sarbanes–Oxley Act ([Dye17]) or Security Management Systems Standards (e.g., ISO/IEC 27,001, Example 4.15). Next in the hierarchy are the application-area-specific security standards, such as the HIPAA (Health Insurance Portability and Accountability Act, [Frew15]) for the health industry or Banking Cybersecurity Standards ([FSI17]) for financial institutions. Many National Data Protection Standards (e.g., [GDPR16] for

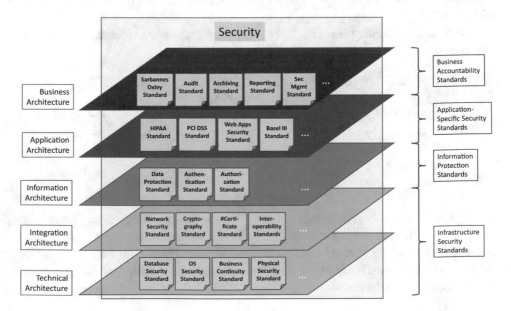

**Fig. 4.14**  Security standards hierarchy

Europe) and information protection technologies are available for the information architecture. The integration architecture benefits from many interoperability standards, e.g., the digital certificate standard X.509, which is the base for trust in electronic transactions (Example 4.13). Last but not least, security standards for the runtime platform, such as database security (e.g., [Malik16]), are available.

Example 4.13 describes a typical *interoperability standard*: The digital certificate standard X.509, which establishes dependable trust relationships in electronic transactions.

---

**Example 4.13: X.509 Digital Certificate Standard**

*Digital certificates* (Definition 4.19) are the foundation of *trust* in electronic transactions (e.g., [Buchmann16], [Vacca19]). The digital certificate standard X.509 (https://tools.ietf.org/html/rfc5280) enforces interoperability between all stakeholders in electronic transactions.

▶ **Definition 4.19: Digital Certificate**
Digital certificates are electronic credentials that bind the identity of the certificate owner to a pair of electronic encryption keys, (one public and one private), that can be used to encrypt and sign information digitally.
https://oit.utk.edu/

The X.509 certificate is a trustworthy data structure (Fig. 4.15), guaranteed by a dependable certification authority (CA).

### 4.4.6.2 Security Policies

In today's hostile environment, security is strongly dictated by a growing number of standards, laws, and regulations. To direct an organization's conformant behavior, the relevant obligations must be concentrated into a set of comprehensive, clearly arranged, actionable, enforceable, and auditable *policies* (Definition 4.8). Both for safety and security, effective policies are essential ([Bacik19], [Williams13])! A condensed form for the information needed in a specific policy, a SOTA policy, is described in Example 4.14.

---

**Quote**
*"Any organization should have policies for its management of information assurance. It should include statements that make it clear that the organization regards risk as a serious issue with it being discussed at all appropriate meeting, with those with the correct authority and responsibility taking an active interest in it".*
Andy Taylor, 2020

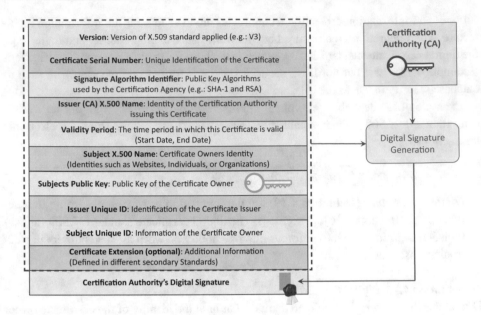

**Fig. 4.15** X.509 digital certificate format

---

**Example 4.14: SOTA Policy**

A modern car (2022) contains up to 100 networked ECU's (Electronic Control Units) and more than 100 Million of Lines of Code (SLOC's)—with a strong upward tendency (see, e.g., [McKinsey19]). This colossal software codebase requires frequent *updates* to add new functionality and correct faults.

Traditionally, the vehicle had to visit a workshop to update the ECU's. The new technology SOTA (Software over-the-air) delivers software, data, or configuration information from the manufacturer directly to a vehicle through wireless technologies (Fig. 4.16, see, e.g., [Möller19], [Schmidt20], Kapitel «Elektronik», S. 152, [Kuppusamy16]).

While software over-the-air updates typically enhance a vehicle's functionality, they can also make it more vulnerable to *cyber-security threats*. The use of third-party provider software, such as the cloud service provider or the distribution platform (Fig. 4.16), increases hackers' opportunities. SOTA is a very typical application of a cyber-physical system that is both safety- and security-critical.

Therefore, SOTA requires a policy to ensure safety and security: The main elements of such a SOTA policy are listed in Table 4.6.

**SOTA Sample Policy Topics** (alphabetically).

Validity Start Date: 1.7.2020/Next Revision: 1.7.2021.

Policy Owner: CIO (Chief Information Officer).

Validity: For all activities related to the SOTA development and operation.

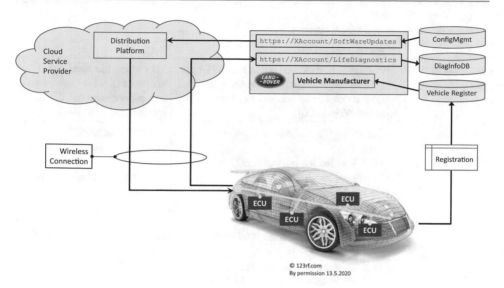

**Fig. 4.16**  Automotive software-over-the-air distribution

## 4.4.7   Governance

Weak, incompetent, or poorly implemented *governance* constitutes a significant risk for any cyber-physical system's security.

Good governance needs an adequate company management structure and unambiguously, traceable, auditable assignments of responsibilities to capable people.

> **Quote**
> *"In many situations, governance is delegated to different functions. This does not relieve top management from their responsibility to give strategic directions, set an example, and provide the necessary resources".*
>   Cesare Gallotti, 2019

## 4.4.8   Security Management System

Developing and evolving cyber-physical systems requires the team effort of many specialists in an organization, often including external partners. The crucial success factor for adequate system security is the satisfactory synchronization of all participants' security-relevant activities. An indispensable prerequisite is the easy availability of extensive information covering the system's security aspects. The essential tool for enabling information access is the *Information Security Management System* (ISMS, Definition 4.20, e.g.: [Gallotti19], [Hintzbergen15], [VanDerWens19], [Williams13]).

**Table 4.6**  Elements of a SOTA Policy

| Policy Topic | Policy Stipulation | Remarks |
|---|---|---|
| Accountability | For any action carried out in the SOTA system, an unquestionably assigned individual is responsible and accountable for that action ([Taylor20]) | Very often, the chief risk officer or the chief information officer, who can delegate specific responsibilities |
| Archiving | The following artifacts must be securely archived:<br>• All downloaded information (All versions)<br>• All recipients<br>• Date of successful download<br>• The vehicle database<br>• The vehicle registration information<br>• All audits<br>• All faults, accidents, recovery actions<br>• All risk assessments & residual risk acceptance<br>• All upload information (diagnostics, etc.) | "Securely archived" means a provably tamper-proof, access-controlled, time-stamped, long-term electronic archive ([EMC12]) |
| Authentication | All connections must be strongly authenticated on a session basis (both endpoints) | Preferably via digital certificates |
| Authorization | All connections must be strongly authorized with the least possible granularity and the least access priviledges | Based on an accurate and timely rights database |
| Availability | All updates, especially safety-critical updates, must be executed to all vehicles in less than xx hours | |
| Client Data | Client data, including all data gathered from registration and vehicle movements, must be adequately protected and managed according to the respective jurisdiction's data protection laws. Explicit personal acceptance of the data usage policy must be obtained from the owner of the vehicle. The data usage policy must be readily available and comprehensive | Respect local jurisdiction |
| Compliance | All SOTA activities must at all times comply with applicable laws and regulations of the affected jurisdictions | Especially the local privacy protection laws |
| Confidentiality | All connections must at all times be protected by strong cryptography | |
| Configuration Management | The SOTA management system must at all times precisely account for all versions of all information downloaded to each individual vehicle | |

(continued)

**Table 4.6**   (continued)

| Policy Topic | Policy Stipulation | Remarks |
| --- | --- | --- |
| Distribution Management | The SOTA system must have full control over the distribution of information to all vehicles. Failures or inability to contact a vehicle must be logged, and appropriate back-up mechanisms to reach the vehicle/owner must be in place | Mandatory for safety-critical updates |
| ECU Recovery | Any ECU may fail and have to be replaced as a complete unit. SOTA must recognize this moment and update the new ECU with the latest versions | |
| Integrity | SOTA must ensure the unconditional integrity of all downloaded information to the vehicle at all times<br>The integrity of the software in the vehicle must be verified at each start-up | Use hash-functions and digital certificates to verify the origin and intactness of all downloaded artifacts, both immediately after download and at each start-up |
| Malicious ECU Replacement | An ECU could be maliciously (physically) replaced in the vehicle and contain harmful software. Ensure that such activity is immediately (before starting the vehicle) recognized and rectified | |
| Provider Audit | Regularly execute safety audits of all the involved providers (Distribution platform, cloud service providers, etc.) | |
| Retention | Define a retention policy, i.e., which data is archived for how long | According to the local jurisdiction |
| Risk Management | Execute a reliable, professional risk assessment and risk mitigation for any activity related to the SOTA | Especially the safety of the downloaded information |
| Safety Testing | Execute thorough testing for the safety of any new software to be downloaded to the vehicles, including recovery/rollback procedures in case of emerging operational problems | Separate safety testing policy (not part of this monograph) |
| Security Testing | Execute thorough testing for the security of any changes to the SOTA platform, including recovery/rollback procedures in case of emerging operational problems | Separate security testing policy (not part of this monograph) |
| Tracing/Audit | Generate a detailed, time-stamped, tamper-proof trail of all SOTA-system activities. Archive the trail for internal and external audit purposes | |

(continued)

**Table 4.6**   (continued)

| Policy Topic | Policy Stipulation | Remarks |
|---|---|---|
| Vehicle Management | Maintain a complete, up-to-date registry of all vehicles served by the SOTA. Retain all registration and vehicle data and their changes over time (archive) | |

▶ **Definition 4.20: Information Security Management System (ISMS)**
That part of the overall management system, based on a business-risk approach, to establish, implement, operate, monitor, review, maintain, and improve Information Security.
  ISO 17799 (https://www.iso.org/standard/39612.html).

> **Quote**
> *"The main principle behind the Information Security Management System is that there should be a 'one-stop shop' for all information pertinent to the assurance of information within an organization. As soon as there is a need to go looking for documentation, policies, practices or anything else to do with assurance, the chances are that someone will not bother and will do their own thing".*
>   Andy Taylor, 2020

Information Security Management Systems are the focus of a number of valuable standards. Of great international importance is the ISO/IEC 27,000 family of standards. Nearly 40 standards of the 27,000-family are planned, of which a multitude has already been published and is regularly revised (e.g., [https://www.itgovernance.co.uk/iso27000-family]).

The ISO/IEC 27000 family of standards offers excellent guidance for the entire field of information security management and provides an invaluable source of well-organized and profound knowledge. A short overview of the standards family is presented in Example 4.15.

---

**Example 4.15: ISO/IEC 27000 Information Security Standards Family**

The *IEC/ISO 27000* family of standards form a towering set of individual, aligned, and focused standards documents (see, e.g., [ITG20], [https://www.iso.org/isoiec-27001-information-security.html]). The 11 standards 27000 – 27010 standardize the basic functionality of a modern ISMS. In addition to these 11 basic standards, a growing number of application- or industry-specific standards are available or are under discussion (Table 4.7).

**Table 4.7**  ISO/IEC 27000 Standards family overview

| ISO/IEC | Topic | Remarks |
|---|---|---|
| *General Standard Parts* | | |
| 27000 | Information security management systems: Overview and vocabulary | Non-normative introduction |
| 27001 | Information technology: Security Techniques— Information security management systems | |
| 27002 | Code of practice for information security controls: A catalog of information security controls that might be managed through the ISMS [Information Security Management System] | Organizations are only required to implement controls that are defined by the risk assessment |
| 27003 | Information security management system: Implementation guidance | |
| 27004 | Information security management: Monitoring, measurement, analysis, and evaluation | |
| 27005 | Information security risk management | |
| 27006 | Requirements for bodies providing audit and certification of information security management systems | |
| 27007 | Guidelines for auditing of information security management systems | |
| 27008 | Guidance for auditors on ISMS controls | |
| 27009 | Internal document for the committee developing sector/industry-specific variants or implementation guidelines for the ISO/IEC 27000 family of standards | |
| 27010 | Information security management for inter-sector and inter-organizational communications | |
| *Application-specific Support Standards: The 27000-family provides many standards focusing on specific industries or applications. Some examples:* | | |
| 27011 | Information security management guidelines for *telecommunications organizations* based on ISO/IEC 27002 | Telecommunications example |
| 27017 | Code of practice for information security controls based on ISO/IEC 27002 for *cloud services* | Cloud services example |
| 27037 | Guidelines for identification, collection, acquisition, and preservation of digital evidence | |
| 27043 | Incident investigation | |
| 27799 | Information security management in health using ISO/IEC 27002 | Advises health organizations on how to protect personal health information using ISO/IEC 27002 |

### 4.4.9  Security Principles

Developers and operators of cyber-physical systems need authoritative guidance for their trusted activities. A colossal body of knowledge and a massive base of literature about IT security is available. This monograph attempts to crystallize this knowledge into a set of *security principles* (Definition 4.21).

▶ **Definition 4.21: Security Principle**
Security principles are rules for the construction, evolution, and operation of secure systems. They are precisely formulated, well-justified, teachable, actionable, and enforceable.
   [Furrer19].

The principles are described in Part 2 of this monograph.

### 4.4.10  Security Implementation

Security is an ongoing, never-ending battle between *threats* (in the form of attackers) and *mitigation measures* (in the form of defense mechanisms or *controls*). The attackers are continuously devising new electronic assaults, and the defenders react with new defensive measures (see, e.g., [Schneier03], [Lusthaus18], [Fazzini19], [Graham20], [Anderson20], [Mukherjee20]). Because more and more cyber-physical systems depend on networks for their operation, the danger of cyber-attacks and potentially hazardous consequences are increasing from day to day. Defending the cyber part of the CPS is, therefore, increasingly essential.

Two basic mechanisms are available for the defense:

1. The thorough risk analysis and the implementation of mitigation mechanisms to eliminate or reduce *known risks* (static defense);
2. The use of algorithms—often based on artificial intelligence (AI) and machine learning (ML)—to detect suspicious system behavior or obviate *unknown threats and risks* (dynamic defense).

For the *static defense* (Fig. 4.17), the risks which may endanger the CPS are identified, evaluated, assessed, and mitigated. The mitigation is done by using adequate *defense mechanisms*. Strong perimeter protection and sufficient, targeted defense mechanisms for all CPS elements are mandatory.

Myriads of specific defense mechanisms have been invented and are documented in a vast body of literature, experience, and analyses (see, e.g., [Taylor20], [Easttom18], [Brotherston17], [Matulevičius17], [Basin11], [Bellovin15], [Stallings18], [Johnson15], [Saydjari18], [Radvanovsky16], [Mohammadi19], [Flammini19], [NIST800-160B],

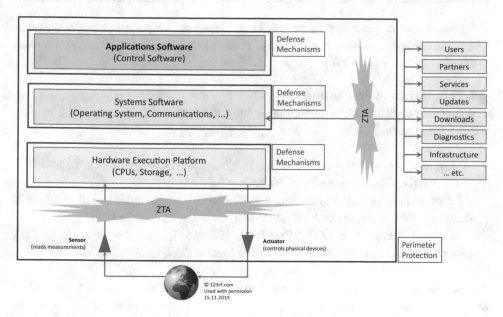

**Fig. 4.17**  CPS cyber-part defense

[Song17], [Diogenes19], [Andress19], [Laing13], [Ackerman17], [Rerup18], [Zongo18], [Loukas15], [Romanovsy17], [Gumzej16]). At this point, only three special topics shall be presented:

a) Security Perimeter Protection;
b) Zero Trust Network Architecture (ZTA);
c) Security Chaos Engineering.

## 4.4.11  Security Perimeter Protection

Any cyber-physical system has a *boundary*, i.e., a *perimeter*. The perimeter delineates between internal and external components or systems. The perimeter is essential because any connection crossing the perimeter is a potential attack path, i.e., a potential *intrusion* route. Unambiguously determining and documenting the perimeter is the first task in securing the system. Note that the perimeter can be a physical/technical boundary, such as a car (Fig. 4.3), or a governance boundary, such as a company (Fig. 7.1). To secure the perimeter, all intrusion routes must be identified, characterized, and secured against intrusion attacks.

*Note*: A public network, such as the Internet or IoT, is always a system boundary, delineating the security perimeter!

The objective of perimeter strength is to withstand the very advanced cyber-attacks, which are rising fast in number on a global scale. Therefore, sound *intrusion prevention* is decisive for cyber-security!

### 4.4.11.1 Intrusion Prevention

The first line of defense of intrusion prevention is to consequently use all the modern technical means, such as:

- A *resilient network architecture* (e.g., [Stallings16], [Brotherston17], [Gallotti19], [ECC16], [Gai20], [Gilman17], [NIST20]);
- All modern *defense technologies* (see, e.g., [Sadiqui20], [Easttom18]), such as Firewalls ([Tegenaw19]), Malware-detection ([Saxe18]), DeMilitarized Zones (DMZ, [Young20]), Zero Trust Architecture (ZTA, [NIST20], [Gilman17]), Cryptography ([Katz21a], [Katz21b], [Stallings16]), Intrusion Detection System (IDS, [Sadiqui20]), etc.

The second line of defense of intrusion prevention is effective, competent *penetration testing* (Definition 4.22, [Weidman14], [Wilhelm13]). There are many intrusion attacks for a complex cyber-physical system, many known, others yet unknown. New functions, products, and technologies still introduce new threats!

▶ **Definition 4.22: Penetration Testing**

Penetration testing, also called "pen testing", ethical hacking, or threat simulation, is the practice of testing a computer system, network, or web application to find security vulnerabilities that an attacker could exploit. Penetration testing can be automated with software applications or performed manually.

Adapted from: https://searchsecurity.techtarget.com/definition/penetration-testing

*Smart homes* ([Juniper18]) are a fast-growing field of CPS application. Smart homes allow the control of many private functions via the smartphone or voice recognition (Fig. 4.19). Often, the connectivity between the sensors, actuators, the control computer, the smartphone, etc., is carried over the Internet or IoT and is therefore at risk of malicious intrusion (Example 4.16).

---

**Example 4.16: Smart Home Intrusion**

Figure 4.19 shows the essential elements of a smart home: A control computer operates all the house's security and comfort functions. Sensors capture environmental parameters, such as temperature, light, etc. The control computer receives commands via smartphones or voice and manages the smart home, very often via the Internet or IoT.

A search for "smart home hacking cases" returns 28'900'000 results (Google, 14.8.2020). Therefore, the control system of a smart home seems to be often vulnerable through many intrusion paths. Figure 4.19 shows the security perimeter of a smart home: The Internet and the IoT are untrusted areas and are thus outside the security perimeter. Figure 4.19 also shows the many attack routes from outside of the security perimeter, i.e., the Internet or the IoT. Especially dangerous is the hacking of sensors and actuators, such as the electronic door lock.

### 4.4.11.2  Penetration Testing

*Penetration testing* has mutated from an automated craft to a very exacting, indispensable art ([Davis20]).

> **Quote**
>
> *"Penetration testing, also called "pentesting", is about more than just getting through a perimeter firewall. The biggest security threats are inside the network, where attackers can rampage through sensitive data by exploiting weak access controls and poorly patched software".*
>   Royce Davies, 2020

Today's *intrusion attacks* distinguish between two categories of targets (Fig. 4.18):

1. Attacks on *random targets*: The attacker scans millions of random IP-addresses. Within this large address set, with certainty, many computers are poorly protected or unpatched and fall victim to the intrusion;
2. Attacks on *specific targets*: The attacker strikes carefully selected targets of high value, such as a financial institution.

Random Targets:

*Random targets* are mostly assailed with widely available hacking tools (e.g., https://www.darknet.org.uk/category/hacking-tools/). Targets are very often *Personal Computers* and their installed software. Myriads of known vulnerabilities are known and documented in various vulnerability databases (see, e.g., https://nvd.nist.gov, https://www.rapid7.com/de/db/, https://www.kb.cert.org/vuls, https://cve.mitre.org/). For the protection against random attacks, standard penetration testing ([Weidman14], [Wilhelm13]) is adequate, although the penetration tools must be continuously updated.

Specific Targets:

*Specific targets*, i.e., targets selected by the attacker according to their properties, are subject to cyberespionage, cyberwar, and cybercrime. The attacker tries to inflict damage to National infrastructures (water supply, electricity distribution, telecommunication networks, etc.), attempts to steal company secrets, or execute blackmail for financial gain ([Loukas15], [Karimipour20], [Ayala16], [Pathan15], [Clark17]).

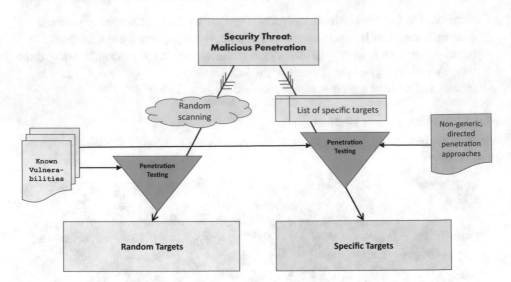

**Fig. 4.18**  Penetration testing

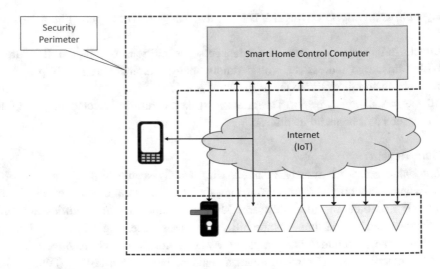

**Fig. 4.19**  Smart home

**Quote**
*"However, if you are specifically targeted by someone with time and resources, you have a problem of an altogether different mangnitude".*
    *Wil Allsopp, 2019*

For directed attacks to specific targets, standard penetration testing is of limited value (However, still required). A new, highly dangerous threat category plays the primary role: *Advanced Persistent Threats* (APT, Definition 4.23, [Allsopp17], [NIST800-172], [Winkler16]).

▶ **Definition 4.23: Advanced Persistent Threat (APT)**
What differentiates an APT from a more traditional intrusion is that it is strongly goal-oriented. The attacker is looking for something (e.g., proprietary data) and is prepared to be as patient and persistent as is necessary to acquire it.
   Will Allsopp, 2017.

Defense against advanced persistent threats is extremely difficult. Many cases of successful APT attacks are documented—e.g., nine alarming cases in [Allsopp17]. No general strategy or recipe against APT attacks is known—each cyber-physical system must be carefully analyzed by competent specialists!

> **Quote**
> *"Conventional penetration testing is next to useless when attempting to protect organizations against a targeted APT attack".*
>    Wil Allsopp, 2019

### 4.4.11.3  Extrusion Prevention
In many cases, only intrusion prevention is considered. However, the reverse way—*extrusion prevention*—is also an effective way to secure a system ([Bejtlich05]). For extrusion prevention, the communication channels are monitored for data crossing the security perimeter from internal to external participants. Many strategies exist to detect unauthorized leaking of data, both on the packet level and the application level.

### 4.4.12  Zero Trust Architecture

"Classical" interconnected computing and data resources in the same network were long considered *implicit trust zone*. The implicit trust zone (Fig. 4.20) was strongly protected at its perimeter, thus guarding against external, malicious access attempts. However, a growing number of successful attacks on protection assets demonstrated the perimeter-protection approach's weaknesses, e.g., its incapability to deal with insider crime.

   In 2017, the US National Institute of Standards and Technology (NIST) researched more secure network architectures. The final result—the *Zero Trust Architecture*

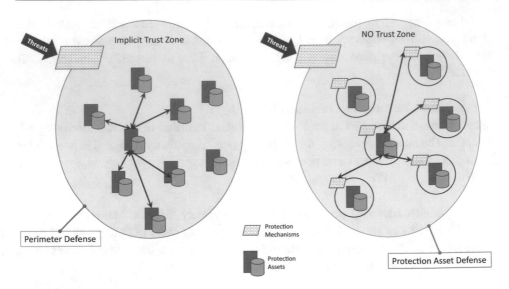

**Fig. 4.20**   Zero trust architecture

(ZTA)—was published in 2020 as [NIST20]. The zero trust architecture (Definition 4.24, [Ghosh21]) introduces a new paradigm: "Do not trust any access attempt, even if it originates within your own network"—i.e., "have zero trust" ([Gilman17], [NIST20], [Kerman20]) (Fig. 4.21).

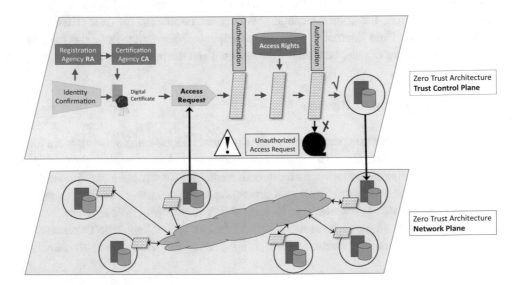

**Fig. 4.21**   ZTA trust plane and network plane

▶ **Definition 4.24: Zero Trust Architecture**

Zero trusts (ZT) is the term for an evolving set of cybersecurity paradigms that move defenses from the static, network-based perimeters to focus on users, assets, and resources. A zero trust architecture (ZTA) uses zero trust principles to plan industrial and enterprise infrastructure and workflows.

https://csrc.nist.gov/publications/detail/sp/800-207/final, 2020.

> **Quote**
> *"Zero Trust Cybersecurity: 'Never Trust, Always Verify'".*
> Alper Kerman (US National Cybersecurity Center of Excellence), 2020

### 4.4.13  Security Chaos Engineering

Testing the resilience of a system traditionally rests on planned, carefully executed testing, e.g., extensive penetration testing. While this is adequate to find causes of failures and vulnerabilities revealed by these selective tests, many failure modes and vulnerability exploitations will not be found. This fact led to the creation of *chaos engineering* (e.g., [Rosenthal20]). The application of chaos engineering to security created the field of *Security Chaos Engineering* (Definition 4.25, [Rinehart18]). In Chaos engineering, random experiments are introduced to test the system's resilience.

Note that security chaos engineering is an enhanced methodology from *fuzzing* (= injection of random inputs and commands into applications (e.g., [Li18c], [Cha12]).

▶ **Definition 4.25: Security Chaos Engineering**

Security Chaos Engineering is the discipline of instrumentation, identification, and remediation of failure within security controls through proactive experimentation to build confidence in the system's ability to defend against malicious conditions in production.

Aaron Rinehart, 2018.

> **Quote**
> *"We spend billions on cybersecurity, but data breaches continuously get bigger and more frequent. Let's try a fresh perspective".*
> Aaron Rinehart, 2018

Unfortunately, no general advice can be given on how to construct secure systems. Modern systems have become so complex and heterogeneous that an uncountable number of vulnerabilities may exist in many cases.

In addition to reducing the systems' complexity, especially the accidental complexity (Definition 4.11), the "weakest link paradigm" should be observed.

### 4.4.14 Weakest Link

Attackers will invest much effort to find the most suitable security vulnerability to strike at systems. A missing or insufficient defense mechanism is often called the "weakest link" (Definition 4.26). Many systems even have considerably more than one single weakest link.

▶ **Definition 4.26: Weakest Link**
The part of a system that is most likely to fail or cause problems.
  https://www.definitions.net

One proven method for improving security is the search for the weakest link. Similar to the *simplification step* in the development process (Fig. 4.6), an explicit discovery step exposing the system's weakest links should be carried out. Methods include specific testing (e.g., chaos engineering, Definition 4.25) or targeted reviews.

> **Quote**
> *"A security system is only as strong as its weakest link".*
>   Niels Ferguson, 2010

Mitigate the *static cyber-defense* targets *known* risks by implementing specific defense mechanisms. The *dynamic cyber-defense* attempts to render the CPS resistant to risk *unknown* at the time of development and deployment of the CPS, i.e., provide dynamic cyber-security (Definition 4.27, [Adams16], [Wang19], [Xu20]).

  Dynamic cyber-security uses adaptive architectures or analysis algorithms—often based on artificial intelligence (AI) and machine learning (ML)—to detect suspicious system behavior or obviate unknown threats and risks (see Chapter 9).

▶ **Definition 4.27: Dynamic Cybersecurity**
Dynamic Cybersecurity is a concept that aims to achieve the modeling, analysis, quantification, operation, and management of cybersecurity from an adaptive, holistic perspective, rather than from a building-blocks perspective.
  Adapted from: Shouhuai Xu, 2020.

### 4.4.14.1 Self-Inflicted Security Issues

Security issues are not always caused by adversary actions: Deficient organizational processes can also be the root of security problems (Example 4.17).

---

**Example 4.17: Expired Digital Certificate**

Digital device certificates (#Cert, Example 4.13) have two popular functions ([Davies10]):

1. The securely confirm the *identity* of a device (Server, Website, IoT-device, etc.), i.e., they authenticate the device;
2. They enable secure communication by *encryption* (SSL-encryption) between the partners.

Digital certificates have a *validity start* and a *validity end* date. Before the start date and after the end date, the digital certificates are not valid and will not be accepted by a client accessing the device or service (Fig. 4.22). Therefore, digital certificates have to be renewed and deployed before their expiry date!

In December 2018, an expired certificate in an Ericsson network management software knocked *32 million UK mobile phones offline* and many more throughout Asia. They could not call or text, nor use 4G connections (Source: https://www.thesslstore. com/blog/expired-certificate-ericsson-o2 [Last accessed: 27.12.2020]).

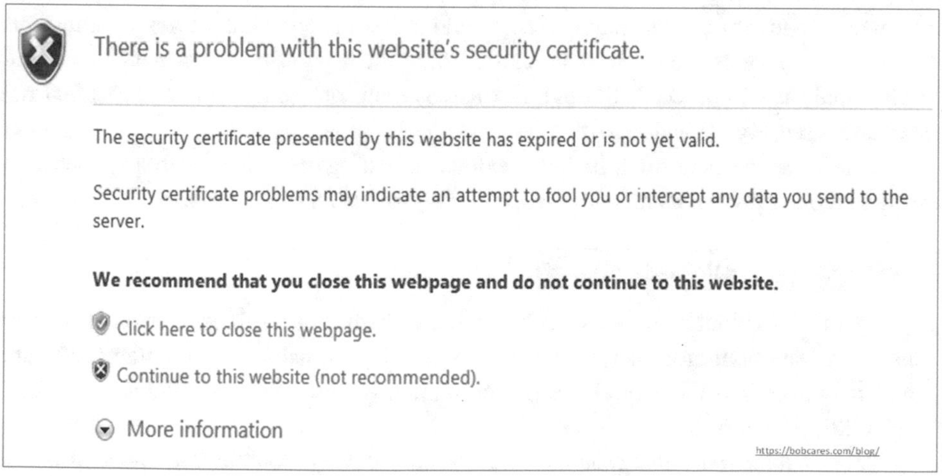

---

There is a problem with this website's security certificate.

The security certificate presented by this website has expired or is not yet valid.

Security certificate problems may indicate an attempt to fool you or intercept any data you send to the server.

**We recommend that you close this webpage and do not continue to this website.**

 Click here to close this webpage.

 Continue to this website (not recommended).

 More information

https://bobcares.com/blog/

**Fig. 4.22** Expired digital certificate

> **Quote**
> *"This episode illustrates the essential role certificates play in keeping IT infra-structure safe and running, and also the risk that enterprises face if they don't have a firm handle on the certificates installed in business-critical systems"*
>      Tim Callan (Sectigo), 2018

### 4.4.15  Security Assessment and Security Audit

Security has two focal points:

1. The *products* and *IT services* offered by the organization;
2. The *processes* used to create, evolve, and operate the products (including all the supporting items, such as the company structure, people behavior, policies, standards, etc.).

The security of the products and services is ensured by periodic *security assessments*, where the security of the processes requires regular *security audits*.

#### 4.4.15.1  Security Assessment

Security is not static: Adding functionality to the products or IT services, changes in the operating environment, modifications in partner or supplier systems, upgrades in laws or regulations often generate new security *vulnerabilities*, many of them implicit and unknown. To identify, understand, assess, and mitigate potential new vulnerabilities, regular security assessments must be carried out. Each security assessment starts with the risk analysis and proceeds through risk assessment, risk mitigation, and residual risk acceptance (see, e.g., [Landoll20]).

A security assessment often includes extensive testing/re-testing, such as penetration testing and other proven techniques (see, e.g., [Wysopal07], Example 4.18).

---

**Example 4.18: Aviation Cyber-Security**

For many 'traditional' products, such as cars, railways, or airplanes, physical security has been paramount for many decades. The growing intrusion of computers, software, and networks into these products has dramatically elevated cyber-attacks risk (e.g., [PTP20], [Haggerty17]).

*"For a long time, the primary security model for airplanes has been physical. Airside security controls are there to prevent access by unauthorized personnel. However, as connectivity has increased (For reasons of efficiency, safety, and passenger convenience), the physical security model has been eroded. While press stories of 'airplane hacking' are often misleading, particularly owing to strong domain segregation, multiple redundant systems, and human pilots in the loop, security of avionics and airborne networks are still essential* (https://www.pentestpartners.com)"

### 4.4.15.2 Security Audit

Security is strongly dependent on people's behavior and their interactions in the organization, with the partners, and the authorities. This is true in all phases of the system life cycle. During the creation, evolution, and maintenance of the systems, people may take shortcuts, ignore policies, violate security principles, neglect security assessments, etc. (e.g., [Dekker11]). During operation, the users can make mistakes, misuse the systems, bypass security controls, leak security-relevant information, etc.

> **Quote**
> *"The heart of any security system is people. This is particularly true in computer security, which deals mainly with technological controls that can usually be bypassed by human intervention"*.
> Matt Bishop, 2019

Enforcing the correct behavior of all people in the organization requires *robust processes* (Definition 4.28, [Dragon19], [Osterwalder20]). These processes guide, constrain, oblige, protect, and make accountable all roles in the organization.

▶ **Definition 4.28: Process**

A process is a specific ordering of work activities across time and space, with a beginning and an end, and clearly defined inputs and outputs, and a structure for action. Processes are the structure by which an organization does what is necessary to produce value, safety, and security for its customers.

Adapted from: Thomas H. Davenport in [Dragon19], 2019.

How does an organization ensure that it has sufficient, adequate, consistent, and effective processes in place? How does it ascertain that the processes are being complied with? How does it make sure that the processes are maintained, i.e., continuously adapted to new requirements?

The instrument to reassure the desired impact of the processes is the *security audit* (Definition 4.29, [Pompon16], [Moeller10], [Mukherjee20], [Dempsey11]). For IT security audits, a number of *audit standards* exist (e.g., [Russo18]).

▶ **Definition 4.29: Security Audit**

A security audit is a structured approach to judging the security measures a company has in place, using a set of defined criteria. During the audit, the auditor will look to identify the policies, standards, and processes that have been defined, then seek evidence that the policies/processes are being followed.

Adapted from: https://www.nexor.com/what-is-a-security-audit/

## 4.4.16  Security Runtime Monitoring

Today most cyber-physical systems are networked, often using the Internet. This fact exposes them to frequent cyber-threats. The *malware growth* is appalling (see, e.g., Example 4.19, [AVTest20]). Even after careful risk analysis and risk mitigation, complex systems will still be left with unknown threats or unidentified vulnerabilities. As the last line of defense, *security runtime monitoring* (Definition 4.30, [Collins17], [Ackerman21], [Ahmed16], [Saxe18]) must be used.

---

**Example 4.19: Malware Growth**

*"Every day, the AV-TEST Institute registers over 350,000 new malicious programs (malware) and potentially unwanted applications (PUA). These are examined and classified according to their characteristics and saved. Visualization programs then transform the results into diagrams that can be updated and produce current malware statistics"* (Source: https://www.av-test.org/).

Total malware, 2018: 856.62 Million.

Total malware, 2019: 1′001.52 Million.

Total malware, 2020: 1′134.06 Million.

https://www.av-test.org/en/statistics/malware/ [Last accessed: 22.12.2020].

▶ **Definition 4.30: Security Runtime Monitoring** Security Runtime Monitoring is a dynamic threat mitigation method that observes the data streams in the system during operation, detects malicious patterns, and discovers operational anomalies in order to trigger automatic or manual interception.

Security runtime monitoring continuously executes six steps (Fig. 4.23):

1. Data/information acquisition from the running system;
2. Data preprocessing/cleaning;
3. Data/information analysis (e.g., Anomaly detection);
4. Anomaly assessment;
5. Automatic or manual anomaly removal/recovery and alerting;
6. Logging and log analysis.

---

**Quote**

*"When we seek to detect cyber-threats, we're analyzing data in the form of files, logs, network packets, and other artifacts"*.

Joshua Saxe, 2018

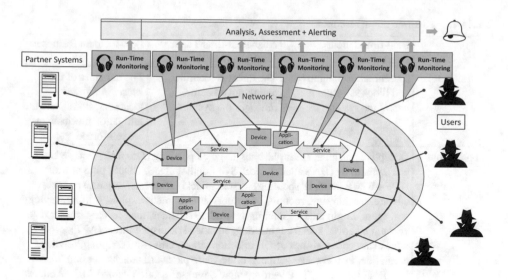

**Fig. 4.23** Security runtime monitoring

Data/information acquisition is obtained from various running system elements, such as the network, the devices, the services, the applications, and communication with partner systems or users (Fig. 4.23). The decisive step is the detection and assessment of *anomalies* (Definition 4.31, [Hawkins14], [Ahmed16]).

▶ **Definition 4.31: Anomaly**
An anomaly is an observation which deviates so much from other observations as to arouse suspicions that it was generated by a different mechanism.
   Hawkins, 1980.

"Anomaly" is a concept from statistics and applied probability. However, it perfectly applies to security runtime monitoring. The security runtime monitoring objective is to detect anomalies, such as virus signatures, suspicious network packets, ominous network addresses, malicious code insertion efforts, unauthorized access attempts, excessive network traffic, increased CPU usage, slowed down response times, strange interface behavior, or crashing, etc.

   Table 4.8 shows the three categories of runtime monitoring (Fig. 4.23) and a choice of widespread threats ([Chio18], [Easttom18], [Ryder19], [Lee20a], [Lee20b], https://www.ibm.com/services/business-continuity/cyber-attack).

   The strongest defense is *real-time threat interception*: The runtime monitoring spots and blocks the threat *before* entering the system. The second line of defense relies on a near-real-time analysis of the data gathered from the system, such as log files: In this case, threats are already in the system but may be detected and removed *before* they

**Table 4.8**  Runtime threat response

|  | Real-time Threat Interception | Near-Real-time Threat Neutralization | Post-Incident Response |
|---|---|---|---|
| Definition | The threat is recognized and blocked before it *enters* the system | The threat is identified and removed before it *damages* the system | The damages of the *successful threat* are located, restored, and the mitigation mechanisms are reinforced |
| Applications Devices Network | Deceptive phishing, spear phishing, targeted phishing (Dropbox, Google Docs, etc.), pharming (DNS attack), DoS, Syn flooding, DoS, DDoS, ping of death, UDP+ICMP flood, IP spoofing, session hijacking, worms, trojans, spyware, adware, ransomware, rootkits, backdoors, bots/botnets, exploit, sniffing, keylogger, spam, LogIn attacks, man-in-the-middle schemes, man-in-the-browser risk, account takeover, social engineering, advanced persistent threat (APT), drive-by attack (insecure websites), SQL injection attack, cross-site scripting, privilege escalation, insider endangerment, brute-force password attacks, session hijacking, DNS tunneling, TCP teardrop attack, ping-of-death, falsified ARP (Address Resolution Protocol) menace, form jacking, crypto-mining = cryptojacking), playback/replay danger, birthday attack on hash, logic bombs, software-defined network attacks, server misconfiguration attacks, malvertising attacks, port redirection, OAuth exploits, shareware attack, 5G-attacks, etc. |  |  |
| Remarks | Perimeter defense | In-time defense | Incident management, cyber-crisis management |

cause damage. Finally, well-implemented incident response and cyber-threat crisis management must be in place to limit the damage of successful assaults.

> **Quote**
> *"There are two types of companies: Those who have been hacked, and those who don't yet know they have been hacked".*
>     John Chambers, former CEO and executive chairman at Cisco.

## 4.4.17  How Much Security is Enough?

Engineering, implementing, and maintaining security is expensive. Therefore, the question "how much security is enough?" must be answered during the engineering process. The *security risk management process* provides the answer: A responsible balance between cost and residual risk must be achieved (Fig. 4.24). The cost curve has an exponential shape—100% security means infinite cost—and is not realizable in any technical system!

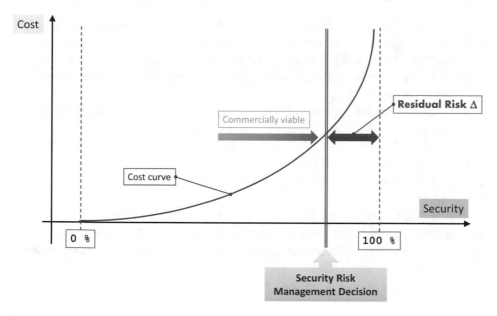

**Fig. 4.24**  Security cost curve

Figure 4.24 shows the critical tradeoff point: It is determined by the product's or the service's commercial viability and the *acceptable residual security risk*. Note that this tradeoff must be carefully evaluated for each *threat*. The explicit and informed acceptance of residual risk is a significant management responsibility.

## 4.5   Convergence of Safety and Security Engineering

Historically, *safety engineering* and *security engineering* were two distinct, mostly unrelated fields. Safety engineering emerged from *mechanical engineering*, and its formal roots were laid in 1950 by the USA Air Force ([Gullo18]). Security engineering is a branch of computer science. It was initiated by the first *computer virus* called "Creeper": An experimental, self-replicating virus that appeared in 1971. Creeper was filling up the computer's hard drive until the computer could not operate any further (https://content. sentrian.com.au/blog/a-short-history-of-computer-viruses).

Today, security vulnerabilities can lead to safety accidents, as has been demonstrated, e.g., by *car hacking* (https://www.wired.com/tag/car-hacking/, [Knight20]) or *aircraft hacking*   (https://www.forbes.com/sites/kateoflahertyuk/2018/08/22/how-to-hack-an-aircraft/ #57a995c641d1,   [Nichols19]) or *infrastructure   hacking*   (https://www.theregister. co.uk/2017/08/16/notpetya_ransomware_attack_cost_us_300m_says_shipping_giant_maersk/).

Therefore, safety and security need to grow into one single engineering discipline: *Resilience engineering* ([Hollnagel06], [Hollnagel13], [Flammini19], [Ganguly18], [Petrenko19], [Heegaard15], [Furrer19]). *Convergence* of safety and security engineering requires an alignment or unification of all elements of safety (Fig. 4.5) and elements of security (Fig. 4.13).

Merging safety culture (Definition 4.6) and security culture (Definition 4.17) into *one* company culture leads to the *dependability culture* (Definition 4.32).

▶ **Definition 4.32: Dependability Culture**

Dependability culture is a combination of safety culture (Definition 4.6) and security culture (Definition 4.17).

Dependability culture is a risk culture (Definition 8.1) in which safety and security are an overriding priority for all levels of the organization (Overriding = Safety and security requirements take priority over other system requirements, such as the functionality or performance).

A dependability culture formalizes the enduring values, attitudes, motivations, and knowledge of an organization in which dependability is prioritized over competing goals in decisions and behaviors.

Adapted from [Gilbert18].

## 4.6   Risk

*Risk* (Definition 3.4) is an inherent part of any technology. It is a constant companion during the development, evolution, application, and use of technology. The possibility of safety accidents and security incidents is just around the corner.

> **Quote**
> *"A safety accident or a security incident are just the tip of the iceberg, a sign of a much larger problem below the surface".*

The engineering teams' indisputable primary responsibility is to build safe and secure systems for their users. Because this expectation often conflicts with the business desire for low development cost and short time-to-market, robust methods and dependable processes must be used in the organization building the CPSs. The fundamental underlying process is the *risk management process*.

## 4.6.1   Risk in Safety and Security

The attitude toward risk in *safety* and *security* has traditionally been separate, using different models, metrics, mindsets, methodologies, and processes. Concerning cyber-physical systems—which are subject to safety and security risks (Example 4.20)—a slow convergence can be seen in research. However, today (2022), the two fields' risk management processes still differ (Table 4.9).

---

**Example 4.20: Water Treatment Plant Attacks**

The responsibility of a *water treatment plant* is to deliver clean, healthy water to its users. A safe water supply is of utmost importance for all people. In most locations, water must be treated before it can be delivered to the users ([Worch19]). Part of the water treatment includes the addition of specific chemicals in very small, strictly controlled quantities.

A water treatment plant's structure is shown in Fig. 4.25: Like all cyber-physical systems, it consists of a *cyber-part* (the controlling computer) and a *physical part* (valves, mixers, pumps, measurement instruments, etc.).

Because water treatment plants are such crucial parts of national infrastructures, they are worthwhile attack targets. In 2016 hackers penetrated the control system and manipulated the programmable logic controllers, which dose the percentage of chemicals used to treat the water. Raising the dose makes the water unsafe to drink or even dangerous to health ([Watertechonline16], [Hassanzadeh19]).

The hackers gained unauthorized access to the controlling computer through a *cyber-part security vulnerability*, allowing unauthorized access. Once they had access, they exploited a *safety vulnerability,* which allowed them to change the dose of added chemicals. By doing so, they could have poisoned a large number of water users.

---

Example 4.20 shows the three most important *risk categories* for a trustworthy water treatment plant:

**Table 4.9**  Risk management in safety and security

|  | Safety | Security |
|---|---|---|
| Objective | Prevent the occurrence of safety accidents | Inhibit the success of security incidents |
| Generic Risk Management Process | Source: Fig. 4.27 | Source: Fig. 4.29 |
| Main Methodologies | FMEA, HAZOP, FTA, ETA | Quantitative risk matrix, qualitative risk matrix, ISRAM, CRAMM, CORA |

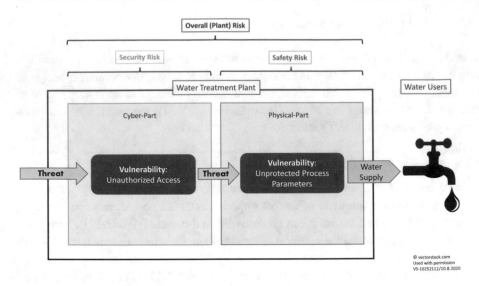

**Fig. 4.25**  Water treatment plant

1. *Overall* plant risk: Example "Delivery of contaminated water to the users";
2. *Security* risk: Example "Unauthorized access to the control computer";
3. *Safety* risk: Example "Malicious manipulation of the process parameters".

This risk list is by far not exhaustive, but exemplary. An extensive *risk assessment* must identify, acknowledge, assess, mitigate, and monitor *all* risks in this application. In the following chapters, the focus lies on safety risks and security risks.

## 4.6.2   Risk Management Process

*Risk* in a complex technical system cannot be avoided—it can only be managed. For risk management (Definition 4.33), proven processes and methodologies exist in many technology areas, as well as for CPS's.

▶ **Definition 4.33: Risk Management**
The identification, analysis, assessment, control, and avoidance, minimization, or elimination of unacceptable risks.
    http://www.businessdictionary.com/definition/risk-management.html

Risk is managed via a *risk management process* (Definition 4.34) based on a *risk management methodology*. From the many known and proven risk management processes and methodologies (e.g., [Hubbard20], [Vélez15], [Hopkin18], [Vellani20],

[Hubbard16], [Kohnke17], [Rasmussen00]), the organizations developing, evolving, and operating the CPS must choose an appropriate process and methodology.

> **Quote**
> *"The entire economy and even human lives are supported in large part by our assessment and management of risk".*
>    Douglas W. Hubbard, 2020

Risk management processes and methodologies greatly vary between the various industries. Some *certification agencies* require the use of specific—often standardized—risk management processes, such as the ISO 31000 standard (Risk management: principles and guidelines: https://risk-engineering.org/ISO-31000-risk-management/) or the ISO/IEC 27,000 family of standards (Information security management: https://www.iso.org/isoiec-27001-information-security.html).

▶ **Definition 4.34: Risk Management Process** An organized, repeatable, effective sequence of steps to identify, analyze, assess, control, and mitigate risks during the development and operation of the CPS, and minimize the negative impact they may have on the CPS, the users, or the involved organizations.

The risk management process is repeated every time:

- A *new project* is initialized (Analyzing the new functionality and the context);
- A *safety accident* or *security incident* has occurred (based on the forensic accident/incident information gained);
- Whenever a *new risk* is identified (e.g., a new cyber-threat);
- Whenever a law, compliance directive, or regulatory decree changes;
- Periodically: Safety and security risk reviews should be executed at regular intervals.

The first and decisive step in the risk management process is identifying and documenting the *protection assets* (Definition 4.35). A protection asset is anything valuable to the organization, i.e., which has an economic value, represents intellectual property, may generate legal responsibilities, uses the organization's products or services, or non-tangible values like reputation, creditworthiness, etc. ([Ruan19]).

▶ **Definition 4.35: Protection Asset**
Any physical or virtual *object* with a value to the organization, its stakeholders, or customers, such as monetary value, intellectual value, strategic value, artistic value or ethical value, etc.

Any *property* of a product or service with the capability to prevent safety accidents or inhibit security incidents, which possibly lead to loss of property or life, damage to the environment, have negative legal or compliance consequences, or degrade the reputation.

> **Quote**
> *"Since institutions can face a broad range of potentially damaging risks, it seems obvious that the truly successful ones will be those that can properly manage their risk exposures".*
> Erik Banks, 2012

Figure 4.26 shows a *generic risk management process* ([Arnold17], [Hopkin18], [Meyer16], [Flammini19], [Hubbard20]): At the core of the process are the *protection assets* (Definition 4.35), which need to be guarded against safety failures and security threats. Once the failures and threats are identified as completely as possible, their specific *impact* on each protection asset is evaluated. The next step is to estimate the *likelihood* (= probability) of each failure and threat.

The risk for each protection asset is then calculated as:

Risk $\Phi\{\Psi_k\} = p$ (Asset $\Psi_k$| Failure/Threat) $\times \Delta$ (Asset $\Psi_k$| Failure/Threat), where:

- $p()$ is the likelihood that a specific failure will occur or a particular threat will materialize (first term);

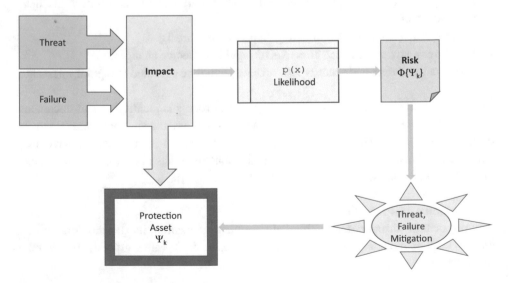

**Fig. 4.26** Generic risk management process

- $\Delta$ ( ) is the measure for the various *damage* levels that the failure or threat inflict on the specific protection asset $\Psi_k$ (second term).

---

**Example 4.21: Risk Calculation**

a) There is a 20% probability that a specific threat $T_n$ will cause 1 Million EUR damage;
b) There is a 10% probability that a specific threat $T_n$ will cause 10 Million EUR damage.

The risk in case a) is $(p=0.2) \times (\Delta = 1'000'000) = \text{Risk } \Phi\{\Psi_k\} = 200'000$.
The risk in case b) is $(p=0.1) \times (\Delta = 10'000'000) = \text{Risk } \Phi\{\Psi_k\} = 1'000'000$.
Real-world mathematics and statistics are definitely more complicated, but this example shows the fundamentals.

### 4.6.3 Risk Analysis and Assessment

The most critical step during risk management is the identification, analysis, and assessment of the risks. Once all possible risks have been identified, their possible impact on the protection assets has to be rated. There are two applicable paradigms:

1. *Qualitative* risk analysis: Example: "There is a medium probability that the protection asset $\Psi_k$ is highly damaged by a successful threat $T_n$";
2. *Quantitative* risk analysis: Example: "There is a 0.27 probability that the protection asset $\Psi_k$ suffers a loss of US\$ 650'000 following a successful threat $T_n$" ([Ruan19]).

Because of the historically different evolution, risk management in safety and security are different. Also, the underlying standards are diverse. Therefore—at the time being (The year 2022)—two separate sections are required, one for safety risk management and one for security risk management.

### 4.6.4 Safety Risk Management

A generic *safety risk management process* is shown in Fig. 4.27: It consists of six steps during system development (Upper part of Fig. 4.27) and four steps during the CPS operation (Lower part of Fig. 4.27).

Numerous safety risk assessment methods are in use (see, e.g., [Bozzano10], [Gullo18], [Ericson15]). The most often used are:

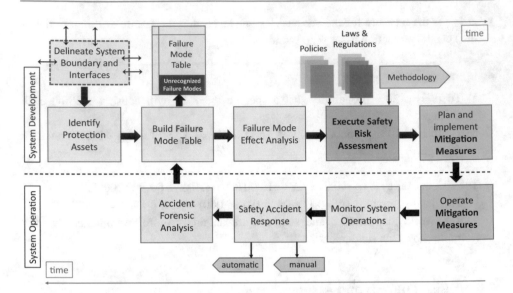

**Fig. 4.27**  Safety risk management process with FMEA

- Fault Tree Analysis (FTA): *FTA* is an analytical technique that starts with an unde-sired state—the *top-level event*. The safety engineer then analyses the system for pos-sible chains of *basic events*, such as system failures, leading to the top-level event. The *fault tree* is the systematic representation of these chains of events. The fault tree uses logical gates, such as AND, OR, NOT, to formalize the logical interrelationship from the basic events to the top-level event.
- Failure Mode and Effects Analysis (FMEA): *FMEA* is a structured process for discov-ering potential failures in a planned CPS. Failure modes are possibilities in which the CPS can fail. Effects are the consequences of the failures, which lead to potentially harmful outcomes in the CPS.
- Hazard and Operability Studies (HAZOP): A *HAZOP* study is the result of a struc-tured process of a CPS under design in order to identify and evaluate possible prob-lems that may cause risk to the stakeholders of the CPS;
- Event Tree Analysis (ETA): *ETA* is a top-down, logical modeling technique for identi-fying possible defects in a CPS under design. It starts with a single initiating event and follows paths to possible failures, assessing outcomes and corresponding probabilities.

**Quote**
*"As is is not possible to achieve absolute safety, designers of a critical system will necessarily be confronted with making decisions about which level of safety may or may not be considered acceptable for the system at hand".*
  *Marco Bozzano, Adolfo Villafiorita, 2011*

One detailed example shall be presented here: The FMEA (Failure Mode and Effect Analysis) for a well-established safety risk assessment methodology (Example 4.22).

**Example 4.22: Failure Mode and Effect Analysis (FMEA)**

The Failure Mode and Effect Analysis (FMEA) is a systematic method of identifying and preventing product and process problems before they occur ([McDermott09]).

The *FMEA process* is shown in Fig. 4.28: It applies to both products and processes. First, the boundary of the system under consideration is clearly delineated. Second, all external dependencies—which are often severe sources of failures—are identified and included in the FMEA.

Next, the possible *failure modes* are discovered, identified, and documented: This is the most critical step—the danger is to miss some failure modes, which then become unrecognized failure modes, representing a severe menace to the system's safety.

The failure hits a vulnerability in the system and has negative consequences on safety. The following activity analyzes each failure mode's effects: Potential damage, probability of occurrence, and the likelihood of detection before the damages materialize.

The FMEA calculates a *Risk Priority Number* (RPN) for each possible failure mode. Each failure mode has a potential effect. Three factors determine the relative risk of failure and its effects:

1. *Severity*: The consequence of the failure, should it occur (Range 1 – 10, 1 = low);
2. *Occurrence*: The probability or frequency of the failure occurring (Range 1 – 10, 1 = low);
3. *Detection*: The probability of the failure being detected before the effect's impact is realized (Range 1 – 10, 1 = low).

The Risk Priority Number (RPN) is calculated as a multiplication: `severity` x `occurrence` x `detection` (Range 1 – 1'000). RPN = 1 means the lowest possible risk, RPN = 1'000 means the highest possible risk.

After calculating the RPN, adequate *risk mitigation measures* (Protection measures, controls) are devised for each failure mode.

The final, decisive step is to judge the residual risk $RPN_{resulting}$ and either explicitly accept it or add additional mitigation measures. If it is not acceptable—the residual risk is still too high—additional mitigation measures are introduced, and the FMEA process for this failure mode is repeated (Fig. 4.28).

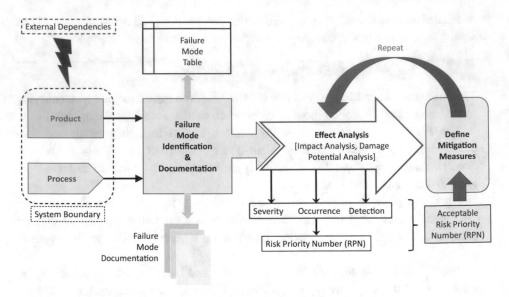

**Fig. 4.28**   FMEA process

> **Quote**
> *"Failures are not limited to problems with the product. Because failures also can occur when the user makes a mistake, those types of failures should also be included in the FMEA".*
>    Robin E. McDermott, 2009

An organization must decide which methodologies and standards it will adopt and enforce. In most cases, the selected methodology must be adapted to the organization's specific needs and processes, whereas domain-specific requirements dictate the applicable standards.

## 4.6.5   Security Risk Management

The skeleton of the *security risk management process* is shown in Fig. 4.29: It encompasses six steps during the development and four steps during the operation of the CPS.

Several *security risk assessment* methods are available. The most common is the *risk matrix* in the form of:

- *Qualitative* risk matrix: The damage potential and the likelihoods are not formalized but described in natural language;
- *Quantitative* risk matrix: The damage potential and the likelihoods are defined by numbers.

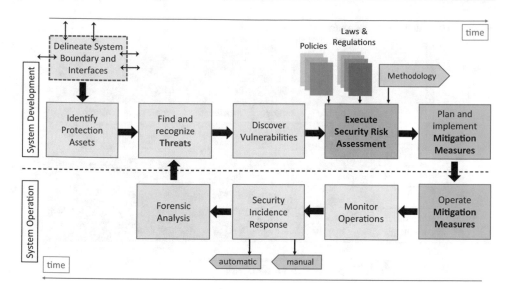

**Fig. 4.29** Security risk management process

### 4.6.5.1 Qualitative Risk Matrix

A standard tool for the risk assessment is the risk matrix (Table 4.10, [Vellani20], [Hubbard20]). The *qualitative risk matrix* is an estimative risk assessment, i.e., it relies not on measured data but on the specialists' expertise in the field.

Once all protection assets have been identified and documented, a qualitative risk matrix is developed for each threat. First, all the known threats are judged concerning their impact as either *catastrophic, major, moderate, minor,* or *negligible* by the experienced team of security experts.

Second, all threats are classified with respect to their probability of occurrence as either *rare, unlikely, possible, likely,* or *almost certain.*

This risk assessment is done *before* the risk mitigation measures (controls) have been defined—the risk matrix represents the *system's unprotected state*! Therefore, this initial qualitative risk matrix will almost certainly present an unacceptable number of red fields.

The matrix is iterated for each threat after any protection measure (control) is introduced, thus rendering the CPS more and more secure. The mitigation measures reduce the risk because either the probability of a successful threat is reduced or the threat's impact is lowered. At some time, the security team may feel that sufficient (or commercially viable) protection measures have been proposed. The team then delivers the risk matrix (Table 4.10) to the formal approval body. Green fields are acceptable, yellow fields represent an acceptable residual risk, whereas red fields should under no circumstance be acceptable.

In rare cases, it could be too expensive or unreasonable to implement additional mitigation measures: A field may remain yellow (never red!). Such cases must be explicitly

**Table 4.10**   Qualitative risk matrix

| | | *Danger*: Threat XYZ / *Target*: Protection Asset $A_n$ | | | | |
|---|---|---|---|---|---|---|
| | | Project: P123 | | | | |
| | | IMPACT | | | | |
| | | negligible | minor | moderate | major | catastrophic |
| LIKELIHOOD | rare | | | | | |
| | unlikely | | | | | |
| | possible | | | | | |
| | likely | | | | | |
| | almost certain | | | | | |

| | |
|---|---|
| | acceptable residual risk |
| | explicit acceptance required |
| | **not** acceptable |

understood and accepted by the *company risk officer*—and the management must be aware of this risk the project introduces! This process is explained in detail in Example 4.23 below.

---

**Example 4.23: Cyberespionage**

*Cyberespionage* ([Buchan18]) is a cyberattack that gains unauthorized access to the intellectual property stored in a company's computer systems. Cyber-espionage is well documented (e.g., https://blog.eccouncil.org/9-latest-cyberespionage-affairs/) and is a severe threat.

---

**Quote**
*"The advent of cyberspace has led to a dramatic increase in state-sponsored political and economic espionage".*
   *Russell Buchan, 2018*

---

In this example, a company is developing a new *torpedo* ([Branfill14]). Their engineers are using a modern, networked design tool with access to the *Internet*. The company's risk officer presented the initial risk matrix for the project shown in Table 4.11. This risk matrix looks somewhat unusual. For the experienced risk officer in this project, only *catastrophic impacts* are identified. The loss of the project data through *cyberespionage* to a

**Table 4.11**   Initial risk matrix for the Torpedo-Project

| | | IMPACT | | | | |
|---|---|---|---|---|---|---|
| | | negligible | minor | moderate | major | catastrophic |
| LIKELIHOOD | rare | × | × | × | × | |
| | unlikely | × | × | × | × | |
| | possible | × | × | × | × | |
| | likely | × | × | × | × | |
| | almost certain | × | × | × | × | |

*Danger*: Threat «Cyber-Espionnage» / *Target*: Design Documents

Project: New Torpedo TP-123A

competitor or a foreign state would be a catastrophe both for the company (loss of intellectual property and future revenues) and for the country (loss of a new naval weapon providing superior advantages)

The initial risk matrix (Table 4.11) showed only unacceptable risks (red fields)! Therefore, serious mitigation measures needed to be devised. The detailed risk analysis resulted in two insights:

1. *Complete separation* of the internal development network from the Internet (so-called "air-gap") must be assured at all times;
2. The torpedo-project must run on its own, isolated development network, with no access to/from other development networks in the company (again: "air-gaps").

The risk matrix *after* implementing the mitigation measures (controls) is shown in Table 4.12.

All foreseeable threats have been successfully mitigated (green entries). The rare and unlikely cases (yellow listings), however, remain critical. Events such as an employee opening a back-door into the system (insider crime) or an unknown access path, e.g., through a maintenance/update channel from a vendor or subcontractor, still pose a—rare or unlikely—risk.

> **Quote**
> *"Risk matrices are among the simplest of the risk assessment methods and this is one reason they are so popular".*
> *Dogulas W. Hubbard, 2020*

**Table 4.12**  Risk matrix for the Torpedo-Project after mitigation measures

| | | IMPACT | | | | |
|---|---|---|---|---|---|---|
| *Danger*: Threat «Cyber-Espionnage» / *Target*: Design Documents | | | | | | |
| Project: New Torpedo TP-123A | | | | | | |
| | | negligible | minor | moderate | major | catastrophic |
| LIKELIHOOD | rare | × | × | × | × | |
| | unlikely | × | × | × | × | |
| | possible | × | × | × | × | |
| | likely | × | × | × | × | |
| | almost certain | × | × | × | × | |

### 4.6.5.2  Quantitative Risk Matrix

Qualitative risk assessment (Table 4.10) strongly depends on the soundness of expert judgment. Therefore, more fact-based risk assessments—especially *quantitative risk assessments*—were developed ([Hubbard20], [Munteanu06], [Aven11], [Leveson16], [Young10]).

> **Quote**
> *"The entire economy and even humn lives are supported in large part by our assessment and management of risks".*
>   Douglas W. Hubbard, 2020

The first step is the *quantitative risk matrix* (Table 4.13). Instead of estimated values for the two dimensions, *numeric* values are assigned (Example in Table 4.14). The choice of criteria and the weight of the two dimensions must be defined according to the specific organization's business model.

Although the criteria and weights are clearly specified, obtaining realistic values for each threat is a real challenge. The impact is relatively easy to appraise. To numeralize the likelihood is more complicated. Here, statistics from various companies available on the Internet are often helpful (e.g., https://dataprot.net/statistics/hacking-statistics/, www.dataconnectors.com).

### 4.6.5.3  Monte Carlo Simulations

Both qualitative and quantitative risk matrix approaches are *static* assessments. For complex situations with significant uncertainties, more elaborate methods must be used.

**Table 4.13** Quantitative risk matrix

| | | Danger: Threat $T_x$ / Target: Protection Asset $A_n$ | | | | |
|---|---|---|---|---|---|---|
| | | Project: $P_n$ | | | | |
| | | IMPACT | | | | |
| | | negligible | minor | moderate | major | catastrophic |
| LIKELIHOOD | rare | | | | | |
| | unlikely | | | | | |
| | possible | | | | | |
| | likely | | | | | |
| | almost certain | | | | | |

One powerful, dynamic method is the Monte Carlo simulation of scenarios (Definition 4.36, [Blom06], [Hubbard16], [Hubbard20], [Barbu20]).

▶ **Definition 4.36: Monte Carlo Risk Analysis**
Monte Carlo simulation performs risk analysis by building models of possible results by substituting a range of values — a probability distribution — for any factor that has inherent uncertainty. It then calculates results over and over, each time using a different set of random values from the probability functions.
  https://www.palisade.com/risk/monte_carlo_simulation.asp

Several methodologies to achieve security- or cyber-risk assessments are in use. They are based on a well-defined model and subsequent simulations to evaluate risk. To be valid, all these methods need a *realistic* model and reliable *data* on the uncertainty of the parameters, such as probability distributions.

> **Quote**
> *"We are ultimately trying to move cybersecurity in the direction of more quantitative risk assessment methods"*.
>   Douglas W. Hubbard, 2016

## 4.6.6 Cyber-Crisis Management

Today, *cyberattacks* are the norm, not the exception. Example 4.19 shows the worrying yearly increase of one specific attack: Malware injections. Some of these attacks are

**Table 4.14**  Metrics for the quantitative risk matrix

| Dimension | Quantifier | Metric |
|---|---|---|
| **LIKELIHOOD** | rare | 1 time in 10 years |
| | unlikely | 1 time in 5 years |
| | possible | 1 ... 5 times/2 years |
| | likely | 1 ... 10 times/year |
| | almost certain | > 10 times per year |
| **IMPACT** | negligible | ▪ No injury<br>▪ Cost < 0,1 % annual turnover of the company<br>▪ Invisible reputation damage<br>▪ No or irrelevant legal or compliance consequences<br>▪ Minimal, localized damage to the environment |
| | minor | ▪ Slight injury<br>▪ Cost < 1 % of the annual turnover of the company<br>▪ Low reputation damage<br>▪ Low legal or compliance consequences<br>▪ Neglectable damage to the environment |
| | moderate | ▪ Moderate injury<br>▪ Cost < 5 % of the annual turnover of the company<br>▪ Annoying reputation damage<br>▪ Endurable legal or compliance consequences<br>▪ Contained local damage to the environment |
| | major | ▪ Invalidity or grave injury<br>▪ Cost < 25 % of the annual turnover of the company<br>▪ Severe reputation damage<br>▪ Grave legal or compliance consequences<br>▪ Severe damage to the environment (contained) |
| | catastrophic | ▪ Loss of life or grave injury<br>▪ Cost > annual turnover of the company cost<br>▪ Irreparable reputation damage<br>▪ Fatal legal or compliance consequences<br>▪ Heavy damage to a large environment |
| *Note*: The impact criteria are OR-criteria, i.e., one or more criteria must apply | | |

*targeted*, e.g., aim at selected targets, such as banks or water treatment plants. Others are *randomized*, i.e., use IP addresses [Fall11] generated randomly to find vulnerabilities.

Some successful malware attacks have disastrous consequences, such as the ransomware attack on a Californian hospital in September 2019 (Example 4.24). Therefore, defense against ransomware is crucial ([NIST1800_20]).

---

**Example 4.24: Hospital Ransomware**

SIMI VALLEY, California—September 18, 2019:

«Wood Ranch Medical ("WRM") was the victim of a ransomware attack that resulted in its patients' personal healthcare information being encrypted. As a result, we were unable to restore patients' healthcare records and will be closing our practice on December 17, 2019. Although there is no indication that any information was accessed, in an abundance of caution, we have taken steps to notify all patients and to provide resources to assist them. The attack encrypted our servers, containing your electronic health records as well as our backup hard drives.»

Source: https://www.woodranchmedical.com/ [Last accessed: 1.7.2020].

Organizations, therefore, must be prepared to deal with a cyber-crisis. This is achieved by using *cyber-crisis management* (Definition 4.37, ([Kaschner20], [Ryder19], [Schneier18], [Stallings18], [Deloitte16]).

▶ **Definition 4.37: Cyber-Crisis Management** Cyber-crisis management is a risk-based process to prevent the success of cyberattacks (= readiness), to minimize the impact of the cyberattack (= response), and to return to regular operation in the shortest possible time (= recovery).

> **Quote**
> *"The risk that cyber crises pose to reputation, brand, operations, and customer and supplier relationships will continue to increase, as will the associated legal and financial effects".*
> *Deloitte, 2016*

The basic process of cyber-crisis management is shown in Fig. 4.30 (e.g., [Stallings18], [Deloitte16], [NIST06]). Four phases are required:

- *Readiness*: Preparation before cyber-incidents occur;
- ⇒ *Cyber-Incident*: The successful cyber-incident occurs;
- *Response*: Behavior during the cyber-incident;
- *Recovery*: Actions after the cyber-incident.

Successful cyber-crisis management critically depends on two factors:

1. *Crisis management group*: A well-balanced, competent, and cooperative team of people—with a "battle-tried", integrative, and unshakeable leader. The crisis management group must periodically be trained in mock situations;
2. *Crisis management process*: During crisis handling, many decisions must be taken under enormous time pressure—often with incomplete or even incorrect factual information. A proven process model must be strictly followed to maximize the decisions' correctness and viability (Example 4.25).

> **Quote**
> *"Incident management and response are the last steps in risk management and the final barrier to what may become an unmitigated disaster".*
> Krag Brotby, 2009

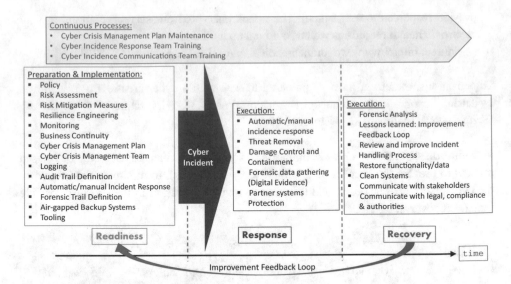

**Fig. 4.30**  Cyber-crisis management phases

## Example 4.25: Crisis Management Process

A proven, effective decision-making process in crises is the FOR-DEC-process (Fig. 4.31, [Soll16]). The acronym FOR-DEC stands for six distinct phases of the decision-making process: Facts, Options, Risks & Benefits, Decision, Execution, and Check (Initially developed and implemented for aeronautical applications). Strictly following the process counteracts the cognitive mechanisms that can adversely affect the quality of the decisions.

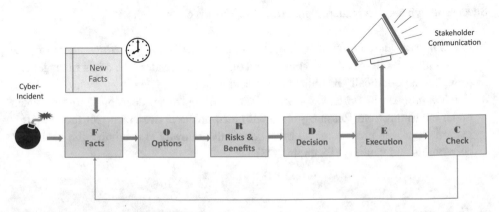

**Fig. 4.31**   Crisis management process model FOR-DEC

FOR-DEC is a repetitive process that is executed whenever new facts become available or at periodic intervals.

A highly relevant part of the crisis management process is timely, accurate, and honest *stakeholder communication*. The content and truthfulness of stakeholder communication during the crisis dictate the relationship's quality after the crisis.

### 4.6.7  Agile Risk Management

All Agile development processes are relatively open to risk introduction into the cyberphysical system. Such risks lead to potentially harmful events in the operation of the CPS. For trustworthy CPS, risk must be reduced by an adequate *extension* of the Agile method. Such a proven extension is the introduction of a *Review/Quality Gate* into the Agile process, as shown in Fig. 4.32);

A process step, "Risk Identification & Evaluation/Mitigation," is added to be executed before each iteration (Sprint), and a checkpoint "Risk Mitigation Assessment" is performed before the release of the new or modified software (Fig. 4.32). Only if the residual risk is acceptable can the software be released into production.

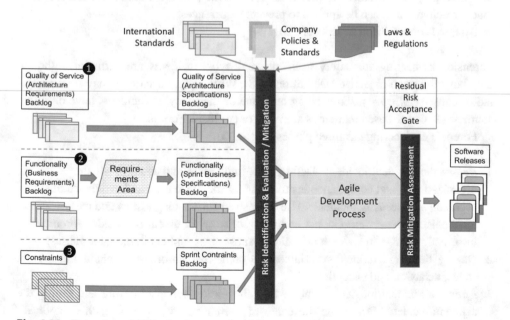

**Fig. 4.32**  Extension of Agile development by risk management

**Quote**

*"Many Agile software development processes at best implicitly tackle risk and those methodologies that lack a risk management framework suffer from deficiencies "*

Alan Moran, 2014

## 4.7    Forensic Engineering

Safety accidents and security incidents will occur despite all care in the cyber-physical systems' development and operations. An essential part of risk management is careful analysis, interpretation, and lessons learned from such unfortunate events using *forensic engineering* (Definition 4.38, [Eloff17], [Chen19], [Easttom17], [Casey11], [ECCouncil16], [Nader20], [Luttgens14], [Reddy19]).

▶ **Definition 4.38: Forensic Engineering**
Forensic Engineering is the detailed investigation of a safety accident or a security incident after they have occurred to determine the specific causes for the accident/incident so that corrective action can be applied to prevent recurrences.
Adapted from: Clifton A. Ericson, 2011

Forensic engineering for safety accidents and cyber-incidents are different, although they have the same objective (Definition 4.38). While safety accidents may have a myriad of causes (*software failures* being only one of them), cyber-incidents have only two sources: (1) Cyber-based malicious activities or (2) Insider crime.

Forensic engineering has three phases (Fig. 4.33):

1. *Before* the accident/incident: Define, test, implement and monitor all the technical, organizational, and compliance instruments for the prevention and avoidance of safety accidents and security incidents. Use extensive runtime diagnostics and early warning mechanisms. *Important*: Collect and securely store all useful preaccident/preincident information and define which information must be assembled in the later phases;
2. *During* the accident/incident: Gather the predefined information about the system states, behavior, and exceptions.
3. *After* the accident/incident: Analyze the information collected before and during the accident/incident. Determine the chain of events and the role of each component. Identify the primary source of the accident/incident and the sequence of events following the primary event. Execute forensic engineering to avoid recurrence of the accident/incident and improve forensic engineering preparation/execution.

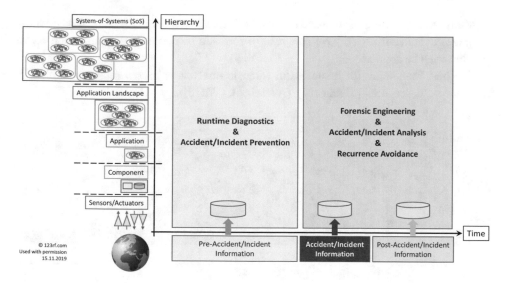

**Fig. 4.33** Runtime diagnostics and forensic engineering

Note: The appropriate instruments must be operated on all software hierarchy levels, i.e., from sensor-/actuator-level to the system-of-systems level). The most important function is to collect sufficient information and store it securely during all phases.

> **Quote**
> *"Despite the best efforts of all involved, accidents and performance failures will continue to happen. The responsible parties need to be identified, liability must be established, and recurrences must be prevented".*
> Adapted from: https://www.ceerisk.com/site/services/forensic-engineering

### 4.7.1 Safety Accident Forensic Engineering of Software

Today (2022), fortunately, few safety accidents are caused by *software failures*. However, while more and more functions of cyber-physical systems are transferred into software implementations (e.g., [Alur15], [Dajsuren19], Example 2.6, Example 4.1), the probability of software failures leading to accidents will continuously rise.

Current safety accident forensic investigation methods, therefore, are not yet focused on software failures (see, e.g., [Wong10], [Fiorentini19], [Bibel08], [Bibel18]). Also, at the present day (2021), the dangerous threat of *cyber-attacks* on cyber-physical systems is not fully recognized but will be a mightful force in the future (see [Knight20], [Gupta19], [Martin20], [Chantzis21], [Karimipour20], [Loukas15], [Li20a], [Li20b], [Colbert16], [Ayala16b], [Liyanage18], [Ayala17]).

Many industries, e.g., aerospace (Example: https://www.icao.int/safety/airnavigation/aig/pages/documents.aspx), have *compulsory* forensic procedures, standards, and methods that *must* be applied.

The best foundation for a successful forensic analysis is *preparation* (Example 2.5, Example 4.26, see, e.g., [Thames17], [Ayala16a], [Wu13]).

---

**Quote**
*"The objective is to plan how best to respond to a cyber-physical attack so you can make decisions quickly and take proper action to mitigate the impact of the attack. When a cyber-physical attack occurs, the last thing you want to do is make things up as you go".*
  Luis Ayala, 2016

---

**Example 4.26: SCADA Forensic Analysis Preparation**

This example presents a checklist for the preparation of a SCADA (Supervisory control and data acquisition system) forensic analysis (Adapted from [Ayala16a]):

**Cyber-Physical Attack Recovery Procedures**

Purpose of the Recovery Procedures
Cyber-Physical Attack Timetable
Recovery Procedures Information
Applicable Directives
Objectives for a Plan
Incident Response Teams (IRT
Recovery Management Team (MGMT)
    General Activities
    Procedures by Phase
Communications Team (CT)
    Stakeholder Communications
    Communication to the Public
    Liaison with the Authorities
Recovery Facilities Team (FAC)
    Procedures by Phase
Recovery Tech Support Team (TECH)
    Procedures by Phase
Recovery Security Team (SEC)

Procedures by Phase
Forensic Evaluation Team (FET)
Recovery Phases
  Phase 1: Detection
  Phase 2: Mitigation
  Phase 3: Recovery
Assumptions
Critical Success Factors
Mission-Critical Systems

## 4.7.2  Cyber-Incident Forensic Engineering of Software

The forensic detective work after a cyber-incident has two missions:

a) Understand the attack, the exploited vulnerability, the impact in order to remedy the weakness and to *prevent reoccurrence* (e.g., [ECCouncil16], [Messier17], [Alleyne18], [Le-Khac20]);
b) Collect *electronic evidence* to identify the culprits and start a prosecution in court (e.g., [Luttgens14], [Akhgar14], [Brown15]).

The first task is an *engineering responsibility*, whereas the second task is assigned to specialized investigators and law enforcement authorities (e.g., https://www.cps.gov.uk/legal-guidance/cybercrime-prosecution-guidance).

Cyber-attacks are executed either through the *network*, by *insiders* (Definition 2.21, [Cappelli12], [CSI20], [Gelles16]), or via *3rd party software* (Example 2.16).

The cyber-attack has one of two objectives (Fig. 4.34):

• Gain unauthorized access to some valuable protection asset (for financial gain or competitive advantage);
• Cause damage to the target.

After the cyber-crisis has been overcome, the forensic analysis relies on data collected before, during, and after the attack. The data collection requires sources on all levels of the system hierarchy (Fig. 4.34). The data must be securely stored *immediately* after its origination. Note that the attacker often erases all traces of the intrusion on the target system. The data is subsequently analyzed to reconstruct the attack path and uncover the system's exploited weakness by a set of specific forensic analysis tools.

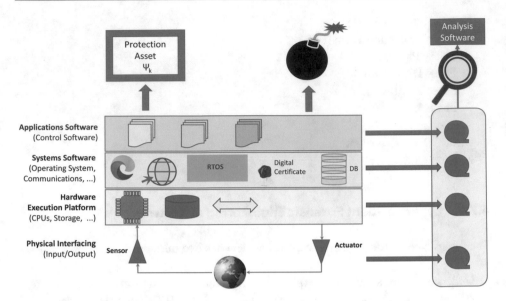

**Fig. 4.34** Cyber-incident forensics

## 4.8    Ethics

Cyber-physical systems can cause considerable harm, either unintentionally (through safety accidents or security incidents) or intentionally, such as CPS's for military use (e.g., [Scharre19], [Clarke12], [Johnson15]). Because the range of functionality and applications of CPS's is almost unlimited, *ethical guidelines* (Definition 4.39) are requested by most nations ([Schoitsch20], [Jasanoff16]).

▶ **Definition 4.39: Ethics**
A system of accepted beliefs that control behavior, especially such a system based on morals for the benefit of people and society.
  https://dictionary.cambridge.org

More and more cyber-physical systems use artificial intelligence, especially *machine learning*, to execute automated decision-*making* (ADM, see, e.g., [EUACM18]). An often-cited, impactful example is an autonomous car that will make many automated decisions ([Artuñedo20], [Ebers20]).

Ethical axioms will probably play a more significant role in future industries ([Boddington17], [Dignum19]), especially for autonomous cyber-physical systems. Some ethical axioms will also gradually become binding law ([Aggarwal19]), such as in Example 4.27.

---

**Example 4.27: Protection from Automated Decision Making**

The European General Data Protection Regulation GDPR ([GDPR16], [ITGP19], [Massey20]) contains the following clause:

*"The data subject should have the right not to be subject to a decision, which may include a measure, evaluating personal aspects relating to him or her which is based solely on automated processing (… continued)."*

(GDPR Legislative Act, § 71).

---

**Quote**

*"Forthcoming applications of automated decision systems, most visibly self-driving cars, raise the possibility of large and potentially lethal physical objects operating under the control of machine-learning models"*.

[EUACM18], 2018

---

## References

[Ackerman17]      Pascal Ackerman: Industrial Cybersecurity - *Efficiently secure critical infrastructure systems*. Packt Publishing, Birmingham, UK, 2017. ISBN 978-1788395151

[Ackerman21]      Pascal Ackerman: Industrial Cybersecurity - *Efficiently monitor the Cybersecurity Posture of your ICS Environment* Packt Publishing, Birmingham, UK, 2nd edition, 2021. ISBN 978-1-800-20209-2

[Adams16]         Niall M. Adams, Nicholas A. Heard: Dynamic Networks and Cyber-Security. World Scientific Publishing (UK) Ltd., London, UK, 2016. ISBN 978-1-783-26904-4

[Ahmed16]         Mohiuddin Ahmed, Abdun Naser Mahmood, Jiankun Hu: *A Survey of Network Anomaly Detection Techniques* Journal of Network and Computer Applications, Elsevier, Amsterdam, Netherlands, Vol. 60, 2016, pp. 19–31. Downloadable from: https://www.gta.ufrj.br/~alvarenga/files/CPE826/Ahmed2016-Survey.pdf [Last accessed: 6.6.2021]

[Alur15]          Rajeev Alur: Principles of Cyber-Physical Systems MIT Press, Cambridge, MA, USA, 2015. ISBN 978-0-262-02911-7

[Allsopp17]       Wil Allsopp: Advanced Penetration Testing - Hacking the World's Most Secure Networks. John Wiley & Sons, Indianapolis, IN, USA, 2017. ISBN 978-1-119-36768-0

[Anderson20]      Ross Anderson: Security Engineering - *A Guide to Building Dependable Distributed Systems* John Wiley & Sons Inc, Indianapolis, IN, USA, 3rd edition, 2020. ISBN 978-1-119-64278-7

[Andress19]       Jason Andress: Foundations of Information Security - *A Straightforward Introduction* No Starch Press, San Francisco, CA, USA, 2019. ISBN 978-1-7185-0004-4

[Antonsen17]        Stian Antonsen: Safety Culture - Theory, Method, and Improvement CRC
                    Press (Taylor & Francis Group), Boca Raton, FL, USA, 2017. ISBN
                    978-1-138-07533-7

[Arnold17]          Rob Arnold: Cybersecurity - *A Business Solution (An Executive
                    Perspective on managing Cyber Risk)* Threat Sketch, LLC, Winston-
                    Salem, NC, USA, 2017. ISBN 978-0-6929-4415-8

[Artuñedo20]        Antonio Artuñedo: Decision-making Strategies for Automated Driving in
                    Urban Environments Springer Nature, Cham, Switzerland, 2020. ISBN
                    978-3-030-45904-8

[Aven11]            Terje Aven, Ortwin Renn: Risk Management and Governance – Concepts,
                    Guidelines, and Applications. Springer Verlag, Berlin, Germany, 2011.
                    ISBN 978-3-642-13926-0.

[AVTest20]          AV-Test Institute: SECURITY REPORT 2018/19 – Facts and Figures
                    AV-Test Institute, Magdeburg, Germany, 2020. Downloadable from:
                    https://www.av-test.org/fileadmin/pdf/security_report/AV-TEST_Security_
                    Report_2018-2019.pdf [Last accessed: 22.12.2020]

[Ayala16]           Luis Ayala: Cybersecurity for Hospitals and Healthcare Facilities - *A
                    Guide to Detection and Prevention* Apress Media LLC, New York, N.Y.,
                    USA, 2016. ISBN 978-1-4842-2154-9

[Ayala16a]          Luis Ayala: Cyber-Physical Attack Recovery Procedures - *A Step-by-Step
                    Preparation and Response Guide* Apress Media LLC, New York, N.Y.,
                    USA, 2016. ISBN 978-1-484-22064-1

[Ayala17]           Luis Ayala: Cyber-Physical Attack Defenses - *Preventing Damage to
                    Buildings and Utilities* CreateSpace Independent Publishing Platform,
                    North Charleston, S.C., USA, 2017. ISBN 978-1-5466-4830-7

[Bacik19]           Sandy Bacik: Building an Effective Information Security Policy
                    Architecture CRC Press, Taylor & Francis Ltd., Boca Raton, FL, USA,
                    2008 (Paperback: 2019). ISBN 978-0-367-38730-3

[Bak09]             Stanley Bak, Deepti K. Chivukula, Olugbemiga Adekunle, Mu Sun, Marco
                    Caccamo, Lui Sha: *The System-Level Simplex Architecture for Improved
                    Real-Time Embedded System Safety* 2009, 15th IEEE Real-Time and
                    Embedded Technology and Applications Symposium, IEEE Press, New
                    York, N.Y., USA, 2009. Downloadable from: https://citeseerx.ist.psu.edu/
                    viewdoc/download?doi=10.1.1.160.7573&rep=rep1&type=pdf      [Last
                    accessed: 23.9.2020]

[Barbu20]           Adrian Barbu, Song-Chun Zhu: Monte Carlo Methods. Springer Nature
                    Singapore, Singapore, Singapore, 2020. ISBN-13: 978-9-811-32970-8

[Bartocci19]        Ezio Bartocci, Yliès Falcone (Editors): Lectures on Runtime Verification
                    - *Introductory and Advanced Topics* Springer Nature, Cham, Switzerland,
                    2019. ISBN 978-3-319-75631-8 (LNCS 10457)

[Basin11]           David Basin, Patrick Schaller, Michael Schläpfer: Applied Information
                    Security - *A Hands-on Approach* Springer Verlag, Berlin, Germany, 2011.
                    ISBN 978-3-642-24473-5

[Bejtlich05]        Richard Bejtlich: Extrusion Detection - *Security Monitoring for Internal
                    Intrusions* Addison-Wesley Professional, Upper Saddle River, NJ, USA,
                    2005. ISBN 978-0-321-34996-5

[Bellovin15]        Steven M. Bellovin: Thinking Security - *Stopping Next Year's Hackers*
                    Addison-Wesley, New York, N.Y., USA, 2015. ISBN 978-0-134-27754-7

[Bibel08]        George Bibel: Beyond the Black Box - *The Forensics of Airplane Crashes* John Hopkins University Press, Baltimore, USA, 2008. ISBN 978-0-8018-8631-7

[Bibel18]        George Bibel: Train Wreck - *The Forensics of Rail Disasters* John Hopkins University Press, Baltimore, USA, 2018. ISBN 978-1-4214-2707-2

[Bishop18]       Matt Bishop: Computer Security - *Art and Science* Addison Wesley, Boston, USA, 2$^{nd}$ edition, 2018. ISBN 978-0-321-71233-2

[Blom06]         Henk A. P. Blom, Sybert H. Stroeve, Hans H. de Jong: Safety Risk Assessment by Monte Carlo Simulation of Complex Safety Critical Operations. In: Redmill, F., Anderson, T. (editors): Developments in Risk-based Approaches to Safety. Springer-Verlag, London, UK, 2006. ISBN 978-1-846-28-447-3_3.

[BMW20]          BMW Group: *SAE Level 3 Automated Driving System Safety Assessment Report*. BMW Group, USA, May 2020. Downloadable from: https://bmw-movement.org/bmw-releases-safety-assessment-report-of-autonomous-vehicle-testing/ [Last accessed: 11.10.2020]

[Boddington17]   Paula Boddington: Towards a Code of Ethics for Artificial Intelligence Springer Nature, Cham, Switzerland, 2017. ISBN 978-3-319-60647-7

[Bozzano10]      Marco Bozzano, Adolfo Villafiorita: Design and Safety Assessment of Critical Systems CRC Press (Taylor & Francis Inc.), Boca Raton, FL, USA, 2010. ISBN 978-1-439-80331-8

[Branfill14]     Roger Branfill-Cook: Torpedo - *The Complete History of the World's Most Revolutionary Naval Weapon* Pen & Sword Books Ltd., Barnsley, UK, 2014. ISBN 978-1-8483-2215-8

[Brotherston17]  Lee Brotherston, Amanda Berlin: Defensive Security Handbook - *Best Practices for Securing Infrastructure* O'Reilly Media Inc., Sebastopol, CA, USA, 2017. ISBN 978-1-491-96038-7

[Brown15]        Cameron S. D. Brown: *Investigating and Prosecuting Cyber Crime - Forensic Dependencies and Barriers to Justice* International Journal of Cyber Criminology, Vol. 9, Issue 1. January – June 2015. Downloadable from: https://www.cybercrimejournal.com/Brown2015vol9issue1.pdf [Last accessed: 12.1.2021]

[Buchan18]       Russell Buchan: Cyber Espionage and International Law Hart Publishing, Oxford, UK, 2018. ISBN 978-1-7822-5734-9

[Buchmann16]     Johannes A. A. Buchmann, Evangelos Karatsiolis, Alexander Wiesmaier: Introduction to Public Key Infrastructures Springer-Verlag, Heidelberg, Germany, 2016. ISBN 978-3-662-52450-3

[Cais14]         Štěpán Cais, Petr Pícha: *Identifying Software Metrics Thresholds for Safety Critical Systems* Proceedings of the Second International Conference on Computer and Communication Technologies (IC3T), 2015, Volume 3. ISBN 978-0-989-1305-8-5 Downloadable from: https://www.academia.edu/8429284/Identifying_Software_Metrics_Thresholds_for_Safety_Critical_System [Last accessed: 21.5.2020]

[Cais15]         Štěpán Cais: Assessing and Improving Quality of Safety-Critical Systems Technische Universität Pilsen, Pilsen, 2015. Downloadable from: https://pdfs.semanticscholar.org/deeb/62e234204c98bbdbaa79c0391e55b4a07d5d.pdf [Last accessed: 21.5.2020]

[Cappelli12]     Dawn M. Cappelli, Andrew P. Moore, Randall F. Trzeciak: The CERT Guide to Insider Threats - *How to Prevent, Detect, and Respond to*

*Information Technology Crimes (Theft, Sabotage, Fraud)* Addison-Wesley, Upper Saddle River, NJ, USA, 2012. ISBN 978-0-321-81257-5

[Casey11] Eoghan Casey: Digital Evidence and Computer Crime - *Forensic Science, Computers, and the Internet* Academic Press (Elsevier), Waltham, MA, USA, 3rd edition, 2011. ISBN 978-0-123-74268-1

[Cassar17] Ian Cassar, Adrian Francalanza, Luca Aceto, Anna Ingólfsdóttir: *A Survey of Runtime Monitoring Instrumentation Techniques* In: Adrian Francalanza, Gordon J. Pace (Editors): Second International Workshop on Pre- and Post-Deployment Verification Techniques (PrePost 2017). EPTCS 254, 2017, pp. 15–28, doi:https://doi.org/10.4204/EPTCS.254.2. Downloadable from: https://arxiv.org/pdf/1708.07229.pdf [Last accessed: 12.10.2020]

[Cha12] S. K. Cha, T. Avgerinos, A. Rebert, D. Brumley: *Unleashing Mayhem on Binary Code* 2012 IEEE Symposium on Security and Privacy, pp. 380–394. IEEE New York, NY, USA, 2012. Downloadable from: https://ieeexplore.ieee.org/stamp/stamp.jsp?tp=&arnumber=6234425 [Last accessed: 23.12.2021]

[Chantzis21] Fotios Chantzis, Ioannis Stais, Paulino Calderon, Evangelos Deirmentzoglou, Beau Woods: Practical IoT Hacking - *The Definitive Guide to Attacking the Internet of Things* No Starch Press, San Francisco, CA, USA, 2021. ISBN 978-1-7185-0090-7

[Chen19] Lei Chen, Hassan Takabi, Nhien-An Le-Khac (Editors): Security, Privacy, and Digital Forensics in the Cloud John Wiley & Sons, Singapore, 2019. ISBN 978-1-119-05328-6

[Chio18] Clarence Chio, David Freeman: Machine Learning and Security - *Protecting Systems with Data and Algorithms* O'Reilly Media Inc., Sebastopol, CA, USA, 2018. ISBN 978-1-491-97990-7

[Clark17] Robert M. Clark, Simon Hakim (Editors): Cyber-Physical Security - *Protecting Critical Infrastructure at the State and Local Level* Springer International Publishing, Cham, Switzerland, 2017. ISBN 978-3-319-32822-5

[Clarke12] Richard A. Clarke, Robert Knake: Cyber War - *The Next Threat to National Security and What to Do About It* Ecco (Harper Collins Publishers), New York, N.Y., USA, Reprint edition 2012. ISBN 978-0-061-96224-0

[Colbert16] Edward J. M. Colbert, Alexander Kott (Editors): Cyber-Security of SCADA and Other Industrial Control Systems Springer International Publishing, Cham, Switzerland, 2016. ISBN 978-3-319-32123-3

[Collins17] Robert Collins: Network Security Monitoring - *Basics for Beginners* CreateSpace Independent Publishing Platform, 2017. ISBN 978-1-9783-0923-4

[CSI20] Cybersecurity Insiders: *2020 Insider Threat Report* Cybersecurity Insiders, USA, 2020. Available at: https://www.cybersecurity-insiders.com/portfolio/2020-insider-threat-report/ [last accessed: 2.5.2020]

[Dajsuren19] Yanja Dajsuren, Mark van den Brand (Editors): Automotive Systems and Software Engineering - *State of the Art and Future Trends* Springer Nature Switzerland, Cham, Sitzerland, 2019. ISBN 978-3-030-12156-3

[Davies10] Joshua Davies: Implementing SSL/TLS Wiley Publishing Inc., Indianapolis, IN, USA, 2010. ISBN 978-0-470-92041-1

| [Davis20] | Royce Davis: The Art of Network Penetration Testing - *Taking Over Any Company in the World* Manning Publications, Shelter Island, New York, N.Y., USA, 2020. ISBN 978-1-617-29682-6 |
| --- | --- |
| [Dearden18] | Harvey T. Dearden: Functional Safety In Practice CreateSpace Independent Publishing, North Charleston, S.C., USA, 2nd edition, 2018. ISBN 978-1-9782-4641-6 |
| [Debouk18] | Rami Debouk: *Overview of the 2nd Edition of ISO 26262: Functional Safety-Road Vehicles* Preprint · August 2018. Downloadable from: https://www.researchgate.net/publication/331650722_Overview_of_the_2nd_Edition_of_ISO_26262_Functional_Safety-Road_Vehicles/link/5c8660fc299bf16918f61f43/download [last accessed: 2.9.2020] |
| [DeFlorio16] | Filippo De Florio: Airworthiness - *An Introduction to Aircraft Certification and Operations* Butterworth-Heinemann (Elsevier), Oxford, UK, 3rd edition, 2016. ISBN 978-0-081-00888-1 |
| [Dekker11] | Sidney Dekker: Drift into Failure - *From Hunting Broken Components to Understanding Complex Systems* CRC Press (Taylor & Francis), Boca Raton, FL, USA, 2011. ISBN 978-1-4094-2221-1 |
| [Deloitte16] | Deloitte Touche Tohmatsu Limited: *Cyber Crisis Management - Readiness, Response, and Recovery* London, UK, 2016, Downloadable from: https://www2.deloitte.com/global/en/pages/risk/articles/cyber-crisis-management.html [Last accessed: 1.7.2020] |
| [DeLong05] | T.A. DeLong, D.T. Smith, B.W. Johnson: *Dependability metrics to assess safety-critical systems* IEEE Transactions on Reliability, New York, N.Y., USA, Volume 54, Issue 3, September 2005, pages 498–505 |
| [Dempsey11] | Kelley Dempsey, Nirali Shah Chawla, Arnold Johnson, Ronald Johnston, Alicia Clay Jones, Angela Orebaugh, Matthew Scholl, Kevin Stine: Information Security Continuous Monitoring (ISCM) for Federal Information Systems and Organizations NIST Special Publication 800-137, Washington, D.C., USA, 2011. Downloadable from: https://nvlpubs.nist.gov/nistpubs/Legacy/SP/nistspecialpublication800-137.pdf [Last accessed: 20.12.2020] |
| [Dignum19] | Virginia Dignum: Responsible Artificial Intelligence - *How to Develop and Use AI in a Responsible Way* Springer Nature Switzerland, Cham, Switzerland, 2019. ISBN 978-3-030-30373-0 |
| [Dimitrakopoulos20] | George J. Dimitrakopoulos, Lorna Uden, Iraklis Varlamis: The Future of Intelligent Transport Systems Elsevier Inc., Amsterdam, Netherlands, 2020. ISBN 978-0-128-18281-9 |
| [Diogenes19] | Yuri Diogenes, Erdal Ozkaya: Cybersecurity – Attack and Defense Strategies: *Counter modern threats and employ state-of-the-art tools and techniques to protect your organization against cybercriminals* Packt Publishing, Birmingham, UK, 2nd edition, 2019. ISBN 978-1-83882-779-3 |
| [Dragon19] | Dave Dragon: I Solve Mysteries - *The Art and Science of Business Process Optimization and Transformation* Silver Tree Publishing, Kenosha, WI, USA, 2019. ISBN 978-1-948238-16-8 |
| [Drusinsky06] | Doron Drusinsky: Modeling and Verification Using UML Statecharts - *A Working Guide to Reactive System Design, Runtime Monitoring, and Execution-based Model Checking* Newnes (Elsevier), Burlington, MA, USA, 2006. ISBN 978-0-7506-7949-7 |

[Dye17]            Jon Dye: A Practical Introduction to Sarbanes-Oxley Compliance
                   CreateSpace Independent Publishing Platform, North Charleston, S.C.,
                   USA, 2017. ISBN 978-1-54324-949-1

[Easttom17]        Chuck Easttom: System Forensics, Investigation, and Response Jones and
                   Bartlett Publishers, Inc., Burlington, MA, USA, 3rd edition 2017. ISBN
                   978-1-284-12184-1

[Easttom18]        William Easttom: Network Defense and Countermeasures - *Principles and*
                   *Practices* Pearson Education (Addison-Wesley), Upper Saddle River, NJ,
                   USA, 3rd edition, 2018. ISBN 978-0-789-75996-2

[Ebers20]          Martin Ebers, Marta Cantero Gamito (Editors): Algorithmic Governance
                   and Governance of Algorithms - *Legal and Ethical Challenges* Springer
                   Nature, Cham, Switzerland, 2020. ISBN 978-3-030-50558-5

[ECC16]            EC-Council: Ethical Hacking and Countermeasures - *Secure Network*
                   *Operating Systems and Infrastructures* EC-Council Press, Boston, MA,
                   USA, 2016. ISBN 978-1-3058-8346-8

[ECCouncil16]      EC-Council: Computer Forensics - *Investigating Network Intrusions*
                   *and Cybercrime* EC-Council Press, Boston, MA, USA, 2016. ISBN
                   978-1-3058-8350-5

[EMC12]            EMC Education Services: Information Storage and Management - *Storing,*
                   *Managing, and Protecting Digital Information in Classic, Virtualized, and*
                   *Cloud Environments* John Wiley &Sons, Indianapolis, USA, 2nd edition,
                   2012. ISBN 978-1-118-09483-9

[Ericson11]        Clifton A. Ericson: System Safety Primer CreateSpace Independent
                   Publishing Platform, North Charleston, S.C., USA, 2011. ISBN
                   978-1-4663-4539-3

[Ericson15]        Clifton A. Ericson: System Safety Engineering – *Design-Based Safety*
                   CreateSpace Independent Publishing Platform, North Charleston, S.C.,
                   USA, 2015. ISBN 978-1-5085-4398-5

[EUACM18]          Informatics Europe & EUACM: *When Computers Decide - European*
                   *Recommendations on Machine-Learned Automated Decision Making*
                   Informatics Europe and ACM Europe Policy Committee, 2018.
                   Downloadable from: https://www.acm.org/binaries/content/assets/public-
                   policy/ie-euacm-adm-report-2018.pdf [last accessed: 19.7.2020]

[EUAR18]           European Union Agency for Railways: *Safety Management System*
                   *Requirements for Safety Certification or Safety Authorisation* Publications
                   Office of the European Union, Luxembourg, 2018. ISBN 978-92-9205-
                   428-1. Downloadable from: https://www.era.europa.eu/sites/default/files/
                   activities/docs/guide_sms_requirements_en.pdf [Last accessed: 26.9.2020]

[FAA19]            FAA (US Federal Aviation Agency): System Safety Handbook US
                   Department of Transportation, Federal Aviation Administration,
                   Washington, DC, USA, 2019. Downloadable from: https://www.faa.gov/
                   regulations_policies/handbooks_manuals/aviation/risk_management/ss_
                   handbook/ [Last accessed 3.4.2020]

[Fall11]           Kevin R. Fall, W. Richard Stevens: TCP/IP Illustrated, Volume 1 - *The*
                   *Protocols* Pearson Education (Addison-Wesley), Upper Saddle River, NJ,
                   USA, 2nd edition, 2011. ISBN 978-0-321-33631-6

[Fazzini19]        Kate Fazzini: Kingdom of Lies - *Adventures in Cybercrime* St. Martin's
                   Press, New York, N.Y., USA, 2019. ISBN 978-1-7860-7637-3

[Finkbeiner19]     Bernd Finkbeiner, Leonardo Mariani (Editors): Runtime Verification Proceedings of the 19th International Conference, RV 2019, Porto, Portugal, October 8–11, 2019. Springer Nature, Cham, Switzerland, 2019. ISBN 978-3-030-32078-2 (LNCS 11757)

[Fiorentini19]     Luca Fiorentini, Luca Marmo: Principles of Forensic Engineering Applied to Industrial Accidents John Wiley & Sons, Hoboken, N.J., USA, 2019. ISBN 978-1-118-96281-7

[Flammini19]       Francesco Flammini (Editor): Resilience of Cyber-Physical Systems - *From Risk Modelling to Threat Counteraction* Springer Nature Switzerland, Cham, Switzerland, 2019. ISBN 978-3-319-95596-4

[Ford21]           Neal Ford, Mark Richards, Pramod Sadalage, Zhamak Dehghani: Software Architecture: The Hard Parts - *Modern Tradeoff Analysis for Distributed Architectures* O'Reilly Media, Inc., Sebastopol, CA, USA, 2021. ISBN 978-1-492-08689-5

[Fowler10b]        Kim Fowler: Mission-Critical and Safety-Critical Systems Handbook - *Design and Development for Embedded Applications* Newnes Publishing (Elsevier), Burlington, MA, USA, 2009. ISBN 978-0-7506-8567-2

[Frew15]           A. C. Frew (Editor): HIPAA Deskbook - *Privacy and Security Regulations with Risk Assessment and Audit Standards* Salient Point Media, Loves Park, IL, USA, 2015. ISBN 978-1-508439-22-6

[Frigerio18]       Alessandro Frigerio, Bart Vermeulen, Kees Goossens: *A Generic Method for a Bottom-Up ASIL Decomposition* SafeComp 2018, September 18–21, 2018, Västeras, Sweden. Downloadable from: http://www.es.ele.tue.nl/~kgoossens/2018-safecomp.pdf [last accessed: 3.9.2020]

[FSI17]            Bank for International Settlements: Cyber-resilience - Range of practices. Basle, Switzerland, 2017. ISBN 978-92-9259-228-8. Downloadable from: https://www.bis.org/bcbs/publ/d454.pdf [Last accessed: 6.6.2022]

[Furrer19]         Frank J. Furrer: Future-Proof Software-Systems – *A Sustainable Evolution Strategy* Springer Vieweg Verlag, Wiesbaden, Germany, 2019. ISBN 978-3-658-19937-1

[Gai20]            Silvano Gai: Building a Future-Proof Cloud Infrastructure - *A Unified Architecture for Network, Security, and Storage Services* Addison-Wesley Educational Publishers Inc., Upper Saddle River, NJ, USA, 2020. ISBN 978-0-136-62409-7

[Gallotti19]       Cesare Gallotti: Information Security - *Risk Assessment, Management Systems, the ISO/IEC 27001 Standard* www.lulu.com, 2019. ISBN 978-0-244-14955-0

[Ganguly18]        Auroop Ratan Ganguly, Udit Bathia, Stephen E. Flynn: Critical Infrastructures Resilience - *Policy and Engineering Principles* Routledge Publishers (Taylor & Francis), Abingdon, UK, 2018. ISBN 978-1-498-75863-5

[GDPR16]           The European Parliament and Council: Regulation (EU) 2016/679 (General Data Protection Regulation - GDPR) Brussels, Belgium, 27 April 2016. Downloadable from: https://eur-lex.europa.eu/legal-content/EN/TXT/PDF/?uri=CELEX:32016R0679 [last accessed: 29.6.2020]

[Gebhardt13]       Vera Gebhardt, Gerhard M. Rieger, Jürgen Mottok, Christian Gießelbach: Funktionale Sicherheit nach ISO 26262 - *Ein Praxisleitfaden zur Umsetzung* dPunkt Verlag, GmbH, Wiesbaden, Germany, 2013. ISBN 978-3-898-64788-5

[Gelles16]          Michael G. Gelles: Insider Threat - *Prevention, Detection, Mitigation, and Deterrence* Butterworth-Heinemann, Kidlington, OX, UK. 2016. ISBN 978-0-128-02410-2

[Gilbert18]         Claude Gilbert, Benoît Journé, Hervé Laroche, Corinne Bieder (Editors): Safety Cultures, Safety Models - Taking Stock and Moving Forward Springer Nature Switzerland AG, Cham, Switzerland, 2018. ISBN 978-3-319-95128-7. Downloadable from: https://link.springer.com/book/10.1007%2F978-3-319-95129-4 [Last accessed: 25.5.2020]

[Gilman17]          Evan Gilman, Doug Barth: Zero Trust Networks - *Building Secure Systems in Untrusted Networks* O'Reilly Media Inc., Sebastopol, CA, USA, 2017. ISBN 978-1-491-96219-0

[Graham20]          Roderick S. Graham: Cybercrime and Digital Deviance Routledge, New York, N.Y., USA, 2020. ISBN 978-0-8153-7631-6

[Gullo18]           Louis J. Gullo, Jack Dixon: Design for Safety John Wiley & Sons, Inc., Hoboken, NJ, USA, 2018. ISBN 978-1-118-97429-2

[Gumzej16]          Roman Gumzej: Engineering Safe and Secure Cyber-Physical Systems - *The Specification PEARL Approach* Springer International Publishing, Cham, Switzerland, 2016. ISBN 978-3-319-28903-8

[Gupta19]           Aditya Gupta: The IoT Hacker's Handbook - *A Practical Guide to Hacking the Internet of Things* Apress Media LLC, New York, N.Y., USA, 2019. ISBN 978-1-4842-4299-5

[Haggerty17]        Al Haggerty: The Failover File Uncial Press, Canada, 2017. ISBN 978-0-6928-3877-8

[Hassanzadeh19]     Amin Hassanzadeh, Amin Rasekh, Stefano Galelli, Mohsen Aghashahi, Riccardo Taormina, Avi Ostfeld, M. Katherine Banks: *A Review of Cybersecurity Incidents in the Water Sector* Journal of Environmental Engineering,· American Society of Civil Engineers, Cincinnati, USA, 11 September, 2019. ISSN: 0733-9372. Downloadable from: https://www.researchgate.net/publication/335753106_A_Review_of_Cybersecurity_Incidents_in_the_Water_Sector [last accessed: 10.8.2020]

[Hawkins14]         Douglas M. Hawkins: Identification of Outliers Springer-Verlag, Dordrecht, Netherlands, 2014. ISBN 978-94-015-3996-8 (Originally published by Chapman & Hall, 1980)

[Hayden10]          Lance Hayden: IT Security Metrics - *A Practical Framework for Measuring Security & Protecting Data* McGraw-Hill (Osborne Media), New York, N.Y., USA, 2010. ISBN 978-0-071-71340-5

[Hayden16]          Lance Hayden: People-Centric Security - *Transforming Your Enterprise Security Culture* McGraw-Hill Education, New York, N.Y., USA, 2016. ISBN 978-0-07-184677-6

[Heegaard15]        Poul Heegaard, Erwin Schoitsch (Editors): *Combining Safety and Security Engineering for Trustworthy Cyber-Physical Systems.* ERCIM News, Nr. 102, July 2015. Free pdf-Download from: https://ercim-news.ercim.eu/en102/special/combining-safety-and-security-engineering-for-trustworthy-cyber-physical-systems [last accessed 10.7.2021]

[Hintzbergen15]     Jule Hintzbergen, Kees Hintzbergen, André Smulders, Hans Baars: Foundations Of Information Security Based on ISO27001 And ISO27002 Van Haren Publishing, Zaltbommel, Nederlands, 3$^{rd}$ edition, 2015. ISBN 978-94-018-0012-9

| [Hobbs20] | Chris Hobbs: Embedded Software Development for Safety-Critical Systems CRC Press (Taylor & Francis), Boca Raton, FL, USA, 2nd edition, 2020. ISBN 978-0-367-33885-5 |
|---|---|
| [Hollnagel06] | Erik Hollnagel, David D. Woods, Nancy Leveson (Editors): Resilience Engineering - *Concepts and Precepts* CRC Press (Taylor & Francis), Boca Raton, FL, USA, 2006. ISBN 978-0-754-64904-5 |
| [Hollnagel13] | Erik Hollnagel, Jean Paries, John Wreathall (Editors): Resilience Engineering in Practice - *A Guidebook* CRC Press (Taylor & Francis), Boca Raton, FL, USA, 2013. ISBN 978-1-472-42074-9 |
| [Hollnagel14] | Erik Hollnagel: Safety-I and Safety-II CRC Press, Francis & Taylor, Boca Raton, FL, USA, 2014. ISBN 978-1-472-42308-5 |
| [Hopkin18] | Paul Hopkin: Fundamentals of Risk Management - *Understanding, Evaluating, and Implementing Effective Risk Management* Kogan Page Ltd., New Delhi, India, 5th edition, 2018. ISBN 978-0-7494-8307-4 |
| [Hubbard16] | Douglas W. Hubbard, Richard Seiersen: How to Measure Anything in Cybersecurity Risk John Wiley & Sons, Inc., Hoboken, N.J., USA, 2016. ISBN 978-1-119-08529-4 |
| [Hubbard20] | Douglas W. Hubbard: The Failure of Risk Management – *Why it's broken and how to fix it.* John Wiley & Sons, Inc., Hoboken, New Jersey, USA, 2nd edition, 2020. ISBN 978-1-119-52203-4 |
| [ICAO13] | International Civil Aviation Organization (ICAO): Safety Management Manual (SMM) ICAO, Montréal, Quebec, Canada, 3rd edition, 2013. Downloadable from : https://www.skybrary.aero/bookshelf/books/644.pdf [Last accessed: 27.9.2020] |
| [IEC61508:10] | The IEC 61508 Association: *What is IEC 61508? - IEC 61508 is the international Standard for Electrical, Electronic, and Programmable Electronic Safety-Related Systems.* https://www.61508.org/knowledge/what-is-iec-61508.php [last accessed: 1.9.2020] |
| [ISO21448:19] | ISO/PAS 21448:2019 Road vehicles - *Safety of the Intended Functionality* International Organization for Standardization, Vernier, Switzerland, 2019. Downloadable from: https://www.iso.org/standard/70939.html [last accessed: 31.8.2020] |
| [ISO 262262-1:2018] | ISO 26262-1:2018: Road Vehicles - Functional Safety International Organization for Standardization, Vernier, Switzerland, 2018. Downloadable from: https://www.iso.org/standard/68383.html [last accessed: 31.8.2020] |
| [ISO262262-9:2018] | ISO 26262-9:2018: Road Vehicles - Functional Safety - Part 9: Automotive Safety Integrity Level (ASIL)-oriented and Safety-oriented Analyses International Organization for Standardization, Vernier, Switzerland, 2018. Downloadable from: https://www.iso.org/standard/68391.html [Last accessed: 2.3.2021] |
| [ISO26262:18] | ISO 26262-1:2018: Road Vehicles - Functional Safety. International Organization for Standardization, Vernier, Switzerland, 2018. Downloadable from: https://www.iso.org/standard/68383.html [last accessed: 31.8.2020] |
| [ITG20] | UK IT Governance: *Information Security and ISO 27001 – An Introduction* UK IT Governance, Ely, UK: Green Paper, January 2020. Downloadable from: https://www.itgovernance.co.uk/green-papers/ |

|                      | information-security-and-iso-27001-an-introduction    [Last    accessed: 30.11.2020] |
| [ITGP19] | IT Governance Privacy Team: **EU General Data Protection Regulation (GDPR) - *An Implementation and Compliance Guide*** IT Governance Publishing, Ely, UK, 3rd edition, 2019. ISBN 978-1-78778-191-7 |
| [Jackson07] | Daniel Jackson, Martyn Thomas, Lynette I. Millett (Editors): Software for Dependable Systems: Sufficient Evidence? US National Research Council, Washington, DC, USA, 2007. ISBN 978-0-309-38450-6. Downloadable from: https://www.nap.edu/download/11923 [last accessed: 9.10.2020] |
| [Jaquith07] | Andrew Jaquith: Security Metrics - *Replacing Fear, Uncertainty, and Doubt* Pearson Education, Boston, MA, USA, 2007. ISBN 978-0-321-34998-9 |
| [Jasanoff16] | Sheila Jasanoff: The Ethics of Invention - *Technology and the Human Future* W W NORTON & CO, New York, N.Y., USA, 2016. ISBN 978-0-393-07899-2 |
| [Jayasri17] | Kotti Jayasri: A Study on Safety Metrics for Safety-Critical Computer Systems - *Software Safety Assessment Methods* LAP LAMBERT Academic Publishing, Riga, Latvia, 2017. ISBN 978-3-3303-4986-5 |
| [Johnson15] | Thomas A. Johnson (Editor): Cybersecurity - *Protecting Critical Infrastructures from Cyber Attack and Cyber Warfare* CRC Press, Taylor & Francis Ltd., Boca Raton, FL, USA, 2015. ISBN 978-1-482-23922-5 |
| [Juniper18] | Adam Juniper: Smart Home Handbook - *Connect, Control, and Secure your Home the Easy Way* Ilex Press, London, UK, 2018. ISBN 978-1-7815-7580-2 |
| [Kanat-Alexander12] | Max Kanat-Alexander: Code Simplicity - *The Fundamentals of Software* O'Reilly and Associates, Sebastopol, CA, USA, 2012. ISBN 978-1-449-31389-0 |
| [Kanat-Alexander17] | Max Kanat-Alexander: Understanding Software - *Max Kanat-Alexander on simplicity and coding* Packt Publishing, Birmingham, UK, 2017. ISBN 978-1-7886-2881-5 |
| [Kane15] | Aaron Kane, Omar Chowdhury, Anupam Datta, Philip Koopman: *A Case Study on Runtime Monitoring of an Autonomous Research Vehicle (ARV) System* Preprint 6th International Conference on Runtime Verification (RV 2015), Vienna, Austria, September 22–25, 2015. Proceedings, Springer International Publishing, Cham, Switzerland, 2015. LNCS 9333. ISBN 978-3-319-23819-7. Downloadable from: https://www.researchgate.net/ publication/300254024_A_Case_Study_on_Runtime_Monitoring_of_an_ Autonomous_Research_Vehicle_ARV_System [Last accessed: 13.5.2021] |
| [Karimipour20] | Hadis Karimipour, Pirathayini Srikantha, Hany Farag, Jin Wei-Kocsis (Editors): Security of Cyber-Physical Systems - *Vulnerability and Impact* Springer Nature Switzerland AG, Cham, Switzerland, 2020. ISBN 978-3-030-45540-8 |
| [Kaschner20] | Holger Kaschner: Cyber Crisis Management - *Das Praxishandbuch zu Krisenmanagement und Krisenkommunikation* Springer Fachmedien Wiesbaden GmbH, Wiesbaden, Germany, 2020. ISBN 978-3-658-27913-4 |
| [Katz21a] | Jonathan Katz, Yehuda Lindell: Introduction to Modern Cryptography CRC Press (Taylor & Francis), Boca Raton, FL, USA, 3rd edition, 2021. ISBN 978-0-815-35436-9 |

[Katz21b]          Jonathan Katz, Vadim Lyubashevsky: Lattice-based Cryptography CRC Press (Taylor & Francis), Boca Raton, FL, USA, 2021. ISBN 978-1-498-76347-)

[Kerman20]         Alper Kerman: Zero Trust Cybersecurity: 'Never Trust, Always Verify' NIST Publication, October 28, 2020. Downloadable from: https://www.nist.gov/blogs/taking-measure/zero-trust-cybersecurity-never-trust-always-verify [Last accessed<>: 10.11.2020]

[Knight20]         Alissa Knight: Hacking Connected Cars - *Tactics, Techniques, and Procedures* John Wiley & Sons, Inc., Hoboken, New Jersey, USA, 2020. ISBN 978-1-119-49180-4

[Kohnke17]         AnneKohnke, Ken Sigler, Dan Shoemaker: Implementing Cybersecurity - *A Guide to the National Institute of Standards and Technology Risk Management Framework* CRC Press, Taylor & Francis, Boca Raton, FL, USA, 2017. ISBN 978-1-498-78514-3

[Kopetz11]         Hermann Kopetz: Real-Time Systems - *Design Principles for Distributed Embedded Applications* Springer Science & Business Media, New York, N.Y., USA, 2nd edition, 2011. ISBN 978-1-461-42866-4

[Kopetz19]         Hermann Kopetz: Simplicity is Complex - *Foundations of Cyber-Physical System Design* Springer Nature, Cham, Switzerland, 2019. ISBN 978-3-030-20410-5

[Krutz16]          Ronald L. Krutz: Industrial Automation and Control System Security International Society of Automation (ISA), Research Triangle Park, N.C., USA, 2nd edition 2016. ISBN 978-1-9415-4682-6

[Kuppusamy16]      Trishank Karthik Kuppusamy, Akan Brown, Sebastien Awwad, Damon McCoy, Russ Bielawski, Cameron Mott, Sam Lauzon, André Weimerskirch, Justin Cappos: *Uptane: Securing Software Updates for Automobiles* 14th escar Europe 2016 Conference, Munich, Germany, November 16 and 17, 2016. Downloadable from: https://ssl.engineering.nyu.edu/papers/kuppusamy_escar_16.pdf [last accessed: 25.10.2021]

[Laing13]          Christopher Laing, Atta Badii, Paul Vickers (Editor): Securing Critical Infrastructures and Critical Control Systems - *Approaches for Threat Protection* Information Science Reference (IGI Global), Hershey, PA, USA, 2013. ISBN 978-1-4666-2659-1

[Landoll16]        Douglas J. Landoll: Information Security Policies, Procedures, and Standards - *A Practitioner's Reference* CRC Press, Taylor & Francis Ltd., Boca Raton, FL, USA, 2016. ISBN 978-1-482-24589-9

[Landoll20]        Douglas Landoll: The Security Risk Assessment Handbook - *A Complete Guide for Performing Security Risk Assessments* CRC Press (Taylor & Francis), Boca Raton, FL, USA, 2nd edition, 2020. ISBN 978-0-367-65929-5

[Lee20a]           Kang B. Lee, Richard Candell, Hans-Peter Bernhard, Dave Cavalcanti, Zhibo Pang, Inaki Val: *Reliable, High-Performance Wireless Systems for Factory Automation* NIST White Paper, NIST, Gaithersburg, MD, USA, 2020. Downloadable from: https://www.nist.gov/publications/reliable-high-performance-wireless-systems-factory-automation [Last accessed: 4.5.2021]

[Lee20b]           Robert M. Lee *2020 SANS Cyber Threat Intelligence (CTI) Survey* SANS Insitute Report, SANS, bethesda, MA, U.S.A., February 2020.

Downloadable from: https://www.domaintools.com/content/SANS_CTI_Survey_2020.pdf [Last accessed: 23.12.2020]

[Le-Khac20] Nhien-An Le-Khac, Kim-Kwang Raymond Choo: Cyber and Digital Forensic Investigations - *A Law Enforcement Practitioner's Perspective* Springer Nature Switzerland, Cham, Switzerland, 2020. ISBN 978-3-030-47130-9

[Leveson16] Nancy G. Leveson: Engineering a Safer World - *Systems Thinking Applied to Safety* MIT Press Ltd., Massachusetts, MA, USA, 2016. ISBN 978-0-262-53369-0

[Li18c] Jun Li, Bodong Zhao and Chao Zhang Fuzzing - *A Survey* Cybersecurity (2018) 1:6, Springer Open, Heidelberg, Germany, 2018. Downloadable from: https://cybersecurity.springeropen.com/track/pdf/https://doi.org/10.1186/s42400-018-0002-y.pdf [Last accessed: 23.12.2021]

[Li20a] Beibei Li, Rongxing Lu, Gaoxi Xiao: Detection of False Data Injection Attacks in Smart Grid Cyber-Physical Systems Springer Nature Switzerland, Cham, Switzerland, 2020. ISBN 978-3-030-58671-3

[Li20b] Kuan-Ching Li, Brij B. Gupta, Dharma B. Agrawal (Editors): Recent Advances in Security, Privacy and Trust for Internet-of-things IoT and Cyber-Physical Systems CRC Press (Taylor & Francis Group), Boca Raton, FL, USA, 2020. ISBN 978-0-367-22065-5

[Liyanage18] Madhusanka Liyanage, Ijaz Ahmad, Ahmed Bux Abro, : A Comprehensive Guide to 5G Security John Wiley & Sons, Hoboken, NJ, USA, 2018. ISBN 978-1-119-29304-0

[Loukas15] George Loukas: Cyber-Physical Attacks - *A Growing Invisible Threat* Butterworth-Heinemann, Kidlington, OX, UK, 2015. ISBN 978-0-1280-1290-1

[Lusthaus18] Jonathan Lusthaus: Industry of Anonymity - *Inside the Business of Cybercrime* Harvard University Press, Cambridge, MA, USA, 2018. ISBN 978-0-674-97941-3

[Luttgens14] Jason T. Luttgens: Incident Response and Computer Forensics McGraw-Hill Education, New York, N.Y., USA, 3rd edition, 2014. ISBN 978-0-071-79868-6

[Malik16] Mubina Malik, Trisha Patel: Database Security – *Attacks and Control Methods* International Journal of Information Sciences and Techniques (IJIST), Vol.6, No.1, March 2016. Downloadable from: https://www.researchgate.net/publication/301277002_Database_Security_-_Attacks_and_Control_Methods/link/5d8c598092851c33e93c6b98/download [Last accessed: 22.11.2020]

[Martin20] Dominic Martin, Amy Martin, Jeremy Martin, Richard Medlin, Nitin Sharma, Jams Ma: Cyber-Secrets: *A Threat Hunting, Hacking, and Intrusion Detection - SCADA, Dark Web, and APTs* Information Warfare Center, Colorado Springs, USA, 2nd edition, 2020. ISBN 979-8-65183-430-3

[Matulevičius17] Raimundas Matulevičius: Fundamentals of Secure System Modelling Springer International Publishing, Cham, Switzerland, 2017. ISBN 978-3-319-87143-1

[McDermott09] Robin E. McDermott, Raymond J. Mikulak, Michael R. Beauregard: The Basics of FMEA CRC Press (Taylor & Francis), Boca Raton, FL, USA, 2nd edition 2009. ISBN 978-1-56327-377-3

[McKinsey19]          McKinsey & Company: *Automotive Software and Electronics 2030 - Mapping the Sector's future Landscape* McKinsey & Company, Munich Office, Munich, Germany, 2019. Downloadable from: https://www.mckinsey.com/~/media/mckinsey/industries/automotive%20and%20assembly/our%20insights/mapping%20the%20automotive%20software%20and%20electronics%20landscape%20through%202030/automotive-software-and-electronics-2030-final.pdf [Last accessed: 26.11.2020]

[Mead16]              Nancy R. Mead, Carol C. Woody: Cyber Security Engineering - *A Practical Approach for Systems and Software Assurance* Addison-Wesley Professional, Boston, USA, 2016. ISBN 978-0-134-18980-2

[Merkow10]            Mark S. Merkow, Lakshmikanth Raghavan: Secure and Resilient Software Development CRC Press (Taylor & Francis), Boca Raton, FL, USA, 2010. ISBN 978-1-439-82696-6

[Messier17]           Ric Messier: Network Forensics John Wiley & Sons, Indianapolis, USA, 2017. ISBN 978-1-119-32828-5

[Meyer16]             Thierry Meyer, Genserik Reniers: Engineering Risk Management Walter De Gruyter GmbH, Berlin, Germany, 2nd edition, 2016. ISBN 978-3-110-41803-3

[Miller20]            Joseph D. Miller: Automotive System Safety - *Critical Considerations for Engineering and Effective Management* John Wiley & Sons, Chichester, UK, 2020. ISBN 978-1-119-57962-5

[Mitchell11]          Melanie Mitchell: Complexity - *A Guided Tour* Oxford University Press, New York, N.Y., USA, 2011. ISBN 978-0-199-79810-0

[Moeller10]           Robert R. Moeller: IT Audit, Control, and Security John Wiley & Sons, Inc., New York, N.Y., USA, 2010. ISBN 978-0-471-40676-1

[Mohammadi19]         Nazila Gol Mohammadi: Trustworthy Cyber-Physical System - *A Systematic Framework towards Design and Evaluation of Trust and Trustworthiness* Springer Vieweg Verlag, Wiesbaden, Germany, 2019. ISBN 978-3-658-27487-0

[Möller19]            Dietmar P.F. Möller, Roland E. Haas: Guide to Automotive Connectivity and Cybersecurity - *Trends, Technologies, Innovations, and Applications* Springer International Publishing, Cham Switzerland, 2019. ISBN: 978-3-319-73511-5

[Mukherjee20]         Aditya Mukherjee: Network Security Strategies - *Protect your Network and Enterprise against advanced Cybersecurity Attacks and Threats* Packt Publishing, Birmingham, UK, 2020. ISBN 978-1-7898-0629-8

[Munteanu06]          Adrian Munteanu: Information Security Risk Assessment - The Qualitative Versus Quantitative Dilemma. Proceedings of the 6th International Business Information Management Association (IBIMA) Conference, Bonn, Germany, pp. 227–232, June 19-21, 2006. Downloadable from: https://papers.ssrn.com/sol3/papers.cfm?abstract_id=917767 [Last accessed: 6.6.2022]

[Nader20]             Mohamed Nader, Jameela Al-Jaroodi, Imad Jawhar: *Cyber-Physical Systems Forensics - Today and Tomorrow* Journal of Sensor and Actuator Networks, Basel, Switzerland, 2020, August 2020, DOI:https://doi.org/10.3390/jsan9030037. Downloadable from: https://www.mdpi.com/2224-2708/9/3/37 [Last accessed: 17.10.2020]

[Nichols19]           R.K. Nichols, JJCH Ryan, H.C. Mumm, W.D. Lonstein, C. Carter, J.P. Hood: Unmanned Aircraft Systems in the Cyber-Domain – *Protecting*

_USA's Advanced Air Assets_ New Prairie Press, Kansas State University, Kansas, USA, 2nd edition, 2019. ISBN 978-1-9445-4823-0

[NIST06]          Karen Kent, Suzanne Chevalier, Tim Grance, Hung Dang: Guide to Integrating Forensic Techniques into Incident Response - _Recommendations of the National Institute of Standards and Technology_ NIST Special Publication 800-86, 2006. Downloadable from:https://nvl-pubs.nist.gov/nistpubs/Legacy/SP/nistspecialpublication800-86.pdf    [last accessed: 6.7.2020]

[NIST1800_20]     Timothy McBride, Michael Ekstrom, Lauren Lusty, Julian Sexton, Anne Townsend: Data Integrity - _Recovering from Ransomware and Other Destructive Events_ NIST SPECIAL PUBLICATION 1800-11B, US National Cybersecurity Center of Excellence, National Institute of Standards and Technology, Gaithersburg, MD, USA, 2020. Downloadable from: https://doi.org/10.6028/NIST.SP.1800-11 [Last accessed: 22.9.2020]

[NIST20]          Scott Rose, Oliver Borchert, Stu Mitchell, Sean Connelly: Zero Trust Architecture NIST Special Publication 800-207 (final), August 2020. Downloadable from: fhttps://doi.org/10.6028/NIST.SP.800-207 or https://csrc.nist.gov/publications/detail/sp/800-207/final   [last · accessed: 12.8.2020]

[NIST800-160B]    Ron Ross, Victoria Pillitteri, Richard Graubart, Deborah Bodeau, Rosalie McQuaid: Developing Cyber Resilient Systems - _A Systems Security Engineering Approach_ NIST Special Publication 800-160, Volume 2, Rev. 1, December 2021. Downloadable from: https://nvlpubs.nist.gov/nistpubs/SpecialPublications/NIST.SP.800-160v2r1.pdf [Last accessed: 09.12.2021]

[NIST800-172]     Ron Ross, Victoria Pillitteri, Gary Guissanie, Ryan Wagner, Richard Graubart, Deb Bodeau : Enhanced Security Requirements for Protecting Controlled Unclassified Information NIST Special Publication 800-172, NIST, Washington, USA, February 2021. Dowloadable from: https://nvl-pubs.nist.gov/nistpubs/SpecialPublications/NIST.SP.800-172.pdf    [Last accessed: 3.2.2021]

[Osterwalder20]   Alexander Osterwalder, Yves Pigneur, Frederic Etiemble, Alan Smith: The Invincible Company - _How to Constantly Reinvent Your Organization with Inspiration From the World's Best Business Models_ John Wiley & Sons, Inc., Hoboken, N.J., USA, 2020. ISBN 978-1-119-52396-3

[Pathan15]        Al-Sakib Khan Pathan: Securing Cyber-Physical Systems CRC Press (Taylor & Francis Inc.), Boca Raton, FL, USA, 2015. ISBN 978-1-498-70098-6

[Petrenko19]      Sergei Petrenko: Cyber Resilience River Publishers, Delft, NL, 2019. ISBN 978-87-70221-16-0

[Pimentel19]      Juan R Pimentel (Editor): The Role of ISO 26262: Book 4 - _Automated Vehicle Safety_ SAE Technical Paper Collection, Society of American Engineers, Warrendale, PA, USA, 2019. ISBN 978-0-7680-0274-4

[Pompon16]        Raymond Pompon: IT Security Risk Control Management - _An Audit Preparation Plan_ Apress Media LLC, New York, N.Y., USA, 2016. ISBN 978-1-48422-139-6

[PTP20]           Pen Test Partners _Securing Airplanes & Airports_ White Paper, 2020. Downloadable       from:       https://www.pentestpartners.com/content/uploads/2020/07/PTP-Aviation-Cyber-Security.pdf    [Last   accessed: 20.12.2020]

[Radvanovsky16]    Robert Radvanovsky, Jacob Brodsky (Editors): Handbook of SCADA/ Control Systems Security CRC Press (Taylor & Francis), Boca Raton, FL, USA, 2nd edition 2016. ISBN 978-1-498-71707-6

[Rasmussen00]     Jens Rasmussen, Inge Svedung: *Proactive Risk Management in a Dynamic Society* Risk & Environmental Department, Swedish Rescue Services Agency, Karlstad, Sweden, 2000. ISBN: 91-7253-084-7. Downloadable from: https://www.msb.se/RibData/Filer/pdf/16252.pdf [Last accessed: 28.11.2020]

[Reddy19]          Niranjan Reddy: Practical Cyber Forensics - *An Incident-Based Approach to Forensic Investigations* Apress Media LLC, New York, N.Y., USA, 2019. ISBN 978-1-484-24459-3

[Rerup18]          Neil Rerup (Autor), Milad Aslaner: Hands-On Cybersecurity for Architects - *Plan and Design robust Security Architectures* Packt Publishing, Birmingham, UK, 2018. ISBN 978-1-78883-026-3

[Rierson13]        Leanna Rierson: Developing Safety-Critical Software - *A Practical Guide for Aviation Software and DO-178C Compliance* CRC Press (Taylor & Francis Inc.), Boca Raton, FL, USA, 2013. ISBN 978-1-439-81368-3

[Ries11]           Eric Ries: The Lean Startup - *How Today's Entrepreneurs Use Continuous Innovation to Create Radically Successful Businesses* Penguin Books, London, UK, 2011. ISBN 978-1-524-76240-7

[Rinehart18]       Aaron Rinehart, Charles Nwatu: *Security Chaos Engineering - A new Paradigm for Cybersecurity* Whitepaper, 2018. Downloadable from: https://opensource.com/article/18/1/new-paradigm-cybersecurity [last accessed: 17.8.2020]

[Rodin99]          Robert Rodin: Free, Perfect, and Now - *Connecting to the Three Insatiable Customer Demands* Touchstone Publishing (Simon & Schuster), New York, N.Y., USA, 1999. ISBN 978-0-684-86312-2

[Roer15]           Kai Roer: Build a Security Culture IT Governance Publishing, Ely, UK, 2015. ISBN 978-1-84928-719-7

[Romanovsky17]     Alexander Romanosky, Fuyuki Ishikawa (Editors): Trustworthy Cyber-Physical Systems Engineering CRC Press, Boca Raton FL, USA, 2017. ISBN 978-1-4978-4245-0

[Rosenthal20]      Casey Rosenthal, Nora Jones: Chaos Engineering - *System Resiliency in Practice* O'Reilly Ltd., Sebastopol, CA, USA, 2020. ISBN 978-1-492-04386-7

[Ross16]           Hans-Leo Ross: Functional Safety for Road Vehicles - *New Challenges and Solutions for E-mobility and Automated Driving* Springer International Publishing, Cham, Switzerland, 2016. ISBN 978-3-319-33360-1

[Roughton19]       James Roughton, Nathan Crutchfield, Michael Waite: Safety Culture - *An Innovative Leadership Approach* Butterworth-Heinemann (Elsevier), Kidlington, UK, 2nd edition, 2019. ISBN 978-0-128-14663-7

[Ruan19]           Keyun Ruan: Digital Asset Valuation and Cyber Risk Measurement - *Principles of Cybernomics* Academic Press (Elsevier), Cambridge, MA, USA, 2019) ISBN 978-0-128-12158-0

[Russo18]          Mark A. Russo: Information Technology Security Audit Guidebook - *NIST SP 800-171* Independently published, 2018. ISBN 978-1-7266-7490-4

[Ryder19]        Rodney D. Ryder, Ashwin Madhavan: Cyber Crisis Management
                 Bloomsbury Publishing India, New Delhi, India, 2019. ISBN
                 978-9-3891-6550-0

[Sadiqui20]      Ali Sadiqui: Computer Network Security ISTE LTD., John Wiley & Sons
                 Inc., Hoboken, N.J. UK, 2020. ISBN 978-1-786-30527-5

[Saxe18]         Joshua Saxe, Hillary Sanders: Malware Data Science - *Attack Detection
                 and Attribution* No Starch Press Inc., San Francisco, USA, 2018. ISBN
                 978-1-5932-7859-5

[Saydjari18]     O. Sami Saydjari: Engineering Trustworthy Systems - *Get Cybersecurity
                 Design Right the First Time* McGraw-Hill Education, New York, N.Y.,
                 USA, 2018. ISBN 978-1-260-11817-9

[Scarfone10]     Karen Scarfone, Dan Benigni, Tim Grance: Cyber Security Standards
                 National Institute of Standards and Technology (NIST), Gaithersburg,
                 Maryland, USA, 2010. Downloadable from: https://tsapps.nist.gov/publi-
                 cation/get_pdf.cfm?pub_id=152153 [Last accessed: 21.11.2020]

[Scharre19]      Paul Scharre: Army of None - *Autonomous Weapons and the Future
                 of War* WW Norton & Company, New York, N.Y., USA, 2019. ISBN
                 978-0-393-35658-8

[Schmidt20a]     Boris Schmidt, Roger Crathorne, Matthias Pfannmüller: Der neue Land
                 Rover Defender Delius Klasing Verlag, GmbH, Bielefeld, Germany, 2020.
                 ISBN 978-3-6671-1662-8

[Schmidt20b]     Herbie Schmidt: *Das Auto wird immer mehr zum Datenkraken* Neue
                 Zürcher Zeitung (NZZ), Zürich, Switzerland, 18. November 2020, p. 27

[Schneier03]     Bruce Schneier: Beyond Fear - *Thinking Sensibly About Security in an
                 Uncertain World* Copernicus Books, New York, N.Y., USA, Corrected 2nd
                 printing 2006. ISBN 978-0-387 02620 6

[Schneier18]     Bruce Schneier: *Artificial Intelligence and the Attack/Defense Balance*
                 IEEE Security & Privacy, New York, NY, USA, Volume 16, Issue 2,
                 March/April 2018. Downloadable from: https://ieeexplore.ieee.org/stamp/
                 stamp.jsp?tp=&arnumber=8328965 [Last accessed: 15.5.2021]

[Schoitsch20]    Erwin Schoitsch: *Machine Ethics* in: ERCIM News, Number 122,
                 July 2020, pp. 4–5. ERCIM EEIG, Sophia Antipolis Cedex, France.
                 Downloadable from: https://ercim-news.ercim.eu/images/stories/EN122/
                 EN122-web.pdf [last accessed: 18.7.2020]

[Sessions08]     Roger Sessions: Simple Architectures for Complex Enterprises Microsoft
                 Press, Redmond, USA, 2008. ISBN 978-0-7356-2578-5

[Sessions09]     Roger Sessions: *The IT Complexity Crisis – Danger and Opportunity*.
                 White Paper, November 2009. Downloadable from: http://www.object-
                 watch.com/whitepapers/ITComplexityWhitePaper.pdf    (last    accessed:
                 8.2.2013)

[Song17]         Houbing Song, Glenn A. Fink, Sabina Jeschke (Editors): Security and
                 Privacy in Cyber-Physical Systems - *Foundations, Principles, and
                 Applications* John Wiley & Sons, Inc., Hoboken, NJ, USA, 2017. ISBN
                 978-1-119-22604-8

[Stallings16]    William Stallings: Network Security Essentials - *Applications and
                 Standards* Pearson Education Ltd., Harlow, Essex, UK, 6th edition, 2016.
                 ISBN 978-1-292-15485-5

[Stallings18]    William Stallings: Effective Cybersecurity Pearson Education (Addison
                 Wesley), Upper Saddle River, N.J., USA, 2018. ISBN 978-0-134-77280-6

[Stolzer15]          Alan J. Stolzer, John J. Goglia: Safety Management Systems in Aviation Ashgate Publishing Ltd., Farnham, UK, 2nd edition, 2015. ISBN 978-1-4724-3175-2

[Taylor20]           Andy Taylor, David Alexander, Amanda Finch, David Sutton: Information Security Management Principles British Computer Society (BCS), Swindon, UK, 3rd edition, 2020. ISBN 978-1-78017-518-8

[Tegenaw19]          Hailu Tegenaw, Mesfin Kifle: A Firewall Architecture to Enhance Performance of Enterprise Network LAP LAMBERT Academic Publishing, Riga, Latvia, 2019. ISBN 978-6-1394-4750-3

[Thames17]           Lane Thames, Dirk Schaefer: Cybersecurity for Industry 4.0 - *Analysis for Design and Manufacturing* Springer International Publishing, Cham, Switzerland, 2017. ISBN 978-3-319-50659-3

[Trim16]             Peter Trim, David Upton: Cyber Security Culture - *Counteracting Cyber Threats Through Organizational Learning and Training* Routledge Publishing (Taylor & Francis), Abingdon, UK, 2016. ISBN 978-1-409-45694-0

[Vacca19]            John R. Vacca: Public Key Infrastructure - Building Trusted Applications and Web Services Routledge Publishers (Taylor & Francis), Abingdon, UK, 2019. ISBN 978-0-367-39432-5

[VanDerWens19]       Cees van der Wens: ISO 27001 Handbook - Implementing and Auditing an Information Security Management System in small and medium-sized Businesses Independently published, 2019. ISBN 978-1-098547-68-4

[Vélez15]            Tony Uceda Vélez, Marco M. Morana: Risk Centric Threat Modeling - *Process for Attack Simulation and Threat Analysis* John Wiley & Sons Inc., Hoboken, NJ, USA, 2015. ISBN 978-0-470-50096-5

[Vellani20]          Karim H. Vellani: Strategic Security Management - *A Risk Assessment Guide for Decision Makers.* CRC-Press, Taylor & Francis Ltd., Boca Raton, FL, USA, 2nd edition, 2020. ISBN 978-1-138-58366-5

[Wang19]             Cliff Wang, Zhuo Lu (Editors). Proactive and Dynamic Network Defense Springer Nature Switzerland, Cham, Switzerland, 2019. ISBN 978-3-030-10596-9

[Watertechonline16]  Watertechonline: *Water Technology (Industrial Water Management)* Fort Atkinson, Wisconsin, USA, 23 March, 2016. Downloadable from: https://www.watertechonline.com/home/article/15550066/hackers-change-chemical-settings-at-water-treatment-plant [last accessed: 10.8.2020]

[Weidman14]          Georgia Weidman: Penetration Testing - *A Hands-On Introduction to Hacking* No Starch Press, San Francisco, CA, USA, 2014. ISBN 978-1-5932-7564-8

[Wilhelm13]          Thomas Wilhelm: Professional Penetration Testing - *Creating and Learning in a Hacking Lab* Syngress (Elsevier), Cambridge, MA, USA, 2013. ISBN 978-1-597-49993-4

[Williams13]         Barry L. Williams: Information Security Policy Development for Compliance: *ISO/IEC 27001, NIST SP 800-53, HIPAA Standard, PCI DSS V2.0, and AUP V5.0* CRC Press (Taylor & Francis), Boca Raton, FL, USA, 2013. ISBN 978-1-466-58058-9

[Winkler16]          Ira Winkler, Araceli Treu Gomes: Advanced Persistent Security - *A Cyberwarfare Approach to Implementing Adaptive Enterprise Protection, Detection, and Reaction Strategies* Syngress (Elsevier), Cambridge, MA, USA, 2017. ISBN 978-0-128-09316-0

[Winner18]          Hermann Winner, Günther Prokop, Markus Maurer (Editors): Automotive Systems Engineering II Springer International Publishing, Cham, Switzerland, 2018. ISBN 978-3-319-61605-6

[Wong10]            W. Eric Wong, Vidroha Debroy, Adithya Surampudi, HyeonJeong Kim, Michael F. Siok: Recent Catastrophic Accidents - *Investigating How Software was Responsible* 2010 Fourth IEEE International Conference on Secure Software Integration and Reliability Improvement, 9–11 June 2010, Singapore, Singapore (DOI https://doi.org/10.1109/SSIRI.2010.38). Downloadable from: https://www.researchgate.net/publication/221502705_Recent_Catastrophic_Accidents_Investigating_How_Software_was_Responsible/link/568d73aa08aead3f42eda330/download [Last accessed: 8.1.2021]

[Wong11]            Caroline Wong: Security Metrics - *A Beginner's Guide* McGraw-Hill (Osborne Media), New York, N.Y., USA, 2011. ISBN 978-0-071-74400-3

[Worch19]           Eckhard Worch: Drinking-Water Treatment - *An Introduction* Walter De Gruyter GmbH, Berlin/London, Germany, 2019. ISBN 978-3-110-55154-9

[Wu13]              Tina Wu, Jules Ferdinand Pagna Disso, Kevin Jones, Adrian Campos: *Towards a SCADA Forensics Architecture* Proceedings of the 1st International Symposium for ICS & SCADA Cyber Security Research, 2013. Downloadable from: https://www.scienceopen.com/document_file/312c507f-84ce-458d-b2cc-453e3fff6655/ScienceOpen/012_Wu.pdf [Last accessed: 11.1.2021]

[Wysopal07]         Chris Wysopal: The Art of Software Security Testing - *Identifying Software Security Flaws* Addison-Wesley Professional, Upper Saddle River, NJ, USA, 2007. ISBN 978-0-321-30486-5

[Xu20]              Shouhuai Xu: Cybersecurity Dynamics - *A Foundation for the Science of Cybersecurity* Preprint, October 2020.

[Young20]           Scott Young: Designing a DMZ SANS Institute White Paper, Bethesda, MA, USA, 2020. Downloadable from: https://www.sans.org/reading-room/whitepapers/firewalls/designing-dmz-950 [last accessed: 12.8.2020]

[Yourdon03]         Edward Yourdon: Death March Prentice-Hall, Upper Saddle River, N.J., USA, 2nd edition, 2003. ISBN 978-0-131-43635-0

[Zongo18]           Phillimon Zongo: The Five Anchors of Cyber-Resilience – *Why some Enterprises are hacked into Bankruptcy while others easily bounce back* Broadcast Books, Australia, 2018. ISBN 978-0-6480078-4-5

# Safe Software and Secure Software

# 5

*In most cyber-physical systems, the functionality is implemented in software. Therefore, the software is responsible for a significant part of safety and security. Because software is infinitely malleable, it is easy to construct new or extend existing software. In both cases, the software must ensure the safety and security of the cyber-physical system. Engineering safe and secure software is a highly challenging but crucial task. The most important precondition for safe and secure software is adequate system architecture. The architecture is the foundation of safe and secure software and determines many of its quality properties and the complexity of implementations.*

## 5.1    Introduction

Most of the functionality of cyber-physical systems is implemented in *software*. Therefore, the software is the most critical part of the system. Numerous potential safety and security *vulnerabilities* are hidden in the large number of programs controlling a CPS (Fig. 5.1).

CPS software has three levels (Fig. 5.2):

I.    The *application software*: This software executes the cyber-physical system's control functions and is application-specific. The application software can either be developed in-house, it can be third-party software, or commercial off-the-shelf software (COTS);

II.   The *systems software*: The system software provides all the functionality to execute basic needs, such as operating systems, communications, databases, vital security functions (e.g., access control), archiving, and much more;

III.  The *execution platform*: Contains all the hardware (CPUs, video integrated circuits, disk drives, memory, etc.) and an enormous amount of *firmware*.

© The Author(s), under exclusive license to Springer Fachmedien Wiesbaden GmbH,     187
part of Springer Nature 2022
F. J. Furrer, *Safety and Security of Cyber-Physical Systems*,
https://doi.org/10.1007/978-3-658-37182-1_5

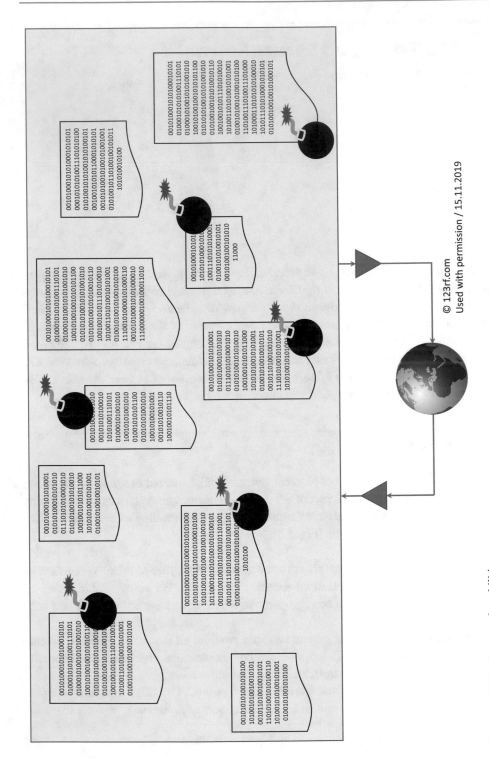

**Fig. 5.1** CPS software vulnerabilities

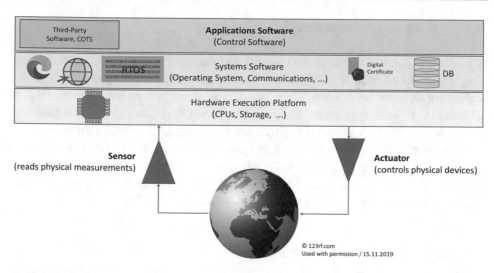

**Fig. 5.2** Software levels

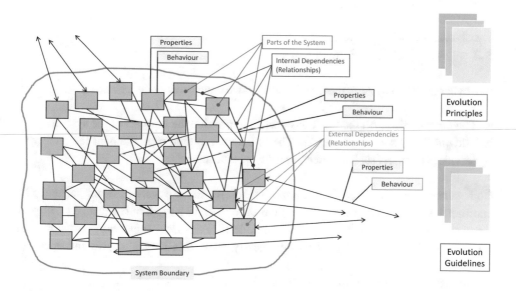

**Fig. 5.3** Software architecture

Vulnerabilities exist on all three levels. Even successful attacks through weaknesses in the processor (CPU) are known (e.g., MELTDOWN and SPECTRE, see: [https://strategynewmedia.com/spectre-vulnerability/], [NAP19], [GUT18]).

## 5.2     Software Architecture[1]

*Software architecting* is certainly *the* essential engineering discipline of our modern world! The complexity of today's large software-systems can only be managed if the system's underlying *software architecture* (Definition 5.1) is sound and remains well maintained ([Murer11], [Furrer19], [Ingeno18], [Ford20], [Bass13], [Rozanski12], [Fairbanks10], [Cervantes16], [Ford17], [Knodel16], [Hohmann03], [Langer20], [Gorton11], [Farley21]).

Of particular interest to cyber-physical systems are *embedded* software architectures ([Lacamera18], [White11], [Marwedel18], [Pohl14], [Kopetz11], [Hobbs20]) and *system-of-systems (SoS) architectures* ([Jamshidi09a], [Jamshidi09b], [Haimes19], [Bondavalli16]), supported by *domain-specific architectures*, such as *automotive architectures* ([Dajsuren19], [Staron17], [Scheid15]).

▶ **Definition 5.1: Software Architecture**
Software architecture is the fundamental organization of a system embodied in its components, their relationships to each other and to the environment, their properties, and the principles guiding its design and evolution.
IEEE Std. 1471–2000 (2000)

## 5.3     Architecture Framework

In a cyber-physical system, there are typically two categories of concerns:

a. The *functionality*: All the functions needed to provide the purpose of the system;
b. The *quality of service properties*: The functions required to implement the quality of service properties, such as safety, security, performance, etc.

Both concerns are realized in *software*. Therefore, the software-systems can become enormous, such as having over 100 million code lines in a modern car (see, e.g., https://www.visualcapitalist.com/millions-lines-of-code/)! A strict separation of concerns (Definition 5.2) is mandatory to keep complexity, changeability, and quality manageable.

▶ **Definition 5.2: Separation of Concerns**
One of the most important principles in software engineering is the separation of concerns (SOC) principle. This principle states that a given problem involves different kinds of concerns, which should be identified and separated to cope with complexity, and to

---

[1] Part of this section has been reused from the authors previous book "Future-Proof Software-Systems", Springer Vieweg Verlag, Wiesbaden, Germany, 2019. ISBN 978-3-658-19,937-1 ([Furrer19]).

achieve the required engineering quality factors such as robustness, adaptability and reusability.

Mehmet Akşit, Bedir Tekinerdoğan, Lodewijk Bergmans, 2001.

Separation of concerns has a long history in software engineering. Starting with Edsger W. Dijkstra ([Dijkstra82]), who coined the term, followed by many more (e.g., [Reade89], [Laplante07], [Akşit01]).

> **Quote**
>
> *"Separation of Concerns is one of the most important concepts that a software architect must internalize—It is also one of the most difficult to describe because it is an abstraction of an abstraction"*
>   Machine Words, 2019

Many schemes for the separation of concerns are presented in the literature. For *mission-critical cyber-physical systems,* the separation between the concern "functionality" and the concern "quality of service properties" (Fig. 5.4) is highly beneficial. Note that the separation of concerns starts with the system requirements and is strictly carried through specifications, architecting, design, and implementation.

To support the separation of concerns, an architectural framework, such as in Fig. 5.5, is inevitable (e.g., [Furrer19], [Murer11]).

The architecture framework in Fig. 5.5 has three axes:

- Horizontal architecture layers: *Functionality.* The components of functionality are uniquely assigned to one of the five layers:
  - *Business* architecture layer: Contains the enterprise architecture ([Kotusev18], [Lankhorst17]), the business processes ([Dumas18]), the business models—such as the domain model and the business object model ([Furrer19])—and the business-IT alignment mechanisms ([DeHaes20]);
  - *Application* architecture layer: Contains the application landscape architecture (= totality of all applications and their interactions), including their models, documentation, code, and evolution strategy;
  - *Information* architecture layer: Contains the definitions, models, and the organization of all information and data of the organization;
  - *Integration* architecture layer: Contains the interaction mechanisms between applications and their environments—such as the communication bus systems ([Chappell04] [DiNatale12], [Tooley13])—all the services and the management of the services and interfaces;
  - *Technical* architecture layer: Contains the elements of the runtime infrastructure, such as computing and communications hardware, ECU's, operating systems, etc.

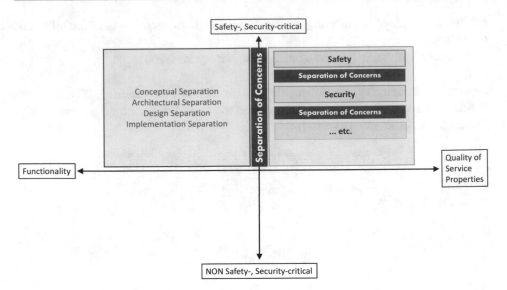

**Fig. 5.4**   Separation of concerns in a mission-critical cyber-physical system

- Vertical architecture layers: *Quality of service properties*. Each quality property should have its own explicit, documented architecture. The implementation of the quality properties (in most cases) affects *all* horizontal, functional layers in Fig. 5.5; they are visualized as vertical architecture layers;
- Hierarchy: *Granularity* of a cyber-physical system-of-systems. This axis organizes the system into the assembly hierarchy: Starting with the lowest elements—the sensor and actuators—up to the highest level—the SoS.

The segmentation into horizontal and vertical architecture layers in Fig. 5.5 is often termed "orthogonal". The orthogonality principle is a strong backing of the separation of concerns (Definition 5.2). It prevents the system designers, e.g., from implementing security functionality directly into the application layer.

An instance of the orthogonality principle is shown in Example 5.1.

---

**Example 5.1: ABS Safety Architecture**

Figure 2.1 shows the *functional architecture* of an automotive Anti-Skid Braking System (ABS). Figure 2.1 does *not* address *safety concerns*. The vertical safety architecture is added in Fig. 5.6.

Figure 5.6 overlays the following safety elements to the functional architecture of Fig. 2.1:

- Dual-redundancy on the physical part of the architecture;
- Dual-redundancy on the cyber part of the architecture;
- Arbiter for the detection of differences in the computation.

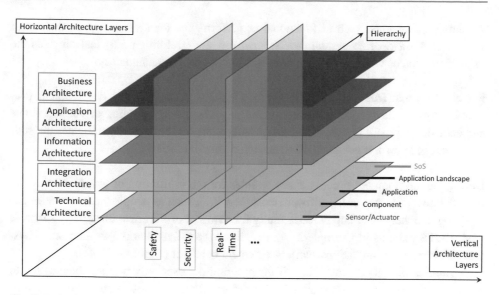

**Fig. 5.5**  Architecture framework for the separation of concerns

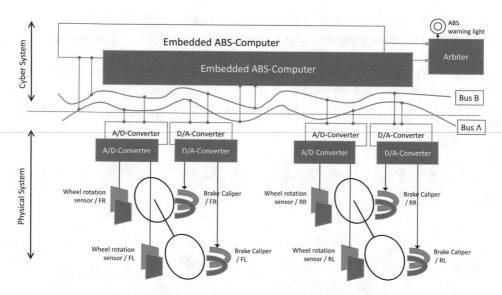

**Fig. 5.6**  ABS safety architecture

## 5.4    **Trustworthy Development Process**

### 5.4.1    **Introduction**

Figure 2.19 introduced a symbolic *safety- and security-aware development process*. It stressed the necessity to weigh safety and security requirements higher than the other systems requirements, such as functional or performance features. However, this is not sufficient: A *trustworthy system development process* (Definition 5.3) that prevents the implementation of vulnerabilities and defects in the system is mandatory.

▶ **Definition 5.3: Trustworthy Development Process** A trustworthy development process is a system development process which prevents the introduction of vulnerabilities and code defects in all phases of the system life cycle, and sufficiently mitigates the consequences of known, unavoidable, and accepted vulnerabilities

During the development process, many *sources* for the infiltration of vulnerabilities and defects into the system or the software exist (Fig. 5.7), such as people (architects, designers, programmers, etc.), buggy tools, faulty frameworks or libraries, third-party software (e.g., COTS, such as in Example 2.16), deficits in the infrastructure (weaknesses in system software, communications, etc.), or emergent behavior (Definition 2.4).

A *vulnerability* is generated, e.g., if the programmer uses only weak authentication to access confidential data, whereas a (code) defect (Definition 5.4) results from the programmer's activity, such as a mistake, omission, or violation of coding standards.

▶ **Definition 5.4: (Code) Defect** A programming mistake or a violation of coding standards that could cause a program to either produce erroneous results, fail, or stop operation

A sample of a *coding defect* is given in Example 5.2, and an example of a deficient infrastructure is described in Example 5.3.

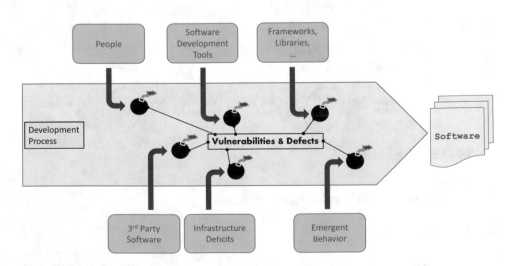

**Fig. 5.7**  Infiltration of vulnerabilities and code defects through the development process

---

**Example 5.2: A Disastrous Line of Code**

On January 15, 1990, AT&T's long-distance telephone switching system crashed: It was a disastrous event. 60'000 people completely lost their telephone service. During the nine long hours of outage, more than 70'000'000 telephone calls went uncompleted.

In early December 1989, technicians had modified the software to speed up the processing of messages. Although the upgraded code had been rigorously tested, a one-line defect was added to the software undetected. The defective software was distributed to all of the 114 switches in the network.

The software defect was a C-program that included a break statement located within an if clause that was nested within a switch clause.

In pseudocode, the responsible program fragment reads as follows:

```
1    while (ring receive buffer not empty
              and side buffer not empty) DO

2       Initialize pointer to first message in side buffer
        or ring receive buffer

3       get copy of buffer

4       switch (message)

5          case (incoming_message):

6              if (sending switch is out of service) DO

7                  if (ring write buffer is empty) DO

8                      send "in service" to status map

9                  else

10                     break

                   END IF

11             process incoming message, set up pointers to
               optional parameters

12                 break
        END SWITCH


13   do optional parameter work
```

When the destination switch received the second of the two closely timed messages while it was still busy with the first (buffer not empty, line 7), the program should have dropped out of the if clause (line 7), processed the incoming message, and set up the pointers to the database (line 11). Instead, the program dropped out of the case statement entirely because of the else clause's break statement (line 10). It began doing optional parameter work, which overwrote the data (line 13). Error correction software detected the overwrite and *shut the switch down* while it could reset. Because every switch contained the same software, the resets cascaded down the network, incapacitating the system.

> **Quote**
> *"Generally, a single line of code is not expected to have such power, and that is why software failures are hard to predict and hard to analyze"*
>     John Knight, 2012

Sources: [https://users.csc.calpoly.edu/~jdalbey/SWE/Papers/att_collapse], [Knight12]

**Example 5.3: Infrastructure Deficit**

January 14, 2021: U.S. safety regulators have asked Tesla to recall 158,000 vehicles over media control unit failures that cause the touchscreen displays to stop working, following a months-long investigation by the National Highway Traffic Safety Administration (https://techcrunch.com/2021/01/13/feds-ask-tesla-to-recall-158000-vehicles-over-failing-touchscreen-displays)

The letter of the US National Highway Traffic Safety Agency (NHTSA) to TESLA of January 13, 2021 (https://static.nhtsa.gov/odi/inv/2020/INRM-EA20003-11321.pdf) states:

*"According to Tesla, for subject vehicles equipped with the NVIDIA Tegra 3 processor with an integrated 8 GB eMMC NAND flash memory device, the eMMC NAND cell hardware will fail when reaching lifetime wear, for which the eMMC controller has no available memory blocks necessary to recover. With this failure mode, the only recovery available is a replacement of the eMMC device, achieved by physical part replacement of either the MCU assembly or visual control module subcomponent. Tesla provided information concerning the effects of MCU failure on vehicle function, including loss of rearview/backup camera and loss of HVAC (defogging and defrosting) setting controls (if the HVAC status was OFF status prior to failure.) The failure also affects the Autopilot advanced driver assistance system (ADAS) and turn signal functionality due to the possible loss of audible chimes, driver sensing, and alerts associated with these vehicle functions.*

*As discussed more fully below, ODI has tentatively concluded that the failure of the media control unit (MCU) constitutes a defect related to motor vehicle safety. Accordingly, ODI requests that Tesla initiate a recall to notify all owners, purchasers, and dealers of the subject vehicles of this safety defect and provide a remedy, in accordance with the requirements of the National Traffic and Motor Vehicle Safety Act, 49 U.S.C. §§ 30,118–30,120".*

This is an example of defective runtime infrastructure: The software design (frequent overwrites) did not respect a restriction of the eMMC NAND flash memory device and its predictable end-of-life span (see e.g., https://www.elinux.org/images/2/2c/Wear_Estimation_for_Devices_with_eMMC_Flash_Memory.pdf).

## 5.4.2   Defect Avoidance and Defect Elimination

Two categories of *deficiencies* endanger safety and security (Fig. 5.8):

1. *Safety and security vulnerabilities*: Vulnerabilities to failures or threats introduced through weak or inadequate risk mitigation measures. This issue is addressed by the safety- and security-aware development process (Definition 2.22);
2. *Code defects*: Potential software execution hazard introduced through programmer mistakes, negligence, or disregard of coding standards, architecture principles, or good programming practices (Definition 5.4);

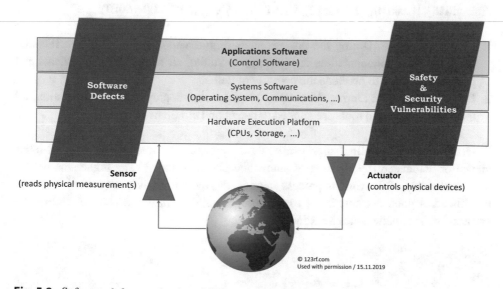

**Fig. 5.8**   Software defects and vulnerabilities

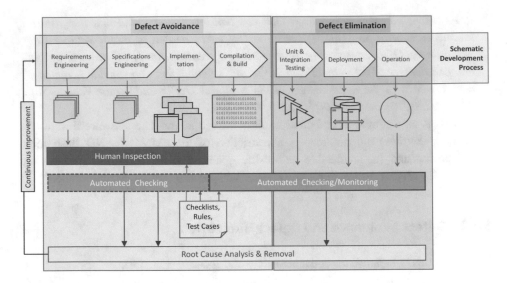

**Fig. 5.9** Defect avoidance and defect elimination

Code defects are hidden risks for the software's execution and can cause failures, misbehavior, or enable successful attacks. Code defects must be eliminated to the highest possible degree to render a cyber-physical system trustworthy.

In addition to the safety- and security-aware development process, the development
process must ensure the minimum (Objective: *Zero*) number of defects in the code produced. Such a process has two phases (Fig. 5.9, [Knight12], [Martin08], [Tornhill15],
[Tornhill18], [Laski10], [Kourie12], [NIST800-160A], [NIST800-160B]):

1. Defect *avoidance*: During the process steps requirements engineering, specification
   management, implementation, and compilation/build, all reasonable instruments and
   practices are used to preclude the *creation* of defects;
2. Defect *elimination*: During the process steps testing, deployment, and operation, all
   the sound means and tools are applied to detect and eliminate defects.

Defect avoidance and elimination can be done manually (by inspection, reviews, interviews) or automated (by all sorts of automated checking tools). The higher the degree
of formalization of the various process steps is, the more effective, automated tools can
be utilized. A good example is the use of strict *coding standards*, which can entirely be
enforced by automatic code checking.

> **Quote**
> *"The best way to have defect-free software is to avoid defects in the first place,*
> *rather then put them in and remove them later"*
>   Bruce Powel Douglass, 2016

### 5.4.3   Coding Standards

A powerful weapon to fight code defects are *coding standards* (Definition 5.5).
Many excellent coding standards exist (e.g., [Seacord13], [Seacord14], [Bagnara18],
[Harrison19], [Marshall09], [Chess07]).

▶ **Definition 5.5: Coding Standard**
Coding standards are collections of coding rules, guidelines, and best practices to support the production of cleaner code.
  https://www.perforce.com/resources/qac/coding-standards

For any organization producing trustworthy cyber-physical systems, it is absolutely
indispensable to use and strictly enforce a suitable coding standard! Also, an automatic
tool must check adherence to the coding standard, and any deviation must be immediately corrected. A well-known coding standard is presented in Example 5.4.

> **Quote**
> *"Commonly exploited software vulnerabilities are usually caused by avoidable software defects"*
>     Robert C. Seacord, 2014

**Example 5.4: CERT C Coding Standard**

*"The SEI CERT C Coding Standard, 2016 Edition provides rules for secure coding
in the C programming language. The goal of these rules and recommendations is
to develop safe, reliable, and secure systems, for example, by eliminating undefined
behaviors that can lead to undefined program behaviors and exploitable vulnerabilities. Conformance to the coding rules defined in this standard is necessary (but not
sufficient) to ensure the safety, reliability, and security of software systems developed
in the C programming language. It is also necessary, for example, to have a safe and
secure design. Safety-critical systems typically have stricter requirements than are
imposed by this coding standard, for example, requiring that all memory be statically
allocated. However, the application of this coding standard will result in high-quality
systems that are reliable, robust, and resistant to attack (Source: CERT-C16)."*

> **Quote**
> *"The most important thing to do with coding style standards is pick one"*
>     Mark Harrison, 2019

The CERT C Coding Standard ([Seacord14]), 2016 edition ([CERT-C16]), has 528 pages of invaluable coding advice. All rules are explicitly formulated and supported by non-compliant and compliant code fragments, such as (extract from [Seacord14], p. 43):

**DCL41-C. Do not declare variables inside a switch statement before the first case label**

If a programmer declares variables, initializes them before the first case statement, and then tries to use them inside any of the case statements, those variables will have scope inside the switch block but will not be initialized and will consequently contain indeterminate values.

**Non-compliant Code Example:**

This *non-compliant code example* declares variables and contains executable statements before the first case label within the switch statement:

```
#include <stdio.h>

extern void f(int i);

void func(int expr) {
      switch (expr) {
            int i = 4;
            f(i);
      case 0:
            i = 17;
            /* Falls through into default code */
      default:
            printf("%d\n" , i);
  }
 }
```

Another well-known, significant coding rule standard is the *MISRA coding standard* ([Bagnara18], https://www.misra.org.uk/Publications/tabid/57/Default.aspx). MISRA C is a formal set of guidelines for developing software using the C programming language. MISRA was initially developed in 1990 for the automotive sector but is now adopted across most safety- and security-critical industry sectors. The Motor Industry Software Reliability Association (MISRA, UK) leads the development of the MISRA C standards. The MISRA language documents ("The Guidelines") are widely used in the development of critical software-systems, especially when the requirements of a safety or security quality standard must be met ([MISRA16]).

> **Quote**
> *"Compliance with MISRA guidelines must be an integral part of the code development phase and compliance requirements need to be satisfied before code is submitted for review or unit testing"*
>   MISRA Guideline, 2016

An essential element of good code quality is *automated coding rules checking*. A Rule Checker is a static program analyzer that automatically checks C or C++code for compliance with MISRA rules, CERT coding standards, or other coding guidelines.

Two powerful instruments to achieve good code quality are modeling and formal languages.

## 5.4.4 Good Programming Practices for Safety and Security

In addition to coding standards—which are language-specific—a large number of good programming practices exist in the literature. These recommendations are language-independent, as shown in Example 5.5.

**Example 5.5: Good Programming Practices**

A few *examples* of good programming practices (Quoted from: [NASA04]) are:

a. Check variables for reasonableness before use*: If the value is out of range, there is a problem—memory corruption, incorrect calculation, hardware problems (if sensor), or other problem (Possible mitigation: Design by contract, contract checking);*

b. Use execution logging, *with independent checking, to find software runaway, illegal functions, or out-of-sequence execution. If the software must follow a known path through the components, a check log will uncover problems shortly after they occur;*

c. Come-from checks*: For safety-critical components, make sure that the correct previous component called it and that it was not called accidentally by a malfunctioning component;*

d. Test for memory leakage*: Instrument the code and run it under load and stress tests. See how the memory usage changes, and check it against the predicted usage;*

e. Use read-backs to check values*: When a value is written to memory, the display, or hardware, another function should read it back and verify that the correct value was written.*

An extensive base of literature on "good practices software recommendations" exists (Google search returns 1′430′000′000 results [14.2.2021]). Many of these recommendations have proven their practical value in many applications. Because of this information treasure's colossal magnitude, it is impossible for the individual programmer to browse and understand it. Therefore, the organization should extract the relevant recommendations and bundle them into a specific, comprehensive *internal coding standard* (e.g., [Bishop18]).

## 5.4.5    Modeling and Formal Languages

### 5.4.5.1  Models

The world of software-systems is rich and complicated. Software systems are, therefore, difficult to understand, build, and evolve. People, especially physicists and engineers, were looking for instruments to document and understand their systems better for a long time. The most powerful instrument to understand systems is *models* (Definition 5.6). Models represent a reduced, focused, and abstracted view of the part of the world of interest. In engineering, models must have a strong syntax and unambiguous semantics. Numerous modeling languages have been introduced, many of them based on a formal logic foundation, and most of them have a graphical representation. Models have become probably the most important and powerful instruments for the systems development process (see, e.g., [Brambilla17], [Volter06], [Pohl12], [Feiler12], [Starr17], [Friedenthal17], [Pastor07], [Borky19], [Micouin14], [Kelly08]).

▶ **Definition 5.6: Model**
A model is an abstraction of an aspect of reality (as-is or to-be) that is built for a given purpose. In engineering, models have well defined syntax and semantics and are often based on a system of logic.
    Benoit Combemale, 2017.

The high value and the enormous impact of models in engineering (see, e.g., [Boerger18], [Lieberman06], [Borky18]) is based on the "4 C's" in Table 5.1.

**Quote**
*"Producing software implementations without the help of models is a considerably more complex task, that is, it leads to significant accidental complexity"*
    Benoit Combemale, 2017

**Table 5.1** Four «C's» of modeling

| # | Model Benefit | Description |
|---|---|---|
| $C_1$ | Clarity | The concepts, relationships, and their attributes are unambiguously defined, described, and understood by all stakeholders |
| $C_2$ | Communication | The model truly and sufficiently represents the real world's critical properties to be mapped into the IT solution. All stakeholders have contributed to and understood the model |
| $C_3$ | Commitment | All stakeholders have accepted the model, its representation, and the consequences (stakeholder's agreement) |
| $C_4$ | Control | The model is used to assess specifications, architecture, design, implementation, reviews, and system evolution. Any deviation from the model must be justified and fed back into the model (round trip engineering) |

A complete model defines:

i.   The unambiguous, precise terminology ([Hewitt19]);
ii.  The structure of the system;
iii. The behavior of the system (including error and fault handling),

and has:

i.   An underlying mathematical construct (e.g., set theory);
ii.  An unambiguous syntax (= Notation);
iii. Well-defined semantics;
iv.  A style guide and modeling guidelines;
v.   Preferably a graphical representation.

Today's software engineering toolbox offers a large number of different modeling techniques, such as *UML* (Example 5.6, [Miles06], [Drusinsky06], [Fowler03], [Lamsweerde09], [Pilone05]), *SysML* ([Holt19], [Weilkiens08], [Delligatti13]), *BPMN* ([Silver17], [Allweyer16], [Walters20]), *Timed Automata* ([Baier07], [Penczek10]), *Petri Nets* ([Penczek10], [Reisig13]), *Temporal Logic* ([Berard10], [Kröger08]), *graphs* ([Heckel20]), and many more.

### Example 5.6: CPSSoS UML Structure Diagram

Figure 5.10 shows the *conceptual structure diagram* of a cyber-physical system-of-systems (Top-level diagram). It depicts all the CPSSoS concepts and their relationships. The modeling language used is pure UML 2.0.

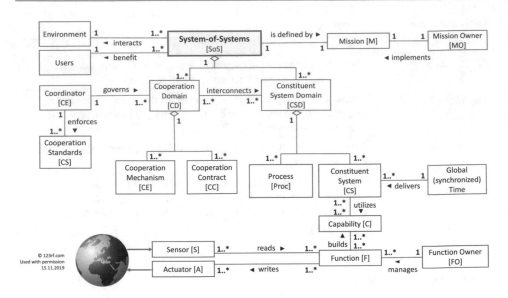

**Fig. 5.10**  Cyber-physical system-of-systems—UML structure diagram

A modeling language and notation specifies syntax and semantics, leaving the modeling engineer unrestricted freedom of expression—which is detrimental to clarity and precision. Therefore, the modeling engineer needs a *style guide* and *modeling guidelines* (Note that these are often company-specific). The style guide and the modeling guidelines restrict and standardize the individual modeling activity and thus significantly improve *model quality* (e.g., [Pitschke20], [Long14], [Ambler05]).

### 5.4.5.2  Model Checking

Contemporary IT systems are huge, complicated, and complex. *Models* provide a powerful instrument to describe, define, communicate, implement, and control these systems. The precondition is that these models are sufficiently complete, consistent, adequately formalized, and undoubtedly convincing—quite a challenge!

An impactful step is *model checking* (Definition 5.7, [Mitra21], [Berard10], [Baier08], [Clarke18a], [Clarke18b]).

> **Quote**
> *A main challenge for the field of computer science is to provide formalisms, techniques, and tools that will enable the efficient design of correct and well-functioning systems despite their complexity".*
> Christel Baier, 2007

▶ **Definition 5.7: Model Checking**

A tool-supported, enabling methodology to ensure the validity and dependability of an information system model. Model checking has two parts:

*Validation*: Validation is the process of determining the degree to which a model and its associated data are an accurate representation of the real world from the perspective of the intended uses of the model, i.e., compared to the requirements and specifications, including completeness and consistency (adapted from: https://www.mitre.org);

*Verification*: Verification is the process to ascertain that the model fully adheres to all the formal constraints, such as the metamodel, the applicable ontology, the relevant domain models, the syntax, the notation, the naming conventions, etc.

Model checking has two parts (Fig. 5.11):

*Validation*: Validation answers the question: "Are we building the right system?". It means conducting a check of the model(s) against the requirements and the specifications. The model(s) must exhaustively and unambiguously incorporate all commitments and explicitly address completeness and consistency;

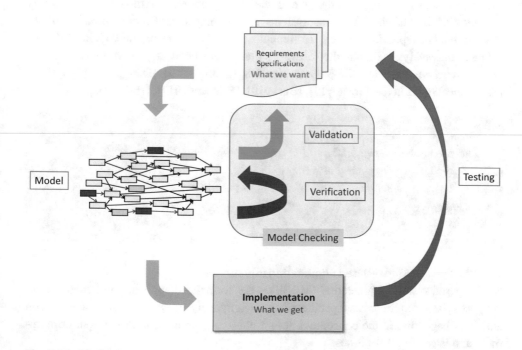

**Fig. 5.11**  Model checking

*Verification*: Verification answers the question: "Are we building the system right?". It checks the model against all formal (or semi-formal) modeling standards applicable to the modeling notation and methodology.

Model checking implementations have many possibilities, ranging from expert reviews to fully automated model checking. The depth and extent of automation of model checking depend on the *degree of formalization*, i.e., on the models' underlying concepts' mathematical rigor. The more expressive the model's mathematical foundation is, the more significant automatic model testing can be done (see, e.g., [Cao20], [Debbabi10]).

Unfortunately, highly formalized modeling is difficult, expensive, and requires highly specialized engineers. Therefore, an organization must carefully choose its modeling techniques, methodologies, and model checking procedures. At the time being (2022), only highly safety-critical systems are constructed based on formal models. Generally, an organization should aim at the highest number, and most in-depth degree of formalization, which it can (and will) afford, especially concerning the safety and security of the cyber-physical systems (e.g., [Cao20], [Cofer14]).

### 5.4.5.3 Model-Based Engineering

The continuously rising complexity and the more and more stringent dependability demands of the stakeholders force systems engineering toward more *formal methods*. Formal methods provide Clarity, Communication, Commitment, and Control (Table 5.1) and massively ameliorate the software's quality attributes. A powerful technology is *model-based engineering* MBSE (Definition 5.8, [Delligatti13], [Douglass15], [Dori16], [Weilkiens08], [Holt19], [Tockey19], [Borky19], [Friedenthal17], [Hart15]).

> **Quote**
> *"The people who design and build houses gave up trying to describe them in natural language over one hundred years ago. What makes you think you can successfully describe something that's orders of magnitude more complex using natural language?"*
> Steve Tockey, 2019

▶ **Definition 5 8: Model-Based Engineering (MBSE)**
Model-based systems engineering (MBSE) is the formalized application of modeling to support system requirements, specifications, design, analysis, verification, and validation activities beginning in the conceptual design phase and continuing throughout development and later life cycle phases.
INCOSE SE Vision 2020 (INCOSE-TP-2004-004-02, Sep 2007)

Model-based engineering relies on a multitude of models, such as in Fig. 5.12. The MBSE uses the *construction models*, which guide the life cycle, and the support models, ensuring specific software quality properties or defining process functionalities (Fig. 5.12).

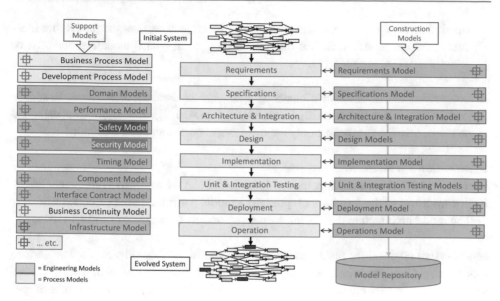

**Fig. 5.12**  MBSE model landscape

**Quote**

*"MBSE's key aim is the separation of concerns into models, so that people can work with models that record all and only the information they need to make their decisions"*
  Perdita Stevens, 2019

The particular *challenges* in MBSE for real, large, and complex systems are:

- Model size;
- Model consistency.

Models of today's cyber-physical systems control software are enormous, both in the number of elements and relationships. This characteristic is known as the *model explosion*. Such large models are extremely difficult to present, understand, and maintain! Therefore, three measures must be taken:

1. *Partitioning*: Divide the cyber-physical system into coherent, manageable *domains* based on a domain model (Definition 2.23). Use maximal cohesion within the domains and minimum cohesion between the domains as partitioning criteria;
2. *Model Hierarchy*: Structure the models hierarchically. Start with a coarse top-level model and refine it stepwise to lower levels (e.g., [Furrer19], [Vykhovanets19]). Large models require massive tool support;

3. *Views:* Provide specific representations of one or more aspects of the model that illustrates how the model addresses one or more concerns—such as safety, security, or performance (Definition 5.11).

Before starting the organization's modeling endeavor, the *model's architecture* should be defined (Definition 5.9, [Rozanski12], [Clements10], [ISO42010]). An early, diligent definition of the model's architecture dramatically improves model quality and maintainability.

▶ **Definition 5.9: Model Architecture**
Hierarchical framework defining the top-level model and its successive refinement levels. Each constituent of the framework represents a model partition, based on coherence.
 The model architecture defines the concept taxonomy (or ontology) and all views.

Model-based engineering makes use of a landscape of models (Fig. 5.12). The various models cover one specific aspect of the total system, such as the architecture model, safety model, security model, deployment model, etc. Therefore, each engineer with specific responsibility can work with his dedicated model and is not distracted by the complete system's overwhelming information.
 Unfortunately, working with different models raises the danger of *model inconsistency*: The contents of the individual models may become different, divergent, redundant, or even contradictory—with possibly very dire consequences for the cyber-physical system! Therefore, maintaining *model consistency* is of most significant importance in MBSE (Definition 5.10, [Miret14], [Stevens20], [Nuseibeha01]).

▶ **Definition 5.10: Model Consistency**
Model consistency means that all model elements, their relationships, and attributes of the different models are at all times correct, synchronized, and never dissenting.
 Eventual consistency is a characteristic of distributed models such that the value for a specific model element will, given enough time without updates, be consistent across all models.

> **Quote**
> *"Based on case studies at NASA, inconsistency management should be a core activity in software development, and the main danger arises not from inconsistency between models but from unrecognised inconsistency"*
>  Bashar Nuseibeha, 2001

Several techniques for ensuring model consistency exist (e.g., [Giese09]). One proven approach is to generate and maintain one single *master model* and provide the specific models via model *views* (Definition 5.11, [Clements10], [Rozanski12], [SEI06], [IEEE1471], [ISO42010]).

▶ **Definition 5.11: View**

A view is a representation of one or more aspects of a model that illustrates how the model addresses one or more concerns—such as safety, security, or performance—used by one or more of the stakeholders.

Adapted from IEEE Standard 1471.

> **Quote**
> *"A complex system is much more effectively described (modelled) by a set of inter-related views, which collectively illustrate its features and quality properties and demonstrate that it meets its goals, than by a single overloaded model"*
> Nick Rozanski, 2012

### 5.4.5.4  Formal Methods and Languages

*Formal methods* are an approach to employ mathematical knowledge to enhance the human-centric system/software development process. Formal methods have a very long history ([Cohen95]). In 1930, formal methods (Definition 5.12, [Nielson19], [O'Regan17], [Monin13], [Fisher11], [Serrano20], [Liu04], [Boulanger12], [Mills09], [Cao20], [Rouff06]) were introduced into computer science. Fortunately, the necessary mathematical frameworks were available at that time.

▶ **Definition 5.12: Formal Method**

Formal methods consist of notations and tools with a mathematical basis that are used to specify the requirements of a computer system unambiguously and that support the proof of properties of this specification and proofs of correctness of an eventual implementation with respect to the specification.

Michael Hinchey, 2006 in [Rouff06].

The arsenal of formal methods is large and growing. Formal methods are an active research area and of increasing importance for safety and security in cyber-physical systems (e.g., [Gnesi12], [Boulanger12], [Nanda19], [Artho16], [Cremers12], [Davenport19], [Kulik18], [Knight12], [Ábrahám16]).

Formal methods can (must) be applied to all phases of the software development process (Fig. 5.13). First, the requirements are formalized, leading to the *formal specification*. After the verification, a formal specification, e.g., in Z-notation, can be transformed into program code using several refinement cycles (e.g., [Spivey89]). Note that correct and complete error, fault, and exception handling is mandatory for safety and security!

The highest benefit of formal methods is undoubtedly in the specification phase. It serves as a single, unambiguous, and dependable source for all stakeholders involved in constructing the software-system. Unfortunately, even a formal specification is no guarantee that the system's requirements are faultless: The requirements engineering

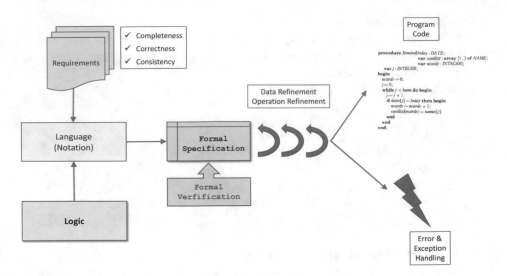

**Fig. 5.13**  Formal method

team must ensure the *completeness*, *correctness*, and *consistency* of the requirements ([Dick17], [Lamsweerde09], [Robertson13], Example 5.7).

---

**Example 5.7: Requirements Error**

A requirements error in an F-16 fighter plane avionics program would cause the plane to flip upside down whenever it crossed the equator. Fortunately, this bug (correctly programmed according to the flawed requirements) was found before it caused an accident. The requirements did not foresee an F-16 crossing the equator.

Adapted from: Edward Yourdon ([Yourdon03]), 2003.

---

Any formal method needs a *formal notation* (or *formal language*). Many formal languages—especially for the formal specification of systems—have emerged in computing history. The industrially most significant are listed in Example 5.8.

---

**Example 5.8: Formal Languages -**

*Ada Language*

---

**Quote**

*"Computers do not make mistakes or so we are told. However, computer software is written by, and hardware systems are designed and assembled by, humans, who certainly do make mistakes"*

Michael Hinchey, 2006 in [Rouff06]

*"Ada is a state-of-the-art programming language that development teams worldwide are using for critical software: From microkernels and small-footprint, real-time embedded systems to large-scale enterprise applications, and everything in between (https://www.adacore.com/about-ada)"*.

"Ada" is not an acronym: It was chosen remembering Augusta Ada Lovelace (1815–1852), a mathematician who is sometimes recognized as the world's first programmer, building on Charles Babbage "Analytical Engine" ([Hollings18]).

The development of the Ada language had several phases:

- Ada was initially developed in the early 1980s at CII-Honeywell-Bull in France as a mandate of the US DoD (Version known as Ada 83);
- Revised and enhanced in an upward compatible release in the early 1990s, by Intermetrics in the U.S. Known as Ada 95;
- Internationally standardized by ISO/IEC. Minor revision to Ada 2005;
- The most recent version of the language standard is Ada 2012, revised 2018 ([ISO8652)].

Ada is a programming language supporting modern software constructs, such as strongly-typed, structured, object-oriented, generic, distributed, and concurrent programming. Also, it has libraries for real time and is supported by mature, certified development tools (Language: [ADACore16], [Barnes14], [Key15] / Applications: e.g., [McCormick11], [ISO15942], [Burns16], [McCormick15]);

**Formal Languages** - *Z Language*
*"Z is a model-oriented, formal specification language, based on Zermelo-Fränkel axiomatic set theory and first-order predicate logic. It is a mathematical specification language, with the help of which natural language requirements can be converted into mathematical form"* ([Ruhela12]).

Z (pronounced "zed") was developed in several steps:

- In 1977, Jean-Raymond Abrial, Steve Schuman, and Bertrand Meyer published the original Z language (in: [McKeag80]);
- Following the 1980s, a number of Z language derivates and extensions were introduced;
- Z was standardized by ISO in 2002 ([ISO13568]).

Z is, in fact, two languages: 1) The Z *specification* notation language and 2) the Z *schema* language. The schema language offers comprehensive, legible structuring of large specification documents, schema composition, and schema reuse (Language: [Jacky97], [Ince93], [Ruhela12], [Woodcock96], [Zeller09] / Applications: e.g., [Spivey89], [ISO13568]).

*Formal Languages - B Method*
*"The B-method provides an Abstract Machine Notation for writing system specifications and rules for refinement of specifications into programs ([Liu04])"*
The *B-method* evolved in several stages:

- Initially developed in the 1980s by Jean-Raymond Abrial ([Abrial05], [Cansell03], [Schneider01]);
- Extended with the central concept of "events", called *Event-B* ([Hoang13], [Abrial10], [Cansell06], [Collazos20]);

The B-method and especially Event-B are used in industrial, safety-critical systems ([Boulanger12], [Boulanger13], [Boulanger14], [Su11], [Dieumegard17], [Damchoom09], [Jarrar18]).

From Example 5.8, it becomes apparent that formal languages are demanding with respect to effort, people's qualifications, and tools. Consequently, formal languages are used only in applications where the software-system's required quality properties justify the time and effort. This is increasingly the case, e.g., in safety-critical applications, real-time applications, applications requiring certification, and communications protocol development.

> **Quote**
> *"Formal methods have presented the most reasonable, rigorous, and controllable approach to software development so far, at least theoretically, but their application requires high skills in mathematical abstraction and proof"*
>     Shaoying Liu, 2004

*Model-Driven Dependability Assessment of Software-Systems*
Ideally, the dependability—including safety and security—would be directly assessable from the system's or software's model. This active research area will soon deliver valuable support for constructing safe and secure software (e.g., [Bernardi13], [Masmali19], [Lucio14]).

## 5.5    Safe Software

The ultimate objective of the systems engineering efforts is a safe system. In this monograph, the focus is on software, i.e., on *safe software* (Definition 5.13, [Boulanger13], [Gullo18], [Ericson13], [Kleidermacher12], [Hobbs20], [NASA04], [Rierson13]).

▶ **Definition 5.13: Safe Software**
Executable software artifact which only causes or contributes to a safety accident with a
known, recognized, documented, and accepted residual risk in case of hardware or software failures.

In case of a hardware or software failure, the executable software artifact will minimize the consequences of the failure (whenever possible go into degraded operation and
finally into a fail-safe state).

Figure 5.14 recalls the context for safe software. It spans:

- A system development process that produces trustworthy *applications software*;
- System *operation processes* that allow the stakeholders to use the systems' functionality and data;
- The application software providing the functionality and data: The application software heavily relies on third-party software (such as COTS, DB management systems,
  authentication/authorization packages, libraries, etc.;
- The *execution platform* required for running the software. It contains all the hardware
  (such as CPUs, video integrated circuits, disk drives, memory, communications, etc.)
  and the system software (Operating systems, device drivers, development tools, version management, etc. and an enormous amount of firmware);
- The *people* creating, evolving, operating, and using the system.

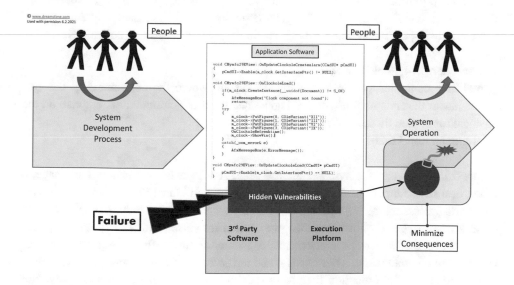

**Fig. 5.14**   Safe software context

Unfortunately, all the elements of Fig. 5.14 in any system will embody *vulnerabilities*—some known (Hopefully correctly mitigated) and some unknown (Hidden vulnerabilities). Any failure may strike a vulnerability and cause a *safety accident*. Therefore, dependable, safe software results from:

1. A trustworthy development process that prevents vulnerabilities;
2. Mitigation measures which confine the impacts of failures during the operation process.

---

**Quote**
*"Software is different from hardware: It has unique properties and characteristics that make it more difficult to thoroughly analyze and evaluate"*
    Clifton A. Ericson, 2013

---

### 5.5.1   Time

In cyber-physical systems, *time* plays an important role. Many cyber-physical systems are *real-time systems*, where most computing has to be completed within stringent, unforgivable timing deadlines (Definition 5.14). Overrunning such a timing deadline generates failures and may cause the system's misbehavior, leading to accidents. Designing, implementing, and operating real-time systems is a challenging task ([Kopetz11], [Furia12], [Manna12], [Gomaa16], [Cooling19], [Carrillo20], [Kopetz22]). Real-time systems require *timing integrity* (Principle 12.7), i.e., their implementations must strictly respect the system's timing deadlines under all operating conditions.

▶ **Definition 5.14: Real-Time System**
A real-time computer system is a computer system where the correctness of the system behavior depends not only on the logical results of the computation, but also on the physical time when these results are produced
    Hermann Kopetz, 2011

Figure 5.15 introduces the *system response time*: A trigger event, such as a message, a timer, or a sensor input change, arrives at the system and launches its reaction. The system starts processing the information and generates the computed output. A specific time elapses from trigger event to reaction—the system response time. The system response time is the sum of *hardware latency* (time consumed in the hardware, e.g., data conversion time or communications bus latency) and the *software execution time*.

   Unfortunately, the execution time for a specific program can vary considerably. Not only does the processor architecture (e.g., multicore processing, caches, pipelines, branch

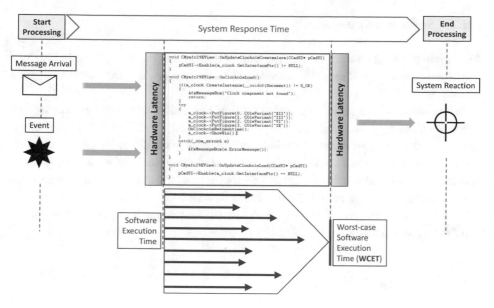

**Fig. 5.15**  System response time

prediction, higher-level interrupts) influence the time the program needs to complete the calculation, but the transversion time through various paths of the program differs massively. Each branch or decision in the program generates a new path. Therefore, the number of possible paths through the program will soon count in millions or more (Fig. 5.15).

Concerning software, a crucial concept is the *worst-case execution time* (WCET, Definition 5.15, [Wilhelm08], [Thiele04], [Erciyeş19], [Buttazzo11], [Kopetz11], [Cooling19], [Gomaa16]). The worst-case execution time is the time the program needs to run through its *longest* path.

▶ **Definition 5.15: Worst-Case Execution Time (WCET)**
Worst-case execution time (WCET) is a software metric that determines the maximum length of time a task or set of tasks requires on a specific hardware platform in a specific application context. The worst-case scenario is a crucial consideration for hard real-time systems, which have a non-negotiable deadline.
    https://searchsoftwarequality.techtarget.com/definition/worst-case-execution-time-WCET

Regrettably, the precise determination of the WCET of large programs is very exacting (e.g., https://ti.tuwien.ac.at/cps/teaching/courses/real-time-systems/slides/rts06_wcet_analysis-1.pdf, Last accessed: 31.1.2021). However, a number of sufficiently reliable methods and tools exist to estimate the WCET (e.g., [Wilhelm08], [Thiele04], [Erciyeş19], [Buttazzo11], [Franke18], [Engblomxy03]).

### 5.5.2   Software Categories

An important categorization for creating safe and secure software is shown in Fig. 5.16: Execution platform, infrastructure services, third-party software, and in-house application software.

In the stone age of software creation (1960), the application software accessed the execution platform—i.e., processor and memory—directly, using assembly language. In the 1980s, direct access to the execution platform was gradually replaced by infrastructure services, and vendors started to offer specialized third-party software. This tendency is continuing, and the execution platform disappears under more and more powerful services, and the usage of specialized third-party software increases rapidly (2022).

The software categorization of Fig. 5.16 substantially impacts the cyber-physical systems' safety and security.

### 5.5.3   In-House Software

For many organizations, their *in-house developed software* (Definition 5.16) is the distinguishing factor on the market, giving them their competitive advantage (e.g., [Harris17]). Therefore, such organizations develop much software with their own staff.

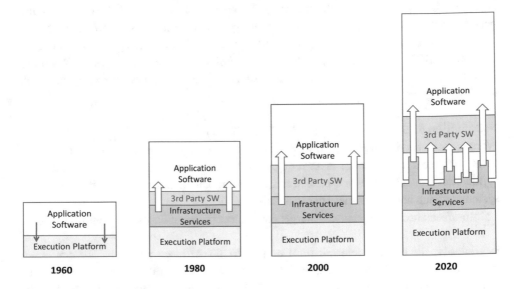

**Fig. 5.16**   Software categories

▶ **Definition 5.16: In-House Software** Software developed under the complete govern-ance of the organization, either by employees or by contractors. Full authority about all specifications (functional and quality properties). Seamless integration into the exist-ing system possible. All policies, principles, coding standards and good programming recommendations can be enforced. Fault correction procedures fully controlled by the organization. All rights belong to the organization.

Concerning *safety*, in-house software is entirely governed by the organization, from requirements to deployment. Therefore, the organization can enforce all the provisions for safe software and ultimately control/accept the resulting residual safety risks.

## 5.5.4   Third-Party Software

For many organizations, the use of *third-party software* (Definition 5.17) is beneficial: Many advantages support this decision, such as cost, maturity, lack of expertise, or fruit-ful collaborations. third-party software is available on all software stack levels, from the lowest level (Execution platform) to the highest level (Application software).

▶ **Definition 5.17: Third-Party Software** Software acquired form a third-party vendor and integrated into the organization's IT system or product. Very limited influence on the functionality, quality properties (i.e., safety and security), and evolution of the third-party software (Restricted to contractual agreements).

For many functions, third-party software is indispensable, such as operating systems, specialized libraries, compilers, database management systems, communications stacks, authentication/authorization software, system management software, etc. Component-based software engineering ([Tiwari20], [Gupta11]) is based on the sourcing of third-party components.
   Third-party software introduces *safety risk,* which is not under the acquiring organiza-tion's control (e.g., Example 2.16, Example 4.20).

> **Quote**
> *"Acquiring software, whether off-the-shelf, previously created, or custom made, carries with it a set of risks and rewards that differ from those related to software development"*
>    [NASA04], 2004

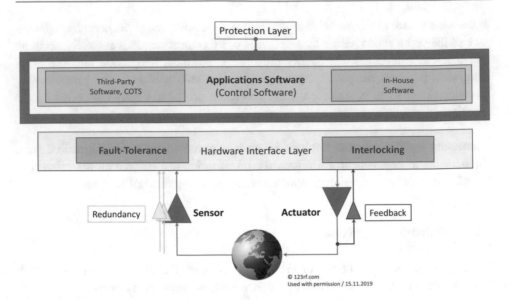

**Fig. 5.17**   Third-party software safety assurance measures

To manage the safety of third-party software, the acquiring organization must (Fig. 5.17):

1. Define and enforce a procurement *policy* (e.g., [NASA98]) and a reliable *procurement process* (e.g., [ITRB99]): The safety considerations and risk assessment are shifted to a large part into the quality of the evaluation process;
2. Use due diligence to assess the compatibility of the third-party software with the existing's systems architecture (e.g., the introduction of unmanaged data redundancy): To force a third-party misfit into an existing, stable system is a sure recipe for software evolution problems and operational difficulties;
3. Protect the *operating environment* of the third-party software: Use all justifiable safeguards, such as perimeter protection, runtime monitoring, crisis management, etc.;
4. Encapsulate the third-party software into safe hardware: Use redundancy, interlocking, fault-tolerance, etc., to absorb malfunctions of the control software.

### 5.5.5   Execution Platform

A safe *execution platform* (Definition 5.18, Fig. 5.8) is indispensable for safety-critical cyber-physical systems. In addition to safe hardware—provided, e.g., through fault-tolerance—*safe operating systems* (https://www.embedded.com/safety-critical-operating-systems), certified system development tools, libraries, compilers, simulators, test environments, requirements traceability tools, etc. are essential ([Hobbs20], [Marwedel18], [Cooling19], [Cooling21a], [Cooling21b]). Many adequately safe execution platforms are provided by industry associations (such as AUTOSAR, e.g., [Scheid15]) or by dependable vendors.

▶ **Definition 5.18: Execution Platform**  Hardware, communications, firmware, and systems software required for running the applications software. It contains all the hardware (such as CPUs, video integrated circuits, disk drives, memory, communications, etc.) and the system software (Operating systems, device drivers, development tools, version management, etc.) and an enormous amount of firmware

---

**Quote**

*"No matter what we do, our programs will sometimes fail anyway. There are bugs in our compilers, operating systems, window managers, and all the other system software on which we depend."*
   Andreas Zeller, 2009

---

However, additional provisions are essential in many applications, either because of safety requirements, requested by industry standards, or demanded by a certification authority, as described in Example 5.9.

---

**Example 5.9: Embedded Systems Infrastructure Safety**

Safety-critical software is often implemented in networked, distributed *embedded systems* ([White11], [Paret07], [Staron17], [Zaman15], [Dajsuren19], [Kopetz11]). This infrastructure consists of microprocessors, memory, networks, interfaces, etc. To mitigate the consequences of infrastructure defects, a number of *safe programming practices* exist (e.g., ([NASA04], [NASA20], [Brown98], [Stewart99]).

The following (incomplete) examples (Quoted from: [NASA04]) list some of the common practices:

a. CPU self-test: *If the CPU becomes partially crippled, it is important for the software to know this. Cosmic Radiation, EMI, electrical discharge, shock, or other effects could have damaged the CPU. A CPU self-test, usually run at boot time, can verify the correct operations of the processor. If the test fails, then the CPU is faulty, and the software can go to a safe state;*

b. Guarding against illegal jumps: *Filling ROM or RAM with a known pattern, particularly a halt or illegal instruction, can prevent the program from operating after it jumps accidentally to unknown memory. On processors that provide traps for illegal instructions (or a similar exception mechanism), the trap vector could point to a process to put the system into a safe state;*

c. ROM tests: *Prior to executing the software stored in ROM (EEPROM, Flash disk), it is important to verify its integrity. This is usually done at power-up, after the*

*CPU self-test, and before the software is loaded. However, if the system has the ability to alter its own programming (EEPROMS or flash memory), then the tests should be run periodically;*

d. Watchdog Timers: *Usually implemented in hardware, a watchdog timer resets (reboots) the CPU if it is not "tickled" within a set period of time. Usually, in a process implemented as an infinite loop, the watchdog is written to once per loop (Example 4.9);*

e. Guard against Variable Corruption: *Storing multiple copies of critical variables, especially on different storage media or physically separate memory, is a simple method for verifying the variables. A comparison is made when the variable is used, using two-out-of-three voting if they do not agree, or using a default value if no two agree. Also, critical variables can be grouped, and a CRC used to verify they are not corrupted;*

f. Stack Checks: *Checking the stack guards against stack overflow or corruption. By initializing the stack to a known pattern, a stack monitor function can be used to watch the available stack space. When the stack margin shrinks to some predetermined limit, an error processing routine can be called that fixes the problem or puts the system into a safe state.*

A rich literature for *good programming practices* for safety exists. Any organization involved in cyber-physical systems needs to understand these practices and distill the relevant ones into the company-specific programming standards.

> **Quote**
> *"Many designers and programmers refuse to listen to the experiences of others, claiming that their applications are different, and of course, much more complicated"*
> *David B. Stewart, 1999*

## 5.6   Secure Software

Most of the severe security incidents in the last years (see e.g., https://cyware.com/category/breaches-and-incidents-news) were due to insecure software. Therefore, an essential goal of cyber-physical systems engineering is to build *secure software* (Definition 5.19).

▶ **Definition 5.19: Secure Software**
Executable software artifact which only causes or contributes to a security incidents with a known, recognized, documented, and accepted residual risk in case of hardware or software failures, or malicious attacks.

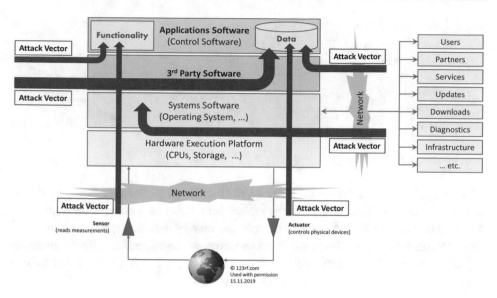

**Fig. 5.18**   Security attack vectors

In case of successful security incidents, the executable software artifact will minimize the consequences of the incident, trigger crisis management, and log sufficient information for forensic analysis and future prevention

Secure software relies on three defenses (Fig. 5.18):

1. Identify, assess, and eliminate or protect the security vulnerabilities (Definition 3.1) in the system: Vulnerabilities are present in all layers of the software-system;
2. Block or intercept the *attack vectors* (Definition 5.20): Regrettably, the modern, very complex systems have myriads of attack vectors. Attack vectors exist through all layers of the cyber-physical system, from attacks through IoT devices, attacks through the system software, attacks via third-party software, and attacks directly to the application software. New attack vectors are found by assailants or security firms every day;
3. Reduce the *impact* of successful malicious activities: Absorb the negative consequences by additional (often runtime) measures.

▶ **Definition 5.20: Attack Vector**
An attack vector is a path or means by which an (external or internal) attacker can gain unauthorized access to networks, computers, or applications.

Attack vectors permit to exploit vulnerabilities, e.g. install malware, access sensitive information, damage the system, or execute unwanted functionality

**Quote**
*"Writing code is work and art, which means there is always a margin of error. We are all humans, and making mistakes is part of the process. But there is a chance that someone will exploit our mistakes for the wrong purposes"*
Ayrunk Publications, 2021

### 5.6.1   Secure by Design

*"Secure by Design"* ([Johnsson19], [Talukder09]) is based on the idea that good design and high-quality programming practices reduce the number of security-related mistakes. The design starts with crystal-clear concept/terminology definitions, ensuring the complete system's conceptual integrity. In many cases, domain-specific software engineering (Definition 5.23) is an excellent methodology choice. The implementation follows proven principles, patterns, and secure code constructs.

### 5.6.2   In-House Software

*In-house developed software* (Definition 5.16) has the highest level of security quality control by the organization: Policies, principles, guidelines, etc., can be decreed, enforced, and audited ([Neumann94], [Sherer92], [NISTIR8151], [Merkow10], [Ousterhout18], [Knight12], [Matulevičius17], [NIST800-160A]). Also, sound knowledge of the application domain is existent in the development team. Therefore, extensive control by the organization over the *application's security* (Definition 5.21) exists.

▶ **Definition 5.21: Applications Security**
Application security describes security measures at the application level that aim to prevent data or code within the application from unauthorized and malicious use. It encompasses the security considerations that happen during application development and design, but it also involves systems and approaches to protect applications after they get deployed.
Adapted from: https://www.vmware.com/topics/glossary/content/application-security

The four primary sources of *vulnerabilities in the applications software* are:

a. *Requirements flaws*: The requirements—both functional and quality requirements (safety and security)—are the origin of all cyber-physical systems development and evolution. If requirements are flawed, the creation of safe and secure software is very difficult. Requirements flaws can be classified into: "missing or incomplete, incorrect information, inconsistent, ambiguous or unclear, misplaced, infeasible or

non-verifiable, redundant or duplicate, typo or formatting, and not relevant or extrane-
ous" ([Alshazly14]). Therefore, *flawless requirements* both for functionality and for
security are indispensable ([Dalpiaz16], [Wiegers13a, b]);

b. *Architecture erosion* and accumulation of *technical debt*: Architecture erosion
   (Definition 9.2, [Li13], [Furrer19], [Fairbanks10], [Lilienthal19]) and accumulation of technical debt (Definition 9.1, [Furrer19], [Janca20], [Moyle20], [Fernandez-
   Buglioni13], [Rerup18], [Wahe11], [Sherwood05], [Rerup18], [Kruchten19]) are a
   major root cause of vulnerabilities. Both are consequences of an inadequate development process and misguided incentives of the stakeholders;

c. *Bad code quality*: Any inadequacy in the program code (Definition 5.4) potentially
   opens up an attack vector ([Tornhill15], [Tornhill18], [VERACODE20], [Seacord13],
   [Seacord14], [Graff03], [Long13], [Martin11], [Janca20], [Kästner17], [Janca20]).

d. *Application programming mistakes*: Application programmers have countless opportunities to introduce insecure code or precarious logic into the applications ([Bell17],
   [Lewis20]). Examples include:

   a. The omission of the authentication/authorization checks before accessing confi-
      dential information;
   b. Neglecting to write to application logs or the archive correctly;
   c. Faulty roll-back procedures in transactions;
   d. Incorrect handling of application-level exceptions or errors;

Assuring the security of the in-house application software is mainly a *process issue*:
Each step of the development/deployment/maintenance/operations *process* must use adequate means and controls, such as (e.g., [NISTIR8151]):

- *Formal methods*:
  - Formal requirements and specifications,
  - Sound static program analysis,
  - Modeling and model checking,
  - Pre- and post-conditions, invariants, aspects, and contracts,
  - Correct by construction and model-based development,
  - Repository of verified tools and verified code,
  - Cyber-retrofitting (modernizing legacy software).
- System-level security:
  - Operating system containers (Resource isolation, Definition 5.22),
  - Microservices.
- *Additive software analysis techniques* (Use of multiple advanced software checking
  tools):
  - Software information expression and exchange standards,
  - Tool development framework or architecture,
  - Strategy to combine analysis results,
  - Technology to combine analysis results,

- *Domain-specific software development frameworks*:
  - Rapid framework adoption,
  - Advanced test methods,
  - Conflict resolution in multi-framework composition,
- Moving target defenses and automatic software diversity:
  - Compile-time techniques,
  - System and network techniques
  - Operating system interface techniques,

### 5.6.2.1 Containers

In the last few years (> 2013), the usage of *containers* (Definition 5.22, [Scholl19], [Faella20]) has massively increased. Apart from delivering cost and time advantages in software development, such as continuous delivery and DevOps ([Bass15]), containers are a strong concept to isolate resources, such as applications or data. Containers can, therefore, significantly improve *security* by providing isolation and thus decoupling deployment ([Rice20], [Candel20], [Sharma20]).

▶ **Definition 5.22: Container**
A container is a software package that contains everything the software needs to run. This includes the executable program, the data, as well as system tools, libraries, and settings. Containers are not installed like traditional software programs, which allows them to be isolated from the other software and the operating system itself.

https://techterms.com/definition/container

**Quote**
*"Container-based isolation can clearly reduce the impact of software vulnerabilities if the isolation is strong enough"*
  [NISTIR8151], 2016

### 5.6.2.2 Domain-Specific Development

Each application domain (banking, aerospace, power grid, health management, etc.) has specific functional requirements and particular safety and security demands. Therefore, researchers started in 2004 ([Evans04]) to include intrinsic domain knowledge into systems engineering: The *domain-specific software engineering* was born and delivered substantial advantages (Definition 5.23, [Wlaschin17], [Combemale16], [Kleppe09], [Voelter13], [Fowler10b]).

Soon after, the power of domain-specific *frameworks* for security was recognized. Such frameworks are elaborated by industry-consortia and condense enormous, proven, domain-specific *security knowledge* into a set of documents, e.g., for the Industrial Internet of Things (IIoT, [IIC16], [IIC19] or [NIST14], [Fowler10a]).

> **Quote**
>
> *"The Industrial Internet Security Framework (IISF) is the most in-depth cross-industry-focused security framework comprising expert vision, experience and security best practices. It reflects thousands of hours of knowledge and experiences from security experts, collected, researched and evaluated for the benefit of all IIoT system deployments"*
>
> https://www.iiconsortium.org/IISF.htm, 2016

Remember that security is a planned, built-in property of the applications, not an afterthought ([NIST80053B], [Morana08], [Horn18], [Johnson20], Fig. 13.2)!

▶ **Definition 5.23: Domain-Specific Software Engineering**
Domain Software Engineering (DSE) is an architectural methodology for creating and evolving a software-system that closely aligns with the corresponding business domain.

A Domain-specific Language (DSL) is a (formal) language that is created to describe and create software systems. DSL's are unique, because of their focus on a certain application domain. A significant part of the domain knowledge is therefore included in a DSL (Anneke Kleppe, 2009).

A Domain-specific Framework is a set of methods and tools for developing a broad range of different IT architectures. It enables IT users to design, evaluate, and build the right architecture for their organization, and reduces the costs of planning, designing, and implementing architectures based on open systems solutions (http://pubs.opengroup.org/architecture)

## 5.6.3   Third-Party Software

Software procured from a third party may present a significant *security risk*. Third-party software acquisition—from components to complete application packages—in many cases makes good business sense. However, assessing and controlling the security properties of third-party software is challenging (Fig. 2.17)—or in many cases, impossible (Example 5.10).

> **Quote**
>
> *"Cybersecurity in the supply chain cannot be viewed as an IT problem only. Cyber supply chain risks touch sourcing, vendor management, supply chain continuity and quality, transportation security and many other functions across the enterprise and require a coordinated effort to address"*
>
> [NISTSCM15], undated

**Example 5.10: BioNTech Hit by Industrial Espionage**

BioNTech: Statement Regarding Cyber Attack on European Medicines Agency (December 9, 2020 / Source: https://investors.biontech.de/node/8886/pdf):

*"Today, we were informed by the European Medicines Agency (EMA) that the agency has been subject to a cyber attack and that some documents relating to the regulatory submission for Pfizer and BioNTech's COVID-19 vaccine candidate, BNT162b2, which has been stored on an EMA server, had been unlawfully accessed. It is important to note that no BioNTech or Pfizer systems have been breached in connection with this incident, and we are unaware that any study participants have been identified through the data being accessed. At this time, we await further information about EMA's investigation and will respond appropriately and in accordance with EU law."*

This is an interesting example of the rising perilousness of *industrial espionage* via cyber-channels. The vulnerable system was *not* BionTech's software but a third-party software installed and used at the European Medicines Agency (EMA, https://www.ema.europa.eu/en/news/cyberattack-european-medicines-agency).

Acquisition and use of third-party software require a *procurement contract* ([Meyers01]) and a trusted relationship. The integration into the organization's system must be carefully planned, and all reasonable protection measures—such as containers, runtime monitoring, incident management—should be taken. A diligent, repeated review of the vendor's processes is advisable (Example 5.11).

**Example 5.11: Third-Party Software Acquisition Questionnaire**

The US National Institute of Standards and Technology recommends reviewing the following questions with the vendor (Source: [NISTSCM15]):

- Is the vendor's software/hardware *design process* documented? Repeatable? Measurable?
- Is the mitigation of known *vulnerabilities* factored into product design (through product architecture, runtime protection techniques, code reviews)?
- How does the vendor stay current on emerging *vulnerabilities*? What are vendor capabilities to address new "zero-day" vulnerabilities?
- What controls are in place to manage and monitor production processes?
- How is configuration management performed? Quality assurance? How is it tested for code quality or vulnerabilities?
- What levels of malware protection and detection are performed?
- What steps are taken to "tamper-proof" products? Are backdoors closed?
- What are physical security measures in place? Documented? Audited?
- What access controls, both cyber and physical, are in place? How are they documented and audited?
  - How do they protect and store customer data?

- How is the data encrypted?
- How long is the data retained?
- How is the data destroyed when the partnership is dissolved?
- What type of employee background checks are conducted, and how frequently?
- What security practice expectations are set for upstream suppliers? How is adherence to these standards assessed?
- How secure is the distribution process?
- Have approved and authorized distribution channels been clearly documented?
- What are the component disposal risk and mitigation strategies?
- How does the vendor assure security through the product life-cycle?

Note: Crisis situations and liability issues must also be discussed and defined with the vendor.

For *open source components*, security assurances are even more difficult to obtain. The use of open source components should be carefully managed. A significant risk arises when programmers download components from Internet sources (e.g., Rogue components [Maggi20], [Synopsys20], [Pittenger20], [Kulkarni20]) or deficient libraries (such as the Apache Log4j Library in December 2021, [NCSC21a, b]). The organization, therefore, should have an *open source policy* and *strategy* governing the sourcing, management, use, and security assessment ([Ahlawat21]). Also, the legal situation (usage rights, licensing conditions) must be respected ([Meeker20]).

---

**Quote**
*"Your developers are using open source — even if you don't know about it"*
   https://www.csoonline.com/article/3191870/how-to-track-and-secure-open-source-in-your-enterprise.html, 2021

---

## 5.6.4   Execution Platform

Figure 5.2 depicts the software layers: The *execution platform* (Definition 5.18) is the underlying foundation for the applications' execution. Unfortunately, a vulnerability in the execution platform has grave consequences because 1) the elements of the execution platform are almost always third-party components, and 2) they are used in myriads of systems worldwide.

The attack vector is often through third-party *infrastructure components* (Operating system browser, database systems, firewall software, etc., Example 5.12). In such cases, the organization must rely on the respective vendors to issue updates (Table 2.1). The organization's responsibility is:

i.   Install and activate the vendor's security *updates* immediately after they become available (Automatic updates recommended);

ii.  Regularly execute professional penetration testing (Definition 4.22): Because penetration testing has become a challenging task, the assistance of specialized companies may be recommended;

iii. Additionally, perform specific *Advanced Persistent Threat* (APT, Definition 4.23) penetration testing.

---

**Example 5.12: SSL/TLS Vulnerability "Heartbleed"**

Many communication channels rely heavily on the *SSL/TLS protocol* ([Davies10], [Ristic14]): It is the standard for Internet security. SSL/TLS provides the essential protection functionality, such as cryptography, authentication, etc. A colossal number of applications run the Open-Source implementation "*OpenSSL*" ([Viega02]).

In 2014 a severe deficiency in the OpenSSL-protocol heartbeat extension was identified (CVE-2014–0160 in the MITRE-database, Table 2.1). The exploitation leads to the leak of memory contents from the server to the client and from the client to the server. This compromises the secret keys used to identify the service providers and encrypt the traffic, the users' names and passwords, and the actual content (https://heartbleed.com).

As long as the vulnerable version of OpenSSL is in use, it can be attacked. A fix has been published (https://www.openssl.org/news/secadv/20140407.txt). However, worldwide, all execution platform software vendors have to adopt the fix and notify their clients. All service providers and users have to install the fix.

Unfortunately, today (2021), still not all OpenSSL-implementations have been fixed, which leaves some channels attackable.

---

## 5.7    Correct-by-Construction

*Correct-by-construction* (CbyC, Definition 5.24, [Amey06], [Knight12], [Hinchey17], [Gamatié10], [Kröger08], [Kourie12], [McGhan15], [Hamiaz14], [Grumberg08], [Thomas17]) is an aspiring *methodology* to build and evolve safe and secure software.

▶ **Definition 5.24: Correct-By-Construction**
Correct-by-construction (CbyC) design and synthesis (development) of software is done in a way so that post-development verification is minimized and correct operation of the systems is maximized. CbyC is not a single, rigid process: It is a framework and a set of principles.

Adapted from: Sandeep Kumar Shukla, 2010

> **Quote**
> *"CbyC is to consider its linguistic near opposite: Construction-by-correction (i.e.,*
> *build and debug), which is still the way the majority of software is developed today.*
>     *In contrast to build and debug, CbyC seeks to produce a product that is initially*
> *correct. Testing becomes a demonstration of correct functionality, rather than the*
> *point where debugging can begin"*
>     https://us-cert.cisa.gov/bsi/articles/knowledge/sdlc-process/correctness-
> by-construction

Many researchers have proposed *CbyC methodologies*. The main elements of a CbyC (Correct-by-Construction) methodology are:

1. Use sound *formal notations* (languages) for all deliverables. For example, use the Z-language, B-method, or Simulink ([Eshkabilov19], [Eshkabilov20]) for the precise formalization and verification of the specifications. Ensure that *all* requirements (functional, safety/security, regulatory, compliance specifications) are adequately covered;
2. Ensure that the specifications are complete, consistent, free of redundancy, and conform to the domain model. This applies not only to the new, incremental specifications but also to the integration into the existing system;
3. Use strong, tool-supported validation methods to validate the deliverables of each development stage;
4. Use a *modular system architecture* (Fig. 5.5) with a strict separation of concerns (Fig. 5.4). Be especially careful in mixed-criticality systems (Fig. 4.3);
5. Apply all modern, state-of-the-art *software constructs*, such as contract-based software construction ([McCormick15], [Ploesch04], [Meyer09], [Mitchell01], [Benveniste12]);
6. Rely on *models*: Model as much as possible and use model-based system engineering (MBSE), also for the *integration* of the new specifications into the existing system;
7. Divide the requirements and specifications into coherent sets (Requirement decomposition, [Minghui19]). Add features to the system build *incrementally*, using small steps and fully test each feature before adding another one (This is one good idea taken from Agile methods, [Meyer14]);
8. Use rigorous, unambiguous notations in the *programming language*, such as ADA or SPARK (Example 5.13, [McCormick15]);
9. *Automate* all processes as much as possible using compulsory, documented, auditable steps (Continuous delivery, DevOps, [Bass15], [Humble10], [Hibbs09]).

**Example 5.13: The SPARK Methodology**

Seminal computer scientists, such as Herman Goldstine, John von Neumann, Alan Turing, and Haskell Curry, realized already in 1947–1949 that it would be necessary to reason about software because merely testing would never be sufficient to produce dependable software ([Thomas17]). In the decade preceding 1970, luminaries like Tony Hoare, Robert Floyd, Edsger W. Dijkstra, Caper Jones, and others had elaborated the computer science foundations for program verification.

This foundational work showed that 1) a high degree of formalization, based on logical groundwork, and 2) a continuous, deep, tool-based verification of all deliverables in the development chain was indispensable. Many methodologies and toolchains were developed, of which SPARK (Definition 5.25, [McCormick15], https://www.adacore.com/about-spark) gained great industrial importance.

SPARK is a strict subset of Ada (the latest version complies with Ada 2012) with formal annotations (contracts) and provides an integrated tool-chain for verification.

▶ **Definition 5.25: SPARK**
SPARK is a software development technology specifically designed for engineering high-reliability applications.

It consists of a programming language, a verification toolset and a design method which, taken together, ensure that ultra-low defect software can be deployed in application domains where high-reliability must be assured, for example where safety and security are key requirements.

    http://www.spark-2014.org/about

Correct by Construction is not a mere technical or process issue: The whole organization (Policies, people, structures, etc.) must be oriented towards this quality objective.

*Construction methods make it possible to guarantee software and to provide high assurance for the safety and cybersecurity of the many systems for which these properties are vital. Software development may, at last, be emerging from its craft stage and becoming a mature engineering discipline"*
  Martyn Thomas, 2017

## 5.8  Importance of People

Last but not least, *people* are of paramount importance for safe and secure software. In addition to technology, methods, standards, etc., ultimately, people's actions and decisions during the creation, evolution, maintenance, and operation of the cyber-physical systems facilitate or avoid *safety accidents* and *security incidents*.

From the lowest level of coding to the highest management level, people's behavior and decisions are strongly influenced by their *work environment*. A favorable work environment is indispensable and consists of a number of elements (Fig. 5.19).

**Quote**
*"Safety is not the sole responsibility of the system safety engineer. Creating a safe system is a team effort and safety is everyone's responsibility"*
  NASA Software Safety Guidebook ([NASA04])

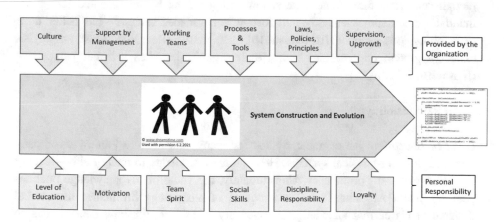

**Fig. 5.19** Work environment for safe and secure software construction

Provided by the organization:

- *Culture*: The risk culture ([Banks12]), safety culture ([Antonsen17]), and security culture ([Hayden16]) form the strong fabric that guides all actions and decisions on all levels of the organization;
- *Management Support*: Conflicts between different goals of the various stakeholders in the cyber-physical system are a fact. Especially the tension between time-to-market/development cost vs. adherence to principles/processes is ever-present ([Hayden16], [Yourdon03], [Murer11]). During such conflicts, the development team must have the full backing of all levels of management, otherwise, the quality properties of the software will be endangered;
- *Working Teams*: The creation and evolution of cyber-physical systems is an enormous team effort. The teams' composition, cooperation, and leadership are vital for performance and quality properties. Therefore, the teams must be carefully assembled, integrated, and managed ([Forsgren18], [Schwartz19], [DeVisch20]);
- *Processes* and *Tools*: Well-founded, forcefully established, consistent and lean processes for all activities of the organization are decisive for safe and secure cyber-physical systems ([Kim18a], [ Kim19]). The teams need support by adequate processes in their work. Software development is very tool-intensive: The organization must provide proven, efficient, up-to-date tools for all functions;
- *Laws, Regulations, Policies, Principles*: Laws and regulations are the backbones of cyber-physical systems' activities. These and additional directives are condensed into policies and principles ([Furrer19]). A sufficient set of consistent, comprehensive, and enforced policies and principles must be available in clearly presented form;
- *Supervision and Upgrowth*: People need guidance, leadership, encouragement, and recognition. Therefore, some supervision and feedback by an accepted leader are mandatory, sometimes embodied in formal performance reviews—but better done as a continuous practice ([Halvorson16], [Singh19], [Rees16]). Most employees expect upgrowth opportunities within their organization, i.e., a career perspective ([Fasano14], [Heller20]).

Personal Responsibility:

- *Level of Education*: An engineer studies for many years before joining a company involved in cyber-physical systems. He acquires knowledge in many fields—leading to a state-of-the-art proficiency. Unfortunately, the value of his knowledge declines year by year ([Arbesman13]). The progress of technology invalidates the market demand of his current knowledge successively. Therefore, life-long education is an important part and responsibility of each engineer!

> **Quote**
> *"Modern estimates place the half-life of an engineering degree at between 2.5 and 5 years, requiring between 10 and 20 hours of study per week. Welcome to the treadmill, where you have to run faster and faster so that you don't fall behind"*
> https://fs.blog/2018/03/half-life

- *Motivation*: Motivation means having the will and perseverance to do the best possible job (e.g., [Haden18]). Motivation can come from inside a person (intrinsic motivation, such as personal goals) or outside forces (extrinsic motivation, such as the working conditions). Motivation is essential for the performance of individuals and teams;
- *Team Spirit*: Successful teamwork is a consequence of 1) an adequate skill mix for the project execution and 2) effective cooperation within the team members (e.g., [Dyer19], [DeVisch20]). The indispensable prerequisite is a driving team spirit—which every member must assist to;
- *Social Skills*: Social skills are essential for personal and professional relationships. They allow trust, tolerance, helpfulness, and many other attributes that render possible people's constructive cooperation (e.g., [Briggs19]). Without sufficient social skills, effective teamwork becomes difficult;
- *Discipline and Sense of Responsibility*: Systems engineering, especially software development, relies on many decisions taken daily by team members. Finally, the sum of these decisions decides about the quality and "fitness for purpose" of the resulting software. An accumulation of careless or unjustifiable choices leads without fail to the system's unmanageability ([Furrer19]). The reasons are, e.g., the accumulation of technical debt (Definition 9.1) or the architecture erosion (Definition 9.2). All team members need to possess a strong discipline and sense of responsibility to keep the quality of their decisions high and in conformance with the applicable policies and principles;
- *Loyalty*: Work relationships are defined in contracts. However, faithful collaboration needs loyalty and allegiance from all employees or contractors (e.g., [Sirota14]).

In addition to Fig. 5.19, an important insight is: Creating, evolving, and operating safe and secure cyber-physical system is crucially dependent on two "soft" capabilities of the organization (e.g., [Gottschalk16]):

1. The effective cooperation of *management* and *engineering* in the organization ([Ricketts20], [Larson21]): Only a trusting, purposeful, and equitable working relationship can eliminate friction and prevent the loss of information—and thus safety and security deficiencies;

2. The productive integration of *project management* and *systems engineering* ([Rebentisch17], [Pinto19]): During the development and deployment of a cyber-physical system, many safety- and security-related design decisions have to be taken. Close cooperation of project management and systems engineering increases the probability of appropriate decisions.

## 5.9   Drift into Vulnerabilities

In today's modern, highly complex systems, *vulnerabilities* are inevitable. Also, *threats* and *failures* are unpreventable. Fortunately, conscientious risk management and accountable mitigation measures reduce the probability of safety accidents and security incidents in most cases to acceptable residual risks.

Fundamentally, seven mechanisms create vulnerabilities (Fig. 5.20):

1. The inherent *complexity*, merciless *change*, and inescapable *uncertainty* of today's cyber-physical systems and their creation/evolution processes ([Murer11], [Furrer19]);
2. The accumulation of *technical debt* ([Kruchten19], [Tornhill18], [Furrer19], [Lilienthal19], [Beine21], [Suryanarayana14]) and the *architecture erosion* ([Fairbanks10], [Erder16], [Ford17], [Furrer19]) during the evolution of the CPS;
3. And (introduced here): The presence of "*far-effects*" i.e., the *law of unintended consequences* ([Mansfield10], [Dekker11]) and *emergence* (Emergent behavior and emergent properties, [Rainey18], [Haimes19], [Mittal18]);

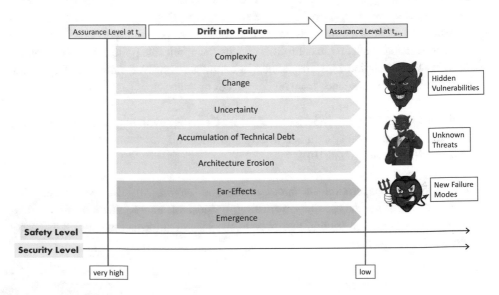

**Fig. 5.20**  Causes of system vulnerabilities

### 5.9.1   The Law of Unintended Consequences

The *law of unintended consequences* (Definition 5.26) was introduced into socio-eco-nomic problems by Robert K. Merton in 1936 ([Merton36]). It was subsequently applied to complex technical systems (e.g., [Mansfield10]).

▶ **Definition 5.26: Law of Unintended Consequences** The law of unintended consequences states that changes in a complex system may have unintended (unforeseen, unexpected) consequences, both in the temporal, functional, and in the spatial domain

Two effects of the law of unintended consequences are shown in Fig. 5.21:

1. A *functional change* is made in a part of the system: The change not only affects the intended part but propagates through the interconnections to other parts of the system and causes unpredicted, unintended effects, such as faulty behavior or unavailability of resources (= "far-effect");
2. *Functionality is added* in the form of new parts to the system. The new functionality interacts in unpredictable, unanticipated ways with the existing functionality to produce unintended, *emerging* functionality or properties (Definition 2.4).

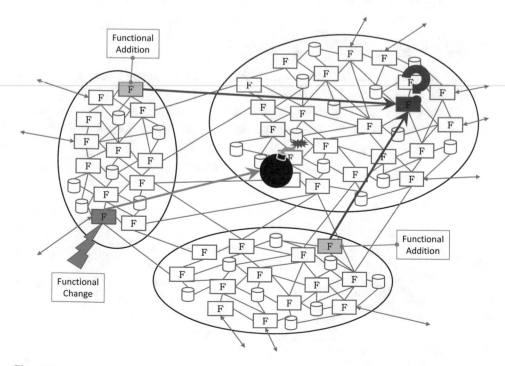

**Fig. 5.21**   Law of unintended consequences

#### 5.9.1.1 Far-Effect

Two instances of far-effects are shown in Example 5.14 and Example 5.15.

---

**Example 5.14: Overlooked Dependency**

In a financial institution at year-end, 40 '000 '000 accounts & interest statements have to be calculated, printed, and distributed to the customers (= End-of-Year processing EoY). When the EoY-processing was started in early January, the software crashed. The resulting panic in the IT department was soon appeased by finding and fixing the fault: The fault was refactoring a component remotely required for EoY-processing: This component was refactored in September tested well. No problems in operations occurred from September until EoY-processing.

The analysis showed that the refactored component's dependency (Fig. 5.22) had been erroneously eliminated (= far-effect). This dependency was not functionally required during EoY-processing but had been in the system for a long time. This was quickly recognized, the dependency was eliminated from the EoY-processing, and the year-end reporting was successfully completed in time.

This example shows that identifying dependencies in components/modules to be refactored to/from other components/modules can be an arduous task. Elimination of a dependency that is still used can crash software in another part of the application landscape at a later time.

---

**Example 5.15: Cross-Platform Vulnerability**

*Facebook Hack* (https://www.nytimes.com/2018/09/28/technology/facebook-hack-data-breach.html [last accessed 3.3.2021]).

In September 2018, unknown hackers gained access to more than 50 million Facebook users' accounts. This was possible due to a *security vulnerability* in a

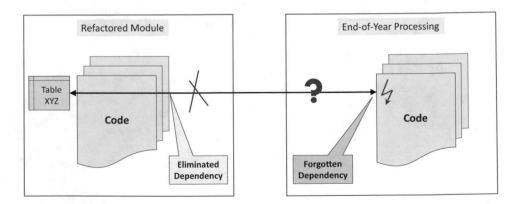

**Fig. 5.22** Accidentally removed dependency during refactoring

relatively minor function: Facebook's "View As" feature, which lets users see how their profile appears on other people's screens. The vulnerability allowed hackers to steal *access tokens*. The stolen access tokens could then be used to take over people's accounts.

The stolen access tokens theoretically also give admission to any other service in which a person uses their Facebook account to register—including applications like Tinder, Spotify, Airbnb, or a niche smartphone game. As a consequence, highly personal information could be viewed and copied.

Facebook temporarily disabled the "View As" feature and executed a thorough security review. This is an example of a relatively small peripheral function implemented with deficient security and massively impacted the whole platform ("Far-effect").

*Far-effects* in huge, complex systems are potentially dangerous because they are often only detected during operation, causing failures, malfunctions, or outages—and possibly safety accidents or security incidents. To a large extent, far-effects can be excluded during system evolution by using an *application portfolio* (Definition 5.27, [Sidhu16], [Mandal14], [Mega20]).

▶ **Definition 5.27: Application Portfolio**
Electronic repository (database) containing all applications, their relationships, dependencies, properties, and the pertaining metadata.

Application portfolio management is the process by which an organization maintains correct and complete view of their existing applications and technologies.

The application portfolio must be kept up-to-date at all times (= Application portfolio management). Fortunately, automated tools exist to support this tracking. When modifying an application (During corrective, adaptive, or predictive maintenance), the project team queries the application portfolio to discover all the respective applications' *dependencies* and takes them into account.

### 5.9.1.2 Emergence
*Emergence* (Definition 2.4, Fig. 5.23) is another cause for a drift into vulnerability. Unexpected effects of the combination of functionality, events, or circumstances may lead to accidents or incidents (Fig. 2.6), as shown in Example 5.16 and Example 5.17.

**Example 5.16: 2016 Tesla Car Accident**

*"The driver of a Tesla car died in Florida in May 2016 after colliding with a tractor-trailer. In a statement, Tesla said it appeared the Model S car was unable to recognize the white side of the tractor-trailer against a brightly lit sky' that had driven across the car's path"* (https://www.bbc.com/news/technology-36680043).

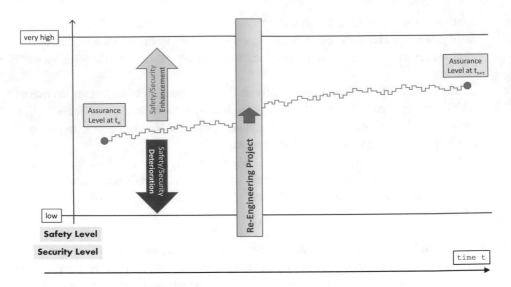

**Fig. 5.23**   Drift into vulnerability

For this accident, the combination of four factors was necessary:

1. The tractor-trailer crossed the path in front of the Tesla car;
2. The trailer had an unpainted, white side: The Tesla image processing algorithm did not sufficiently distinguish the white trailer against the brightly lit sky behind the trailer;
3. The trailer had a high ride height: The Tesla's forward radar's view angle did not capture the obstacle in front;
4. The car was in autonomous mode, and the driver ignored the traffic condition.

The *emerging consequence* was the collision. No guilt (with the possible exception of the driver) can be assigned: This emergence was not foreseeable, i.e., this was a 'normal accident' ([Perrow99]) contributing to an increase of the *technology readiness level* of autonomous driving (TRL, [Engel12]).

---

**Example 5.17: Vehicular Traffic Flow**

Vehicular traffic, e.g., on motorways, is crucial for the economy. An increasingly larger number of cars and trucks is using the traffic infrastructure. Unfortunately, this leads to the phenomenon of *traffic breakdowns*. Often, these disruptions of the traffic flow are due to accidents or construction work. However, many traffic breakdowns (traffic jams) develop without such a deterministic cause.

In a kilometer-long, multilane, continuously flowing traffic, a traffic jam may suddenly appear and develop into a severe traffic breakdown. Traffic flow dynamics ([Treiber13]) explained by models that this is a consequence of many vehicles and drivers' interplay. Collective behavior that does not depend on individual cars or drivers' details causes complex, dynamic patterns of acceleration, deceleration, and standstill of groups of vehicles.

The resulting traffic jam is an *emergent property* of a large number of vehicles and the laws of traffic flow dynamics.

### 5.9.1.3  Drift into Vulnerability

During the cyber-physical system's creation and evolution, many decisions have to be made by the development team (Fig. 5.23). Some—hopefully, most—decisions *enhance* the safety or security assurance level. Unfortunately, also decisions are taken which deteriorate the safety or security assurance level. Most vulnerabilities are the consequence of many detrimental architectures and damaging design decisions: Each of them slightly weakens the system—leading to a slow *drift into vulnerability*! This slow, imperceptible progress is known as *system erosion* due to *entropy* ([Hubert02], [Furrer19]). To keep the cyber-physical system safe and secure, two practices must be performed (Fig. 5.23):

i.  The *gradient* of the trajectory of all accumulated decisions must be positive, i.e., continuously improve the safety and security of the system—in step with the increasing complexity;

ii. The organization must regularly execute *re-engineering/refactoring projects* to eliminate technical debt, compensate for architecture erosion, and simplify the system.

> **Quote**
> *"Although each decision was locally rational, making sense for decision-makers in their time and place, the global picture became one of drift into failure"*
>    Sidney Dekker, 2011

### References

[Ábrahám16]        Erika Ábrahám, Marcello Bonsangue, Einar Broch Johnsen (Editors): **Theory and Practice of Formal Methods** Springer International Publishing, Cham, Switzerland, (Lecture Notes in Computer Science, Volume 9660), 2016. ISBN 978-3-319-30733-6

[Abrial05]         Jean-Raymond Abrial: The B-Book - Assigning Programs to Meanings.Cambridge University Press, Cambridge, UK, 2005. ISBN 978-0-521-02175-3

[Abrial10]          Jean-Raymond Abrial: Modeling in Event-B - System and Software
                    Engineering. Cambridge University Press, Cambridge, UK, 2010. ISBN
                    978-0-521-89556-9

[ADACore16]         Various authors: Learning Ada (Release 2022-05). AdaCore Inc., New
                    York, NY, USA, New version 2022. Downloadable from: https://learn.
                    adacore.com/pdf_books/learning-ada.pdf [Last accessed: 8.6.2022]

[Ahlawat21]         Pranay Ahlawat, Johannes Boyne, Dominik Herz, Florian Schmieg,
                    Michael Stephan: *Why You Need an Open Source Software Strategy*
                    BCG (Boston Consulting Group), Boston, MA, USA, April 16, 2021.
                    Accessible    at:    https://www.bcg.com/publications/2021/open-source-
                    software-strategy-benefits [Last accessed: 22.6.2021]

[Akşit01]           Mehmet Aksit (Editor): **Software Architectures and Component
                    Technology. Springer International Series in Engineering and
                    Computer Science**, Band 648), Springer-Verlag, Heidelberg, 2001.
                    ISBN 978-1-461-35286-0

[Allweyer16]        Thomas Allweyer: BPMN 2.0 - Introduction to the Standard for
                    Business Process Modeling. BoD – Books on Demand, Norderstedt,
                    Germany, 2016. ISBN 978-3-8370-9331-5

[Alshazly14]        Amira A. Alshazly, Ahmed M.Elfatatry, Mohamed S.Abougabal:
                    *Detecting Defects in Software Requirements Specifications* Alexandria
                    Engineering Journal, Volume 53, Issue 3, September 2014, pages
                    513–527. Available at: https://doi.org/10.1016/j.aej.2014.06.001 [Last
                    accessed: 19.2.2021]

[Ambler05]          Scott W. Ambler: **The Elements of UML™ 2.0 Style** Cambridge
                    University Press, New York, NY, USA, 2005. ISBN 978-0-521-61678-2

[Amey06]            Peter Amey: *Correctness by Construction* US Cybersecurity and
                    Infrastructure Security Agency (CISA), Washington, DC, USA, 2006
                    (Last revised: 2013). Downloadable from: https://us-cert.cisa.gov/bsi/arti-
                    cles/knowledge/sdlc-process/correctness-by-construction [Last accessed:
                    3–3–2021]

[Antonsen17]        Stian Antonsen 2017 Safety Culture - Theory, Method, and
                    Improvement CRC Press (Taylor & Francis Group) FL, USA Boca
                    Raton 978-1-138-07533-7

[Arbesman13]        Samuel Arbesman 2013 The Half-Life of Facts - Why Everything We
                    Know Has an Expiration Date Current (Penguin Group) NY, USA New
                    York 978-1-591-84651-2

[Artho16]           Cyrille Artho, Peter Csaba Ölveczky (Editors): **Formal Techniques for
                    Safety-Critical Systems**: 4th International Workshop, FTSCS 2015,
                    Paris, France, November 6–7, 2015. Springer International, Cham,
                    Switzerland, 2016. ISBN 978-3-319-29509-1

[Bagnara18]         Roberto Bagnara, Abramo Bagnara, Patricia M. Hill: **The MISRA C
                    Coding Standard and its Role in the Development and Analysis of
                    Safety- and Security-Critical Embedded Software** Arxiv-Preprint
                    Nr. 1809.00821, 4.9.2018. Downloadable from: https://arxiv.org/
                    pdf/1809.00821.pdf [Last accessed:2.11.2020]

[Baier07]           Christel Baier, Joost-Pieter Katoen: Principles of Model Checking.
                    MIT Press, The MIT Press, Cambridge, MA, USA, 2007. ISBN
                    978-0-262-02649-9

| [Baier08] | Christel Baier Joost-Pieter Katoen 2008 Principles of Model Checking The MIT Press MA, USA Cambridge 978-0-262-02649-9 |
|---|---|
| [Banks12] | Erik Banks: **Risk Culture - *A Practical Guide to Building and Strengthening the Fabric of Risk Management*** Palgrave Macmillan, London, UK, 2012. ISBN 978-1-137-26371-1 |
| [Barnes14] | John Barnes: **Programming in Ada 2012** Cambridge University Press, Cambridge, UK, 2014. ISBN 978-1-107-42481-4 |
| [Bass13] | Len Bass Paul Clements Rick Katzman 2013 Software Architecture In Practice Pearson Education, SEI-Series, (Addison-Wesley) 3 Upper Saddle River NJ, USA 978-9-332-50230-7 |
| [Bass15] | Len Bass Ingo Weber Liming Zhu 2015 DevOps - A Software Architect's Perspective SEI Series in Software Engineering Addison Wesley Upper Saddle River, NJ, USA 978-0-134-04984-7 |
| [Beine21] | Gerritt Beine 2021 Technical Debts - Economizing Agile Software Architecture De Gruyter Oldenbourg Oldenbourg Germany 978-3-110-46299-9 |
| [Bell17] | Laura Bell, Michael Brunton-Spall, Rich Smith, Jim Bird: **Agile Application Security - *Enabling Security in a Continuous Delivery Pipeline*** O'Reilly Ltd., Farnham, UK, 2017. ISBN 978-1-491-93884-3 |
| [Benveniste12] | Albert Benveniste, Benoît Caillaud, Dejan Nickovic, Roberto Passerone, Jean-Baptiste Raclet, Philipp Reinkemeier, Alberto Sangiovanni-Vincentelli, Werner Damm, Tom Henzinger, Kim Larsen: *Contracts for Systems Design* INRIA RESEARCH REPORT, N° 8147, November 2012 ISSN 0249–6399 Downloadable from: http://hal.inria.fr/docs/00/75/85/14/PDF/RR-8147.pdf [last accessed: 23.9.2017] |
| [Berard10] | B. Berard, M. Bidoit, A. Finkel, F. Laroussinie, A. Petit, L. Petrucci, Ph. Schnoebelen, P. McKenzie : **Model-Checking Techniques and Tools** Springer Verlag, Berlin, Germany, 2010 (Softcover reprint of Hardcover 1st edition 2001). ISBN 978-3-642-07478-3 |
| [Bernardi13] | Simona Bernardi, José Merseguer, Dorina Corina Petriu: **Model-Driven Dependability Assessment of Software** Springer Verlag, Heidelberg, Germany, 2013 (reprinted 2016). ISBN 978-3-662-50942-5 |
| [Bishop18] | Matt Bishop 2018 Computer Security - Art and Science Addison Wesley 2 Boston USA 978-0-321-71233-2 |
| [Boerger18] | Egon Börger, Alexander Raschke: **Modeling Companion for Software Practitioners** Springer-Verlag, Berlin, Germany, 2018. ISBN 978-3-662-56639-8 |
| [Bondavalli16] | Andrea Bondavalli, Sara Bouchenak, Hermann Kopetz (Editors): **Cyber-Physical Systems of Systems: Foundations - A Conceptual Model and Some Derivations: The AMADEOS Legacy** Springer Lecture Notes in Computer Science, Heidelberg, Germany, 2016. ISBN 978-3-319-47589-9 |
| [Borky19] | John M. Borky, Thomas H. Bradley: **Effective Model-Based Systems Engineering** Springer Nature Switzerland AG, Cham, Switzerland, 2019. ISBN 978-3-319-95668-8 |
| [Boulanger12] | Jean-Louis Boulanger (Editor): **Formal Methods - *Industrial Use from Model to the Code*** ISTE and John Wiley & Sons, Inc., Hoboken, NJ, USA, ISBN 978-1-848-21362-3 |

[Boulanger13]        Jean-Louis Boulanger: **Safety Management of Software-Based Equipment** John Wiley & Sons, Inc., Hoboken, NJ, USA, 2013 (published jointly with ISTE Ltd., London, UK). ISBN 978-1-848-21452-1

[Boulanger14]        Jean-Louis Boulanger (Editor): **Formal Methods Applied to Complex Systems** - *Implementation of the B Method* ISTE and John Wiley & Sons, Inc., Hoboken, NJ, USA, 2014. ISBN 978-1-848-21709-6

[Brambilla17]        Marco Brambilla Jordi Cabot Manuel Wimmer 2017 Model-Driven Software Engineering in Practice Morgan & Claypool Publishers 2 San Rafael CA, USA 978-1-681-73233-6

[Briggs19]           Roy Briggs: **Improve Your Social Skills** - *How to Talk to Anyone. The Ultimate Guide to Improve Your Conversations and Your People Skills* Independently published, 2019. ISBN 978-1-0791-0125-6

[Brown98]            Doug Brown: *Solving the Software Safety Paradox* Embedded Systems Programming, Volume 11, Nr. 13, December 1998.

[Burns16]            Alan Burns, Andy Wellings: **Analysable Real-Time Systems** - *Programmed in Ada* CreateSpace Independent Publishing Platform, North Charleston, S.C., USA, 2016. ISBN 978-1-5302-6550-3

[Buttazzo11]         Giorgio C Buttazzo: **Hard Real-Time Computing Systems** - *Predictable Scheduling Algorithms and Applications* Springer Science + Business Media, New York, N, USA, 3$^{rd}$ edition, 2011. ISBN 978-1-461-40675-4

[Candel20]           Jose Manuel Ortega Candel 2020 DevOps and Containers Security - Security and Monitoring in Docker Containers BPB Publications Ne Delhi India 978-9-38942-353-2

[Cansell03]          Dominique Cansell, Dominique Mery: *Foundations of the B-Method* Computing and Informatics, Vol. 22, 2003, pp. 1-31. Downloadable from: https://www.researchgate.net/publication/230688722_Foundations_of_the_B_Method/link/02e7e53b6936f68fa6000000/download        [Last accessed: 5.2.2021]

[Cansell06]          Dominique Cansell, Dominique Méry: *Tutorial on the event-based B method* IFIP FORTE 2006, Paris, France, 2006. Downloadable from:    https://cel.archives-ouvertes.fr/inria-00092846/document    [Last accessed: 5.2.2021]

[Cao20]              Zongyu Cao, Wanyou Lv, Yanhong Huang, Jianqi Shi, Qin Li: *Formal Analysis and Verification of Airborne Software based on DO-333* Electronics 2020, Vol. 9, Nr. 2, (Design and Applications of Software Architectures),    February    2020).    Downloadable    from:    https://doi.org/10.3390/electronics9020327 [Last accessed: 23.1.2021]

[Carrillo20]         Miguel Carrillo, Vladimir Estivill-Castro, David A. Rosenblueth: *Verification and Simulation of Time-Domain Properties for Models of Behaviour* In: Hammoudi S., Pires L.F., Selić B. (editors): Model-Driven Engineering and Software Development. MODELSWARD 2020. Communications in Computer and Information Science, Vol 1361. Springer International Publishing, Cham, Switzerland, 2020

[CERT-C16]           Carnegie Mellon University: **SEI CERT C Coding Standard** - *Rules for Developing Safe, Reliable, and Secure Systems* Software Engineering Institute, Carnegie Mellon University, Hanscom, MA, USA, 2016. Downloadable from: https://resources.sei.cmu.edu/downloads/secure-coding/assets/sei-cert-c-coding-standard-2016-v01.pdf [Last accessed: 3.11.2020]

[Cervantes16]    Humberto Cervantes, Rick Kazman: **Designing Software Architectures - *A Practical Approach*** Addison Wesley, Upper Saddle River, NJ, USA, 2016. ISBN 978-0-134-39078-9

[Chappell04]    David A. Chappell: **Enterprise Service Bus** O'Reilly and Associates, Sebastopol, CA, USA, 2004. ISBN 978-0-596-00675-4

[Chess07]    Brian Chess, Jacob West: **Secure Programming with Static Analysis - *Getting Software Security Right with Static Analysis*** Addison-Wesley, Upper Saddle River, NJ, USA, 2007. ISBN 978-0-321-42477-8

[Clarke18a]    Edmund M. Clarke, Thomas A. Henzinger, Helmut Veith, Roderick Bloem (Editors): **Handbook of Model Checking** Springer International Publishing, Cham, Switzerland, 2018. ISBN 978-3-319-10574-1

[Clarke18b]    M Edmund 2018 Clarke, Orna Grumberg, Daniel Kroening, Doron Peled, Helmut Veith: Model Checking The MIT Press 2 Cambridge MA, USA 978-0-262-03883-6

[Clements10]    Paul Clements, Felix Bachmann, Len Bass, David Garlan, James Ivers, Reed Little, Paulo Merson, Robert Nord, Judith Stafford: **Documenting Software Architectures - *Views and Beyond*** SEI Series in Software Engineering, Addison-Wesley Education, Upper Saddle River, NJ, USA, 2010. ISBN 978-0-321-55268-6

[Cofer14]    Darren Cofer, Steven P. Miller: *NASA/CR-2014-218244 Formal Methods Case Studies for DO-333* National Aeronautics and Space Administration, Langley Research, Hampton, Virginia, USA, 2014. Downloadable from: https://ntrs.nasa.gov/citations/20140004055 [Last accessed: 23.1.2021]

[Cohen95]    Bernie Cohen: *A Brief History of Formal Methods* Formal Aspects of Computing, Springer Verlag, Heidelberg, Germany, January 1995. Downloadable from: https://www.researchgate.net/publication/233960390 [Last accessed: 26.1.2021]

[Collazos20]    Néstor Cataño Collazos: **Java Software Development With Event B - *A Practical Guide*** Morgan & Claypool Publishers, San Rafael, CA, USA, 2020. ISBN 978-1-681-73689-1

[Combemale16]    Benoit Combemale Robert B France Jean-Marc Jézéquel Bernhard Rumpe Jim Steel Didier Vojtisek 2016 Engineering Modeling Languages - Turning Domain Knowledge into Tools CRC Press Inc FL, USA Boca Raton 978-1-4665-8373-3

[Cooling19]    Jim Cooling: **Software Engineering for Real-Time Systems - *A Software Engineering Perspective toward designing real-time Systems*** Packt Publishing, Birmingham, UK, 2019. ISBN 978-1-839-21658-9

[Cooling21a]    Jim Cooling 2021 Real-time Operating Systems Book 1 - The Theory Lindentree Associates 2 Markfield UK 978-1-79534-065-6

[Cooling21b]    Jim Cooling 2021 Real-time Operating Systems Book 2 - The Practice: Using STM Cube, FreeRTOS and the STM32 Discovery Board Lindentree Associates 2 Markfield UK 978-1-97340-993-9 (Engin

[Cremers12]    Cas Cremers, Sjouke Mauw: **Operational Semantics and Verification of Security Protocols** Springer Verlaag, Berlin, Germany, 2012. ISBN 978-3-540-78635-1

[Dajsuren19]    Yanja Dajsuren, Mark van den Brand (Editors): **Automotive Systems and Software Engineering - *State of the Art and Future Trends*** Springer Nature Switzerland, Cham, Sitzerland, 2019. ISBN 978-3-030-12156-3

[Dalpiaz16]            Fabiano Dalpiaz Elda Paja Paolo Giorgini 2016 Security Requirements
                      Engineering - Designing Secure Socio-Technical Systems The MIT
                      Press MA, USA Cambridge 978-0-262-03421-0

[Damchoom09]          Kriangsak Damchoom, Michael Butler: *Applying Event and Machine
                      Decomposition to a Flash-Based Filestore in Event-B* Brazilian
                      Symposium on Formal Methods (SBMF 2009), Formal Methods:
                      Foundations and Applications, pp. 134–152. Springer Verlag, Berlin,
                      Germany, LNCS, Volume 5902. ISBN 978-3-642-10451-0

[Davenport19]         J.H. Davenport: *Formal Methods and Cyber-Security* Preprint,
                      Department of Computer Science, University of Bath, Bath, UK,
                      2019. Downloadable from: https://arxiv.org/pdf/1909.03325.pdf [Last
                      accessed: 26.1.2021]

[Davies10]            Joshua Davies: **Implementing SSL/TLS** Wiley Publishing Inc.,
                      Indianapolis, IN, USA, 2010. ISBN 978-0-470-92041-1

[Debbabi10]           Mourad Debbabi, Fawzi Hassaïne Yosr Jarraya, Andrei Soeanu, Luay
                      Alawneh: **Verification and Validation in Systems Engineering
                      - Assessing UML/SysML Design Models** Springer Verlag, Berlin,
                      Germany, 2010. ISBN 978-3-642-42316-1

[DeHaes20]            Steven Haes De Wim Grembergen Van Anant Joshi Tim Huygh
                      2020 Enterprise Governance of Information Technology - Achieving
                      Alignment and Value in Digital Organizations Springer Nature 3 Cham
                      Switzerland 978-3-030-25917-4

[Dekker11]            Sidney Dekker 2011 Drift into Failure - From Hunting Broken
                      Components to Understanding Complex Systems CRC Press (Taylor &
                      Francis) FL, USA Boca Raton 978-1-4094-2221-1

[Delligatti13]        Lenny Delligatti 2013 SysML Distilled - A Brief Guide to the Systems
                      Modeling Language Addison-Wesley Professional Upper Saddle River
                      USA 978-0-321-92786-6

[DeVisch20]           Jan DeVisch, Otto Laschke: **Practices of Dynamic Collaboration - *A
                      Dialogical Approach to Strengthening Collaborative Intelligence in
                      Teams*** Springer Nature Switzerland, Cham, Switzerland, 2020. ISBN
                      978-3-030-42548-7

[Dick17]              Jeremy Dick Elizabeth Hull Ken Jackson 2017 Requirements
                      Engineering Springer International Publishing 4 Cham Switzerland
                      978-3-319-86997-1

[Dieumegard17]        Arnaud Dieumegard, Ning Ge, Eric Jenn: *Event-B at Work - Some
                      Lessons Learnt from an Application to a Robot Anti-collision
                      Function* 9th International Symposium on NASA Formal Methods
                      (NFM 2017), May 2017, Moffet Field, CA, USA, pp.312–341.
                      Downloadable from: https://hal.archives-ouvertes.fr/hal-01535060/doc-
                      ument [Lastaccessed: 5.2.2021]

[Dijkstra82]          Edsger W. Dijkstra: *On the Role of Scientific Thought* in: **Selected
                      Writings on Computing - *A Personal Perspective*** Springer-Verlag,
                      Berlin, Germany, 1982. ISBN 0-387-90652-5. Downloadable from:
                      https://www.cs.utexas.edu/users/EWD/ewd04xx/EWD447.PDF      [Last
                      accessed: 9.7.2020]

[DiNatale12]          Marco Natale Di Haibo Zeng Paolo Giusto Arkadeb Ghosal 2012
                      Understanding and Using the Controller Area Network Communication
                      Protocol - Theory and Practice Springer Science & Business Media NY,
                      USA New York 978-1-461-40313-5

[Dori16]        Dov Dori 2016 Model-Based Systems Engineering with OPM and SysML Springer Verlag N.Y., USA New York 978-1-493-932948

[Douglass15]    Bruce Powel Douglass: **Agile Systems Engineering** Morgan Kaufmann (Elsevier), Waltham, MA, USA, 2015. ISBN 978-0-128-02120-0f

[Drusinsky06]   Doron Drusinsky: **Modeling and Verification Using UML Statecharts - *A Working Guide to Reactive System Design, Runtime Monitoring, and Execution-based Model Checking*** Newnes (Elsevier), Burlington, MA, USA, 2006. ISBN 978-0-7506-7949-7

[Dumas18]       Marlon Dumas Marcello Rosa La Jan Mendling Hajo A Reijers 2018 Fundamentals of Business Process Management Springer Verlag 2 Berlin Germany 978-3-662-56508-7

[Dyer19]        W. Gibb Dyer, Jeffrey H. Dyer: **Beyond Team Building - *How to Build High Performing Teams and the Culture to Support Them*** John Wiley &Sons, Inc., Hoboken, USA, 2019. ISBN 978-1-119-55140-9

[Engblomxy03]   Jakob Engblomxy, Andreas Ermedahlx, Mikael Sjödin, Jan Gustafsson, Hans Hansson: *Worst-case execution-time analysis for embedded real-time systems* International Journal on Software Tools for Technology Transfer, Vol. 4, August 2003, pp. 437–455. Downloadable from: https://www.researchgate.net/publication/220643605_Worst-case_execution-time_analysis_for_embedded_real-time_systems/link/09e415088c899bb9f4000000/download [Last accessed: 31.1.2021]

[Engel12]       D.W. Engel, A.C. Dalton, K. Anderson, C. Sivaramakrishnan, C. Lansing: *Development of Technology Readiness Level (TRL) Metrics and Risk Measures* Technical Report, U.S. Department of Energy, Pacific Northwest National Laboratory, Richland, USA, 2012. Downloadable from: https://core.ac.uk/download/pdf/190816293.pdf [Last accessed: 7.3.2021]

[Erciyeş19]     Kayhan Erciyeş: **Distributed Real-Time Systems - *Theory and Practice*** Springer Nature Switzerland, Cham, Switzerland, 2019. ISBN 978-3-030-22569-8

[Erder16]       Murat Erder, Pierre Pureur: **Continuous Architecture - *Sustainable Architecture in an Agile and Cloud-Centric World*** Morgan Kaufmann (Elsevier), Waltham, MA, USA, 2016. ISBN 978-0-12-803284-8

[Ericson13]     Clifton A. Ericson: **Software Safety Primer** CreateSpace Independent Publishing Platform, North Charleston, S.C., USA, 2013. ISBN 978-1-4909-7598-6

[Eshkabilov19]  L Sulaymo 2019 Eshkabilov: Beginning MATLAB and Simulink - From Novice to Professional Apress Media LLC N.Y., USA New York 978-1-484-25060-0

[Eshkabilov20]  L Sulaymon 2020 Eshkabilov: Practical MATLAB Modeling with Simulink - Programming and Simulating Ordinary and Partial Differential Equations Apress Media LLC N.Y., USA New York 978-1-4842-5798-2

[Evans04]       Eric Evans: **Domain-Driven Design - *Tackling Complexity in the Heart of Software*** Pearson Education, Addison-Wesley, Boston, USA, 2004. 7th printing 2006. ISBN 978-0-321-12521-5

[Faella20]      Marco Faella: **Seriously Good Software - *Code that works, survives, and wins*** Manning Publications, Shelter Island, NY, USA, 2020. ISBN 978-1-617-29629-1

[Fairbanks10]            George Fairbanks: **Just Enough Software Architecture - *A Risk-Driven Approach*** Marshall & Brainerd Publications, Boulder, CO, USA, 2010. ISBN 978-0-984-61810-1

[Farley21]              David Farley: **Modern Software Engineering - *Doing What Really Works to Build Better Software Faster*** Addison Wesley, Upper Saddle River, NJ, USA, 2021. ISBN 978-0-137-31491-1

[Fasano14]              Anthony Fasano 2014 Engineer Your Own Success - 7 Key Elements to Creating an Extraordinary Engineering Career IEEE Press John Wiley & Sons Inc. Hoboken, NJ, USA 978-1-118-65964-9

[Feiler12]              Peter H. Feiler, David P. Gluch: **Model-based Engineering with AADL** Addison-Wesley Longman (SEI Series in Software Engineering), Amsterdam, NL, 2012. ISBN 978-0-321-88894-5

[Fernandez-Buglioni13]  Eduardo Fernandez-Buglioni **Security Patterns in Practice: *Designing Secure Architectures Using Software Patterns*** John Wiley & Sons, USA, 2013. ISBN 978-1-119-99894-5

[Fisher11]             Michael Fisher: **An Introduction to Practical Formal Methods Using Temporal Logic** John Wiley & Sons., Ltd., Chichester, UK, 2011. ISBN 978-0-470-02788-2

[Ford17]               Neal Ford, Rebecca Parsons, Patrick Kua: **Building Evolutionary Architectures - *Support Constant Change*** O'Reilly, Farnham, UK, 2017. ISBN 978-1-491-98636-3

[Ford20]               Neal Ford, Mark Richards: **Fundamentals of Software Architecture: *An Engineering Approach - A Comprehensive Guide to Patterns, Characteristics, and Best Practices*** O'Reilly, Farnham, UK, 2020. ISBN 978-1-492-04345-4)

[Forsgren18]           Nicole Forsgren, Jez Humble), Gene Kim: **Accelerate: The Science of Lean Software and DevOps - *Building and Scaling High Performing Technology Organizations*** IT Revolution Press, Portland, OR, USA, 2018. ISBN 978-1-9427-8833-1

[Fowler03]             Martin Fowler 2003 UML Distilled - A Brief Guide to the Standard Object Modeling Language Addison-Wesley (Object Technology Series) 3 Boston USA 978-0-321-19368-1

[Fowler10a]            Martin Fowler, Rebecca Parsons: **Domain-Specific Languages** Addison-Wesley, Upper Saddle River, NJ, USA, 2010. ISBN 978-0-321-71294-3

[Fowler10b]            Kim Fowler: **Mission-Critical and Safety-Critical Systems Handbook - *Design and Development for Embedded Applications*** Newnes Publishing (Elsevier), Burlington, MA, USA, 2009. ISBN 978-0-7506-8567-2

[Franke18]             Björn Franke: **Embedded Systems.** *Lecture 11: Worst-Case Execution Time* University of Edinburgh, Edinburgh, Scotland, 2018. Downloadable from: http://www.inf.ed.ac.uk/teaching/courses/es/PDFs/lecture_11.pdf [Last accessed: 31.1.20121]

[Friedenthal17]        Sanford Friedenthal, Christopher Oster: **Architecting Spacecraft with SysML - *A Model-based Systems Engineering Approach*** CreateSpace Independent Publishing Platform, North Charleston, S.C., USA, 2017. ISBN 978-1-5442-8806-2

[Furia12]      Carlo A. A. Furia, Dino Mandrioli, Angelo Morzenti, Matteo Rossi: **Modelling Time in Computing** Springer Verlag, Berlin, Germany, 2012. ISBN 978-3-642-43136-4

[Furrer19]     Frank J. Furrer: **Future-Proof Software-Systems** - *A Sustainable Evolution Strategy* Springer Vieweg Verlag, Wiesbaden, Germany, 2019. ISBN 978-3-658-19937-1

[Gamatié10]    Abdoulaye Gamatié: **Designing Embedded Systems with the SIGNAL Programming Language** - *Synchronous, Reactive Specification* Springer Science & Business Media, New York, NY, USA, 2010. ISBN 978-1-489-98512-5

[Giese09]      Holger Giese, Stephan Hildebrandt: **Efficient Model Synchronization of Large-Scale Models** Technical Report 28, Hasso Plattner Institute at the University of Potsdam, 2009. Downloadable from: https://publishup. uni-potsdam.de/opus4-ubp/frontdoor/deliver/index/docId/2883/file/ tbhpi28.pdf [Last accessed: 24.1.2021]

[Gnesi12]      Stefania Gnesi, Tiziana Margaria: **Formal Methods for Industrial Critical Systems** - *A Survey of Applications* John Wiley & Sons, Hoboken, NJ, USA, 2012. ISBN 978-0-470-87618-3

[Gomaa16]      Hassan Gomaa 2016 Real-Time Software Design for Embedded Systems Cambridge University Press N.Y., USA New York 978-1-1107-04109-7

[Gorton11]     Ian Gorton 2011 Essential Software Architecture Springer Verlag 2 Heidelberg Germany 978-3-642-19175-6

[Gottschalk16] Melissa M. Gottschalk: **The Soft Side of Project Management** - *How to Encourage, Empower and Cultivate a Thriving Team* Independently published, 2016. ISBN 978-1-51472-691-4

[Graff03]      Mark G. Graff, Kenneth R. van Wyk: **Secure Coding** - *Principles & Practices* O'Reilly & Associates, Sebastopol, CA, USA, 2003. ISBN 978-0-596-00242-8

[Grumberg08]   Orna Grumberg, Tobias Nipkow, Christian Pfaller: **Formal Logical Methods for System Security and Correctness** IOS Press, Amsterdam, Netherlands, 2008. Downloadable from: https://epdf.pub/ formal-logical-methods-for-system-security-and-correctness.html [Last accessed: 3.3.2021]

[Gullo18]      Louis J. Gullo, Jack Dixon: **Design for Safety** John Wiley & Sons, Inc., Hoboken, NJ, USA, 2018. ISBN 978-1-118-97429-2

[Gupta11]      Ratneshwer Gupta, Anil Tripathi: **Component-Based Software Engineering** - *Dependability and Software Process Issues* LAP LAMBERT Academic Publishing, Riga, Latvia, 2011. ISBN 978-3-84338-680-7

[GUT18]        Graz University of Technology: *Meltdown and Spectre -Vulnerabilities in Modern Computers leak Passwords and Sensitive Data* GUT Website "Meltdown and Spectre", 2018. Accessible via: https://spec-treattack.com/ [Last accessed: 21.12.2020]

[Haden18]      Jeff Haden: **The Motivation Myth** - *How High Achievers Really Set Themselves Up to Win* Portfolio (Penguin), New York, NY, USA. ISBN 978-0-399-56376-8

[Haimes19]             Yacov Y. Haimes: **Modeling and Managing Interdependent Complex Systems of Systems** John Wiley & Sons, Inc. (Wiley - IEEE), Hoboken, NJ, USA, 2019. ISBN 978-1-119-17365-6

[Halvorson16]          Chad Halvorson: **People Management -** *Everything you need to know about managing and leading People at Work* CreateSpace Independent Publishing Platform, North Charleston, S.C., USA, 2016. ISBN 978-1-5229-7235-8

[Hamiaz14]             Mounira Kezadri Hamiaz, Benoit Combemale, Marc Pantel, Xavier Thirioux: **Correct-by-Construction Model Composition -** *Application to the Invasive Software Composition Method* In: B. Buhnova, L. Happe, J. Kofron (Editors): *Formal Engineering - Approaches to Software Components and Architectures* Electronic Proceedings in Theoretical Computer Science (EPTCS), Waterloo, Australia, 2014, pp. 108-122. Downloadable from: https://www.researchgate.net/publication/261368803_Correct-by-Construction_model_composition_Application_to_the_Invasive_Software_Composition_method          [Last accessed: 3.3.2021]

[Harris17]             Michael D. S. Harris: **The Business Value of Software** CRC Press (Taylor & Francis Inc.), Boca Raton, FL, USA, 2017. ISBN 978-1-498-78286-9

[Harrison19]           Mark Harrison: **The CTO's Guide to Code Quality (PHP Edition)** Independently published, 2019. ISBN 978-1-6935661-8-9

[Hart15]               Laura E. Hart: *Introduction To Model-Based System Engineering MBSE) and SysML [Presentation]* Delaware Valley INCOSE Chapter Meeting, July 30, 2015. Downloadable from: https://www.incose.org/docs/default-source/delaware-valley/mbse-overview-incose-30-july-2015.pdf [Last accessed: 18.1.2021]

[Hayden16]             Lance Hayden 2016 People-Centric Security - Transforming Your Enterprise Security Culture McGraw-Hill Education N.Y., USA New York 978-0-07-184677-6

[Heckel20]             Reiko Heckel, Gabriele Taentzer: **Graph Transformation for Software Engineers -** *With Applications to Model-Based Development and Domain-Specific Language Engineering* Springer Nature Switzerland, Cham, Switzerland, 2020. ISBN 978-3-030-43918-7

[Heller20]             Daniel Heller 2020 Building a Career in Software - A Comprehensive Guide to Success in the Software Industry Apress Media LLC N.Y., USA New York 978-1-4842-6146-0

[Hewitt19]             Eben Hewitt: **Semantic Software Design -** *A New Theory and Practical Guide for Modern Architects* O'Reilly Media Inc., Sebastopol, CA, USA, 2019. ISBN 978-1-492-04595-3

[Hibbs09]              Curt Hibbs, Steve Jewett, Mike Sullivan: **The Art of Lean Software Development -** *A Practical and Incremental Approach* O'Reilly and Associates Ltd., Sebastopol, CA, USA, 2009. ISBN 978-0 596 51731 1

[Hinchey17]            Mike Hinchey Jonathan P Bowen Ernst-Rüdiger Olderog Eds 2017 Provably Correct Systems NASA Monographs in Systems and Software Engineering Springer International Publishing Cham, Switzerland 978-3-319-48627-7

[Hoang13]              Thai Son Hoang: *An Introduction to the Event-B Modelling Method* In "Industrial Deployment of System Engineering Methods",

Springer-Verlag, Berlin, Germany, 2013. http://www.springer.com/computer/swe/book/978-3-642-33169-5. Downloadable from: https://www.researchgate.net/publication/259929812_An_Introduction_to_the_Event-B_Modelling_Method [Last accessed: 5.2.2021]

[Hobbs20]        Chris Hobbs 2020 Embedded Software Development for Safety-Critical Systems CRC Press (Taylor & Francis) 2 Boca Raton FL, USA 978-0-367-33885-5

[Hohmann03]     Luke Hohmann 2003 Beyond Software Architecture - Creating and Sustaining Winning Solutions Addison-Wesley Professional Upper Saddle River USA 978-0-201-77594-5

[Hollings18]     Cristopher Hollings Ursula Martin Adrian Rice 2018 Ada Lovelace Byron - The Making of a Computer Scientist University of Chicago Press IL, USA Chicago 978-1-851-24488-1

[Holt19]         Jon Holt Simon Perry 2019 SysML for Systems Engineering - A Model-Based Approach Institution of Engineering and Technology 3 London UK 978-1-78561-554-2

[Horn18]         Christopher Horn, Anita D'Amico: *Measuring Application Security* US Department of Homeland Security, Washington, USA, 2018. Downloadable      from:      https://securedecisions.com/wp-content/uploads/2018/08/Measuring_Application_Security.pdf [Last accessed: 20.2.2021]

[Hubert02]       Richard Hubert 2002 Convergent Architecture John Wiley & Sons N.Y., USA New York 978-0-471-10560-0

[Humble10]       Jez Humble, David Farley: **Continuous Delivery** - *Reliable Software Releases Through Build, Test, and Deployment Automation* Addison Wesley, Upper Saddle River, NJ, USA, 2010. ISBN 978-0-321-60191-9

[IEEE1471]       Institute of Electrical and Electronics Engineers: **IEEE Recommended Practice for Architectural Description of Software-Intensive Systems** IEEE Standard 1471-2000, Institute of Electrical and Electronics Engineers, New York, NY, USA, 2000. ISBN 0-7381-2518-0

[IIC16]          Industrial Internet Consortium (IIC): **Industrial Internet of Things - Volume G4: Security Framework** Industrial Internet Consortium (IIC), Needham, Massachusetts, USA, 2016. IIC:PUB:G4:V1.0:PB:20160926. Downloadable from: https://www.iiconsortium.org/IISF.htm [Last accessed: 24.2.2021]

[IIC19]          Industrial Internet Consortium (IIC): **The Industrial Internet of Things - *Volume G1: Reference Architecture*** Industrial Internet Consortium (IIC), Needham, Massachusetts, USA, Version 1.9, 2019. Downloadable from: https://www.iiconsortium.org/IIRA.htm [Last accessed: 24.2.2021]

[Ince93]         DC Ince 1993 An Introduction to Discrete Mathematics, Formal System Specification, and Z Oxford University Press 2 New York NY, USA 978-0-198-53836-3

[Ingeno18]       Joseph Ingeno: **Software Architect's Handbook -** *Become a successful Software Architect by implementing effective Architecture Concepts* Packt Publishing, Birmingham, UK, 2018. ISBN 978-1-7886-2406-0

[ISO13568]       ISO/IEC JTC 1/SC 22: **Information technology - Z formal Specification Notation - Syntax, Type System and Semantics** ISO/

IEC 13568:2002 Standard (revised 2013), Geneva, Switzerland, 2002. Available at: https://www.iso.org/standard/21573.html [Last accessed: 2.2.2021]. Free download from: https://standards.iso.org/ittf/PubliclyAvailableStandards/c021573_ISO_IEC_13568_2002(E).zip [Last accessed: 4.2.2021]

[ISO15942]      ISO: *Information Technology - Programming Languages - Guide for the use of the Ada Programming Language in High Integrity Systems* Technical Report ISO/IEC TR 15942:2000(E), 1st edition, March 2000. ISO, Geneva, Switzerland. Downloadable from: http://www.dit.upm.es/~str/ork/documents/adahis.pdf [Last accessed: 2.2.2021]

[ISO42010]      ISO, IEC JTC 1, SC 7: Systems and Software Engineering - Architecture Description ISO, IEC, IEEE 42010:2011, Standard, revised, 2017 INTERNATIONAL ORGANIZATION FOR STANDARDIZATION (ISO) Switzerland Geneva 2017

[ISO8652]       ISO/IEC JTC 1/SC 22: **ISO/IEC 8652:2012 Standard: Information technology - Programming Languages - ADA** ISO, Geneva, Switzerland, 2018. Available from: https://www.iso.org/standard/61507.html [Last accessed: 2.2.2021]

[ITRB99]        Information Technology Resources Board (ITRB): *Assessing the Risks of Commercial-Off-The-Shelf Applications* ITRB White Paper, 1999. Downloadable from: https://govinfo.library.unt.edu/npr/howto/cots819.pdf [Last accessed: 15.2.2021]

[Jacky97]       Jonathan Jacky: **The Way of Z - Practical Programming with Formal Methods** Cambridge University Press, Cambridge, UK, 1997 (Reprinted 2001). ISBN 978-0-521-55976-8

[Jamshidi09a]   Mo Jamshidi (Editor): **Systems of Systems Engineering - *Principles and Applications*** CRC Press, Taylor & Francis Group, Boca Raton, USA, 2009. ISBN 978-1-4200-6588-6

[Jamshidi09b]   Mo Jamshidi (Editor): **Systems of Systems Engineering - *Innovations for the 21st Century*** John Wiley & Sons Inc., Hoboken, New Jersey, USA, 2009. ISBN 978-0-470-19590-1

[Janca20]       Tanya Janca: **Alice and Bob Learn Application Security** John Wiley & Sons, Inc., Indianapolis, IN, USA, 2020. ISBN 978-1-119-68735-1

[Jarrar18]      Abdessamad Jarrar, Youssef Balouki: *Formal Reasoning for Air Traffic Control System Using Event-B Method* In: Computational Science and Its Applications (ICCSA 2018), Springer Nature Switzerland AG, Cham, Switzerland, 2018. ISBN 978-3-319-95164-5

[Johnson20]     Patricia Johnson: *Top 10 Application Security Best Practices* White Paper, WhiteSource In., New York, USA, October 22, 2020. Accessed at: https://resources.whitesourcesoftware.com/blog-whitesource/application-security-best-practices [Last accessed: 20.2.2021]

[Johnsson19]    Dan Bergh Johnsson, Daniel Deogun, Daniel Sawano: **Secure By Design** Manning Publications, Shelter Island, NY, USA, 2019. ISBN 978-1-617-29435-8

[Kästner17]     Daniel Kästner, Laurent Mauborgne, Christian Ferdinand: *Detecting Safety-and Security-Relevant Programming Defects by Sound Static Analysis* Conference Paper, CYBER 2017: The Second International Conference on Cyber-Technologies and Cyber-Systems, Barcelona, Spain, November 12 - 16, 2017. Downloadable from: https://www.

|            | researchgate.net/publication/330311935_Detecting_Safety-and_ Security-Relevant_Programming_Defects_by_Sound_Static_Analysis/ link/5c3868eb458515a4c71cd2b0/download [Last accessed: 20.5.2021] |
|---|---|
| [Kelly08] | Steven Kelly, Juha-Pekka Tolvanen: Domain-Specific Modeling - Enabling Full Code Generation. Wiley-IEEE Computer Society Press, Los Alamitos, CA, USA, 2008. ISBN 978-0-470-03666-2 |
| [Key15] | Sam Key: **ADA Programming Success In A Day - *Beginner's Guide to fast, easy, and efficient learning of ADA programming*** CreateSpace Independent Publishing Platform, North Charleston, S.C., USA, 2015. ISBN 978-1-5153-7132-8 |
| [Kim18a] | Gene Kim, Kevin Behr, George Spafford: **The Phoenix Project - *A Novel About IT, DevOps, and Helping Your Business Win*** IT Revolution Press, Portland, OR, USA, 2018. ISBN 978-1-9427-8829-4 |
| [Kim19] | Gene Kim: **The Unicorn Project - *A Novel about Developers, Digital Disruption, and Thriving in the Age of Data*** IT Revolution Press, Portland, OR, USA, 2019. ISBN 978-1-9427-8876-8 |
| [Kleidermacher12] | David Kleidermacher, Mike Kleidermacher: **Embedded Systems Security - *Practical Methods for Safe and Secure Software and Systems Development*** Newnes (Elsevier), Burlington, MA, USA, 2012. ISBN 978-0-123-86886-2 |
| [Kleppe09] | Anneke Kleppe: **Software Language Engineering - *Creating Domain-Specific Languages using Metamodels*** Addison-Wesley, Upper Saddle River, N.J., USA, 2009. ISBN 978-0-321-55345-4 |
| [Knight12] | John Knight 2012 Fundamentals of Dependable Computing for Software Engineers CRC Press (Taylor & Francis) FL, USA Boca Raton 978-1-439-86255-1 |
| [Knodel16 | ]Jens Knodel, Matthias Naab: **Pragmatic Evaluation of Software Architectures** Springer International Publishing, Cham, Switzerland, 2016. ISBN 978-3-319-34176-7 |
| [Kopetz11] | Hermann Kopetz 2011 Real-Time Systems - Design Principles for Distributed Embedded Applications Springer Science & Business Media 2 New York N.Y., USA 978-1-461-42866-4 |
| [Kotusev18] | Svyatoslav Kotusev: **The Practice of Enterprise Architecture - *A Modern Approach to Business and IT Alignment*** SK Publishing, Melbourne, Australia, 2018. ISBN 978-0-6483-0983-3 |
| [Kourie12] | Derrick G. Kourie, Bruce W. Watson: **The Correctness-by-Construction Approach to Programming** Springer Verlag, Berlin, Germany, 2012. ISBN 978-3-642-27918-8 |
| [Kröger08] | Fred Kröger, Stephan Merz: **Temporal Logic and State Systems** Springer Verlag, Berlin, Germany, 2008. ISBN 978-3-540-67401-6 |
| [Kruchten19] | Philippe Kruchten, Robert Nord: **Managing Technical Debt - *Reducing Friction in Software Development*** Addison-Wesley, Upper Saddle River, NJ, USA, 2019. ISBN 978-0-135-64593-2 |
| [Kulik18] | Tomas Kulik, Peter Gorm Larsen: ***Towards formal Verification of Cyber Security Standards*** Proceedings ISP RAS, Vol. 30, Issue 4, 2018, pp. 79-94. DOI: https://doi.org/10.15514/ISPRAS2018-30(4)-5 https://www.researchgate.net/publication/327878060_ Towards_Formal_Verification_of_Cyber_Security_Standards/ link/5baade8a299bf13e604c91ea/download |

[Kulkarni20]      Dhanashree C. Kulkarni: *Managing Open Source Library Risks*
                  OWASP White Paper, Open Web Application Security Project,
                  Maryland, USA, 2020. Downloadable from: https://owasp.org/www-
                  pdf-archive//Managing_Open_Source_Library_Risks.pdf      [Last
                  accessed: 25.2.2021]

[Lacamera18]      Daniele Lacamera: **Embedded Systems Architecture - *Explore
                  Architectural Concepts, pragmatic Design Patterns, and best Practices
                  to produce robust Systems*** Packt Publishing, Birmingham, UK, 2018.
                  ISBN 978-1-7888-3250-2

[Lamsweerde09]    Axel van Lamsweerde: **Requirements Engineering - *From System
                  Goals to UML Models to Software Specifications*** John Wiley & Sons
                  Inc., Chichester, UK, 2009. ISBN 978-0-470-01270-3

[Langer20]        Arthur M. Langer: **Analysis and Design of Next-Generation Software
                  Architectures - *5G, IoT, Blockchain, and Quantum Computing***
                  Springer Nature Switzerland, Cham, Switzerland, 2020. ISBN
                  978-3-030-36898-2

[Lankhorst17]     Marc Lankhorst 2017 Enterprise Architecture at Work - Modelling,
                  Communication, and Analysis Springer Verlag 4 Berlin Germany
                  978-3-662-53932-3

[Laplante07]      A Philip 2007 Laplante: What Every Engineer Should Know about
                  Software Engineering CRC Press (Taylor & Francis) FL, USA Boca
                  Raton 978-0-849-37228-5

[Larson21]        Will Larson: **Staff Engineer - *Leadership beyond the Management
                  Track*** Independently published, 2021. ISBN 978-1-73641-791-1

[Laski10]         Janusz Laski William Stanley 2010 Software Verification and Analysis
                  - An Integrated Hands-On Approach Springer Verlag London, UK
                  978-1-849-96829-4

[Lewis20]         Elijah Lewis: **Clean Architecture (3 in 1) - Beginner's Guide +
                  Tips and Tricks + Advanced and Effective Strategies using Clean
                  Architecture Principles** Independently published, 2020. ISBN
                  979-8-6689-1899-7

[Li13]            Huxi Li: **The Myth of Enterprise System Pollutions - *The Hidden
                  Demons*** CreateSpace Independent Publishing Platform, 2013. ISBN
                  978-1-4812-8050-1

[Lieberman06]     Benjamin A. Lieberman: **The Art of Software Modeling** Auerbach
                  Publishers Inc. (Taylor & Francis), Boca Raton, FL, USA, 2006. ISBN
                  978-1-420-04462-1

[Lilienthal19]    Carola Lilienthal 2019 Sustainable Software Architecture - Analyze and
                  Reduce Technical Debt Dpunkt Verlag GmbH 2 Heidelberg Germany
                  978-3-8649-0673-2

[Liu04]           Shaoying Liu: **Formal Engineering for Industrial Software
                  Development - *Using the SOFL Method*** Springer Verlag, Heidelberg,
                  Germany, 2004. ISBN 978-3-540-20602-6

[Long13]          Fred Long: **Java Coding Guidelines - *75 Recommendations for
                  Reliable and Secure Programs*** Pearson Education (Addison-Wesley),
                  Upper Saddle River, NJ, USA, 2013. ISBN 978-0-321-93315-7

[Long14]          John Long: **Process Modeling Style** Morgan Kaufmann (Elsevier),
                  Waltham, MA, USA, 2014. ISBN 978-0-128-00959-8

## References

[Lucio14]  Levi Lucio, Qin Zhang, Phu H. Nguyen, Moussa Amrani, , Jacques Klein, Hans Vangheluwe, Yves Le Traon: **Advances in Model-Driven Security** Chapter in Advances in Computers, February 2014. Downloadable from: https://www.researchgate.net/publication/261361604_Advances_in_Model-Driven_Security [Last accessed: 15.01.2022]

[Maggi20]  Federico Maggi, Marcello Poglian: *Rogue Automation - Vulnerable and Malicious Code in Industrial Programming* Trend Micro Research White Paper, Irving, TX, USA, 2020. Downloadable from: https://documents.trendmicro.com/assets/white_papers/wp-rogue-automation-vulnerable-and-malicious-code-in-industrial-programming.pdf [Last accessed: 21.2.2021]

[Mandal14]  Kamales Mandal, Tapodhan Sen: *A New Approach to Application Portfolio Assessment for New-Age BusinessTechnology Requirements* White Paper, Cognizant Inc., Teaneck, NJ, USA, 2014. Downloadable from: https://www.cognizant.com/whitepapers/A-New-Approach-to-Application-Portfolio-Assessment-for-New-Age-Business-Technology-Requirements-codex939.pdf [Last accessed: 27.2.2021]

[Manna12]  Zohar Manna 2012 Temporal Verification of Reactive Systems - Safety Springer Verlag N.Y., USA New York 978-1-461-28701-8

[Mansfield10]  John Mansfield: **The Nature of Change or the Law of Unintended Consequences** - *An Introductory Text to Designing Complex Systems and Managing Change* Imperial College Press, London, UK, 2010. ISBN 978-1-84816-540-3

[Marshall09]  Donis Marshall, John Bruno: **Solid Code - Optimizing the Software Development Life Cycle** Microsoft Press, Redmond, USA, 2009 (Developer Best Practices). ISBN 978-0-7356-2592-1

[Martin08]  Robert Martin: **Clean Code - *A Handbook of Agile Software Craftsmanship*** Prentice-Hall, Upper Saddle River, N.J., USA, 2008. ISBN 978-0-132-35088-4

[Martin11]  Robert C. Martin: **The Clean Coder** - *A Code of Conduct for Professional Programmers* Addison-Wesley, Upper Saddle River, NJ, USA, 2011. ISBN 978-0-137-08107-3

[Marwedel18]  Peter Marwedel 2018 Embedded System Design - Embedded Systems Foundations of Cyber-Physical Systems, and the Internet of Things Springer International Publishing 3 Cham Switzerland 978-3-319-56043-4

[Masmali19]  Omar Masmali, Omar Badreddin: **Model-Driven Security - *A Systematic Mapping Study*** Software Engineering, 2019, 7(2), pp. 30-38. Downloadable from: https://www.researchgate.net/publication/335822278_Model_Driven_Security_A_Systematic_Mapping_Study [Last accessed: 15.01.2022]

[Matulevičius17]  Raimundas Matulevičius: **Fundamentals of Secure System Modelling** Springer International Publishing, Cham, Switzerland, 2017. ISBN 978-3-319-87143-1

[McCormick11]  John W. McCormick, Frank Singhoff, Jérôme Hugues: **Building Parallel, Embedded, and Real-Time Applications with Ada** Cambridge University Press, Cambridge, UK, 2011. ISBN 978-0-521-19716-8

[McCormick15]        John W. McCormick: **Building High Integrity Applications with SPARK** Cambridge University Press, Cambridge, UK, 2015. ISBN 978-1-107-65684-0

[McGhan15]           Catharine L. R. McGhan, Richard M. Murray: *Application of Correct-by-Construction Principles for a Resilient Risk-Aware Architecture* White Paper, California Institute of Technology, Pasadena, CA, USA, 2015. Downloadable from: https://authors.library.caltech.edu/65888/1/application.pdf [Last accessed: 3.3.2021]

[McKeag80]           R. M. McKeag, A. M. Macnaghten (Editors): **On the Construction of Programs - *An Advanced Course*** Cambridge University Press, Cambridge, UK, 1980. ISBN 0-521-23090-X

[Meeker20]           Heather Meeker: **Open (Source) for Business - *A Practical Guide to Open Source Software Licensing*** Independently published, 3$^{rd}$ edition, 2020. ISBN 979-8-61820-177-3

[Mega20]             Mega Inc.: *Application Portfolio Management -Key Principles and Best Practices* White Paper, Mega Inc., Raynham MA, USA, 2020. Downloadable    from:    https://www.mega.com/en/confirmation/application-portfolio-management-key-principles-best-practices        [Last accessed: 27.2.2021]

[Merkow10]           S Mark 2010 Merkow, Lakshmikanth Raghavan: Secure and Resilient Software Development CRC Press (Taylor & Francis) FL, USA Boca Raton 978-1-439-82696-6

[Merton36]           Robert K Merton 1936 The Unanticipated Consequences of Purposive Social Action American Sociological Review 1 6 894 904

[Meyer09]            Bertrand Meyer: **A Touch of Class - *Learning to Program Well with Objects and Contracts*** Springer-Verlag, Berlin, 2009. ISBN 978-3-540-92144-5

[Meyer14]            Bertrand Meyer: **Agile! - The Good, the Hype, and the Ugly** Springer International Publishing, Cham, Switzerland, 2014. ISBN 978-3-319-05154-3

[Meyers01]           B. Craig Meyers, Patricia Oberndorf: **Managing Software Acquisition - *Open Systems and COTS Products*** Pearson Education (Taylor & Francis), Boca Raton, FL, USA, 2001 (SEI Series in Software Engineering). ISBN 978-0-201-70454-9

[Micouin14]          Patrice    Micouin:    **Model-Based    Systems    Engineering    - *Fundamentals and Methods*** Wiley-ISTE, London, UK, 2014. ISBN 978-1-848-21469-9

[Miles06]            Russ Miles, Kim Hamilton: **Learning UML 2.0 - *A Pragmatic Introduction to UML*** O'Reilly Media, Sebastopol, CA, USA, 2006. ISBN 978-0-596-00982-3

[Mills09]            Bruce Mills 2009 Practical Formal Software Engineering - Wanting the Software You Get Cambridge University Press NY, USA New York 978-0-521-87903-3

[Minghui19]          Sun Minghui, Georgios Bakirtzis, Hassan Jafarzadeh, Cody Fleming: *Correct-by-Construction:    a    Contract-Based    semi-automated Requirement Decomposition Process* arXiv Preprint, 4. September 2019. Downloadable    from:    https://arxiv.org/abs/1909.02070    [Last accessed: 3.3.2021]

[Miret14]         Leticia Pascual Miret: *Consistency Models in modern distributed Systems. An Approach to Eventual Consistency.* Master's Thesis in Parallel and Distributed Computing, 09, 2014. Universitat Politecnica de Valencia, Valencia, Spain. Downloadable from : https://riunet.upv.es/ bitstream/handle/10251/54786/TFMLeticiaPascual.pdf [Last accessed: 24.1.2021]

[MISRA16]         Motor Industry Software Reliability Association (MISRA): **Achieving Compliance with MISRA Coding Guidelines** MISRA Guideline (MISRA Compliance:2016), UK, 2016. ISBN 978-1-906400-13-2. Downloadable from: https://misra.org.uk/LinkClick.aspx?fileticket=w_ Syhpkf7xA%3D&tabid=57 [Last accessed: 3.11.2020]

[Mitchell01]      Richard Mitchell, Jim McKim: **Design by Contract, by Example** Addison-Wesley, Indianapolis, IL, USA, 2001. ISBN 978-0-201-634600

[Mitra21]         Sayan Mitra 2021 Verifying Cyber-Physical Systems - A Path to Safe Autonomy The MIT Press MA, USA Cambridge 978-0-262-04480-6

[Mittal18]        Saurabh Mittal, Saikou Diallo, Andreas Tolk (Editors): **Emergent Behaviour in Complex Systems - *A Modeling and Simulation Approach*** John Wiley & Sons, Inc., Hoboken, NJ, USA, 2018. ISBN 978 - 1-119-37886-0

[Monin13]         Jean François Monin: **Understanding Formal Methods** Springer Verlag, London, UK, 2013. ISBN 978-1-852-33247-1

[Morana08]        Marco Morana: *Building Security into Applications* OWASP Chapter Lead Presentation, Blue Ash, USA, July 30th 2008. Downloadable from. https://owasp.org/www-pdf-archive/Build_Security_Into_Applications_ Short.pdf [Last accessed: 20.2.2021]

[Moyle20]         Ed Moyle, Diana Kelley: **Practical Cybersecurity Architecture - *A Guide to creating and implementing robust Designs for Cybersecurity Architects*** Packt Publishing, Birmingham, UK, 2020. ISBN 978-1-8389-8992-7

[Murer11]         Stephan Murer, Bruno Bonati, Frank J. Furrer: **Managed Evolution - *A Strategy for Very Large Information Systems*** Springer Verlag, Berlin, Germany, 2011. ISBN 978-3-642-01632-5

[Nanda19]         Manju Nanda, Yogananda Jeppu (Editors): **Formal Methods for Safety and Security - *Case Studies for Aerospace Applications*** Springer Nature Singapore Pte Ltd., Singapore, Singapore (Softcover Reprint of the original 1st edition 2018). ISBN 978-9-811-35054-2

[NAP19]           NAP (US National Academy of Sciences): **Beyond Spectre - *Confronting New Technical and Policy Challenges*** Forum on Cyber-Resilience - Proceedings of a Workshop, NAP, Washington, D.C., USA, 2019. ISBN 978-0-309-49146-4. Downloadable from: https://www.nap. edu/download/25418 [Last accessed: 5.12.2020]

[NASA04]          NASA Glenn Research Center: **NASA Software Safety Guidebook** US National Aeronautics and Space Administration (NASA), Washington, DC, USA, 2004. Document NASA-GB-8719.13. Downloadable from: https://standards.nasa.gov/standard/osma/nasa-gb-871913 [Last accessed: 9.2.2021]

[NASA20]          NASA Office of Safety and Mission Assurance (OSMA): **NASA Software Assurance and Safety Standard** US National Aeronautics and Space Administration (NASA), Washington, DC, USA, 2020.

Document NASA-STD-8739.8A. Downloadable from: https://standards. nasa.gov/standard/osma/nasa-std-87398 [Last accessed: 9.2.2021]

[NASA98]        US National Aeronautics and Space Administration: **NASA Software Policies** NASA Directive NPD 2820.1A, 1998 (Revalidated 5/29/04). Downloadable from: http://everyspec.com [Last accessed: 15.2.2021]

[NCSC21a]       NCSC: *NCSC Alert: Critical Vulnerability in Apache Log4j Library* (CVE-2021-44228, Update 2, 15. December 2021). Ireland National Cyber Security Centre, Dublin, Ireland. Downloadable from: https:// www.ncsc.gov.ie/pdfs/apache-log4j-101221.pdf        [Last        accessed: 19.12.2021]

[NCSC21b]       NCSC: *Software Supply Chain Attacks* UK National Cyber Security Centre (NCSC), London, UK, 2021. Downloadable from: https://www. dni.gov/files/NCSC/documents/supplychain/Software_Supply_Chain_ Attacks.pdf [Last accessed: 20.07.2021]

[Neumann94]     Peter  G.  Neumann:  **Computer-Related  Risks**  Addison-Wesley Professional,  Upper  Saddle  River,  NJ,  USA,  1994.  ISBN 978-0-201-55805-0

[Nielson19]     Flemming  Nielson,  Hanne  Riis  Nielson:  **Formal  Methods  -  *An Appetizer*** Springer  Nature  Switzerland,  Cham,  Switzerland,  2019. ISBN 978-3-030-05155-6

[NIST14]        US  National  Institute  of  Standards  and  Technology:  **Framework for  Improving  Critical  Infrastructure  Cybersecurity**  NIST, Gaithersburg, MD, USA, 2014. Downloadable from: https://www.nist. gov/system/files/documents/cyberframework/cybersecurity-frame-work-021214.pdf [Last accessed: 24.2.2021]

[NIST800–160A]  Ron  Ross,  Michael  McEvilley,  Janet  Oren:  **Systems  Security Engineering  Considerations  for  a  Multidisciplinary  Approach in  the  Engineering  of  Trustworthy  Secure  Systems**  NIST  Special Publication 800–160, Volume 1, November 2016. Downloadable from: from:  https://doi.org/10.6028/NIST.SP.800-160v1  or  https://nvlpubs. nist.gov/nistpubs/SpecialPublications/NIST.SP.800-160.pdf        [Last accessed: 12.11.2020]

[NIST800–160B]  Ron  Ross,  Victoria  Pillitteri,  Richard  Graubart,  Deborah  Bodeau, Rosalie McQuaid: **Developing Cyber Resilient Systems - *A Systems Security Engineering Approach*** NIST Special Publication 800–160, Volume 2, Rev. 1, December 2021. Downloadable from: https://nvlpubs. nist.gov/nistpubs/SpecialPublications/NIST.SP.800-160v2r1.pdf    [Last accessed: 09.12.2021]

[NIST80053B]    NIST  JOINT  TASK  FORCE:  **Control  Baselines  for  Information Systems and Organizations** NIST Special Publication 800–53B, NIST Washington, USA, October 2020. Downloadable from: https://doi. org/10.6028/NIST.SP.800-53B [Last accessed: 30.10.2020]

[NISTIR8151]    Paul  E.  Black,  Lee  Badger,  Barbara  Guttman,  Elizabeth  Fong: **Dramatically  Reducing  Software  Vulnerabilities  -  *Report  to  the White House Office of Science and Technology Policy*** US National Institute of Standards and Technology, NISTIR 8151, November 2016. Downloadable    from:    https://nvlpubs.nist.gov/nistpubs/ir/2016/NIST. IR.8151.pdf [Last accessed: 23.8.2020]

[NISTSCM15]    NIST: *Best Practices in Cyber Supply Chain Risk Management* ConferenceMaterials, US National Institute of Standards and Technology, Workshop October 1 - 2, 2015, Washington, DC, USA. Downloadable from: https://csrc.nist.gov/CSRC/media/Projects/Supply-Chain-Risk-Management/documents/briefings/Workshop-Brief-on-Cyber-Supply-Chain-Best-Practices.pdf [Last accessed: 21.2.2021]

[NIST800-160A]    Ron Ross, Michael McEvilley, Janet Oren: Systems Security Engineering Considerations for a Multidisciplinary Approach in the Engineering of Trustworthy Secure Systems. NIST Special Publication 800-160, Volume 1, November 2016. Downloadable from: https://doi.org/10.6028/NIST.SP.800-160v1 or https://nvlpubs.nist.gov/nistpubs/SpecialPublications/NIST.SP.800-160.pdf [Last accessed: 12.11.2020]

[NIST800-160B]    Ron Ross, Victoria Pillitteri, Richard Graubart, Deborah Bodeau, Rosalie McQuaid: Developing Cyber Resilient Systems - A Systems Security Engineering Approach. NIST Special Publication 800-160, Volume 2, Rev. 1, December 2021. Downloadable from: https://nvlpubs.nist.gov/nistpubs/SpecialPublications/NIST.SP.800-160v2r1.pdf [Last accessed: 09.12.2021]

[Nuseibeha01]    Bashar Nuseibeha, Steve Easterbrook, Alessandra Russo: *Making Inconsistency Respectable in Software Development* Journal of Systems and Software, Volume 58, Issue 2, September 2001, pp. 171–180. https://doi.org/10.1016/S0164-1212(01)00036-X

[O'Regan17]    Gerard O'Regan: **Concise Guide to Formal Methods - *Theory, Fundamentals, and Industry Applications*** Springer International Publishing, Cham, Switzerland, 2017. ISBN 978-3-319-64020-4

[Ousterhout18]    John Ousterhout: **A Philosophy of Software Design** Yakniam Press, Palo Alto, CA, USA, 2018. ISBN 978-1-73210-220-0

[Paret07]    Dominique Paret 2007 Multiplexed Networks for Embedded Systems - CAN, LIN, FlexRay Safe-by-Wire Wiley & Sons Inc. Chichester, UK 978-0-470-03416-3

[Pastor07]    Oscar Pastor, Juan Carlos Molina: **Model-Driven Architecture in Practice - *A Software Production Environment Based on Conceptual Modeling*** Springer-Verlag, Berlin, Germany, 2007. ISBN 978-3-540-71867-3

[Penczek10]    Wojciech Penczek, Agata Pólrola: **Advances in Verification of Time Petri Nets and Timed Automata - *A Temporal Logic Approach*** Springer Verlag, Berlin, Germany, 1st edition 2006 (Softcover reprint of Hardcover 1st edition, 2010). ISBN 978-3-642-06942-0

[Perrow99]    Charles Perrow 1999 Normal Accidents - Living with High-Risk Technologies Princeton University Press Princeton USA, Updated Edition 978-0-691-00412-9

[Pilone05]    Dan Pilone, Neil Pitman: **UML 2.0 in a Nutshell - *A Desktop Quick Reference*** O'Reilly Media, Inc., Sebastopol, USA, 2005. ISBN 978-0-596-00795-9

[Pinto19]    K Jeffrey 2019 Pinto: Project Management - Achieving Competitive Advantage Pearson Education Limited 5 Harlow UK 978-1-292-26914-6

[Pitschke20]        Juergen   Pitschke:   **Model-Based-Business-Engineering  -  *Style***
                    ***Guide and Modeling Policies*** Independently published, 2020. ISBN
                    979-8-67240-078-5

[Pittenger20]       Mike Pittenger: ***Open Source Software  -  Security Risks and***
                    ***Best    Practices***   White    Paper,   RogueWave   Software,   Seattle,
                    USA,   2020.   Downloadable   from:   https://dsimg.ubm-us.net/enve-
                    lope/352333/369442/1421334278_Open_source_software-Security_
                    risks_and_best_practices_white_paper.pdf [Last accessed: 25.2.2021]

[Ploesch04]         Reinhold   Ploesch:   **Contracts,   Scenarios,   and   Prototypes  -  *An***
                    ***Integrated Approach to High-Quality Software*** Springer Verlag, Berlin,
                    Germany, 2004. ISBN 978-3-540-43486-3

[Pohl12]            Klaus   Pohl,   Harald   Hönninger,   Reinhold   Achatz,   Manfred   Broy
                    (Editors):  **Model-Based  Engineering  of  Embedded  Systems  -  *The***
                    ***SPES 2020 Methodology*** Springer Verlag, Heidelberg, Germany, 2012.
                    ISBN 978-3-642-43992-6

[Pohl14]            Klaus   Pohl,   Harald   Hönninger,   Reinhold   Achatz,   Manfred   Broy
                    (Editors):  **Model-Based  Engineering  of  Embedded  Systems  -  The**
                    **SPES   2020   Methodology**.   Springer   Verlag,   Heidelberg,   Germany,
                    2012. ISBN 978-3-642-43992-6

[Rainey18]          Larry B Rainey Mo Jamshidi Eds 2019 Engineering Emergence - A
                    Modeling and Simulation Approach CRC Press (Taylor & Francis) FL,
                    USA Boca Raton 978-1-138-04616-0

[Reade89]           Chris   Reade:   **Elements   of   Functional   Programming**   Addison
                    Wesley   Longman,   Upper   Saddle   River,   NJ,   USA,   1989.   ISBN
                    978-0-201-12915-1

[Rebentisch17]      Eric Rebentisch: **Integrating Program Management and Systems**
                    **Engineering  -  *Methods,  Tools,  and  Organizational  Systems  for***
                    ***Improving Performance*** John Wiley & Sons, Inc., Hoboken, NJ, USA,
                    2017. ISBN: 978-1-119-25892-6

[Rees16]            Gary   Rees,   Raymond   French   (Editors):   **Leading,   Managing,   and**
                    **Developing  People**  CIPD,  London,  UK,  5th  edition,  2016.  ISBN
                    978-1-8439-8412-2

[Reisig13]          Wolfgang   Reisig   2013   Understanding   Petri   Nets  -  Modeling
                    Techniques Case Studies Springer Verlag, Berlin Analysis Methods
                    978-3-642-33277-7

[Rerup18]           Neil Rerup (Autor), Milad Aslaner: **Hands-On Cybersecurity for**
                    **Architects  -  *Plan  and  Design  robust  Security  Architectures*** Packt
                    Publishing, Birmingham, UK, 2018. ISBN 978-1-78883-026-3

[Rice20]            Liz Rice: **Container Security  -  *Fundamental Technology Concepts***
                    ***that   Protect   Containerized   Applications*** O'Reilly   Media   Inc.,
                    Sebastopol, CA, USA, 2020. ISBN 978-1-492-05670-6

[Ricketts20]        John   Ricketts:   **Exceeding   the   Goal  -  *Adventures   in   Strategy,***
                    ***Information Technology, Computer Software, Technical Services, and***
                    ***Goldratt's Theory of Constraints*** Industrial Press, Inc., South Norwalk,
                    Connecticut, USA, 2020. ISBN 978-0-8311-3656-7

[Rierson13]         Leanna Rierson: **Developing Safety-Critical Software  -  *A Practical***
                    ***Guide  for  Aviation  Software  and  DO-178C  Compliance*** CRC
                    Press   (Taylor   &   Francis   Inc.),   Boca   Raton,   FL,   USA,   2013.   ISBN
                    978-1-439-81368-3

[Ristic14]            Ivan Ristic: **Bulletproof SSL and TLS - *Understanding and Deploying SSL/TLS and PKI to Secure Servers and Web*** Feisty Duck Ltd., London, UK, 2014. ISBN 978-1-90711-704-6

[Robertson13]         Suzanne Robertson James Robertson 2013 Mastering the Requirements Process - Getting Requirements Right Addison-Wesley 3 Upper Saddle River NJ, USA 978-0-321-81574-3

[Rouff06]             Christopher Rouff, Michael Hinchey, James Rash, Walter Truszkowsi, Diana Gordon-Spears (Editors): **Agent Technology from a Formal Perspective** NASA Monographs in Systems and Software Engineering, Springer Verlag, London, UK, 2006. ISBN 978-1-852-33947-0. Downloadable from: https://www.researchgate.net/publication/227133578_Introduction_to_Formal_Methods [Last accessed: 25.1.2021]

[Rozanski12]          Nick Rozanski Eoin Woods 2012 Software Systems Architecture Addison-Wesley 2 Upper Saddle River NJ, USA 978-9-332-54795-7

[Ruhela12]            Vishal Ruhela *Z Formal Specification Language - An Overview* International Journal of Engineering Research & Technology (IJERT), Vol. 1, Issue 6, August 2012. Downloadable from: https://www.ijert.org/research/z-formal-specification-language-an-overview-IJERTV1IS6492.pdf [Last accessed: 3.2.2021]

[Scheid15]            Oliver Scheid: **AUTOSAR Compendium, Part 1 - *Application & RTE*** CreateSpace Independent Publishing Platform, North Charleston, S.C., USA, 2015. ISBN 978-1-502-75152-2

[Schneider01]         Steve Schneider: **The B-Method - *An Introduction*** Red Globe Press McMillan International), London, UK, 2001. ISBN 978-0-3337-9284-1

[Scholl19]            Boris Scholl, Trent Swanson, Peter Jausovec: **Cloud Native - *Using Containers, Functions, and Data to build next-generation Applications*** O'Reilly Media Inc., Sebastopol, CA, USA, 2019. ISBN 978-1-492-05382-8

[Schwartz19]          Mark Schwartz: **War and Peace and IT - *Business Leadership, Technology, and Success in the Digital Age*** IT Revolution Press, Portland, OR, USA, 2019. ISBN 978-1-9427-8871-3

[Seacord13]           C Robert 2013 Seacord: Secure Coding in C and C++ Addison-Wesley Professional (SEI Series in Software Engineering) 2 Upper Saddle River N.J., USA 978-0-321-82213-0

[Seacord14]           C Robert 2014 Seacord: CERT® C Coding Standard - 98 Rules for Developing Safe, Reliable, and Secure Systems Addison-Wesley Professional (SEI Series in Software Engineering) 2 Upper Saddle River N.J., USA 978-0-321-98404-3

[SEI06]               SEI: **"Views and Beyond" Architecture Documentation Template** Carnegie Mellon University, Software Engineering Institute (SEI), Pittsburgh, PA, USA, 2006. Downloadable from: https://resources.sei.cmu.edu/library/asset-view.cfm?assetid=519381 [Last accessed: 24.1.2021]

[Serrano20]           Eleazar Jiménez Serrano: **System and Software Verification using Petri Nets - *For functional Safety of ISO-26262 Requirements*** Independently published, 2020. ISBN 979-8-5506-9610-1

[Sharma20]            Nitin Sharma: **Securing Docker - *The Attack and Defense Way*** Independently published, Information Waarfare Center, 2020. ISBN 979-8-5545-5938-9

[Sherer92]        Susan A. Sherer: **Software Failure Risk - *Measurement and
                  Management*** Springer Verlag, Heidelberg, Germany, 1992. ISBN
                  978-1-461-36316-3

[Sherwood05]      John Sherwood Andrew Clark David Lynas 2005 Enterprise Security
                  Architecture - A Business-Driven Approach CRC Press (Taylor &
                  Francis) FL, USA Boca Raton 978-1-578-20318-5

[Sidhu16]         Baljit Singh Sidhu, Kirti Gupta: *Application Portfolio Management -
                  An Approach to Overcome IT Management Challenges* International
                  Journal of Trend in Research and Development (www.ijtrd.com),
                  Volume 3, 2016. Downloadable from: https://www.researchgate.
                  net/publication/312378740_Application_Portfolio_Management_
                  An_Approach_to_Overcome_IT_Management_Challenges/
                  link/587cb69e08aed3826aef9dca/download [Last accessed: 27.2.2021]

[Silver17]        Bruce Silver: **BPMN Quick and Easy Using Method and Style -
                  *Process Mapping Guidelines and Examples Using the Business
                  Process Modeling Standard*** Cody-Cassidy Press, Altadena, CA, USA,
                  2017. ISBN 978-0-9823-6816-9

[Singh19]         Arun Singh, Mike Mister: **How to Lead Smart People - *Leadership
                  for Professionals*** Profile Books Ltd., London, UK, 2019. ISBN
                  978-1-7881-6154-1

[Sirota14]        David Sirota Douglas A Klein 2014 The Enthusiastic Employee -
                  How Companies Profit by Giving Workers What They Want Pearson
                  Education (Addison-Wesley) 2 Upper Saddle River NJ, USA
                  978-0-134-05759-0

[Spivey89]        Michael Spivey: *An Introduction to Z and Formal Specifications* ACM
                  Software Engineering Journal, January 1989, pp. 40-50. Association
                  for Computing Machinery, New York, NY, USA. Downloadable from:
                  https://people.csail.mit.edu/dnj/teaching/6898/papers/spivey-intro-to-z.
                  pdf [Last accessed: 2.2.2021]

[Staron17]        Miroslaw Staron: **Automotive Software Architectures - *An
                  Introduction*** Springer International Publishing AG, Cham, Switzerland,
                  2017. ISBN 978-3-319-58609-0

[Starr17]         Leon Starr, Andrew Mangogna, Stephen Mellor: **Models to Code -
                  *With No Mysterious Gaps*** Apress Publishing, New York, NY, 2017.
                  ISBN 978-1-4842-2216-4

[Stevens20]       Perdita Stevens: *Maintaining Consistency in Networks of Models
                  - Bidirectional Transformations in the Large* Software Systems
                  Modeling, Nr. 19, 2020, pp. 39-65. Downloadable from: https://link.
                  springer.com/article/https://doi.org/10.1007/s10270-019-00736-x#Sec11.
                  [Last accessed: 24.1.2021]

[Stewart99]       David B. Stewart: *30 Pitfalls for Real-Time Software Developers*
                  Embedded Systems Programming, October 1999, Downloadable from:
                  https://docplayer.net/9366096-30-pitfalls-for-real-time-software-devel-
                  opers-part-1.html [Last accessed: 13.2.2021]

[Su11]            Wen Su, Jean-Raymond Abrial, Runlei Huang, Huibiao Zhu: *From
                  Requirements to Development: Methodology and Example* White
                  Paper, Software Engineering Institute, East China Normal University,
                  China, 2011. Downloadable from: http://deploy-eprints.ecs.soton.
                  ac.uk/316/1/Modes_version_55.pdf [Last accessed: 5.2.2021]

[Suryanarayana14]   Girish   Suryanarayana,   Ganesh   Samarthyam,   Tushar   Sharma: **Refactoring for Software Design Smells** - *Managing Technical Debt* Morgan Kaufmann, (Elsevier), Waltham, MA, USA 2014. ISBN 978-0-128-01397-7

[Synopsys20]   Synopsys, Inc.: *2020 Open Source Security and Risk Analysis Report* Synopsys, Inc., San Francisco, CA,USA, 2020. Downloadable from: www.synopsys.com/software [Last accessed: 25.2.2021]

[Talukder09]   K Asoke 2009 Talukder, Manish Chaitanya: Architecting Secure Software Systems CRC Press (Talor & Francis) FL, USA Boca Raton 978-0-367-38618-4

[Thiele04]   Lothar Thiele, Reinhard Wilhelm: *Design for Timing Predictability* Real-Time   Systems   (The   International   Journal   of   Time-Critical Computing   Systems),   28.   November   2004,   pp.   157-177.   Springer Nature Switzerland, Cham, Switzerland

[Thomas17]   Martyn Thomas: *Making Software 'Correct by Construction'* Transcript of a Lecture, Tuesday, 2 May 2017, Museum of London. Downloadable   from:   https://www.gresham.ac.uk/lectures-and-events/ making-software-correct-by-construction [Last accessed: 3.3.2021]

[Tiwari20]   Umesh   Kumar   Tiwari   2020   Santosh   Kumar:   Component-Based Software Engineering - Methods and Metrics CRC Press (Taylor & Francis) FL, USA Boca Raton 978-0-367-35488-6

[Tockey19]   Steve Tockey: **How to Engineer Software** - *A Model-Based Approach* IEEE Computer Society & John Wiley & Sons, Inc., Hoboken, NJ, USA, 2019. ISBN 978-1-119-54662-7

[Tooley13]   Mike Tooley 2013 Aircraft Digital Electronic and Computer Systems Routledge Publishers 2 Abingdon UK 978-0-415-82860-4

[Tornhill15]   Adam Tornhill: **Your Code As a Crime Scene** - *Use Forensic Techniques to Arrest Defects, Bottlenecks, and Bad Design in Your Programs* The Pragmatic Bookshelf, Raleigh, NC, USA, 2015. ISBN 978-1-68050-038-7 (The Pragmatic Programmers)

[Tornhill18]   Adam Tornhill: **Software Design X-Rays** - *Fix Technical Debt with Behavioral Code Analysis* The Pragmatic Bookshelf, Raleigh, NC, USA, 2018. ISBN 978-1-68050-272-5

[Treiber13]   Martin Treiber Arne Kesting 2013 Traffic Flow Dynamics - Data Models   and   Simulation   Springer   Verlag   Heidelberg,   Germany 978-3-642-32459-8

[VERACODE20]   VERACODE Inc.: **Secure Coding Best Practices Handbook** - *A Developer's Guide to Proactive Controls* VERACODE Inc., Burlington, MA, USA, 2020. Downloadable from : https://www.veracode.com/sites/ default/files/pdf/resources/guides/secure-coding-best-practices-hand-book-veracode-guide.pdf [Last accessed: 22.11.2020]

[Viega02]   John Viega, Matt Messier, Pravir Chandra: **Network Security with OpenSSL** - *Cryptography for Secure Communications* O'Reilly & Associates, Sebastopol, CA, USA, ISBN 978-0-596-00270-1

[Voelter13]   Markus Voelter: **DSL Engineering** - *Designing, Implementing, and Using Domain-Specific Languages* CreateSpace Independent Publishing   Platform,   North   Charleston,   S.C.,   2013.   ISBN 978-1-4812-1858-0

[Volter06]          Markus Volter 2006 Model-Driven Software Development - Technology, Engineering Management John Wiley & Sons Ltd. Chichester, UK 978-0-470-02570-3

[Vykhovanets19]     V.S. Vykhovanets: *Large-Scale Information Systems based on Conceptual Models* IEEE 2019 Twelfth International Conference "Management of large-scale system development" (MLSD), Moscow, Russia, 1-3 October 2019. DOI: https://doi.org/10.1109/ MLSD.2019.8911106

[Wahe11]            Stefan Wahe: **Open Enterprise Security Architecture (O-ESA) - A Framework and Template for Policy-Driven Security** Van Haren Publishing, Zaltbommel, NL, 2011. ISBN 978-9-0875-3672-5

[Walters20]         Ed Walters: **Using BPMN to model Business Processes - Handbook for Practitioners** Independently published, 2020. ISBN 978-1-67230-127-5

[Weilkiens08]       Tim Weilkiens: **Systems Engineering with SysML/UML** - *Modeling, Analysis, Design* Morgan Kaufmann (Elsevier), Waltham, MA, USA, 2008. ISBN 978-0-123-74274-2

[White11]           Elecia White: **Making Embedded Systems** - *Design Patterns for Great Software* O'Reilly and Associates, Sebastopol, CA, USA, 2011. ISBN 978-1-449-30214-6

[Wiegers13a]        Karl Wiegers Joy Beatty 2013 Software Requirements Microsoft Press 3 Redmond Washington, USA 978-0-7356-7966-5

[Wiegers13b]        Karl Wiegers, Joy Beatty: **Software Requirements** Microsoft Press, Redmond, USA, 3$^{rd}$ edition, 2013. ISBN 978-0-735-67966-5

[Wilhelm08]         Reinhard Wilhelm, Jakob Engblom, Andreas Ermedahl, Niklas Holsti, Stephan Thesing, David Whalley, Guillem Bernat, Christian Ferdinand, Reinhold Heckmann, Tulika Mitra, Frank Mueller, Isabelle Puaut, Peter Puschner, Jan Staschulat, Per Stenström: *The Worst-Case Execution Time Problem - Overview of Methods and Survey of Tools* ACM Transactions on Embedded Computing Systems, Vol. 7, Nr. 3, May 2008. Preprint downloadable from: http://www.cs.fsu.edu/~whalley/ papers/tecs07.pdf [Last accessed: 31.1.2021]

[Wlaschin17]        Scott Wlaschin: **Domain Modeling made Functional** The Pragmatic Bookshelf, Raleigh, NC, USA 2017. ISBN 978-1-6805-0254-1

[Woodcock96]        Jim Woodcock, Jim Davies: **Using Z** - *Specification, Refinement, and Proof* Prentice-Hall, Upper Saddle River, N.J., USA, 1996. ISBN 978-0-139-48472-8. Downloadable from: http://www.usingz.com/usingz.pdf [Last accessed: 4.2.2021]

[Yourdon03]         Edward Yourdon 2003 Death March Prentice-Hall 2 Upper Saddle River N.J., USA 978-0-131-43635-0

[Zaman15]           Najamuz Zaman: **Automotive Electronics Design Fundamentals** Springer Verlag, Germany, 2015. ISBN 978-3-319-17583-6

[Zeller09]          Andreas Zeller: **Formal Specification with Z** Lecture Software Engineering, Saarland University, Germany, 2009. Downloadable from: https://www.st.cs.uni-saarland.de/edu/se/2009/slides/11-Zed-1.pdf [Last accessed: 4.2.2021]

# The Future

**6**

*There is no doubt that cyber-physical systems will affect life, work, business, science, and the environment more and more. Autonomy—based on artificial intelligence and machine learning—will gain importance. Cooperation between intelligent machines and humans will become commonplace. This chapter presents some thoughts on the possible future of safety and security in cyber-physical systems.*

## 6.1    The Rise of the Three Devils

The *three devils of safety and security* in cyber-physical systems have been introduced earlier (Fig. 6.1).

- The *first* devil: *Vulnerabilities*. Vulnerabilities are weaknesses of the system which allow a failure to cause damage or which generate an entry point for malicious activities;
- The *second* devil: *Threats*. A threat is an intentional or unintentional danger aimed at the CPS in order to generate operational disruptions, such as security breaches, malfunctions, or unavailability;
- The *third* devil: *Failures*. Failure causes a part of a system (hardware, software, communications, or infrastructure) to become inoperable or unavailable to the services depending on this part.

© The Author(s), under exclusive license to Springer Fachmedien Wiesbaden GmbH, part of Springer Nature 2022
F. J. Furrer, *Safety and Security of Cyber-Physical Systems*,
https://doi.org/10.1007/978-3-658-37182-1_6

Devil 1: Vulnerabilities    Devil 2: Threats    Devil 3: Failures

**Fig. 6.1**  Three devils of safety and security

The influence of these three devils of safety and security, unfortunately, is on the rise in the development and operation of cyber-physical systems, and they make the assurance of *safety* and *security* increasingly challenging. The reason is *another* set of three devils: The three devils *of systems engineering* (Introduced in: [Furrer19]):

1. *Complexity*: Software-systems are highly complex constructs. During their lifecycles, the complexity increases continuously. Complexity makes the systems very difficult to understand, to predict their behavior, and to evolve them—thus establishing the *first devil* of systems engineering;
2. *Change*: Software-systems are forced to change continuously. They must be adapted to new requirements, respond to environmental changes, and undergo corrective and predictive maintenance. Change cycles may be very short. This continuous, rapid change makes the software-systems difficult to manage and to ensure their conceptual integrity;
3. *Uncertainty*: Uncertainty is an established companion in the modern world. In the systems engineering process, uncertainty is encountered in all phases, from requirements to the operation. This fact forces some decisions that may not be well-founded and require assumptions, which may prove wrong later.

> **Quote**
> *"It is impossible to design a physical device that will not fail eventually, nor is it possible to develop a complex software component without a design error"*
>   Stanley Bak, 2009

*Complexity* is the reason for *vulnerabilities*: The more parts, interconnections, dependencies, products, processes, lines of code, etc., a system has, the more potential vulnerabilities the system contains. At the same time, also, the number of attack points and failure modes increases. Because systems incessantly grow in complexity, vulnerabilities, threat opportunities, and failure potential also augment.

*Change*—in the form of the rapid rate of transformation of the cyber-physical system—is the cause of pressure on CPS's development and deployment activities. Unfortunately, in many cases, this results in carelessness and negligence during the construction or extension of the CPS. Again, this leads to considerable vulnerabilities, threat opportunities, and failure potential.

*Uncertainty* is another characteristic of today's cyber-physical systems, both in all phases of the development and operational environment. For many decisions, not enough information is available, or the information is ambiguous. This may lead to flawed choices of options or behavior—generating unexpected vulnerabilities, opening unanticipated threat opportunities, or enabling surprising failure potential.

## 6.2   Safety: Autonomy

The future of safety has many facets. Currently (2021), the most important and challenging one is *autonomy*. In an autonomous system, observing the operating environment, possible reactions and decisions increasingly shift from humans to machines (or software). The trend from partly autonomous to fully autonomous CPSs is irresistible (e.g., [Burns18], [Gao16], [Liu18], [Scharre19], [Michel21]).

One formidable current use of autonomy is the NASA Mars Rover "Perseverance" (Example 6.1). A sufficient degree of autonomy is a precondition to the success of this mission (https://mars.nasa.gov/mars2020/).

---

**Example 6.1: Mars Rover «Perseverance»**

Source: [Ackerman21]: *"18 February 2021 at 20:55 UTC NASA's Perseverance rover successfully landed on Mars. Like its predecessor, Curiosity, which has been exploring Mars since 2012, Perseverance is a semi-autonomous mobile science platform the size of a small car. It is designed to spend years roving the red planet, looking for (among other things) any evidence of microbial life that may have thrived on Mars in the past.*

*This mission to Mars is arguably the most ambitious one ever launched, combining technically complex science objectives with borderline craziness that includes the launching of a small helicopter. Over the next two days, we will be taking an in-depth look at both that helicopter and how Perseverance will be leveraging autonomy to explore farther and faster than ever before."*

In the last decade, *autonomy* has become an engineering discipline of its own. The theory, mathematics, modeling, etc. are getting more and more powerful (e.g., [Ivancevic17], [Haimes19], [Mitra21], [SASWG19], [Zivic20]). Also, the range of applications is expanding every year, and the technology readiness level rises continuously.

> **Quote**
> *"Soon, many of us will no longer need to own or drive a car. Instead, we will rely on services that safely and conveniently use autonomous vehicles to take us where we want to go"*
>    Lawrence D. Burns, 2018

Autonomous systems provide many advantages over the present-day automatic (= rigidly programmed) systems and enable many promising applications. However, autonomy introduces an unknown extent of *risk* (e.g., [Lawless17], [Mitra21], [Hersch20], [Knight20])—which is a great topic for adequate *risk management*.

Autonomy poses new challenges for the safety of cyber-physical systems, as shown in Table 6.1.

**Table 6.1**  Autonomous systems risks

| Phase | Risk | Risk Management Measures |
|---|---|---|
| Development | Autonomy introduces a new element of uncertainty into risk management: All possible autonomous decisions in all potential operating conditions should enter the risk assessment | For managing autonomy risks, powerful penetrative approaches are needed: (a) Extensive modeling, (b) formal verification of all development artifacts, and (c) large-scale simulations ([Haimes19], [Mitra21], [Zeigler17], [Zeigler18], [Bossel14], [Zhang19]) |
| Operation | The autonomous control programs of the CPS can make wrong decisions during operation (e.g., Example 2.6, Example 5.16, which can lead to dangerous occurrences | After all the due diligence has been applied during the development process, the last line of defense against safety accidents is the running system's supervision with *runtime monitors* (Definition 4.13, Fig. 4.7). A specific implementation of a runtime monitor is the *safety monitor* (e.g., [Haupt19], [Brukman11], [Kane15], [Harrison20], [Bartocci18], [Bartocci19]) |

An additional obstacle to the large-scale adoption of autonomous systems is the *legal uncertainty*: Lawmakers all over the world are slow to put into force corresponding legislature ([Aggarwal19], [Maurer16]).

> **Quote**
>
> *"Safety monitors at runtime complement traditional safety assurance as it facilitates the system in finding design time as well as runtime defects that could potentially occur in software or hardware due to unexpected environmental conditions or runtime faults"*
>
> Nikita Bhardwaj Haupt, 2019

## 6.3 Security: Cryptography Apocalypse

### 6.3.1 Cryptographic Algorithms

Today's protection of *digital assets* heavily relies on *cryptography* (e.g., [Paar10], [Hoffstein14]). Also, secure communications—and other security techniques, such as digital signatures, hash functions, etc.—rely on an encryption/decryption pair of *cryptographic algorithms* (Definition 6.1, Fig. 6.2). The mathematical strength of the cryptographic algorithm used determines the security of the information.

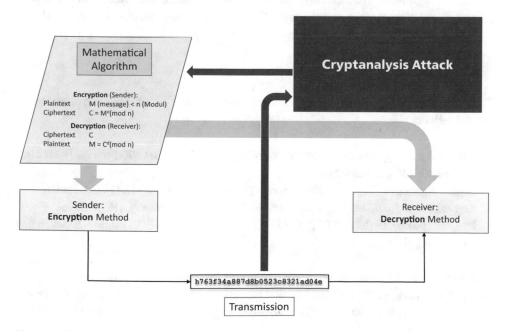

**Fig. 6.2** Cryptography context

Today, several highly secure cryptographic algorithms exist, which guard our information and communications, particularly the *RSA* (named after its inventors: Rivest–Shamir–Adleman, ([Coutinho99]) and *elliptic curves* ([Shemanske17], [Hankerson13]). These algorithms provide unquestionable defense against today's (= year 2020) *cryptanalysis attacks* (Fig. 6.2).

▶ **Definition 6.1: Cryptographic Algorithm**  A mathematical method to protect the confidentiality and integrity of digital information during communications and in storage.

## 6.3.2   Cryptographic Attacks

The history of cryptography is marked by an epic fight between *cryptographers* (devising cryptographic algorithms) and *cryptanalysts* (trying to break the cryptographic algorithms) ([Dooley18], [Bauer13]). A cryptographic attack (Fig. 6.2) either attacks the encrypted artifact or the mathematical algorithm. Cryptanalysis today is a highly developed, challenging field (e.g., [Hinek09], [Stamp07], [Joux09]).

Fortunately, modern cryptographic algorithms are resistant to all known *classical* cryptographic attacks. They rely on very difficult mathematical foundations. However, this comfortable situation changes dramatically with the emergence of *quantum computing* (e.g., [Bernhardt20], [Sutor19], [Grumbling19]).

Quantum computing creates an entirely new, dangerous situation for the existing crypto-algorithms ([Benatti10]). Even the proven workhorses RSA ([Coutinho99], [Shor97]) and elliptic curve cryptography ([Shemanske17]) are seriously threatened!

## 6.3.3   Cryptography Apocalypse

Today (= year 2022), most communication protocols rely on four *core cryptographic* functionalities:

- Public-key encryption ([e.g., http://dx.doi.org/10.6028/nist.sp.800-56br1], [Katz21a], [Hankerson13], [Shemanske17]);
- Digital signatures ([e.g., http://dx.doi.org/10.6028/nist.fips.186-4], [Buchmann16]);
- Key exchange ([e.g., http://dx.doi.org/10.6028/nist.sp.800-56ar2], [Boyd20]);
- Internet protocols like TLS, SSH, IKE, IPsec, DNSSEC, https:, Blockchains ([Farrel04], [Kiong20]).

These crypto-workhorses are seriously threatened by emerging quantum computers and quantum cryptanalysis algorithms. The first quantum prime-factoring algorithm was developed by Peter W. Shor ([Shor97], [Yan10])—before quantum computers were realized!

To break an *RSA-cryptosystem* with $n = 2'048$ bit key length, a quantum computer with 4'099 stable Qubits ($2 \times n + 3$) is required (Note: Improvements of Shor's algorithm will reduce this number).

> **Quote**
>
> *"Shor's algorithm solves in polynomial time:*
>
> - *Integer factorization: RSA is dead;*
> - *The discrete logarithm problem in finite fields: DSA is dead;*
> - *The discrete logarithm problem on elliptic curves: ECDHE is dead;*
>
> *This breaks all current public-key cryptography on the Internet!"*
> Daniel J. Bernstein, 2017

The looming end of secure, classical cryptography is called "Cryptography Apocalypse" ([Grimes19], Definition 6.2) because its impact will be genuinely apocalyptic for the whole existing worldwide cryptography infrastructure ([Furrer20], [Edwards20]).

▶ **Definition 6.2: Cryptography Apocalypse** The situation, where twentieth century cryptography becomes unsecure because of advances in quantum cryptanalysis

Quantum computer processing will become publicly available as a service over the Internet in the Cloud—this leads to "quantum computing for everybody", and all types of malicious cryptanalysts will have easy access (see, e.g., https://www.ibm.com/quantum-computing/technology/experience)! The progress of quantum computer development is frighteningly fast ([Ball21]).

> **Quote**
>
> *"The post-quantum world will be full of weakened and utterly broken cryptography"*
> Roger A. Grimes, 2019

The foreseeable global disaster following *quantum cryptanalytic successes* ([NIST16], [Vermeer20], [Shor20]) is a tremendous challenge for the worldwide IT/communications industry—in fact, a real *quantum crypto catastrophe* is approaching. The risk not only lies with confidentiality/availability/integrity/non-repudiation of information generated and used *after* the quantum crypto break but also with information generated *before*. This *retroactive risk* includes all crypto artifacts acquired and stored before the crypto break—they will become insecure and readable as soon as sufficient quantum computing power becomes widely available. Organizations protecting information with classical cryptography, which should stay secure for a very long time, do already face significant risks from information by now captured and subject to the coming advances in quantum computing ([Vermeer20]).

### 6.3.4   Post-Quantum Cryptography

The escape from the Cryptography Apocalypse is *post-quantum cryptography* (Definition 6.3, [NIST16], [Bernstein08], [Chaubey20], [Takagi21], [Vermeer20], [Yan13], [ENISA21], [Zhang21a]).

▶ **Definition 6.3: Post-Quantum Cryptography (Also called Quantum-Resistant Cryptography)**
Cryptographic algorithms that are secure against both quantum and classical computer decryption and can interoperate with existing communications protocols and networks.
http://dx.doi.org/10.6028/NIST.IR.8105

**Quote**
*"When the forthcoming quantum crypto break happens, the world will change forever. There will be the world's history before and the world's incredible future after"*
   Roger A. Grimes, 2019

The US National Institute of Standards and Technology (NIST) has, for a long time, firmly pointed out the necessity to prepare the world's crypto-infrastructure for the post-quantum apocalypse (e.g., [Chen16]). NIST started a PQC-standardization effort with a call for submissions for quantum-safe cryptographic algorithms (e.g., [NIST8309_2_20]). NIST expects the selection and standardization of a worldwide post-quantum cryptographic algorithm in 2021. As an example of one of the finalists, the NTRU algorithm is presented (Example 6.2).

**Example 6.2: Post-Quantum Cryptographic Algorithm «NTRU»**

In July 2020, the US NIST selected seven proposed crypto-algorithms as third-round finalists ([NIST83092_20]). One of these seven is the *NTRU cryptosystem* (Definition 6.4).

▶ **Definition 6.4: NTRU Cryptosystem**
NTRU is a lattice-based public key cryptosystem from Security Innovation and the leading alternative to RSA and Elliptic Curve Cryptography (ECC) due to its higher performance

and resistance to attacks from quantum computers. NTRU was developed in 1996 as a visionary solution to cyber security challenges for the twenty-first century. NTRU is based on a mathematical problem called the "Approximate close lattice vector problem".

https://github.com/NTRUOpenSourceProject/ntru-crypto

*Lattices* are long-known mathematical structures ([Grätzer13]). Since 1997 lattices have been used as a source of computationally hard problems to build cryptographic algorithms. NTRU is based on the shortest/closest vector problem on integer lattices and provides post-quantum public-key cryptography and digital signatures ([Regev06], [Pipher02], [Katz21b]).

Once quantum cryptanalysis algorithms become widely available, an attacker's retroactive confidentiality risk from classically encrypted information already captured and stored becomes real. This recorded, encrypted information can then be decrypted and used at any time in the future. Therefore, the transition to PQC (Post Quantum Cryptography) should be done at the earliest possible time.

Protecting the *integrity* of information stored within the organization can be done by *cryptographic wrapping* (Fig. 6.3): The classically encrypted information is encrypted a second time using a post-quantum cryptographic algorithm. This envelope then safeguards the confidentiality and integrity of the information—such as legally required archives—into the post-quantum world.

The Cryptography Apocalypse triggered by quantum cryptanalysis algorithms may not be the last *cryptographic catastrophe*. Progress in mathematical cryptanalysis (e.g., [Jordan18]) may generate other unpleasant consequences. To be prepared for such future situations, *security architectures* that ensure cyber-resilience (= ability to respond in time to future cryptographic apocalypses) and cryptographic agility (= separation of cryptographic implementations from protocols, applications, etc.).

**Fig. 6.3** Cryptographic wrapping

> **Quote**
> *"New PQC-systems should aim for (1) future-compatibility with the expected evolution of standards and more demanding requirements of PCQ, and (2) modularity that would allow rapid and inexpensive cryptographic adaptation as new threats or vulnerabilities are discovered"*
>     Michael J. D. Vermeer, 2020

## 6.4    Artificial Intelligence in Safety and Security

*Artificial intelligence* ([Boden18], [Chowdhary20], [Kulkarni15], [Russell17], [Yampolskiy18]), especially *machine learning* (ML, [Stone19], [Goodfellow16], [Burkov20]) started to find applications in both safety and security engineering in the early 2000's. The coming impact of AI and ML will be tremendous—both as positive and negative forces!

### 6.4.1    AI in Safety

*Artificial intelligence* (AI)—especially machine learning (ML)—is unpreventably invading many safety–critical cyber-physical systems.

The big challenge—and risk—for safety are *autonomous systems* (Definition 2.7, [Liu18], [Kulkarni15]). Autonomous systems monitor the environment, react to changes and obstacles, and make autonomous decisions. Many of the autonomous systems are based on *artificial intelligence* (AI, [Brockman19], [Lawless17], [Burns18]), especially *machine learning* (ML, [Alpaydin16], [Kelleher19]).

> **Quote**
> *"Any system simple enough to be understandable will not be complicated enough to behave intelligently, while any system complicated enough to behave intelligently will be too complicated to understand"*
>     George Dyson, 2019

Artificial intelligence will have different boundary conditions for *civil applications* and *military applications*. Not only the policies and the ethical foundation are different, but in military applications, most of the laws and regulations are not applicable.

### 6.4.1.1  Artificial Intelligence in Civil Applications

For civilian applications, the safety—and the provable *safety case*—for autonomous systems based on machine learning is an indispensable and significant challenge of the (near) future (e.g., [Amodei16]).

AI for the safety of cyber-physical systems will have three primary applications (Fig. 6.4):

1. AI and ML will be increasingly used to implement the *on-board functionality* of the CPS, e.g., of a car, such as image processing, sensor fusion, route optimization, battery range maximization (e.g., Example 6.3);
2. AI and ML gain more and more power for safety *accident prevention*, accident handling, and mitigation, both during development and during operation:
3. AI and ML improve the CPS behavior by active integration into the *operating context*, i.e., its environment, such as accident prediction/evasion.

---

**Example 6.3: Traffic Intention Assistant**

Any traffic participant's driving mistake may initiate or produce an accident in today's very dense and dynamic traffic. It is challenging for any driver to observe all other actors and recognize or guess their intended behavior. Often, an accident could be prevented if drivers perceived an unforeseen or illegal maneuver of another vehicle before it is executed.

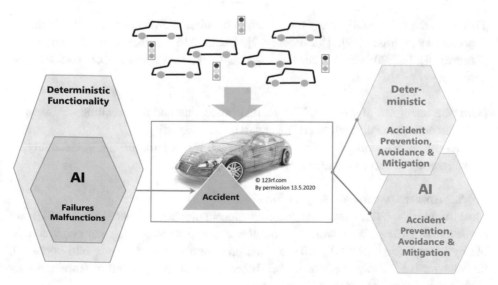

**Fig. 6.4**  AI in safety

A *traffic intention assistant* is an ML-based cyber-physical system that monitors the vehicles and their movements around their own vehicle. Based on its many sensors, it records and analyzes their behavior. The traffic assistant system is ML-trained to interpret the behavior. It not only interprets the apparent behavior, such as a lane-change indicator or braking, but is also trained to take account of minimal signs which possibly forecast a coming movement, such as a spurious steering move, a slight left or right drift of the car, a head movement of the driver, a short brake moment, the specific traffic position, etc.

Based on these indications, the intention assistant warns the driver of an observed car's possible forthcoming maneuver. The car in question will be highlighted in red, and a symbol will indicate its forecasted maneuver in the head-up display. The driver can then preventively react to the possible danger.

In addition to the risks of autonomous systems (Definition 2.7, Table 6.1), AI and especially ML have one additional, considerable risk: The potentially unpredictable risks of ML-based *decision algorithms* (Note: Only technical risks, no societal risk is in focus).

> **Quote**
> *"Machine Learning has many specificities that makes its behavior prediction and assessment much different from explicitly programmed software-systems"*
>      José M. Faria, 2018

Therefore, the *safety of AI* during the operation of mission-critical systems is of primary concern (e.g., [Frtunikj19], [McCloskey20], [Amodei16], [Juliano16], [Fehlmann20], [Szegedy14], [Elçi20]). Two recognized *threats* to the safe operation of AI-based systems are:

(1) *Intransparent ML algorithms*: The internal processing and the resulting decisions of the algorithms are difficult to understand, retrace, and justify;
(2) *Adversarial ML examples*: The trained, operational algorithm can accidentally or maliciously be deceived and lead to erroneous decisions.

### 6.4.1.2  Intransparent ML Algorithms

Most of today's (2022) machine learning algorithms (e.g., [Shalev-Shwartz14]) are *intransparent*: Their "inner working" and the resulting decisions are not comprehensible by humans (e.g., [Rai20]). After a fixed programmed system is put into operation, its behavior is relatively predictable: ML-based systems are intended to learn and may change their behavior over time ([Lance16]).

If these incomprehensible techniques and their findings are utilized in mission-critical systems—such as autonomous cars—an unpleasant number of questions regarding safety, security, liability, conformity to laws and regulations, and trustworthiness arise. The safety of ML-based systems is an emerging research field (e.g., [Tamboli19], [Varshney16], [Hernandez-Orallo20], [Faria18], [Faulkner20]).

> **Quote**
> *"Not knowing what you don't know is a more significant risk than any known risks"*
>    Anand Tamboli, 2019

As research avenues for the *mitigation* of intransparent ML algorithm *risks*, two promising fields are emerging: (1) *Explainable Artificial Intelligence* (XAI, Definition 6.6), and (2) *AI Guardian Angel Bots* (Definition 6.7).

### 6.4.1.3 Adversarial ML Examples

As with any networked technical system, machine learning systems can and will also be *attacked* (e.g., [Warr19], [Kurakin17]). For ML, three attack paths exist (Fig. 6.5):

1. Attack the machine learning algorithm (either during training or during operations, e.g., [Bajo21]) and introduce malicious code;
2. Attack the training data (Poisoning): Force falsified information to the training process;
3. Attack the operations data (Adversarial examples or adversarial inputs, Definition 6.5), first discovered by [Szegedy14]): Misdirect the classification algorithm by altering the ML algorithm input.

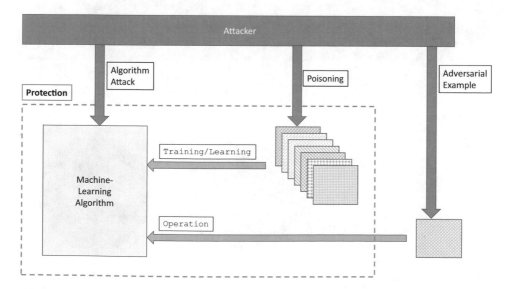

**Fig. 6.5**  Machine learning attacks

Defending against attacks to the algorithms and data poisoning requires system security, i.e., the use of all adequate security protection measures.

Defense against *adversarial examples* (Definition 6.5, Example 6.4), i.e., misdirection attempts during operation, is a challenge of its own ([Joseph19], [Brachman18], [Morgulis19], [Sitawarin18], [Evtimov17], [Juliano16], [Finlayson19]).

▶ **Definition 6.5: Adversarial ML Examples**

Adversarial examples are slight manipulations that cause machine learning algorithms to misclassify inputs, such as images, while going unnoticed to the human eye.

Adapted from: https://bdtechtalks.com/2020/07/15/machine-learning-adversarial-examples/

---

**Example 6.4: Graffiti on a STOP Sign**

Source [Evtimov17]: Fig. 6.6 shows two STOP signs. Left the original STOP sign and right a STOP sign slightly defaced by four graffiti (= Adversarial patch, [Warr19]). In a drive-by test with an autonomous car, the subtle *image attack* causes the right-hand STOP sign to be misclassified as a *Speed Limit 45* sign in 100% of the extracted video frames (37 out of 37).

Input to an ML vision system may be maliciously (intentional, Example 6.4) or inadvertently (damage, wear, decay, corrosion) adversarial.

---

**Quote**

*"While trained models tend to be very effective in classifying benign inputs, recent work shows that an adversary is often able to manipulate the input so that the model produces an incorrect output"*

Aleksander Madry, 2019

**Fig. 6.6**  Malicious graffiti on a STOP sign

> *"Algorithmic defenses against adversarial examples remain an extremely open and challenging problem"*
> Samuel G. Finlayson, 2019

Defense against attacks by *adversarial input* during the operation of the AI/ML system is tricky. Many defense techniques are under consideration (Table 6.2, e.g., [Li18a], [Casola19]).

---

**Example 6.5: Use of Context Information for ML-Decisions**

Example 6.4 showed a graffiti-defaced STOP sign which the ML algorithm misclassified as a speed limit sign. The ML algorithm solely used the disfigured STOP sign as input data.

Figure 6.7 illustrates additional information that the classification algorithm can use:

a) Possible traffic sign redundancy: The STOP sign is present twice (Redundancy);
b) The white road markings unequivocally indicate a STOP situation (Context);
c) The road layout—a large main street ahead—unambiguously refers to a potentially dangerous situation (Visual context+GPS navigation map information);
d) The auxiliary visual information, i.e., (1) the distance indication below the STOP sign, points to danger ahead and (2) the blue signposts characterize a major road with unconditional right of way (in Switzerland);
e) The on-board GPS navigation system delivers additional information: The general speed limit "50 km/h" (Switzerland) and the intersection ahead.

Combining this extended information is sufficient to correct the misguided decision of the ML algorithm.

---

**Quote**
*"It is important to consider the presence of attackers when thinking about machine learning. History reveals that attackers always follow in the footsteps of technology development or sometimes even lead it. The stakes are even higher with AI!"*
Dawn Song, 2019 (in [Casola19])

### 6.4.1.4 Explainable AI (XAI)

Two very promising approaches for ensuring the trustworthiness of AI/ML systems in mission-critical applications are:

**Table 6.2** Adversarial input defense

| Defense Mechanism | Description | References |
|---|---|---|
| Develop and use robust models and algorithms | The robustness of an algorithm is its sensitivity to discrepancies between the assumed model and the reality ([Schain15], https://www.robust-ml.org/). To render ML algorithms resilient against malicious or unintended input, new technologies are under development | [Carlini17], [Warr19], [Madry19], [Ozdag18], [Frtunikj19], [Li18a], [Schain15], [Uesato18] |
| Inhibit 3rd party access to the ML model knowledge | Understanding the ML model architecture and model parameters (such as confidence scores for each predicted class) can be very helpful while generating adversarial examples. However, keeping the internals of the ML-system secret is not a defense in itself | [Bajo21], [Bhagoji18], [Warr19] |
| Adversarial training | During the training of the ML model, targeted adversarial examples are generated and used | [Bajo21], [Raghunathan18], [Goodfellow15], [Tramèr20], [Carlini17] |
| Integrity Indicators | • *Digital integrity indicators*: Are pixels/representations inconsistent? E.g., blurred edges from object insertion, different camera placements, replicated pixels, flawed compression; <br> • *Physical integrity indicators*: Are the laws of physics violated? E.g., inconsistent shadows and lighting, faulty perspectives; | [Casola19] |
| Include Domain Knowledge | Specific application domains often have particular information characteristics, e.g., commercial data (from Internet videos) does not look like military data (from airborne on-board videos),such as resolution, lightning, stability | [Casola19] |
| Redundancy (Multi-channel decisions) | Use not only the camera input for the classification decisions (Example 6.4) but also additional information sources, such as the map information from the GPS navigation system (Example 9.5) | |
| Incorporate context information | Make use of the context information, e.g., the road markings (Example 6.5). Make allowance for bias factors, e.g., a model trained on roads' images during sunny weather can fail to recognize the same situation in snow-covered features | [Casola19] |
| Input preprocessing | Remove or correct adversarial data in the broader processing chain *before* it is submitted to the ML algorithm | [Warr19], [Qiu20] |
| Err on the safe side | All decision mechanisms err on the safe side: The action taken following an ML decision protects all assets | [Warr19], [Fehlmann20], [Frtunikj19] |

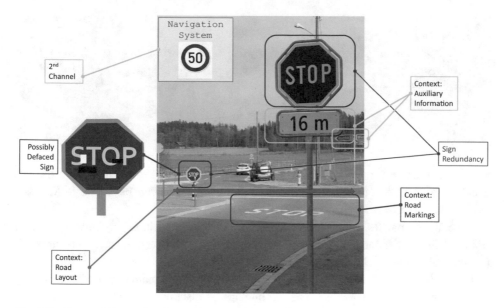

**Fig. 6.7** Context information for road sign recognition

a) *Explainable Artificial Intelligence* (XAI, Definition 6.6, e.g., [Samek19], [Goebel18]);
b) *AI Guardian Angel Bots* (Definition 6.7, [Eliot16]).

▶ **Definition 6.6: Explainable Artificial Intelligence (XAI)**
The goal of Explainable Artificial Intelligence (XAI) are machine learning techniques that produce explainable models that, when combined with effective explanation techniques, enable end users to understand, appropriately trust, and effectively manage the emerging generation of Artificial Intelligence (AI) systems
    Adapted from: https://www.darpa.mil/attachments/DARPA-BAA-16-53.pdf

Classification decisions outputted by machine-learning-based algorithms, especially deep learning with artificial neural networks, are opaque. It is, therefore, difficult for humans to trust these decisions. If such decisions pilot, e.g., an autonomous car, a sufficient degree of comprehension and trust is compulsive. To establish this trust, the ML-based decisions must become *transparent, explainable,* and *comprehensible.*

An active field of research—the *Explainable Artificial Intelligence* (XAI, [Goebel18], [Samek19], [NISTIR8312], [Vilone20], [Longo20], [Hamon20])—tries to accomplish this mission by making the models, results, and decisions comprehensible by humans. The soundness of classification methods is a long-discussed problem—investigated by Charles S. Peirce in logical classification systems already in 1878 ([Peirce1878]).

However, research for explainable AI/ML systems has gained much interest in the last decade, e.g., the US National Institute of Standards published *four principles* of XAI as a draft in 2020 (Table 6.3, [NISTIR8312]).

Several approaches to XAI are currently examined, such as *explainable models* (Fig. 6.8), model induction, profound explanation, knowledge graphs, etc. ([Turek21], [Molnar20], [Beyerer21], [Samek19]).

In Fig. 6.8, the ML learning model is augmented to an explainable model, including an explanation interface—which may be human- or machine-readable (e.g., [Turek21], [Geyik19], [PwC18]). In Example 6.6, the ML algorithm delivers the decision characteristics for the classification "Cat".

---

**Example 6.6: Explainable ML**

The ML algorithm has been trained with 10'000's images of cats of all breeds, in all colors, and in many situations. When presented with the scene in Table 6.4, the algorithm classified «Cat» and delivered the decision characteristics.

**Table 6.3**  NIST-XAI principles

| NIST-XAI Principles | |
| --- | --- |
| Principle | Content |
| *Explanation* | AI systems deliver accompanying evidence or reason(s) for all outputs (results & decisions) |
| *Meaningful* | AI systems provide explanations that are understandable to individual users |
| *Explanation Accuracy* | The explanation correctly reflects the AI system's process for generating the output(s) |
| *Knowledge Limits* | The AI system only operates under conditions for which it was designed or when the AI system reaches sufficient confidence in its output |

**Source**: NISTIR8312/August 2020 ([NISTIR8312])

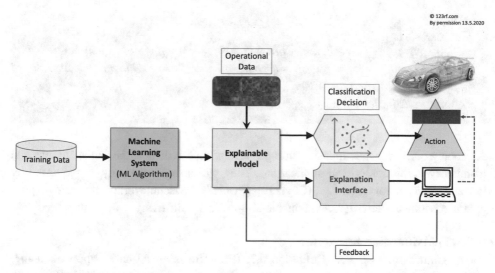

**Fig. 6.8** Explainable AI/ML model

**Table 6.4** ML explanation interface

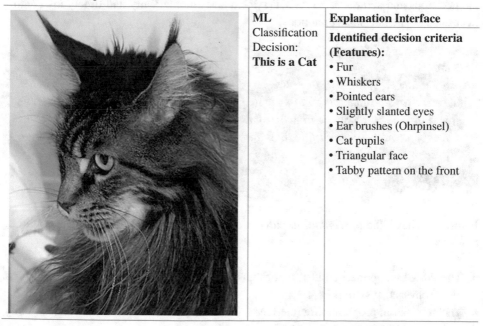

| | ML Classification Decision: **This is a Cat** | **Explanation Interface** |
|---|---|---|
| | | **Identified decision criteria (Features):** <br> • Fur <br> • Whiskers <br> • Pointed ears <br> • Slightly slanted eyes <br> • Ear brushes (Ohrpinsel) <br> • Cat pupils <br> • Triangular face <br> • Tabby pattern on the front |

> **Quote**
> *"The importance of explainability does not just depend on the degree of functional opacity caused by the complexity of your machine learning models but also the impact of the decisions they make"*
>     PricewaterhouseCoopers LLP, 2018

As soon as XAI reaches industry-strength levels, they will become indispensable and mandatory in mission-critical applications. They will also increasingly be requested by law and the regulators (see, e.g., [Geyik19], [PwC18], [Wischmeyer20], https://enterprisersproject.com/article/2019/5/explainable-ai-4-critical-industries).

### 6.4.1.5  AI Guardian Angel Bot

An *AI Guardian Angel Bot* (Definition 6.7, [Eliot16], is an AI-based application-specific implementation of a *runtime monitor* (Definition 4.13, [Amodei16], [Kane15], [Goodloe10], [Schwenger21], [Brukman11], [Uesato18], [Braband20], [Horneman19]).

▶ **Definition 6.7: AI Guardian Angel Bot** Runtime software supervising and analyzing the decisions and actions of an AI/ML system and automatically and timely intervening in case of unsafe decisions or actions.

> **Quote**
> *"Due to the complexity of the system and uncertainty of the environment, it is not adequate to determine all possible system behaviors at design time. Thus, in addition to traditional safety approaches, these systems necessitate runtime safety techniques that are capable of handling unpredictable operating conditions, evaluate system behavior and adapt accordingly"*
>     Nikita Bhardwaj Haupt, 2019

Figure 6.9 shows the *conceptual architecture* of an AI Guardian Angel Bot. Its building blocks are:

- The AI/ML-supported control algorithms with their sensors and actuators, i.e., the cyber-physical system (Fig. 1.1);
- The application programming interface (API) enables access to the data, functionality, and control of the AI/ML system;
- The AI safety angel bot includes intervention mechanisms: Actuators inhibition and forcing transition into a safe state (emergency procedures).

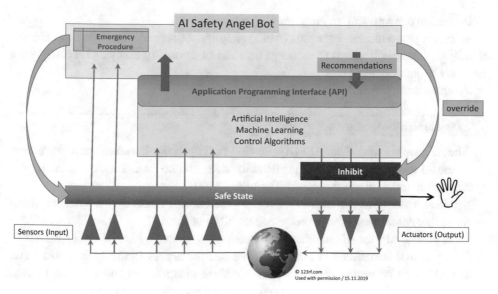

**Fig. 6.9** AI guardian angel bot conceptual architecture

The AI safety angel bot has three levels of intervention following its analysis and decisions:

1. Issuing *recommendations* impacting planned actions of the AI/ML algorithm: The recommendations should influence the AI/ML CPS control algorithm and guide it toward safety;
2. Inhibiting or correcting a possibly harmful output of an actuator to the physical part of the CPS;
3. Moreover—as a last resort—forcing the CPS into a *safe state*: Often, the safe state removes the CPS from dangerous operating conditions and possibly hands control back to the supervising human. Note that the AI safety angel bot may use additional sensory input, such as context information.

The two software-systems: (1) AI/ML CPS control algorithms and (2) the AI safety angel bot need a suitable *communications method*. Good software engineering practice dictates an *application programming interface* (API) for access to the AI safety angel bot (Definition 6.8, [Jacobson11]). Many forms of such an API are conceivable. One possible such API is given in Example 6.7. To render possible a wide use of AI safety angel bots, an independent standard for AI safety angel bots API must evolve and gain broad acceptance ([Eliot16]).

▶ **Definition 6.8: Application Programming Interface (API)**

Application programming interfaces (API's) simplify software development and innovation by enabling applications to exchange data and functionality easily and securely via a formally defined contact point

Adapted from: https://www.ibm.com/cloud/learn/api

---

**Example 6.7: AI Guardian Angel Bot API**

The *AI guardian angel bot API* (Application Programming Interface) provides access to the AI guardian angel bot's functionality, data, and control. The API is the implementation of the interface contract (Definition 2.27).

Various solutions for useful APIs exist—the choices made by the software architects, therefore, impact the coverage and usability of the API.

A generic API for AI guardian angel bots was presented by Lance Eliot ([Eliot16]): This proposal is revealed in Fig. 6.10. Note that the access to the AI guardian angel bot API must be *protected*, i.e., dependable authentication and authorization mechanisms must be in place, and any remote access channel must be encrypted.

The API offers seven services:

1. Connect: Establishes a secure (= authenticated, authorized, and possibly encrypted) bidirectional communication channel between the AI safety angel bot software and the CPS control system software;
2. Status: Retrieval of status information, such as time-out inquiries;

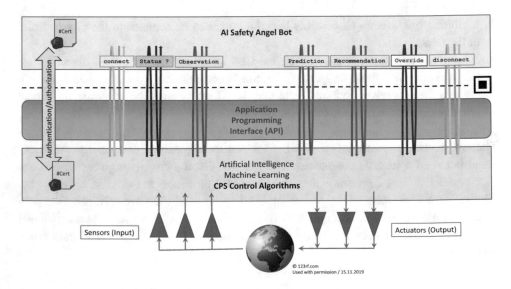

**Fig. 6.10** AI guardian angel bot API

3. `Disconnect`: Terminate the connection and close all pending requests. Before a new request can be issued, `Connect` must be executed again.

   The API is based on the «Observation–Prediction-Recommendation-Override» pattern ([Eliot16]). Therefore, the four operational API calls are:

4. `Observation`: The AI guardian angel bot requests specific observations from the AI/ML system, such as the current speed or direction of an autonomous car and other sensory input data;

5. `Prediction`: The AI guardian angel bot uses the information received from «Observation» to compute valid predictions, i.e., forecasting what could happen next;

6. `Recommendation`: The AI guardian angel bot requests a recommendation for the intended actions of the AI/ML system, compares it to its deduction in a specific situation, and then either accept the AI/ML recommendation or responds with its adjusted recommendation;

7. `Override`: If the AI guardian angel bot detects anything which the AI/ML system is doing or not doing for which the predicted outcome may be harmful, the AI guardian angel bot attempts an override. The override is the most controversial act.

Note that the AI safety angel bot may also have additional access to context information, additional sensors, or other safety bots' information. Furthermore, the AI safety angel bot may have direct access to hardware functions of the CPS (not through the API), such as inhibiting actuator settings or forcing the system into a safe state (Fig. 6.9).

---

**Quote**

*"AI safety is going to determine whether Deep AI is sustainable or not"*
   http://aiguardianangelbots.org/, 2021

---

### 6.4.1.6  Artificial Intelligence in Military Applications

Artificial intelligence applications have an enormous impact on people, work, and civil society—and this trend is accelerating. However, many experts expect that the implications of applying artificial intelligence to *military usage* will have even more consequences on nations and humanity. Military artificial intelligence usage ([IntroBooks20]) will completely change the art of war and the global power balance (e.g., [DelMonte18], [Tangredi21], [Scharre19]). Therefore, military artificial intelligence applications—especially for autonomous systems—are a very strong driver of theoretical and applied research.

Nations are developing *military artificial intelligence strategies*, such as the US National Security Commission on Artificial Intelligence ([NSCAI21]). All of them strongly emphasizes the decisive role of military AI.

> **Quote**
> *"We will not be able to defend against AI-enabled threats without ubiquitous AI capabilities and new warfighting paradigms"*
>     Eric Schmidt, 2021 [NSCAI21]

*Militarized artificial intelligence* (MAI) leads to all kinds of new weapons, such as *warbots* ([Michelson20]) and *Lethal Autonomous Weapons* (LAWS, Definition 6.9, e.g., [Liljefors19], [Hageback17]).

▶ **Definition 6.9: Lethal Autonomous Weapons (LAWS)**
Lethal autonomous weapons systems (LAWS), sometimes called "killer robots", are weapon systems that use artificial intelligence to identify, select, and kill targets without human control. This means that the decision to kill a human target is no longer made by humans but by algorithms.
    https://autonomousweapons.org, 2021

Military, autonomous, AI-powered weapon systems have become the new arms race for the next decades ([Garcia21]). Such systems are currently developed for land, sea, air (Example 6.8), space, and cyberspace.

---

**Example 6.8: Heron AlphaDogfight**

In August 2020, an unmanned, AI-powered drone defeated a top human pilot in the US DARPA AlphaDogfight Trials (Source: https://www.janes.com/defence-news/news-detail/heron-systems-ai-defeats-human-pilot-in-us-darpa-alphadogfight-trials and https://www.newsweek.com/artificial-intelligence-raspberry-pi-pilot-ai-475291).

An artificial intelligence algorithm (Heron Systems, https://heronsystems.com/) defeated a human US Air Force (USAF) pilot 5–0 on 20 August in the final round of the United States Defense Advanced Research Projects Agency's (DARPA's) AlphaDogfight Trials.

The real, physical dogfight was the summit of three rounds of AI algorithms created by contractors controlling *simulated* Lockheed Martin *F-16* Fighting Falcons in aerial combat (dogfights). The teams flew against each other in a computer-simulated round-robin tournament in the second round. In the third round, the top four teams competed in a single-elimination simulated tournament for the right to fly against the *human pilot*.

This is an example of an AI-powered cyber-physical system (the military drone) supremacy against a human-controlled cyber-physical system (the fighter plane).

---

Mostly the same safety and security principles apply to military cyber-physical systems and civilian cyber-physical systems. However, four new, specific risks emerge, especially for fully autonomous systems:

1. *Jamming* the communications control channel of the military CPS, thus interrupting the remote control, e.g., using a *drone jammer* that sends powerful electromagnetic noise at radio frequencies to override the same radio and GPS signals the drone uses to operate. Therefore, *anti-jamming techniques* must be implemented ([Li18b], [Graham10], [Suojanen18]);
2. *Blinding* the military CPS, e.g., by spoofing the satellite navigation system or perturbing the sensors, such as by direct laser interference. Emergency procedures must be in place, such as automatic evasion;
3. *"Friend or Foe" identification*: Before engaging in a destructive action, the autonomous CPS must ensure that the envisaged target is an adversary, e.g., by evaluating its `friend-or-foe signal` (IFF signal, [Scharre19]) or other unambiguous characteristics;
4. *Mission abort*: Many external circumstances may force the military CPS to abort its mission, e.g., jamming, blinding, weather conditions, component or sensor failures. In such cases, the CPS must seek a safe state, such as autonomously returning to base or self-destructing.

The future will see many breakthrough innovations in military artificial intelligence for combat use, ranging from small killer drones to fully autonomous, target-seeking rockets (e.g., [Bistron21], [Sayler20], [Beyerer20]). These will also strongly influence civilian applications.

> **Quote**
> *"Technology has brought us to a crucial threshold in humanity's relationship with war. In future wars, machines may make life-and-death engagement decisions all on their own"*
>    Paul Scharre, 2018

In many worldwide communities, there is great concern about the *reconcilability* of lethal autonomous weapons systems with human ethics and current law (e.g., [Schroeder17], [Heinegg18], [Boddington17], [Schoitsch20], [AI-HLEG19], [Scharre19]).

## 6.4.2   AI in Security

Artificial intelligence (AI, [Cylance17], [Johnson19], [ENISA20]) has become the *third player* in the cybersecurity field (Fig. 6.11). Attackers and defenders use AI to enhance their success rate and devise new, startling methods and techniques for cybercrime and cyberwar. This monograph primarily applies artificial intelligence to the cyber-*defense*, i.e., *security AI* (Definition 6.10), is considered.

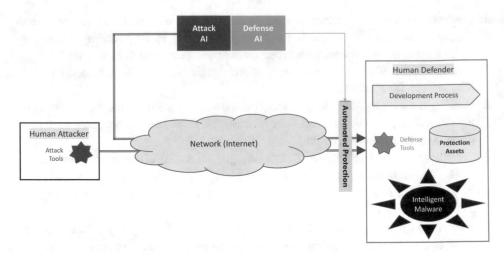

**Fig. 6.11**  Artificial intelligence in security

> **Quote**
> *"The future of security and safety online is going to be defined by the ability of defenders to deploy machine learning to find and stop malicious activity at Internet scale and speed".*
>   *Alex Stamos in* [Chio18]*, 2018*

Today's (2022) rising success rate, increasing destruction power (Example 6.9), and higher frequency of cyber-attacks demonstrate the difficulty of "classical" (i.e., non-AI mechanisms) for cyber-defense to protect the systems, especially large CPSoS's!

▶ **Definition 6.10: Security AI (Artificial Intelligence)**
Cyber AI is a self-learning technology—like the human immune system, it learns 'on the job', from the data and activity that it observes in situ. This means making billions of probability-based calculations in light of evolving evidence. Security AI used for cyber-attacks is malicious security AI. The one used for defense is beneficial AI.
    Adapted from: www.darktrace.com

---

**Example 6.9: US Colonial Pipeline Attack**

*"On May 8, 2021, the* Colonial Pipeline *Company announced that it had halted its pipeline operations due to a ransomware attack (*[IST21]*), disrupting critical supplies of gasoline and other refined products throughout the East Coast of the United States of America"* ([Parfomak21]).

The Colonial Pipeline is a 5500-mile tube for gasoline, diesel fuel, and other refined products to the US East Coast, running from the Gulf Coast to Linden, N.J. The Colonial Pipeline supplies about 45% of the fuel consumed on the East Coast.

Thousands of gas stations across several southeastern states ran dry during the week of May 13, 2021. Drivers waited in long lines for fuel. Consequently, the average U.S. price of a gallon of gasoline rose to US$ 3.03 a gallon.

> **Quote**
>
> *"In 2020, thousands of businesses, hospitals, school districts, city governments, and other institutions in the U.S. and around the world were paralyzed as their digital networks were held hostage by malicious actors seeking payouts. The immediate physical and business risks posed by ransomware are compounded by the broader societal impact of the billions of dollars steered into criminal enterprises, funds that may be used for the proliferation of weapons of mass destruction, human trafficking, and other virulent global criminal activity".*
>
> Institute for Security and Technology, 2021 [IST21]

After paying nearly US$ 5 Million to the group, identified as "DarkSide", the systems were restored, and pipeline operations resumed. Obviously, this was a commercial attack with the sole objective of collecting ransom money (cyber criminality). If this had been a hostile attack (cyberwar), the consequences would have been devastating ([Johnson15], [HomelandSecurity17])!

*June 8, 2021*: The FBI recovered a huge chunk of the Colonial Pipeline ransom by secretly gaining access to DarkSide's bitcoin wallet password (US$ 2.3 million worth of Bitcoin out of the US$ 4.4 million ransom) [2021 Insider Inc. (www.insider.com)].

Cybersecurity is a long-lasting, epic battle between attacker and defender. It started in historical times as an intellectual fight between *defenders* (= cryptographers, cryptographic algorithm developers) and *attackers* (= cryptanalysts, cryptographic algorithm breakers) of information security (e.g., [Kahn96], [Hinek09], [Dooley18]).

Unfortunately, in modern Internet times, the attackers have a much better position than the defenders—Attackers (cybercrime, cyberwar, cybersabotage) win more and more confrontations. The result is a dramatically rising rate of successful cybercrime (e.g., [FBI21], [CPST21], [Kurtz21]). Many industry experts forecast that this trend will even be accelerated in the future. The main reason is the colossal *complexity* of modern software-systems and their known/unpatched or unknown cyber vulnerabilities. Moreover: *"The attacker needs to get it right only once to score, while defenders need to defend successfully 24/7/365"* ([BCG18]).

When the use of *artificial intelligence in cybersecurity* started (e.g., [BCG18], [Bonfanti20], [Szychter18], [Schneier18]), immediately, both the attackers and the defenders tried to apply the power of AI/ML to cybersecurity, thus creating *malicious security AI* and *beneficial security AI*.

Security AI is already having a massive impact on the cybersecurity landscape and will be the dominant technology for the next decades (e.g., [Fiehler20], [Dilek15], [Elçi20], [Thomas20]).

> **Quote**
> *"Artificial intelligence technologies have the potential to upend the longstanding advantage that attack has over defense on the Internet. This has to do with the relative strengths and weaknesses of people and computers, how those all interplay in Internet security, and where AI technologies might change things".*
> Bruce Schneier, 2018

### 6.4.2.1 Malicious Security Artificial Intelligence

Cyber-attackers use *malicious security Artificial Intelligence* (Definition 6.11, [Szychter18], [Kubovič20]) to increase their success rate substantially. Malicious security AI is applied:

- To all phases of the attack process (Fig. 2.15, [Zouave20]);
- As *intelligent malware* infiltrated into the target system;
- As machine-learning-based *malware intrusion* detection and interception (e.g., [Rains20]).

▶ **Definition 6.11: Malicious Security AI (Artificial Intelligence)** Artificial intelligence, especially machine learning, used by a cyber-attacker to identify and exploit vulnerabilities in a target system or to be installed in a target system.

---

**Example 6.10: Malware Detection Evasion**

Malware injection (e.g., [Mohanta20], [Saxe18]) is a perilous cyber-activity with significant, large-scale damage potential. Tremendous efforts are invested every year in malware defense developments. Malware protection systems are an indispensable part of cyber-defense.

Malware infiltration practices (= cyber-attacker) and malware prevention techniques (= cyber-defender) are engaged in an unending battle. Both contenders attempt to use artificial intelligence/machine learning to gain the upper hand. One battlefield is the race to detect/disarm malware versus the evasion of malware identification ([Anderson18], [Anderson17]), both making use of ML (Fig. 6.12). Two paths can be seen in Fig. 6.12:

1. The *attack path* (marked with 1): This path starts with the malware source module, probably detected by the malware defense. The malware evasion ML system modifies the module using a format- and function-preserving transformation, generating

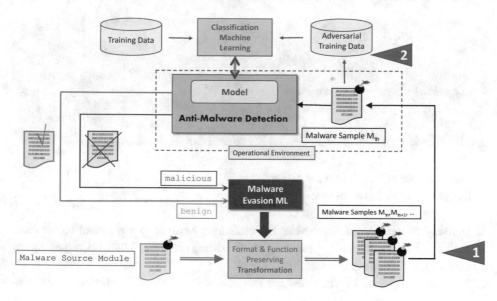

**Fig. 6.12**  Malware detection evasion via machine learning

    mutated copies, i.e., malware samples. These are then fed into the anti-malware detection system, receiving a response as "malicious" or "benign". If the result is "benign", an evading form of the malware has been created and can be used to attack numerous systems protected by the same anti-malware protection;

2. The *defense path* (marked with 2): Whenever the anti-malware detection system recognizes a mutation as "malicious", it is used as additional training data and thus improves the capability of the anti-malware detection system.

A latent fear in cybersecurity is the emerging danger of *intelligent malware*, i.e., malware powered by artificial intelligence (Definition 6.12, [Kaloudi20], [Tolido20], [Truong19]). AI/ML-powered malware poses a significant threat to existing and future CPS's.

▶ **Definition 6.12: Intelligent Malware**  Malware powered by artificial intelligence, leading to greater detection evasion capabilities and higher damage potential.

AI/ML can be used in many ways in future malware (e.g., [Claus20], [Kujawa18]). One application is presented in Example 6.11.

---

**Example 6.11: Intelligent Malware "DeepLocker"**

*DeepLocker* established a class of AI/ML-powered, evasive malware, which conceals its purpose until it reaches a particular victim with specific characteristics ([Kirat18], [Stoecklin18]).

**Quote**

*"DeepLocker is designed to be stealthy. It flies under the radar, avoiding detection until the precise moment it recognizes a specific target. This AI-powered malware is particularly dangerous because, like nation-state malware, it could infect millions of systems without being detected"*

 Marc Stoecklin, 2018

The architecture of DeepLocker is presented in Fig. 6.13.

 DeepLocker has the elements:

1. Delivery carrier: DeepLocker hides the *malicious payload* in a practical application, such as a video conferencing tool or a photo editing program. When downloading the application, the malware is installed;
2. Payload: The payload is the malware. The payload is encrypted to obfuscate any signature known to the antivirus and malware scanners (detection evasion);
3. Decryption & Execution Software and Decryption key: The payload is decrypted and installed as an executable, dormant program;
4. Evasive checks & measures: Before installing, the malware checks for protection precautions of the target system, such as sandboxes. Installation is aborted if any effective defense mechanisms are detected;
5. Machine-based target identification and execution trigger: This is the impressive new part. A trained machine learning model is included in the package: The *ML model*

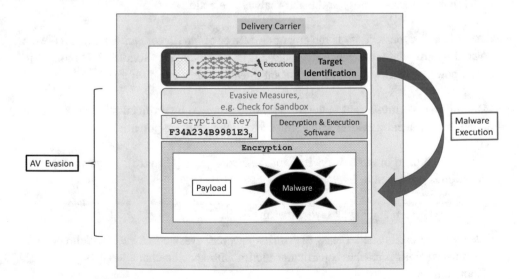

**Fig. 6.13**  DeepLocker malware

recognizes specific input patterns, such as visual or audio, a combination of words, a system configuration, or traffic patterns on the communication links or the Internet. If such a condition is encountered, the malware is activated and executed.

The presence of a trained ML model as the trigger mechanism makes it highly challenging for malware analysts to discover the trigger conditions and the malware payload.

### 6.4.2.2  Beneficial Security Artificial Intelligence

Cyber-defenders use *beneficial security AI* (Definition 6.13, [ENISA20], [Odayan21], [Parisi19], [Chio18], [Szychter18], [Dilek15], [Rao19], [Rubinoff20], [Balamurugan18]) to enhance their defense massively. Beneficial security AI is applied:

- During the development process to identify and eliminate vulnerabilities or mitigate their impact;
- Installed in the target system as online, real-time protection against assaults;
- During the operation to harden the system, e.g., AI/ML-supported penetration testing;
- Return the system to full capability after a security incident.

▶ **Definition 6.13: Beneficial Security AI (Artificial Intelligence)** Artificial intelligence, especially machine learning, used by a cyber-defender to identify and eliminate vulnerabilities in a target system during development or to be installed in a target system as online, real-time protection against assaults

Figure 6.14 shows the architecture of an *AI/ML-powered cyber-defense*: It repeats the possible attack vectors of Fig. 5.18. Artificial intelligence and machine learning support:

- The local intrusion interception and malware defense (Example 6.12);
- The log file analysis and exposure assessment;
- The preservation of healthy system configuration files;
- The anomaly detection and interpretation.

> **Quote**
> *"Only AI-powered defenses can withstand AI-powered attacks"*
>    Max Heinemeyer, 2021 (www.darktrace.com)

**Example 6.12: ML-based IDS (Intrusion Detection System)**

*Intrusion Detection Systems* (IDS) are cyberattack protection systems that try to identify malicious traffic by inspecting the network activities ([Scarfone12], [Fung17]). IDS inspects the traffic gathered from two sources: (1) All packets moving across the network after the firewall, and (b) the inbound/outbound traffic of each node/host

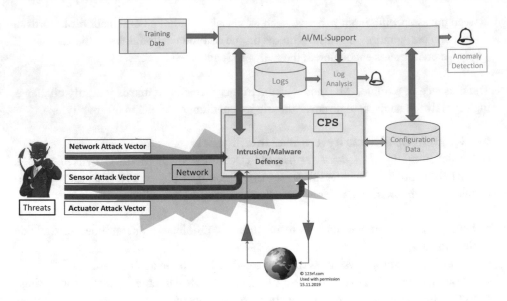

**Fig. 6.14**  AI/ML-powered cyber-defense

(Fig. 6.15). "Classical" IDS's mainly compare the bitstream with stored malware signatures and raise an alarm when a match is found (Left side of Fig. 6.15). The signature database is regularly updated by specialized companies.

This method works well with known malware, i.e., identified signatures. However,
new malware with yet unknown signatures will not be detected.

"*Identifying* unknown *attacks is one of the big challenges in network Intrusion
Detection Systems (IDS) research*" (Muhamad Erza Aminanto, 2016).

Newer research uses *machine learning* for the identification of novel malware,
working independently of recognized signatures (Right side of Fig. 6.15, e.g.,
[Aminanto16], [Kim18b], [Sengupta20], [Dasgupta20]).

One specific, successful ML-IDS technique is *feature extraction* or feature engineering (e.g., [Aminanto16], https://deepai.org/machine-learning-glossary-and-terms/
feature-extraction).

One significant promise of AI/ML-powered cyber-defense is *autonomous cyber-attack
response* (Definition 6.14, [Parisi19], [Zhang21c], [Macas20], [Sarker20], Example 6.13).

▶ **Definition 6.14: Autonomous Cyber-Attack Response**
Real-time, automated threat detection, overcoming the shortcomings of legacy tools, and
cutting through the noise in live, complex networks to accurately identify and neutralize
threatening anomalies.

Adapted from: https://www.information-age.com/autonomous-response-force-
multiplier-human-security-teams-123469048/

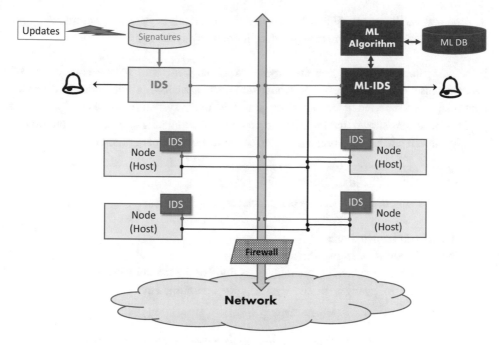

**Fig. 6.15**  Machine learning intrusion detection system

> **Quote**
> *"The need for automated, scalable, machine-speed vulnerability detection and patching is large and growing fast as more and more systems—from household appliances to major military platforms—get connected to and become dependent upon the internet. At the 2016 Defense Advanced Research Projects Agency (DARPA) Grand Cyber Challenge, machines were challenged to find and patch within seconds—not the usual months—flawed code that was vulnerable to being hacked and find their opponents' weaknesses before the defending system did".*
>    http://www.darpa.mil/news-events/2016-08-04. 2016

Basically, three methods of cyber-attack response are used:

1. *Open Loop Response* (Fig. 6.14, Part [a]): The attack is discovered after its incurrence, often days, months, or years after the infiltration—sometimes not at all! Reasons for the detection may be damage done, warnings by anti-malware specialists, log analysis, or forensic activities;
2. *Closed-Loop Response* (Fig. 6.14, Part [b]): The attack triggers an alarm, and a manual or semi-automated reaction is launched. The reaction may limit the impact of the assault and activate further defense mechanisms;

3. *Autonomous Response* (Fig. 6.14, Part [c]): The attack is detected, neutralized, and reported *before* it can harm the system or hide in any system part.

Effective, autonomous, and real-time *cyber-attack response* will be the culmination of cyber-defense: Any attack—even previously unknown assaults—is automatically detected, neutralized, and reported before any damage is done to the system (Example 6.13).

Experience has shown that the damage potential of an attack increases with the time it remains unmitigated. Therefore, a rapid, effective response is paramount (Fig. 6.16).

---

**Example 6.13: Autonomous Cyber-Attack Response**

In today's violent cyberspace, cyber-attack response requires:

a)  Disarm all known threats;
b)  Neutralize new, unknown threats;
c)  React in real-time, i.e., before damage is done or the threat hides in the system;
d)  Self-adapt to changes in the system, the operating environment, or the threat landscape.

The appropriate technology is the *autonomous cyber attack response* (e.g., [Creese20], [Sarker20], [Dressler04], [DarkTrace21a], [DarkTrace21b], [Garcia-Teodoroa09]). In addition to knowledge-based defense (such as rule-based detection), autonomous response heavily relies on *anomaly detection* ([Roberts17]) at all system levels. The

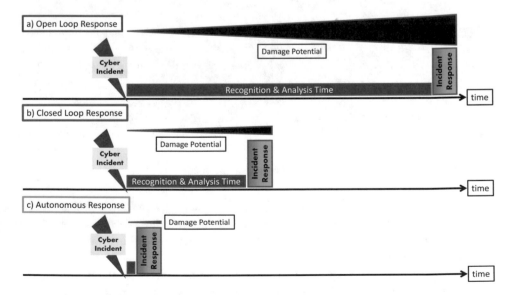

**Fig. 6.16**  Cyber-attack response modes

timely discovery of anomalies in the system requires massive use of artificial intelligence and machine learning (e.g., [Mehrotra17], [Pang20], [Dunning14], [Chandola07],
[Ahmed16], [Wu17]).

The autonomous cyber-attack response system architecture collects real-time information from all available sources in the system (Fig. 6.17), extracts anomalies by
machine learning models, aggregates the automated findings, and transmits them to
the decision AI (Fig. 6.18, 2.2). The decision AI triggers the immediate and appropriate *incident response*.

The various sources of data for anomaly detection are (Fig. 6.17):

- The network traffic;
- All origins of error or exception messages;
- The interconnections between the applications (Message and control flow);
- The behavior of the applications;
- The flow of data and control between the constituent systems of the CPSoS;
- The input and output of the CPS (Sensor inputs and actuator outputs);
- All human input activities;
- etc.

Because nature and the anomaly detection algorithms are different for the respective
data sources, specific ML models are required—each trained for its specific task. The
outcomes of the individual ML models are transmitted to the decision AI, which can

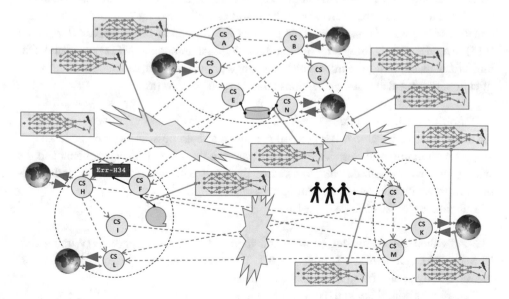

**Fig. 6.17**  Anomaly detection from all system sources

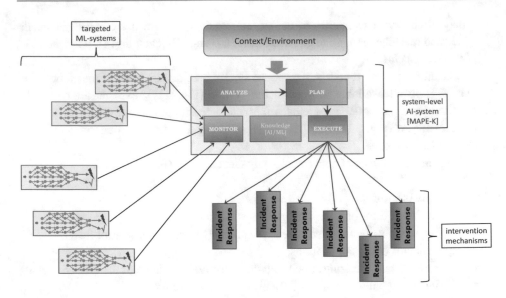

**Fig. 6.18**  Autonomous cyber-attack response system architecture

be the MAPE-K architecture (Example 2.2). The decision AI analyzes and combines the information, decides on actions, and immediately initiates the protection activities.

### 6.4.2.3  Collaborating Agents and Threat Intelligence Aggregation

AI/ML constitutes an indispensable, powerful defense against cyber threats. However, the effectiveness of cyber-defense can be significantly raised by combining the *threat and threat response information*. This requires the formation of *ensembles* (Definition 6.15). An ensemble is formed by linking the individual autonomic ML systems and enabling formalized, real-time information exchange about detected anomalies, threat identification, defense decisions, and actions taken (Fig. 6.19, Part [a]).

▶ **Definition 6.15: Ensemble**

An autonomous ensemble is a set of components that both possess and exchange knowledge. An ensemble achieves awareness by observing the state of its components and the state of its environment, deriving knowledge from this collected information, and deciding how to act on this knowledge through reasoning

Lubomir Bulej, 2015 in [Wirsing15a]

The individual autonomic ML systems are modeled as software agents ([Wirsing15a], [Wirsing15b], [Omicini01], [Boudaoud00]). They cooperate via the exchange of information and by coordinating functional activities.

Two factors contribute to the payoff of AI in security:

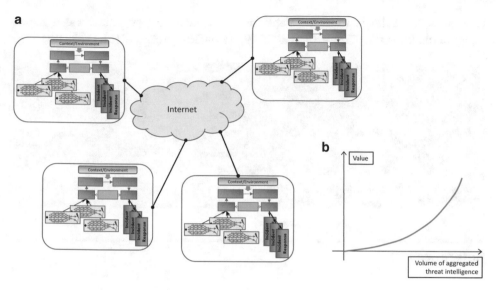

**Fig. 6.19**  Collaborating agents and threat intelligence aggregation

1. Data quality ([Chio18]);
2. Value of the aggregated data (Fig. 6.19, Part [b], [Dehghantanha18], [Bou-Harb20], [Skopik17], [Robertson17], [Soldatos20]);

> **Quote**
> *"The quality of a machine learning system's input data dictates its success or failure"*
>    Clarence Chio, 2018

Risks for the *data quality* are (a) Bias in datasets, (b) Missing data, (c) Label inaccuracy, and (d) Adversarial activity. All four risks must be solidly mitigated during training, deployment, operation, and maintenance ([Chio18]). In a figurative sense, also *model quality* forms part of the data quality—and the model quality must be assessed and maintained at all times (e.g., [Raj21], [ModelOp21]).

## 6.5    AI Conclusions

The potency of artificial intelligence/machine learning heralded a new dimension of the fight for *supremacy in the cyber-world* (e.g., [Rugge19], [Creese20], [West19], [Allen17]).

Individual actors—such as independent organizations—will not be able to handle the imminent cyber threats. Solid, organized, and disciplined cooperation of state actors, legislators, the military, research establishments, academia, companies, and organizations are fundamental ([Creese20], [OECD19], [King20], [Heegaard15], [Veith19]).

> **Quote**
> *"Only AI can fight AI, and the best algorithms will win. Cyber defense is an ongoing battle, and cyber AI is leading the charge, allowing the human responders to take stock and strategize from behind the front line. A new age in cyber defense is just beginning, and the effect of AI on this battleground is already proving fundamental."*
> Adapted from: DarkTrace, 2021 in [DarkTrace21c]

## References

[Ackerman21]        Pascal Ackerman 2021 Industrial Cybersecurity - Efficiently monitor the Cybersecurity Posture of your ICS Environment Packt Publishing 2 Birmingham UK 978-1-800-20209-2

[Aggarwal19]        Nikita Aggarwal, Horst Eidenmüller, Luca Enriques, Jennifer Payne, Kristen van Zwieten: Autonomous Systems and the Law. C.H.Beck Verlag, Munich, Germany, 2019. ISBN 978-3-4067-3683-4

[Ahmed16]           Mohiuddin Ahmed, Abdun Naser Mahmood, Jiankun Hu: *A Survey of Network Anomaly Detection Techniques* Journal of Network and Computer Applications, Elsevier, Amsterdam, Netherlands, Vol. 60, 2016, pp. 19–31. Downloadable from: https://www.gta.ufrj.br/~alvarenga/files/CPE826/Ahmed2016-Survey.pdf [Last accessed: 6.6.2021]

[AI-HLEG19]         AI-HLEG: **Ethics Guidelines for Trustworthy AI** European Commission High-Level Expert Group on Artificial Intelligence, Brussels, Belgium, 2019. Downloadable from: https://digital-strategy.ec.europa.eu/en/library/ethics-guidelines-trustworthy-ai [Last accessed: 24.5.2021]

[Allen17]           John R. Allen, Amir Husain: *On Hyperwar* Proceedings of the US Naval Institute, Vol. 143, Nr. 7, July 2017. Available from: https://www.usni.org/magazines/proceedings/2017/july/hyperwar [Last accessed: 2.6.2021]

[Alpaydin16]        Ethem Alpaydin 2016 Machine Learning – The New AI The MIT Press MA, USA Cambridge 978-0-262-52951-8

[Aminanto16]        Muhamad Erza Aminanto, Kwangjo Kim: **Deep Learning in Intrusion Detection System-An Overview** Technical Report, School of Computing, KAIST, Korea, 2016. Downloadable from: https://caislab.kaist.ac.kr/publication/paper_files/2016/IRCET16_AM.pdf [Last accessed: 22.5.2021]

[Amodei16]          Dario Amodei, Chris Olah, Jacob Steinhardt, Paul Christiano, John Schulman, Dan Mané: **Concrete Problems in AI Safety** Preprint, arXiv, July 2016. Downloadable from: https://arxiv.org/pdf/1606.06565.pdf [Last accessed: 28.3.2021]

[Anderson17]        Hyrum S. Anderson, Anant Kharkar, Bobby Filar, Phil Roth: *Evading Machine Learning Malware Detection* Black Hat USA 2017 Conference, Las Vegas, NV, USA, July 22–27, 2017. Downloadable from: https://paper.bobylive.com/Meeting_Papers/BlackHat/USA-2017/us-17-Anderson-Bot-Vs-Bot-Evading-Machine-Learning-Malware-Detection-wp.pdf [Last accessed: 23.5.2021]

[Anderson18]        Hyrum S. Anderson, Anant Kharkar, Bobby Filar, David Evans, Phil Roth: *Learning to Evade Static PE Machine Learning Malware Models via Reinforcement Learning* arXiv Preprint, 30 Jan 2018. Downloadable from: https://arxiv.org/pdf/1801.08917.pdf [Last accessed: 23.5.2021]

[Bajo21]            Pau Labarta Bajo: *Adversarial Machine Learning: How to Attack and Defend ML Models* Tutorial, Toptal LLC, New York, NY, USA, 2021. Downloadable from: https://www.toptal.com/machine-learning/adversarial-machine-learning-tutorial [Last accessed: 2.4.2021]

[Balamurugan18]     S. Balamurugan: **Principles of Artificial Intelligence for Information Security - *Concepts, Applications, and Case Studies*** Scholars' Press (International Book Market Service Ltd.), Rīgā, Latvia, 2018. ISBN 978-620231-363-6

[Ball21]            Philip Ball November 2021 First 100-Qubit Quantum Computer enters crowded race NATURE New York, NY, USA 599 25 542

[Bartocci18]        Ezi Bartocci, Jyotirmoy Deshmukh, Alexandre Donzé, Georgios Fainckos, Oded Maler, Dejan Ničković: **Specification-Based Monitoring of Cyber-Physical Systems - *A Survey on Theory, Tools, and Applications*** In: Bartocci E., Falcone Y. (Editors): Lectures on Runtime Verification. Lecture Notes in Computer Science, Vol 10457. Springer International Publishing, Cham, Switzerland, 2018. ISBN 978-3-319-75632-5

[Bartocci19]        Ezio Bartocci, Yliès Falcone (Editors): **Lectures on Runtime Verification - *Introductory and Advanced Topics*** Springer Nature, Cham, Switzerland, 2019. ISBN 978-3-319-75631-8 (LNCS 10457)

[Bauer13]           P Craig 2013 Bauer: Secret History-The Story of Cryptology CRC Press (Taylor & Francis) FL, USA Boca Raton 978-1-4665-6186-1

[BCG18]             Ryan Goosen, Anna Rontojannis, Stefan Deutscher, Juergen Rogg, Walter Bohmayr, David Mkrtchian: *Artificial Intelligence is a Threat to Cybersecurity. It's also a Solution.* Technical Memorandum, Boston Consulting Group, Boston, MA, USA, 2018. Downloadable from: https://image-src.bcg.com/Images/BCG-Artificial-Intelligence-Is-a-Threat-to-Cyber-Security-Its-Also-a-Solution-Nov-2018_tcm9-207468.pdf [Last accessed: 16.5.2021]

[Benatti10]         Fabio Benatti, Mark Fannes, Roberto Floreanini, Dimitri Petritis (Editors): **Quantum Information, Computation, and Cryptography - *An Introductory Survey of Theory, Technology, and Experiments*** Springer Verlag, Berlin, Germany, 2010 (Lecture Notes in Physics, Volume 808), 2010. ISBN 978-3-642-11913-2

[Bernhardt20]       Chris Bernhardt : **Quantum Computing for Everyone** MIT Press, Cambridge, MA, USA, 2020. ISBN 978-0-262-53953-1

[Bernstein08]        Daniel J. Bernstein, Johannes Buchmann, Erik Dahmen (Editors): **Post-Quantum Cryptography** Springer Verlag, Berlin, Germany, 2008. ISBN 978-3-540-88701-0

[Beyerer20]          J. Beyerer, P. Martini (Editors): *Rise of Artificial Intelligence in Military Weapon Systems* Position Paper, Fraunhofer IOSB, Ettlingen, Germany, May 2020. Downloadable from: https://www.fraunhofer.de/content/dam/zv/de/forschungsthemen/schutz-sicherheit/rise-of-intelligent-systems-in-military-weapon-systems-position-paper-fraunhofer-vvs.pdf [Last accessed: 25.5.2021]

[Beyerer21]          Jürgen Beyerer, Alexander Maier, Oliver Niggemann (Editors): **Machine Learning for Cyber-Physical Systems - Selected Papers from the International Conference ML4CPS 2020.** Springer Vieweg, Wiesbaden, Germany, 2021. ISBN 978-3-662-62745-7. Downloadable from: https://www.dbooks.org/machine-learning-for-cyber-physical-systems-3662627469/ [Last accessed: 22.3.2021]

[Bhagoji18]          Arjun Nitin Bhagoji, Warren He, Bo Li, Dawn Song: *Practical Black-box Attacks on Deep Neural Networks using Efficient Query Mechanisms* Proceedings of the European Conference on Computer Vision, ECCV 2018, Munich, Germany, September 8–14, 2018. Downloadable from: https://openaccess.thecvf.com/ECCV2018 and https://openaccess.thecvf.com/content_ECCV_2018/papers/Arjun_Nitin_Bhagoji_Practical_Black-box_Attacks_ECCV_2018_paper.pdf [Last accessed: 2.4.2021]

[Bistron21]          Marta Bistron, Zbigniew Piotrowski: *Artificial Intelligence Applications in Military Systems and Their Influence on Sense of Security of Citizens* Electronics 2021, 10(7), 871. Downloadable from: https://doi.org/10.3390/electronics10070871 [Last accessed: 25.5.2021]

[Boddington17]       Paula Boddington: **Towards a Code of Ethics for Artificial Intelligence** Springer Nature, Cham, Switzerland, 2017. ISBN 978-3-319-60647-7

[Boden18]            Margaret A. Boden: **Artificial Intelligence - A Very Short Introduction** Oxford University Press, Oxford, UK, 2018. ISBN 978-0-199-60291-9

[Bonfanti20]         Matteo E. Bonfanti, Kevin Kohler: *Artificial Intelligence for Cybersecurity* CSS Analyses in Security Policy, No. 265, June 2020, ETH Zurich, Zurich, Switzerland, 2020. Downloadable from: https://css.ethz.ch/content/dam/ethz/special-interest/gess/cis/center-for-securities-studies/pdfs/CSSAnalyse265-EN.pdf [Last accessed: 16.5.2021]

[Bossel14]           Hartmut Bossel: **Modeling and Simulation** A.K. Peters, Ltd., Wellesley, MA, USA, 2014. ISBN 978-3-663-10823-8

[Boudaoud00]         K. Boudaoud, H. Labiod, R. Boutaba, Z. Guessoum : *Network Security Management with Intelligent Agents* 2000 IEEE/IFIP Network Operations and Management Symposium (NOMS 2000), 'The Networked Planet: Management Beyond 2000'. Downloadable from: http://rboutaba.cs.uwaterloo.ca/Papers/Conferences/2000/Boudaoud00.pdf [Last accessed: 8.6.2021]

[Bou-Harb20]         Elias Bou-Harb, Nataliia Neshenko: **Cyber Threat Intelligence for the Internet of Things** Springer Nature Switzerland, Cham, Switzerland, 2020. ISBN 978-3-030-45857-

[Boyd20]          Colin Boyd Anish Mathuria Douglas Stebila 2020 Protocols for
                  Authentication and Key Establishment Springer Verlag 2 Heidelberg
                  Germany 978-3-662-58145-2

[Braband20]       Jens Braband, Hendrik Schäbe: *On Safety Assessment of Artificial
                  Intelligence* Preprint February 2020. Downloadable from: https://www.
                  researchgate.net/publication/339642339_On_Safety_Assessment_of_
                  Artificial_Intelligence [Last accessed: 20.9.2020]

[Brachman18]      Ronald Brachman, Peter Stone, Francesca Rossi, Yevgeniy Vorobeychik
                  (Editors): **Adversarial Machine Learning** Morgan & Claypool
                  Publishers, San Rafael, CA, USA, 2018. ISBN 978-1-681-73397-5

[Brockman19]      John Brockman Eds 2019 Possible Minds-Twenty-Five Ways of
                  Looking at AI Penguin Press (Random House) N.Y., USA New York
                  978-0-525-55799-9

[Brukman11]       Olga Brukman, Shlomi Dolev: *Recovery-oriented Programming -
                  Runtime Monitoring of Safety and Liveness* International Journal on
                  Software Tools for Technology Transfer, Springer Nature Switzerland
                  AG, Cham, Switzerland, August 2011. Downloadable from: https://
                  www.researchgate.net/publication/220643413_Recovery_ori-
                  ented_programming_Runtime_monitoring_of_safety_and_liveness/
                  link/59c8e210458515548f3d9769/download [Last accessed: 10.3.2021]

[Buchmann16]      Johannes A. A. Buchmann, Evangelos Karatsiolis, Alexander
                  Wiesmaier: **Introduction to Public Key Infrastructures** Springer-
                  Verlag, Heidelberg, Germany, 2016. ISBN 978-3-662-52450-3

[Burkov20]        Andriy Burkov: **Machine Learning Engineering** True Positive Inc.,
                  Quebec City, Canada, 2020. ISBN 978-1-77700-546-7

[Burns18]         D Lawrence 2018 Burns, Christopher Shulgan: Autonomy: The Quest to
                  Build the Driverless Car - and how it will Reshape our World Harper
                  Collins Publishers N.Y., USA New York 978-0-062-66112-8

[Carlini17]       Nicholas Carlini, David Wagner: *Adversarial Examples Are Not Easily
                  Detected: Bypassing Ten Detection Methods* AISec'17, November 3,
                  2017, Dallas, TX, USA. Downloadable from: https://nicholas.carlini.com/
                  papers/2017_aisec_breakingdetection.pdf [Last accessed: 16.4.2021]

[Carlini17]       Nicholas Carlini, David Wagner: *Towards Evaluating the Robustness of
                  Neural Networks* Preprint, arXiv, March 22, 2017. Downloadable from:
                  https://arxiv.org/pdf/1608.04644.pdf [Last accessed: 2.4.2021]

[Casola19]        Linda Casola, Dionna Ali (Rapporteurs): **Robust Machine Learning
                  Algorithms and Systems for Detection and Mitigation of Adversarial
                  Attacks and Anomalies** Proceedings of a Workshop. The National
                  Academies Press, Washington, DC, USA, 2019. ISBN 978-0-309-
                  49609-4. Downloadable from: https://www.nap.edu/download/25534
                  [Last accessed: 8.4.2021]

[Chandola07]      Varun Chandola, Arindam Banerjee, Vipin Kumar: **Anomaly Detection
                  - A Survey** Technical Report TR 07-017, Department of Computer
                  Science and Engineering, University of Minnesota, Minneapolis, USA.
                  Downloadable       from:      https://conservancy.umn.edu/bitstream/han-
                  dle/11299/215731/07-017.pdf?sequence=1 [Last accessed: 6.6.2021]

[Chaubey20]       Nirbhay Chaubey, Bhavesh Prajapati: **Quantum Cryptography and
                  the Future of Cyber Security** IGI Global Publishing, Hershey, PA,
                  USA, 2020. ISBN 978-1-7998-2254-7

[Chen16]          Lily Chen Stephen Jordan Yi-Kai Liu Dustin Moody Rene Peralta Ray Perlner Daniel Smith-Tone 2016 Report on Post-Quantum Cryptography: NiSTIR 8105 National Institute of Standards and Technology (NIST), Gaithersburg, MD, USA, 2016 CreateSpace Independent Publishing Platform North Charleston, S.C., USA 978-1-9798-7930-9. Downloadable from: doi: https://doi.org/10.6028/NIST. IR.8105 [Last accessed: 14.9.2020]

[Chio18]          Clarence Chio, David Freeman: **Machine Learning and Security - Protecting Systems with Data and Algorithms** O'Reilly Media Inc., Sebastopol, CA, USA, 2018. ISBN 978-1-491-97990-7

[Chowdhary20]     KR Chowdhary 2020 Fundamentals of Artificial Intelligence Springer Nature India New Delhi India 978-8-132-23970-3

[Claus20]         Isabell Claus (Publisher): *IT SECURITY 2025 - This is what experts are expecting for the future* RadarServices, Cybersecurity World, Vienna, Austria, 2020. Downloadable from: https://www.radarcs.com/ wp-content/uploads/2020/04/IT-Security-Magazine-Issue-1-EN.pdf [Last accessed: 30.5.2021]

[Coutinho99]      S. C. Coutinho: **The Mathematics of Ciphers** - *Number Theory and RSA Cryptography* A K Peters Publishers, Natick, MA, USA, 1999. ISBN 978-1-5688-1082-9

[CPST21]          Check Point Software Technologies Ltd (CPST): **Cyber Security Report 2020** Check Point Software Technologies Ltd., Tel Aviv, Israel, 2021. Downloadable from: https://www.ntsc.org/assets/pdfs/cyber-secu-rity-report-2020.pdf [Last accessed: 16.5.2021]

[Creese20]        Sadie Creese, Jamie Saunders, Louise Axon, William Dixon, et al. : **Cybersecurity, Emerging Technology, and Systemic Risk** Insight Report, World Economic Forum, Cologny/Geneva, Switzerland, 2020. Downloadable from: http://www3.weforum.org/docs/WEF_Future_ Series_Cybersecurity_emerging_technology_and_systemic_risk_2020. pdf [Last accessed: 2.6.2021]

[Cylance17]       Cylance Data Science Team: **Introduction to Artificial Intelligence for Security Professionals** The Cylance Press, Irvine, CA, USA, 2017. ISBN 978-0-9980169-2-4

[DarkTrace21a]    DarkTrace: *DarkTrace AI: Combining Unsupervised and Supervised Machine Learning* Technical White Paper, Darktrace Inc., Cambridge, UK, 2021. Downloadable from: https://www.darktrace.com/en/ resources/wp-machine-learning.pdf [Last accessed: 3.6.2021]

[DarkTrace21b]    DarkTrace: *Darktrace Antigena: The Future of AI-Powered Autonomous Response* Technical White Paper, Darktrace Inc., Cambridge, UK, 2021. Downloadable from: https://www.darktrace.com/ en/resources/wp-antigena.pdf [Last accessed: 3.6.2021]

[DarkTrace21c]    DarkTrace: *AI-Augmented Attacks and the Battle of the Algorithms* Technical White Paper, Darktrace Inc., Cambridge, UK, 2021. Downloadable from: http://newtech.mt/wp-content/uploads/2020/09/ AI-Augmented-Attacks-and-the-Battle-of-the-Algorithms.pdf     [Last accessed: 6.6.2021]

[Dasgupta20]       Dipankar Dasgupta, Zahid Akhtar, Sajib Sen: *Machine Learning in Cybersecurity - A comprehensive Survey* Journal of Defense Modeling and Simulation: Applications, Methodology, Technology, Newbury Park, CA, USA, September 2020. Downloadable from: https://www.research-gate.net/publication/344374053_Machine_learning_in_cybersecurity_a_comprehensive_survey [Last accessed: 3.6.2021]

[Dehghantanha18]   Ali Dehghantanha, Mauro Conti, Tooska Darhari (Editors): **Cyber Threat Intelligence** Springer International Publishing, Cham, Switzerland, 2018. ISBN 978-3-319-739502

[DelMonte18]       Louis A. Del Monte: **Genius Weapons - *Artificial Intelligence, Autonomous Weaponry, and the Future of Warfare*** Prometheus Books, Amherst, NY, USA, 2018. ISBN 978-16338-8452-6

[Dilek15]          Selma Dilek, Hüseyin Çakır, Mustafa Aydın: *Applications of Artificial Intelligence Techniques to Combating Cyber Crimes – A Review* International Journal of Artificial Intelligence & Applications (IJAIA), Vol. 6, No. 1, January 2015. Downloadable from: https://arxiv.org/ftp/arxiv/papers/1502/1502.03552.pdf [Last accesse d: 17.5.2021]

[Dooley18]         John F. Dooley: **History of Cryptography and Cryptanalysis - *Codes, Ciphers, and Their Algorithms*** Springer International Publishing, Cham, Switzerland, 2018. ISBN 978-3-31990442-9

[Dressler04]       Falko Dressler, Gerhard Münz, Georg Carle: *Attack Detection using Cooperating Autonomous Detection Systems (CATS)* Technical Report, University of Tübingen, Tübingen, Germany, 2004. Downloadable from: https://www.net.in.tum.de/fileadmin/RI/members/carle/publications/papers/wac04.pdf [Last accessed: 5.6.2021]

[Dunning14]        Ted Dunning, Ellen Friedman: **Practical Machine Learning - *A New Look at Anomaly Detection*** O'Reilly and Associates, Sebastopol, CA, USA, 2014. ISBN 978-1-491-91160-0

[Edwards20]        Simon Edwards: **Quantum Computing and Modern Cryptography** Independently published (by author), 2020. ISBN 979-8-624095-32-8

[Elçi20]           Atilla Elçi, Ashish Kumar Luhach (Editors): **Artificial Intelligence Paradigms for Smart Cyber-Physical Systems** IGI Global Publisher, Engineering Science Reference, Hershey, PA, USA. 2020. ISBN 978-1-7998-5846-1

[Eliot16]          Lance Eliot 2016 AI Guardian Angel Bots for Deep AI Trustworthiness - Practical Advances in Artificial Intelligence (AI) and Machine Learning LBE Press Publishing CA, USA San Francisco 978-0-6928-0061-4

[ENISA20]          ENISA20: **AI Cybersecurity Challenges - *Threat Landscape for Artificial Intelligence*** Technical Report, European Union Agency for Cybersecurity (ENISA), Maroussi, Attiki, Greece, 2020. ISBN 978-92-9204-462-6. Downloadable from: https://www.enisa.europa.eu/publications/artificial-intelligence-cybersecurity-challenges [Last accessed: 23.5.2021]

[ENISA21]          ENISA21: **Post-Quantum Cryptography - *Current State and Quantum Mitigation*** Technical Report, European Union Agency for Cybersecurity (ENISA), Maroussi, Attiki, Greece, 2021. ISBN 978-92-9204-468-8. Downloadable from: https://www.enisa.europa.eu/

publications/post-quantum-cryptography-current-state-and-quantum-mitigation [Last accessed: 23.5.2021]

[Evtimov17]     Ivan Evtimov, Kevin Eykholt, Earlence Fernandes, Tadayoshi Kohno, Bo Li, Atul Prakash, Amir Rahmati, Dawn Song: *Robust Physical-World Attacks on Deep Learning Models* White Paper, University of California, Berkeley, USA, 2017. Downloadable from: http://techpolicylab.uw.edu/wp-content/uploads/2017/12/Robust-Physical-World-Attacks-on-Deep-Learning-Modules.pdf Last accessed: 14.3.2021]

[Faria18]       José M. Faria: *Machine Learning Safety - An Overview* Safe Perspective Ltd. 2018. White Paper, published by the Safety-Critical Systems Club, Bristol, UK, 2018. Downloadable from: https://www.researchgate.net/profile/Jose-Faria-3/publication/320567319_Machine_Learning_Safety_An_Overview/links/59f37efda6fdcc075ec34986/Machine-Learning-Safety-An-Overview.pdf [Last accessed: 12.3.2021]

[Farrel04]      Adrian Farrel : **The Internet and Its Protocols - *A Comparative Approach*** Morgan Kaufmann Publishers (Elsevier), Amsterdam, Netherlands, 2004. ISBN 978-1-558-60913-6

[Faulkner20]    Alastair Faulkner, Mark Nicholson: *The Emergence of Accidental Autonomy* In: Mike Parsons, Mark Nicholson, (Editors): **Assuring Safe Autonomy** Proceedings of the 28th Safety-Critical Systems Symposium (SSS'20), York, UK, 11–13 February 2020. ISBN 978-1-713305-66-8

[FBI21]         US Federal Bureau of Investigation (FBI): **Internet Crime Report 2020** Annual Report, FBI Internet Complaint Center (IC3), New York, NY, USA, 2021. Downloadable from: https://www.ic3.gov/Media/PDF/AnnualReport/2020_IC3Report.pdf [Last accessed: 16.5.2021]

[Fehlmann20]    Thomas Michael Fehlmann: **Autonomous Real-Time Testing - *Testing Artificial Intelligence and Other Complex Systems*** Logos Verlag, Berlin, Germany, 2020. ISBN 978-3-8325-5038-7

[Fiehler20]     Kyle Fiehler: **Smoke and Mirrors - *Do AI and Machine Learning Make a Difference in Cybersecurity?*** WebRoot Company, Broomfield, CO, USA, 2020. Downloadable from: https://mypage.webroot.com/ai-ml-survey-report-2020-access.html [Last accessed: 17.5.2021]

[Finlayson19]   Samuel G. Finlayson, Hyung Won Chung, Isaac S. Kohane, Andrew L. Beam: *Adversarial Attacks Against Medical Deep Learning Systems* Paper Preprint, February 2019. Downloadable from: https://arxiv.org/pdf/1804.05296.pdf [Last accessed: 18.3.2021]

[Frtunikj19]    Jelena Frtunikj, Simon Fürst: *Engineering Safe Machine Learning for Automated Driving Systems* In: Mike Parsons, Tim Kelly (Editors): **Assuring Safe Autonomy** Proceedings of the 27th Safety-Critical Systems Symposium (SSS'19), Bristol, UK, 5–7 February 2019. ISBN 978-1-7293-6176-4

[Fung17]        Carol Fung Raouf Boutaba 2017 Intrusion Detection Networks - A Key to Collaborative Security CRC Press (Taylor & Francis) FL, USA Boca Raton 978-1-138-19889-0

[Furrer19]      Frank J. Furrer: **Future-Proof Software-Systems – *A Sustainable Evolution Strategy*** Springer Vieweg Verlag, Wiesbaden, Germany, 2019. ISBN 978-3-658-19937-1

[Furrer20]      Frank J. Furrer Roger A. Grimes. **Cryptography Apocalypse - *Preparing for the Day When Quantum Computing Breaks Today's***

|  | *Crypto* (Book Review) Informatik Spektrum, Germany, Volume 43, Pages 70 – 72, January 2020. Available from: https://link.springer.com/article/10.1007/s00287-020-01239-6 [Last accessed: 11.9.2020] |
| [Gao16] | Yang Gao (Editor): **Contemporary Planetary Robotics – *An Approach Toward Autonomous Systems*** Wiley-VCH Verlag GmbH & Co. KGaA, Weinheim, Germany, 2016. ISBN-13: 978-3-527-41325-6 |
| [Garcia21] | Denise Garcia May 2021 Stop the emerging AI cold War Nature London, UK 593 13 169 |
| [Garcia-Teodoroa09] | P. Garcia-Teodoroa, J. Diaz-Verdejoa, G. Macia-Fernandeza, E. Vazquez: ***Anomaly-based Network Intrusion Detection - Techniques, Systems and Challenges*** Computers & Security, Elsevier, Amsterdam, Netherlands, Vol. 28, 2009, pp. 18–28. Downloadable from: http://dtstc.ugr.es/~jedv/descargas/2009_CoSe09-Anomaly-based-network-intrusion-detection-Techniques,-systems-and-challenges.pdf [Last accessed: 5.6.2021] |
| [Geyik19] | Sahin Cem Geyik, Krishnaram Kenthapadi & Varun Mithal: **Explainable AI in Industry** KDD 2019 Tutorial, 25th ACM SIGKDD Conference on Knowledge Discovery and Data Mining, Anchorage, Alaska, August 4–8, 2019. Downloadable from: https://de.slideshare.net/KrishnaramKenthapadi/explainable-ai-in-industry-kdd-2019-tutorial [Last accessed: 10.5.2021] |
| [Goebel18] | Randy Goebel, Ajay Chander, Katharina Holzinger, Freddy Lecue, Zeynep Akata, Simone Stumpf, Peter Kieseberg, Andreas Holzinger: ***Explainable AI: the new 42?*** 2nd International Cross-Domain Conference for Machine Learning and Knowledge Extraction (CD-MAKE), August 2018, Hamburg, Germany. Downloadable from: https://hal.inria.fr/hal-01934928/document [Last accessed: 7.5.2021] |
| [Goodfellow15] | Ian J. Goodfellow, Jonathon Shlens, Christian Szegedy: ***Explaining and Harnessing Adversarial Examples*** Preprint arXiv, March 20, 2015, Published as a conference paper at ICLR 2015, San Diego, CA, USA, May 7—9, 2015. Downloadable from: https://arxiv.org/abs/1412.6572 [Last accessed: 17.4.2021] |
| [Goodfellow16] | Ian Goodfellow Yoshua Bengio Aaron Courville 2016 Deep Learning MIT Press MA, USA Cambridge 978-0-262-03561-3 |
| [Goodloe10] | Alwyn E. Goodloe, Lee Pike: ***Monitoring Distributed Real-Time System - A Survey and Future Directions*** NASA Technical Memorandum (NASA/CR–2010-216724), NASA Center for AeroSpace Information, Hanover, MD, USA, July 2010. Downloadable from: https://ntrs.nasa.gov/api/citations/20100027427/downloads/20100027427.pdf [Last accessed: 13.5.2021] |
| [Graham10] | Adrian Graham 2010 Communications 2 Radar and Electronic Warfare John Wiley & Sons Ltd. Chichester, UK 978-0-470-68871-7 |
| [Grätzer13] | George Grätzer 2013 General Lattice Theory Birkhäuser Verlag 2 Basel Switzerland 978-3-764-36996-5 |
| [Grimes19] | Roger A. Grimes: **Cryptography Apocalypse - *Preparing for the Day When Quantum Computing Breaks Today's Crypto*** John Wiley & Sons, Inc., Hoboken, NJ, USA, 2019. ISBN 978-1-119-61819-5 |
| [Grumbling19] | Emily Grumbling, Mark Horowitz (Editors): ***Quantum Computing - Progress and Prospects (A Consensus Study Report)*** THE NATIONAL |

|                      | ACADEMIES PRESS, Washington, DC, USA, 2019. ISBN 978-0-309-47969-1. Downloadable from: https://www.nap.edu/download/25196 [Last accessed: 15.9.2020] |

[Hageback17]            Niklas Hageback Daniel Hedblom 2017 AI for Digital Warfare Taylor & Francis Ltd FL, USA Boca Raton 978-1-032-04871-0

[Haimes19]              Yacov Y. Haimes: **Modeling and Managing Interdependent Complex Systems of Systems** John Wiley & Sons, Inc. (Wiley - IEEE), Hoboken, NJ, USA, 2019. ISBN 978-1-119-17365-6

[Hamon20]               Ronan Hamon, Henrik Junklewitz, Ignacio Sanchez: **Robustness and Explainability of Artificial Intelligence -** *From technical to Policy Solutions* European Commission, JRC Technical Report, Brussels, Belgium, 2020. Downloadable from: https://ec.europa.eu/jrc/en/publication/robustness-and-explainability-artificial-intelligence [Last accessed: 21.3.2021]

[Hankerson13]           Darrel Hankerson, Alfred Menezes, Scott Vanstone: **Guide to Elliptic Curve Cryptography** Springer Science+Business Media, New York, N.Y., USA, 2013 (Softcover reprint of the original $1^{st}$ edition 2004. ISBN 978-1-441-92929-7

[Harrison20]            Lee Harrison: *How to use Runtime Monitoring for automotive Functional Safety* TechDesignForum White Paper, May 2020. Downloadable from: https://www.techdesignforums.com/practice/technique/how-to-use-runtime-monitoring-for-automotive-functional-safety/ [Last accessed: 11.3.2021]

[Haupt19]               Nikita Bhardwaj Haupt, Peter Liggesmeyer: *A Runtime Safety Monitoring Approach for Adaptable Autonomous Systems* In: Romanovsky A., Troubitsyna E., Gashi I., Schoitsch E., Bitsch F. (Editors): Computer Safety, Reliability, and Security. SAFECOMP 2019. Lecture Notes in Computer Science, Vol 11699. Springer International Publishing, Cham, Switzerland, 2019. Downloadable from: https://www.researchgate.net/publication/335557336_A_Runtime_Safety_Monitoring_Approach_for_Adaptable_Autonomous_Systems/link/5dd2b7b0299bf1b74b4e14f7/download [last accessed: 12.5.2021]

[Heegaard15]            Poul Heegaard, Erwin Schoitsch (Editors): *Combining Safety and Security Engineering for Trustworthy Cyber-Physical Systems*. ERCIM News, Nr. 102, July 2015. Free pdf-Download from: https://ercim-news.ercim.eu/en102/special/combining-safety-and-security-engineering-for-trustworthy-cyber-physical-systems [last accessed 10.7.2021]

[Heinegg18]             Wolff Heintschel von Heinegg, Robert Frau, Tassilo Singer (Editors): **Dehumanization of Warfare -** *Legal Implications of New Weapon Technologies* Springer International Publishing, Cham, Switzerland, 2018. ISBN 978-3-319-88402-8

[Hernandez-Orallo20]    Jose Hernandez-Orallo, Fernando Martınez-Plumed, Shahar Avin, Jess Whittlestone, Sean O hEigeartaigh: *AI Paradigms and AI Safety - Mapping Artefacts and Techniques to Safety Issues* $4^{th}$ European Conference on Artificial Intelligence - ECAI 2020, Santiago de Compostela, Spain, 2020. Downloadable from: https://ecai2020.eu/papers/1364_paper.pdf [Last accessed: 12.3.2021]

[Hersch20]        Jack J. Hersch: **The Dangers of Automation in Airliners - *Accidents Waiting to Happen*** Air World (Pen & Sword Books Ltd.), Barnsley, UK, 2020. ISBN 978-1-5267-7314-2

[Hinek09]         M. Jason Hinek: **Cryptanalysis of RSA and Its Variants** CRC Press (Taylor & Francis), Boca Raton, FL, USA, 2009. ISBN 978-1-420-07518-2

[Hoffstein14]     Jeffrey Hoffstein Jill Pipher Joseph H Silverman 2014 An Introduction to Mathematical Cryptography Springer Science & Business Media 2 New York N.Y., USA 978-1-493-91710-5

[HomelandSecurity17]  U.S. Homeland Security: **Emerging Cyber Threats to the United States** Security Technologies of the Committee on Homeland Security House of Representatives Subcommittee on Cybersecurity (Infrastructure Protection), CreateSpace Independent Publishing Platform, North Charleston, S.C., USA,2017. ISBN 978-1-5464-8512-4

[Horneman19]      Angela Horneman, Andrew Mellinger, Ipek Ozkaya: **AI Engineering: 11 Foundational Practices - *Recommendations for Decision Makers from Experts in Software Engineering, Cybersecurity, and applied Artificial Intelligence*** White Paper DM19-0624, 06.06.2019. CARNEGIE MELLON UNIVERSITY, Software Engineering Institute (SEI), Pittsburgh, PA, USA, 2019. Downloadable from: https://resources.sei.cmu.edu/asset_files/WhitePaper/2019_019_001_634648.pdf [last accessed: 19.8.2021]

[IntroBooks20]    IntroBooks. **Artificial Intelligence in Military** IntroBooks, Independently published, 2020. ISBN 979-8-6020-7476-5

[IST21]           IST: **Combating Ransomware - *A Comprehensive Framework for Action: Key Recommendations from the Ransomware Task Force*** Institute for Security and Technology, San Francisco Bay Area, CA, USA, 2021. Downloadable from: https://securityandtechnology.org/wp-content/uploads/2021/04/IST-Ransomware-Task-Force-Report.pdf [Last accessed: 15.5.2021]

[Ivancevic17]     Vladimir G. Ivancevic, Darryn J Reid, Michael J Pilling: **Mathematics of Autonomy - *Mathematical Methods for Cyber-Physical-Cognitive Systems*** World Scientific Publishing Ltd, Singapore, 2017. ISBN 978-9-813-23038-5

[Jacobson11]      Daniel Jacobson, Greg Brail, Dan Woods: **APIs: A Strategy Guide - *Creating Channels with Application Programming Interfaces*** O'Reilly and Associates, Sebastopol, CA, USA, 2011. ISBN 978-1-449-30892-6

[Johnson15]       Thomas A Johnson Eds 2015 Cybersecurity - Protecting Critical Infrastructures from Cyber Attack and Cyber Warfare CRC Press Taylor & Francis Ltd. Boca Raton, FL, USA 978-1-482-23922-5

[Johnson19]       Anne Johnson, Emily Grumbling (Rapporteurs): **Implications of Artificial Intelligence for Cybersecurity – *Proceedings of a Workshop*** The National Academies Press (NAP), Washington, DC, USA, 2019. ISBN 978-0-309-49450-2. Downloadable from: https://www.nap.edu/download/25488 [Last accessed: 15.9.2020]

[Jordan18]        Stephen P. Jordan, Yi-Kai Liu: ***Quantum Cryptanalysis - Shor, Grover, and Beyond*** IEEE Security & Privacy, New York, N.Y., USA, September/October 2018, pp. 14–21, vol. 16. DOI: https://doi.org/10.1109/MSP.2018.3761719

[Joseph19]          Immanual Joseph: **The Fifth Revolution - *Reinventing Workplace Happiness, Health, and Engagement through Compassion*** Independently Publisher, 2019. ISBN 978-0-5785-7789-0

[Joseph19]          Anthony D. Joseph, Blaine Nelson, Benjamin I. P. Rubinstein, J. D. Tygar: **Adversarial Machine Learning** Cambridge University Press, Cambridge, UK, 2019. ISBN 978-1-107-04346-6

[Joux09]            Antoine Joux 2009 Algorithmic Cryptanalysis CRC Press (Taylor & Francis) FL, USA Boca Raton 978-1-420-07002-6

[Juliano16]         Dustin Juliano : **AI Security** CreateSpace Independent Publishing Platform, North Charleston, S.C., USA, 2016. ISBN 978-1-5351-1900-9 Undine Press, Fort Myers, FL, USA, 2016, ISBN 978-1-5351-1900-9 Read online: www. http://aisecurity.org/ [Last accessed: 16.3.2021]

[Kahn96]            David Kahn 1996 The Codebreakers - The Comprehensive History of Secret Communication from Ancient Times to the Internet Scribner Publishing NY, USA, Revised Edition New York 978-0-684-83130-5

[Kaloudi20]         Nektaria Kaloudi, Jingue Li: *The AI-Based Cyber Threat Landscape - A Survey* Technical Report, Norwegian University of Science and Technology, Trondheim, Norway, 2020. Downloadable from: https:// ntnuopen.ntnu.no/ntnu-xmlui/handle/11250/2642553   [Last   accessed: 30.5.2021]

[Kane15]            Aaron Kane, Omar Chowdhury, Anupam Datta, Philip Koopman: *A Case Study on Runtime Monitoring of an Autonomous Research Vehicle (ARV) System* Preprint 6th International Conference on Runtime Verification (RV 2015), Vienna, Austria, September 22–25, 2015. Proceedings, Springer International Publishing, Cham, Switzerland, 2015. LNCS 9333. ISBN 978-3-319-23819-7. Downloadable from: https://www.researchgate.net/publication/300254024_A_Case_Study_ on_Runtime_Monitoring_of_an_Autonomous_Research_Vehicle_ ARV_System [Last accessed: 13.5.2021]

[Kane15]            Aaron Kane: **Runtime Monitoring for Safety-Critical Embedded Systems** Ph.D.Thesis, Carnegie Mellon University, Pittsburgh, PA, USA, February 2015. Downloadable from: https://users.ece.cmu. edu/~koopman/thesis/kane.pdf [Last accessed: 12.10.2020]

[Katz21a]           Jonathan   Katz   Yehuda   Lindell   2021   Introduction   to   Modern Cryptography CRC Press (Taylor & Francis) 3 Boca Raton FL, USA 978-0-815-35436-9

[Katz21b]           Jonathan Katz Vadim Lyubashevsky 2021 Lattice-based Cryptography CRC Press (Taylor & Francis) FL, USA Boca Raton 978-1-498-76347-)

[Kelleher19]        D John 2019 Kelleher: Deep Learning MIT Press MA, USA Cambridge 978-0-262-53755-1

[Kim18b]            Kwangjo Kim, Muhamad Erza Aminanto, Harry Chandra Tanuwidjaja: **Network Intrusion Detection using Deep Learning - *A Feature Learning Approach*** Springer Nature Singapore PTE Ltd., Singapore, Singapore, 2018. ISBN 978-9-811-31443-8

[King20]            Thomas C. King, Nikita Aggarwal, Mariarosaria Taddeo, Luciano Floridi: *Artificial Intelligence Crime - An Interdisciplinary Analysis of Foreseeable Threats and Solutions* Science and Engineering Ethics, SpringerLink, Springer Nature, Cham, Switzerland, 2020.

[Kiong20]          Liew Voon Kiong: **Blockchain and Cryptocurrency - *A Blockchain and Cryptocurrency Guidebook for Everyone*** Independently published, 2020. ISBN 979-8-6504-8548-3

[Kirat18]          Dhilung Kirat, Jiyong Jang, Marc Ph. Stoecklin: *DeepLocker - Concealing Targeted Attacks with AI Locksmithing* Black Hat Conference Paper, Las Vegas, USA, 2018. Downloadable from: https://i.blackhat.com/us-18/Thu-August-9/us-18-Kirat-DeepLocker-Concealing-Targeted-Attacks-with-AI-Locksmithing.pdf [Last accessed: 30.5.2021]

[Knight20]         Alissa Knight: **Hacking Connected Cars - *Tactics, Techniques, and Procedures*** John Wiley & Sons, Inc., Hoboken, New Jersey, USA, 2020. ISBN 978-1-119-49180-4

[Kubovič20]        Ondrej Kubovič, Peter Košinár, Juraj Jánošík: *Can Artificial Intelligence power future Malware?* ESET White Paper, ESET Deutschland GmbH, Jena, Germany, 2020. Downloadable from: https://www.welivesecurity.com/wp-content/uploads/2018/08/Can_AI_Power_Future_Malware.pdf [Last accessed: 23.5.2021]

[Kujawa18]         Adam Kujawa: *Under the Radar – The Future of Undetected Malware* Technical Brief, Malwarebytes Ltd., Cork, Ireland, 2018. Downloadable from: https://resources.malwarebytes.com/files/2018/12/Malwarebytes-Labs-Under-The-Radar-US.pdf [Last accessed: 30.5.2021]

[Kulkarni15]       Parag Kulkarni Prachi Joshi 2015 Artificial Intelligence - Building Intelligent Systems PHI Learning Ltd Delhi India 978-81-203-5046-5

[Kurakin17]        Alexey Kurakin, Ian J. Goodfellow, Samy Bengio: *Adversarial Machine Learning at Scale* Conference paper at ICLR 2017, Toulon, France, April 24 - 26, 2017. Downloadable from: https://bengio.abracadou-dou.com/publications/pdf/kurakin_2017_iclr_scale.pdf [Last accessed: 2.4.2021]

[Kurtz21]          George Kurtz: *CrowdStrike 2020 Global Threat Report* Technical Report, CrowdStrike, Inc., Sunnyvale, CA, USA, 2021. Downloadable from: https://www.crowdstrike.com/resources/reports/2020-crowdstrike-global-threat-report/ [Last accessed: 17.5.2021]

[Lance16]          Lance Eliot: **Pioneering Advances for AI Driverless Cars - *Practical Innovations in Artificial Intelligence and Machine Learning***. LBE Press Publishing, Open Library, San Francisco, CA, USA, New edition 2018. ISBN 978-0-6921-9669-4

[Lawless17]        W.F. Lawless, Ranjeev Mittu, Donald Sofge, Stephen Russell (Editors): **Autonomy and Artificial Intelligence - *A Threat or Savior?*** Springer International Publishing, Cham, Switzerland, 2017. ISBN 978-3-319-59718-8

[Li18a]            Jerry Zheng Li: **Principled Approaches to Robust Machine Learning and Beyond** PhD Thesis, Massachusetts Institute of Technology (MIT), Cambridge, MA, USA, 2018. Downloadable from: https://dspace.mit.edu/bitstream/handle/1721.1/120382/1084485589-MIT.pdf?sequence=1 [Last accessed: 8.4.2021]

[Li18b]            Tongtong Li, Tianlong Song, Yuan Liang: **Wireless Communications under Hostile Jamming - *Security and Efficiency*** Springer Nature Singapore, Singapore, 2018. ISBN 978-9-811-34509-8

[Liljefors19]      Max Liljefors, Gregor Noll, Daniel Steuer: **War and Algorithm** Rowman & Littlefield International, London, UK, 2019. ISBN 978-1-7866-1364-6

[Liu18]            Shaoshan Liu, Liyun Li, Jie Tang: **Creating Autonomous Vehicle Systems** Morgan & Claypool Publishers, San Rafael, CA, USA, 2018. ISBN 978-1-681-73007-3

[Longo20]          Luca Longo, Randy Goebel, Freddy Lecue, Peter Kieseberg: *Explainable Artificial Intelligence: Concepts, Applications, Research Challenges and Visions* IFIP International Federation for Information Processing 2020. Published by Springer Nature Switzerland, Cham, Switzerland 2020 in: A. Holzinger et al. (Editors): LNCS 12279, pp. 1–16, 2020. Downloadable from: https://www.researchgate.net/publication/343751247_Explainable_Artificial_Intelligence_Concepts_Applications_Research_Challenges_and_Visions/link/5f45149992851cd302296018/download [Last accessed: 21.3.2021]

[Macas20]          Mayra Macas, Chunming Wu: *Review: Deep Learning Methods for Cybersecurity and Intrusion Detection Systems* IEEE Latin-American Conference on Communications (LATINCOM), Virtual Conference, 18–20 November 2020. Downloadable from: https://arxiv.org/pdf/2012.02891.pdf [Last accessed: 3.6.2021]

[Madry19]          Aleksander Madry, Aleksandar Makelov, Ludwig Schmidt, Dimitris Tsipras, Adrian Vladu: *Towards Deep Learning Models Resistant to Adversarial Attacks* Paper Preprint, September 2019. Dowloadable from: https://arxiv.org/pdf/1706.06083.pdf [Last accessed: 18.3.2021]

[Maurer16]         Markus Maurer, J. Christian Gerdes, Barbara Lenz, Hermann Winnder (Editors): **Autonomous Driving – *Technical, Legal and Social Aspects*** Springer Verlag, Germany, 2016. ISBN 978-3-662-48845-4

[McCloskey20]      James McCloskey, Rose Gambon, Chris Allsopp, Thom Kirwan-Evans, Richard Maguire: *Generating the Evidence necessary to support Machine Learning Safety Claims* In: Mike Parsons, Mark Nicholson, (Editors): **Assuring Safe Autonomy** Proceedings of the 28th Safety-Critical Systems Symposium (SSS'20), York, UK, 11–13 February 2020. ISBN 978-1-713305-66-8

[Mehrotra17]       Kishan G. Mehrotra, Chilukuri K. Mohan, HuaMing Huang: **Anomaly Detection Principles and Algorithms (Terrorism, Security, and Computation)** Springer International Publishing, Cham, Switzerland, 2017. ISBN 978-3-319-67524-4

[Michel21]         Arthur Holland Michel: **Known Unknowns – Data Issues and Military Autonomous Systems** UNIDIR Technical Report SecTec/21/AI1, United Nations Institute for Disarmament Research, Geneva, Switzerland, 2021. Downloadable from: https://unidir.org/known-unknowns [Last accessed: 31.5.2021]

[Michelson20]      Brian M. Michelson: **Warbot 1.0 - *AI Goes to War*** War Planet Press (Ethan Ellenberg Literary Agency), New York, NY, USA, 2020. ISBN 978-1-6806-8205-2

[Mitra21] Sayan Mitra 2021 Verifying Cyber-Physical Systems - A Path to Safe Autonomy The MIT Press MA, USA Cambridge 978-0-262-04480-6

[ModelOp21] ModelOp: *Model Monitoring - The Path to Reliable AI* eBook, ModelOp Corporation, Chicago, IL, USA, 2021. Downloadable from: https://www.modelop.com/wp-content/uploads/2021/04/ebook-Model-Monitoring-The-Path-to-Reliable-AI-1.pdf [last accessed: 27.10.2021]

[Mohanta20] Abhijit Mohanta Anoop Saldanha 2020 Malware Analysis and Detection Engineering - A Comprehensive Approach to Detect and Analyze Modern Malware Apress Media LLC N.Y., USA New York 978-1-4842-6192-7

[Molnar20] Christoph Molnar: **Interpretable Machine Learning – *A Guide for Making Black-Box Models Interpretable*** www.lulu.com, 2020. ISBN 978-0-244-76852-2

[Morgulis19] Nir Morgulis, Alexander Kreines, Shachar Mendelowitz, Yuval Weisglas: *Fooling a Real Car with Adversarial Traffic Signs* White Paper, Harman International, Automotive Security Business Unit, Stamford, Connecticut, USA, 2019. Downloadable from: https://arxiv.org/ftp/arxiv/papers/1907/1907.00374.pdf [Last accessed: 14.3.2021]

[NIST16] Lidong Chen, Stephen P. Jordan, Yi-Kai Liu, Dustin Moody, Rene C. Peralta, Ray A. Perlner, Daniel C. Smith-Tone: **Report on Post-Quantum Cryptography** US National Institute of Standards and Technology, Gaithersburg, MD, USA, 2016. Downloadable from: https://nvlpubs.nist.gov/nistpubs/ir/2016/NIST.IR.8105.pdf [last accessed: 18.4.2020]

[NIST8309_2_20] Gorjan Alagic, Jacob Alperin-Sheriff, Daniel Apon, David Cooper, Quynh Dang, John Kelsey, Yi-Kai Liu, Carl Miller, Dustin Moody, Rene Peralta, Ray Perlner, Angela Robinson, Daniel Smith-Tone: *NISTIR 8309 - Status Report on the Second Round of the NIST Post-Quantum Cryptography Standardization Process* U.S. Department of Commerce, National Institute of Standards and Technology (NIST), Gaithersburg, MD, USA. Downloadable from: https://doi.org/10.6028/NIST.IR.8309 [Last accessed: 18.9.2020]

[NISTIR8312] US National Institute of Standards and Technology: **Four Principles of Explainable Artificial Intelligence** NIST, Gaithersburg, MD, USA, Draft NISTIR 8312, August 2020. Downloadable from: https://doi.org/10.6028/NIST.IR.8312-draft [Last accessed: 20.3.2021]

[NSCAI21] NSCAI: **Recommendations on Defending America and Winning the Technology Competition in the AI Era** US National Security Commission on Artificial Intelligence, Final Report, Washington, DC, USA, May 2021. Downloadable from: https://www.nscai.gov/wp-content/uploads/2021/03/Full-Report-Digital-1.pdf [Last accessed: 24.5.2021]

[Odayan21] Kershlin Odayan: **Artificial Intelligence controlling Cyber Security** Independently published, 2021. ISBN 979-8-5919-8207-2

[OECD19] OECD: Roles and Responsibilities of Actors for Digital Security OECD (Organisation for Economic Co-operation and Development), Digital Economy Papers, No. 286, July 2019, Paris, France. Downloadable from: https://www.oecd-ilibrary.org/science-and-technology/roles-and-responsibilities-of-actors-for-digital-security_3206c421-en [Last accessed: 6.6.2021]

[Omicini01]      Andrea Omicini, Franco Zambonelli, Matthias Klusch, Robert Tolksdorf (Editors): **Coordination of Internet Agents - *Models, Technologies, and Applications*** Springer Verlag Berlin Heidelberg, Germany, 2010 (Softcover reprint of hardcover 1$^{st}$ edition, 2001). ISBN 978-3-642-07488-2

[Ozdag18]      Mesut Ozdag: *Adversarial Attacks and Defenses Against Deep Neural Networks - A Survey* Cyber Physical Systems and Deep Learning Conference, CAS 2018, 5–7 November 2018, Chicago, Illinois, USA. Dowloadable from: https://www.researchgate.net/publication/327655401_Adversarial_Attacks_and_Defenses_Against_Deep_Neural_Networks_A_Survey/link/5bcff4de92851c1816bc8fed/download [Last accessed: 18.3.2021]

[Paar10]      Christof Paar, Jan Pelzl: **Understanding Cryptography - *A Textbook for Students and Practitioners*** Springer-Verlag, Berlin, Germany, 2010. ISBN 978-3-642-04100-6

[Pang20]      Guansong Pang, Chunhua Shen, Longbing Cao, Anton van den Hengel: *Deep Learning for Anomaly Detection - A Review* arXiv Preprint, 5 Dec 2020. Downloadable from: https://arxiv.org/pdf/2007.02500.pdf [Last accessed: 5.6.2021]

[Parfomak21]      Paul W. Parfomak, Chris Jaikaran: *Colonial Pipeline: The DarkSide Strikes* US Congressional Research Service (CRS) Report, May 11, 2021. Downloadable from: https://crsreports.congress.gov/product/pdf/IN/IN11667 [Last accessed: 15.5.2021]

[Parisi19]      Alessandro Parisi : **Hands-On Artificial Intelligence for Cybersecurity - *Implement smart AI Systems for preventing Cyber-Attacks and detecting Threats and Network Anomalies*** Packt Publishing, Birmingham, UK, 2019. ISBN 978-1-7898-0402-7

[Peirce1878]      Charles S. Peirce: *Illustrations of the Logic of Science - Sixth Paper: Deduction, Inductions and Hypothesis* The Popular Science Monthly, Vol 13, May 1878, pp. 470–482, New York, NY, USA, 1878. Downloadable from: https://books.google.ch/books?id=u8sWAQAAIAAJ&pg=PA472&redir_esc=y#v=onepage&q&f=false [Last accessed: 10.5.2021]

[Pipher02]      Jill Pipher: *Lectures on the NTRU Encryption Algorithm and Digital Signature Scheme* Brown University, Providence, RI, USA, June 2002 (Grenoble Lectures). Downloadable from: http://www.math.brown.edu/~jpipher/grenoble.pdf [Last accessed: 19.9.2020]

[PwC18]      PwC: *Explainable AI* Technical Brief, PricewaterhouseCoopers LLP, London, UK, 2018. Downloadable from: https://www.pwc.co.uk/audit-assurance/assets/explainable-ai.pdf [Last accessed: 9.5.2021]

[Qiu20]      Han Qiu, Yi Zeng, Tianwei Zhang, Yong Jiang, Meikang Qiu: *FENCEBOX: A Platform for Defeating Adversarial Examples with Data Augmentation Techniques* Preprint arXiv, December 2, 2020. Downloadable from: https://arxiv.org/pdf/2012.01701.pdf [Last accessed: 17.4.2021]

[Raghunathan18]      Aditi Raghunathan, Jacob Steinhardt, Percy Lian, Aditi Raghunathan, Jacob Steinhardt, Percy Liang: *Certified Defenses against Adversarial Examples* Conference Paper at ICLR 2018, Vancouver, Canada,

2018. Downloadable from: https://arxiv.org/pdf/1801.09344.pdf [Last accessed: 18.3.2021]

[Rai20]        Arun Rai: *Explainable AI: from Black Box to Glass Box* Journal of the Academy of Marketing Science, Vol. 48, 2020, pp. 137–141. Downloadable from: https://link.springer.com/article/https://doi.org/10.1007/s11747-019-00710-5 [Last accessed: 21.3.2021]

[Rains20]      Tim Rains: **Cybersecurity Threats, Malware Trends, and Strategies - *Mitigate Exploits, Malware, Phishing, and other Social Engineering Attacks*** Packt Publishing Ltd., Birmingham, UK, 2020. ISBN 978-1-800-20601-4

[Raj21]        Emmanuel Raj: **Engineering MLOps - *Rapidly build, test, and manage production-ready machine learning life cycles at scale*** Packt Publishing, Birmingham, UK, 2021. ISBN 978-1-8005-6288-2

[Rao19]        Joysula Rao: *Detection and Mitigation of Adversarial Attacks and Anomalies* In [Casola19], pp. 13–18

[Regev06]      Oded Regev: **Lattice-based Cryptography** crypto2006, August 20–24, 2006, Santa Barbara, California, USA. Downloadable from: https://www.iacr.org/archive/crypto2006/41170129/41170129.pdf [Last accessed: 18.9.2020]

[Roberts17]    Scott J. Roberts, Rebekah Brown: **Intelligence-Driven Incident Response - *Outwitting the Adversary*** O'Reilly Media, Inc, Sebastopol, CA, USA, 2017. ISBN 978-1-491-93494-4

[Robertson17]  John Robertson, Ahmad Diab, Ericsson Marin, Eric Nunes, Vivin Paliath, Jana Shakarian, Paulo Shakarian: **Darkweb Cyber Threat Intelligence Mining** Cambridge University Press, Cambridge, UK, 2017. ISBN 978-1-107-18577-7

[Rubinoff20]   Shira Rubinoff: **Cyber Minds - *Insights on Cybersecurity across the Cloud, Data, Artificial Intelligence, Blockchain, and IoT to keep your Cyber safe*** Packt Publishing, Birmingham, UK, 2020. ISBN 978-1-7898-0700-4

[Rugge19]      Fabio Rugge (Editor): **The Global Race for Technological Superiority** Ledizioni LediPublishing, Milano, Italy, 2019. ISBN 978-8-8552-6143-2. Downloadable from: https://www.ispionline.it/sites/default/files/pubblicazioni/ispi_cybsec_2019_web2.pdf [Last accessed: 2.6.2021]

[Russell17]    Stuart J. Russell, Peter Norvig: **Artificial Intelligence** Prentice-Hall International, Upper Saddle River, N.J., USA, 3rd revised edition, 2017. ISBN 978-1-292-15396-4

[Samek19]      Wojciech Samek, Grégoire Montavon, Andrea Verdaldi, Lars Kai Hansen, Klaus-Robert Müller (Editors): **Explainable AI - *Interpreting, Explaining, and Visualizing Deep Learning*** Springer Nature Switzerland, Cham, Switzerland, 2019. ISBN 978-3-030-28953-9

[Sarker20]     Iqbal H. Sarker, Yoosef B. Abushark, Fawaz Alsolami, Asif Irshad Khan: *IntruDTree: A Machine Learning-Based Cyber Security Intrusion Detection Model* Symmetry, MDPI, Basel, Switzerland, 6 May 2020. Downloadable from: https://www.mdpi.com/2073-8994/12/5/754?type=check_update&version=1

[SASWG19]      Safety of Autonomous Systems Working Group (SASWG): **Safety Assurance Objectives for Autonomous Systems** Independently published, SASWG, 2019. ISBN 978-1-7904-2122-0

[Saxe18]            Joshua Saxe, Hillary Sanders: **Malware Data Science - *Attack Detection and Attribution*** No Starch Press Inc., San Francisco, USA, 2018. ISBN 978-1-5932-7859-5

[Sayler20]          Kelley M. Sayler: *Artificial Intelligence and US National Security* US Congressional Research Service, Report R45178, Washington, DC, USA, updated November 10, 2020. Downloadable from: https://fas.org/sgp/crs/natsec/R45178.pdf [Last accessed: 25.5.2021]

[Scarfone12]        Karen Scarfone, Peter Mell: **Guide to Intrusion Detection and Prevention Systems (IDPS)** NIST Special Publication 800-94 Revision 1 (Draft), National Institute of Standards and Technology, Washington, CD, USA, July 2012. Downloadable from: https://csrc.nist.gov/CSRC/media/Publications/sp/800-94/rev-1/draft/documents/draft_sp800-94-rev1.pdf [Last accessed: 21.5.2021]

[Schain15]          Mariano Schain, Yishay Mansour: **Machine Learning Algorithms and Robustness** PhD Thesis, Tel Aviv University, Tel Aviv, Israel, January 2015. Downloadable from: https://www.tau.ac.il/~mansour/students/Mariano_Scain_Phd.pdf [Last accessed: 16.4.2021]

[Scharre19]         Paul Scharre 2019 Army of None - Autonomous Weapons and the Future of War WW Norton & Company N.Y., USA New York 978-0-393-35658-8

[Schneier18]        Bruce Schneier: *Artificial Intelligence and the Attack/Defense Balance* IEEE Security & Privacy, New York, NY, USA, Volume 16, Issue 2, March/April 2018. Downloadable from: https://ieeexplore.ieee.org/stamp/stamp.jsp?tp=&arnumber=8328965 [Last accessed: 15.5.2021]

[Schneier18]        Bruce Schneier: **Click Here to Kill Everybody - *Security and Survival in a Hyper-connected World*** W.W. Norton & Company, Inc., New York, USA, 2018. ISBN 978-0-393-60888-5

[Schoitsch20]       Erwin Schoitsch: *Machine Ethics* in: ERCIM News, Number 122, July 2020, pp. 4–5. ERCIM EEIG, Sophia Antipolis Cedex, France. Downloadable from: https://ercim-news.ercim.eu/images/stories/EN122/EN122-web.pdf [last accessed: 18.7.2020]

[Schroeder17]       Ted W. Schroeder: **Lethal Autonomous Weapon Systems in Future Conflicts** Independently published, 2017. ISBN 978-1-5207-0240-7

[Schwenger21]       Maximilian Schwenger: *Monitoring Cyber-Physical Systems - From Design to Integration* Project Paper, European Research Council (ERC), Brussels, Project OSARES, Grant No. 683'300, 2021 (https://www.react.uni-saarland.de/research/osares.html) Downloadable from: https://arxiv.org/pdf/2012.08959.pdf [Last accessed: 22.3.2021]

[Sengupta20]        Nandita Sengupta, Jaya Sil: **Intrusion Detection - *A Data Mining Approach*** Springer Nature Singapore, Singapore, 2020. ISBN 978-9-811-52718-0

[Shalev-Shwartz14]  Shai Shalev-Shwartz Shai Ben-David 2014 Understanding Machine Learning - From Theory to Algorithms Cambridge University Press N.Y., USA, New Edition New York 978-1-107-05713-5

[Shemanske17]       R Thomas 2017 Shemanske: Modern Cryptography and Elliptic Curves American Mathematical Society Rhode Island USA 978-1-47043582-0

[Shor20]            Peter Shor: *Quantum Computer Pioneer warns of Complacency over Internet Security* NATURE, Vol. 587, Issue 7833, 12 November 2020, p. 189

[Shor97]    Peter W. Shor: *Polynomial-Time Algorithms for Prime Factorization and Discrete Logarithms on a Quantum Computer* SIAM Journal on Computing, Volume 26, Nr. 5, 1997, pp. 1884–1509. Downloadable from: https://arxiv.org/abs/quant-ph/9508027 [last accessed: 20.12.2019].

[Sitawarin18]    Chawin Sitawarin, Arjun Nitin Bhagoji, Arsalan Mosenia, Prateek Mittal, Mung Chiang: *Rogue Signs: Deceiving Traffic Sign Recognition with Malicious Ads and Logos* White Paper, Department of Electrical Engineering Princeton University, Princeton, USA, 2018. Downloadable from: https://arxiv.org/pdf/1801.02780.pdf [Last accessed: 14.3.2021]

[Skopik17]    Florian Skopik **Collaborative Cyber Threat Intelligence - *Detecting and Responding to Advanced Cyber Attacks at the National Level*** CRC Press (Taylor & Francis Ltd.), Boca Raton, FL, USA, 2017. ISBN 978-1-138-03182-1

[Soldatos20]    John Soldatos, James Philpot, Gabriele Giunta (Editors): **Cyber-Physical Threat Intelligence for Critical Infrastructures Security - *A Guide to Integrated Cyber-Physical Protection of Modern Critical Infrastructures*** now publishers Inc., Boston, USA, 2020. ISBN 978-1-6808-3686-8

[Stamp07]    Mark Stamp, Richard M. Low: **Applied Cryptanalysis - *Breaking Ciphers in the Real World*** John Wiley & Sons, Inc., Hoboken, NJ, USA, 2007. ISBN 978-0-470-11486-5

[Stoecklin18]    Marc Ph. Stoecklin, Jiyong Jang, Dhilung Kirat: *DeepLocker - How AI Can Power a Stealthy New Breed of Malware* Security Intelligence, August 8, 2018. Access: https://securityintelligence.com/deeplocker-how-ai-can-power-a-stealthy-new-breed-of-malware/ [Last accessed: 30.5.2021]

[Stone19]    James V. Stone: **Artificial Intelligence Engines - *A Tutorial Introduction to the Mathematics of Deep Learning*** Sebtel Press, Sheffield, UK, 2019. ISBN 978-0-9563-7281-9

[Suojanen18]    Marko Suojanen: **Military Communications in the Future Battlefield** Artech House Publishers, Norwood, MA, USA, 2018. ISBN 978-1-630-81333-8

[Sutor19]    Robert S. Sutor: **Dancing with Qubits - *How Quantum Computing works and how it can change the World*** Packt Publishing, Birmingham, UK, 2019. ISBN 978-1-8388-2736-6

[Szegedy14]    Christian Szegedy, Wojciech Zaremba, Ilya Sutskever, Joan Bruna, Dumitru Erhan, Ian Goodfellow, Rob Fergus: *Intriguing Properties of Neural Networks* Preprint, 19. February 2014. Downloadable from: https://arxiv.org/abs/1312.6199 [Last accessed: 31.3.2021]

[Szychter18]    Avi Szychter, Hocine Ameur, Antonio Kung, Hervé Daussin: *The Impact of Artificial Intelligence on Security - A Dual Perspective* CESAR Conference Paper, 2018. Downloadable from: https://www.cesar-conference.org/wp-content/uploads/2018/11/articles/C&ESAR_2018_J1-03_A-SZYCHTER_Dual_perspective%20_AI_in_Cybersecurity.pdf [Last accessed: 16.5.2021]

[Takagi21]    Tsuyoshi Takagi, Kirill Morozov: **Mathematics of Post-Quantum Cryptography** Springer Verlag, Japan, 2021. ISBN 978-4-431-55015-0

[Tamboli19]        Anand Tamboli 2019 Keeping Your AI Under Control - A Pragmatic Guide to Identifying, Evaluating, and Quantifying Risks Apress Media LLC N.Y., USA New York 978-1-4842-5466-0

[Tangredi21]       Sam J. Tangredi, George V. Galdorisi (Editors): **AI at War - *How Big Data Artificial Intelligence and Machine Learning Are Changing Naval Warfare*** Naval Institute Press, Annapolis, MD, USA, 2021. ISBN 978-1-6824-7606-2

[Thomas20]         Tony Thomas, Athira P. Vijayaraghavan, Sabu Emmanuel: **Machine Learning Approaches in Cyber Security Analytics** Springer Nature Singapore, Singapore, 2020. ISBN 978-9-811-51705-1

[Tolido20]         Ron Tolido et al.: *Reinventing Cybersecurity with Artificial Intelligence - The new Frontier in Digital Security* CAP Gemini Research Insitute, Technical Report, Paris, France, 2020. Downloadable from: https://www.capgemini.com/wp-content/uploads/2019/07/AI-in-Cybersecurity_Report_20190711_V06.pdf [Last accessed: 30.5.2021]

[Tramèr20]         Florian Tramèr, Alexey Kurakin, Nicolas Papernot, Ian Goodfellow, Dan Boneh, Patrick McDaniel: *Ensemble Adversarial Training: Attacks and Defenses* Preprint arXiv, April 26, 2020, Published as a conference paper at ICLR 2018, April 30 -May 3, Vancouver, Canada, 2018. Downloadable from: https://arxiv.org/abs/1705.07204 [Last accessed: 17.4.2021]

[Truong19]         Cong Thanh Truong, Ivan Zelinka: *A Survey on Artificial Intelligence in Malware as Next-Generation Threats* Mendel Soft Computing Journal, Brno University of Technology, Brno, Czech Republic, 25(2) pp. 27–34, December 2019. Downloadable from: https://www.research-gate.net/publication/338099216_A_Survey_on_Artificial_Intelligence_in_Malware_as_Next-Generation_Threats [Last accessed: 30.5.2021]

[Turek21]          Matt Turek: *Explainable Artificial Intelligence (XAI)* DARPA Tutorial, Defense Advanced Research Projects Agency, Washington, USA, 2021. Downloadable from: https://www.darpa.mil/program/explainable-artificial-intelligence [Last accessed: 31.3.2021]

[Uesato18]         Jonathan Uesato, Ananya Kumar, Csaba Szepesvari, Tom Erez, Avraham Ruderman, Keith Anderson, Krishmamurthy Dvijotham, Nicolas Heess, Pushmeet Kohli: *Rigorous Agent Evaluation: An Adversarial Approach to Uncover Catastrophic Failures* arXiv Preprint, December 4, 2018. Downloadable from: https://arxiv.org/abs/1812.01647 [Last accessed: 16.4.2021]

[Varshney16]       Kush R. Varshney: *Engineering Safety in Machine Learning* White Paper, IBM Thomas J. Watson Research Center, Yorktown Heights, NY, USA, 2016. Downloadable from: https://arxiv.org/pdf/1601.04126.pdf [Last accessed: 12.3.2021]

[Veith19]          Eric M.S.P. Veith, Lars Fischer, Martin Tröschel, Astrid Nieße: *Analyzing Cyber-Physical Systems from the Perspective of Artificial Intelligence* arXiv Preprint, September 2, 2019. Downloadable from: https://arxiv.org/abs/1908.11779 [last accessed 10.7.2021]

[Vermeer20]        Michael J. D. Vermeer, Evan D. Peet: *Securing Communications in the Quantum Computing Age - Managing the Risks to Encryption* RAND Corporation (Research Report), Santa Monica, CA, USA, 2020. Downloadable from: https://www.rand.org/pubs/research_reports/RR3102.html [Last accessed: 16.9.2020]

| | |
|---|---|
| [Vilone20] | Giulia Vilone, Luca Longo: *Explainable Artificial Intelligence: A Systematic Review* Preprint submitted to Elsevier, October 13, 2020. Downloadable from: https://arxiv.org/pdf/2006.00093.pdf [Last accessed: 21.3.2021] |
| [Warr19] | Katy Warr: **Strengthening Deep Neural Networks - *Making AI Less Susceptible to Adversarial Trickery*** O'Reilly, Farnham, UK, 2019. ISBN 978-1-492-04495-6 |
| [West19] | Bruce J. West, Chris Arney: *Nonsimplicity – The Warrior's Way* United States Military Academy, US Army Cyber Institute, West Point, USA, 2019. Downloadable from: https://cyberdefensereview.army.mil/Portals/6/Nonsimplicity_The_Warriors_Way_West_Arney.pdf [Last accessed: 2.6.2021] |
| [Wirsing15a] | Martin Wirsing, Matthias Hölzl, Nora Koch, Philip Mayer (Editors): **Software Engineering for Collective Autonomic Systems - *The ASCENS Approach*** Springer International Publishing, Cham, Switzerland, 2015. ISBN 978-3-319-16309-3 |
| [Wirsing15b] | Martin Wirsing (Coordinator): *Software engineering for self-aware, self-adaptive, self-expressive, open-ended, highly parallel, collective and interactive distributed systems* Final Project Brochure, ASCENS Project, Munich, Germany, 2015. Downloadable from: http://ascens-ist.eu/images/ascens/ascens_broshure_final.pdf [Last accessed: 8.6.2021] |
| [Wischmeyer20] | Thomas Wischmeyer, Timo Rademacher (Editors): **Regulating Artificial Intelligence** Springer Nature Switzerland, Cham, Switzerland, 2020. ISBN 978-3-030-32360-8 |
| [Wu17] | Xuanfan Wu: *Metrics, Techniques, and Tools of Anomaly Detection - A Survey* Technical Report, Washington University in St. Louis, St. Louis, MO, USA, 2017. Downloadable from: https://www.cse.wustl.edu/~jain/cse567-17/ftp/mttad.pdf [Last accessed: 6.6.2021] |
| [Yampolskiy18] | Roman V. Yampolskiy: **Artificial Intelligence Safety and Security** CRC Press (Taylor & Francis Inc.) Boca Raton, FL, USA, 2018. ISBN 978-0-815-36982-0 |
| [Yan10] | YY Song 2010 Yan: Primality Testing and Integer Factorization in Public-Key Cryptography Springer Science and Business Media 2 New York N.Y., USA 978-1-441-94586-0 |
| [Yan13] | Y Song 2013 Yan: Quantum Attacks on Public-Key Cryptosystems Springer Verlag N.Y., USA New York 978-1-441-97721-2 |
| [Zeigler17] | Bernard P Zeigler Hessam S Sarjoughian 2017 Guide to Modeling and Simulation of Systems of Systems Springer International Publishing 2 Cham Switzerland 978-1-447-16933-8 |
| [Zeigler18] | P Bernard 2018 Zeigler, Alexandre Muzy, Ernesto Kofman: Theory of Modeling and Simulation - Discrete Event & Iterative System Computational Foundations Academic Press (Elsevier) 3 London UK 978-0-128-13370-5 |
| [Zhang19] | Lin Zhang, Bernard P. Zeigler, Yuanjun LaiLi (Editors): **Model Engineering for Simulation** Academic Press (Elsevier), London, UK, 2019. ISBN 978-0-128-13543-3 |
| [Zhang21a] | Lei Zhang, Andriy Miranskyy, Walid Rjaibi, Greg Stager, Michael Gray, John Peck: *Making Existing Software Quantum-Safe - Lessons Learned* Preprint arXiv:2110.08661v1[cs.SE], 16 Oct 2021. Downloadable from: https://arxiv.org/abs/2110.08661 [Last accessed: 13.11.2021] |

[Zhang21c]        Zhimin Zhang, Huansheng Ning, Feifei Shi, Fadi Farha, Yang Xu, Jiabo
                  Xu, Fan Zhang, Kim-Kwang, Raymond Choo: *Artificial intelligence in
                  Cyber Security - Research Advances, Challenges, and Opportunities*
                  Artificial Intelligence Review, Springer Verlag, Heidelberg, Germany,
                  13 March 2021. Available at: https://link.springer.com/article/https://doi.
                  org/10.1007/s10462-021-09976-0 [Last accessed: 3.6.2021]
[Zivic20]         Natasa Zivic, Obaid Ur-Rehman: **Security in Autonomous Driving** De
                  Gruyter Oldenbourg, Oldenbourg, Germany, ISBN 978-3-110-62707-7
[Zouave20]        Erik Zouave, Marc Bruce, Kajsa Colde, Margarita Jaitner, Ioana Rodhe,
                  Tommy Gustafsson: **Artificially intelligent Cyberattacks** Technical
                  Report  FOI-R-4947-SE,  FOI  Totalförsvarets  Forskninsinstitut,
                  Stockholm, Sweden, 2020. Downloadable from: https://www.statsvet.
                  uu.se/digitalAssets/769/c_769530-l_3-k_rapport-foi-vt20.pdf       [Last
                  accessed: 23.5.2021]

# Part II
# Principles

# Principle-Based Engineering

**7**

*The essential knowledge about safety, security, and risk has been presented in the previ-*
*ous chapters. The questions coming up at this point are: 1) How are good safety and*
*security defined? 2) How is good safety and security formalized? 3) How is good safety*
*and security taught? 4) How is good safety and security enforced?*

*This monograph's answers are: By defining, formalizing, strictly applying, and*
*enforcing safety and security principles. Safety and security principles represent proven,*
*long-tested, reliable knowledge about successful architecting.*

*The following chapters introduce, explain, and justify an actual number of safety and*
*security principles. For this approach to architecting safe and secure systems, the term*
*"Principle-Based Engineering" has been introduced.*

## 7.1    Risk-Based Engineering

*Risk* is the prevalent notion while developing and operating cyber-physical systems.
Every decision taken during development and operation can have severe consequences in
the physical world, such as *safety accidents* or *security incidents*. Therefore, risk awareness, assessment, and mitigation must be permanent in all phases. An organization faces
*multiple risks* (see, e.g., [Lam17], [Olson20], [Hutchins18], [ISO31000]), such as financial risks, operational risks, reputation risks, legal risks, etc.

F. J. Furrer, *Safety and Security of Cyber-Physical Systems*,
https://doi.org/10.1007/978-3-658-37182-1_7

Serious consideration and management of all risks is the domain of *risk-based engi-neering* (Definition 7.1, Fig. 3.6, e.g., [Varde18], [Fairbanks10], [Meyer16]).

▶ **Definition 7.1: Risk-Based Engineering** Engineering process which continuously and consequently includes risk identification, risk awareness, risk assessment, risk mitiga-tion, and risk monitoring for multiple risks (not only safety and security) in all phases of development and operation.

A multiple-risk sample is given in Example 7.1: The *ransomware attack* on the company GARMIN™.

> **Quote**
> *"The risk-based design approach enables a design process that considers risk as the major parameter driver or focus of the design"*
>    Prabhakar V. Varde, Michael G. Pecht, 2018

**Example 7.1: GARMIN Ransomware (July 2020)**

On July 23, 2020, the company GARMIN™ was successfully attacked by a ransom-ware attack. Ransomware (https://www.malwarebytes.com/ransomware/) encrypts some or all of the files accessible by the infected computer. This also hits backup or recovery files if an air-gap does not protect them (Fig. 7.1). The encrypted data are useless for processing. In many cases, the initiator of the attack requests ransom pay-ment, usually in untraceable bitcoins.

In the GARMIN™ case (see: https://www.wired.com/story/garmin-outage-ran-somware-attack-workouts-aviation/), most of the online services were closed down for approximately five days. The damage for GARMIN was:

- Reputation damage;
- Decrease in customer trust;
- Loss of five business days;
- Potential customers are abandoning GARMIN services.

*Therefore, risk-based engineering is* fundamental for safety and security in CPS's (Principle 12.2). Risk-based engineering addresses all possible risks, whereas safety and security engineering consider specific failures and threats. In this monograph, the focus is on safety and security.

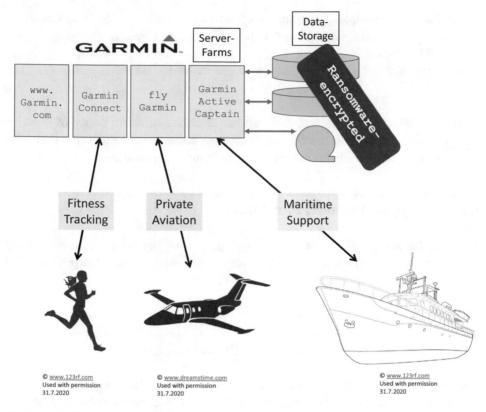

**Fig. 7.1** GARMIN™ Ecosystem as a CPS

## 7.2   Principle-Based Engineering[1]

### 7.2.1   Principles in Science

*Principles* (Definition 7.2) have long been used in science to formalize funda-mental insights in a specific area of scientific knowledge (see, e.g., [Devons1923], [Russell1903], [Dirac1930], [Stark1910], [Graupe13], [Born1999], [Harms00], [Alur15], [Hertz19], [Morley94]).

---

[1] Part of this section has been reused from the authors previous book "Future-Proof Software-Systems", Springer Vieweg Verlag, Wiesbaden, Germany, 2019. ISBN 978-3-658-19937-1 ([Furrer19]).

▶ **Definition 7.2: Principle**  A principle is a fundamental truth or proposition that serves
as the foundation for a system of belief, or behaviour, or for a chain of reasoning.
   Oxford Dictionary.
   **Note** [by author]: An engineering principle must be teachable, actionable, and
enforceable.

Whenever a scientific or technical discipline has reached a sufficient degree of maturity,
the underlying principles become apparent and can be formalized. Scientific principles
represent an objective truth that has been verified in many applications.

## 7.3    Safety and Security Principles

Principles for *safety* and *security* are a subset of general engineering principles. They
have been distilled from a long history of theoretical work, practical experience, and
reflections on failed projects. The following five chapters present an essential set of these
principles for safety and security.

## 7.4    Principle-Based Engineering Process

Many processes and methods exist to create, maintain, and evolve safe and secure cyber-
physical systems. In this monograph, the *principle-based engineering process* is applied
(Definition 7.3, Fig. 7.2, [Furrer19]).

▶ **Definition 7.3: Principle-Based Engineering**

Principle-based engineering is a process, where the knowledge used to create, maintain
and evolve a cyber-physical system is contained in principles and patterns. These principles and patterns are consistently applied and enforced throughout all phases of the CPS
life cycle.

Figure 7.2 shows the principle-based engineering process: The principles as such are not
directly implementable—they require a context-dependent interpretation (which is the
responsibility of the chief safety or security engineer). The operationalization is forced
and constrained by the *applicable laws* and regulations, governed by the *relevant industry standards*, and guided by the *company policies*.

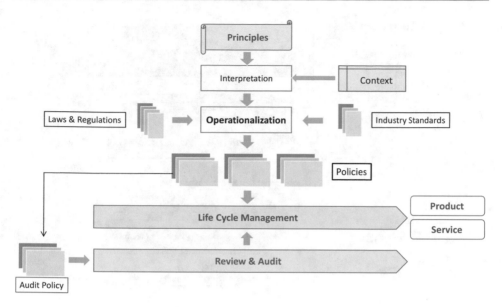

**Fig. 7.2**   Principle operationalization

## 7.5    Safety and Security Patterns

Although safety and security principles are clear, precise, and enforceable, their actual implementation may not be evident during the development. One great help for the development teams is *patterns*.

Many existing patterns support the application of the principles in more detail for particular problems. Patterns live in a hierarchy, as shown in Fig. 7.3.

*Patterns* (Definition 7.4) form a precious treasure of proven knowledge in many scientific disciplines. Christopher Alexander introduced the idea of a pattern as reusable expert knowledge in 1977 for the support of building houses and towns ([Alexander1977], [Alexander1980]). Since then, patterns have been identified and made available in many disciplines, also for safety and security.

▶ **Definition 7.4: Pattern**

A pattern is a proven, generic solution to recurring architectural or design problems, which can be adapted to the task at hand. Safety and security patterns specifically address the trustworthiness of CPCs.

A rich literature about safety and security patterns exists (e.g., [Buschmann96], [Butler07], [Cloutier08], [Fernandez-Buglioni13], [Hanmer07], [Perroud13], [Schumacher05], [Schumacher06], [Steel05], [Schumacher08], [Blackwell16], [Meier03], [Beckers15]). An example for a typical *security pattern* is given in Example 7.2.

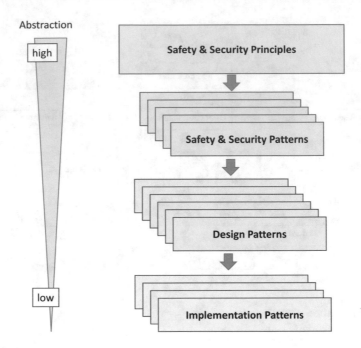

**Fig. 7.3**  Principle-pattern hierarchy

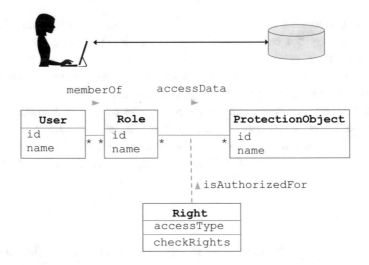

**Fig. 7.4**  RBAC security pattern

> **Example 7.2: RBAC Pattern [Fernandez, ISBN 978-1-119-99894-5]**

The *role-based access control* is the standard protection method in modern commercial systems. It is based on a standardized pattern and is implemented in many products. The situation is that a user wants to access protected/confidential information and needs to be authorized (Fig. 7.4).

The `User` and `Role` classes describe registered users and their predefined roles. Users are assigned roles. Roles are given rights according to their functions. Access is granted or denied based on the right of the role and the ID of the user.

The following chapters introduce five categories of principles:

1. Principles for business and organization;
2. General principles;
3. Principles for safety;
4. Principles for security;
5. Principles for risk.

They form the framework for safe and secure cyber-physical systems.

# References

| [Alexander1977] | Christopher Alexander: A Pattern Language - Towns, Buildings, Construction. Oxford University Press, New York, USA, 1977. ISBN 978-0-195-01919-3 |
|---|---|
| [Alexander1980] | Christopher Alexander: The Timeless Way of Building. Oxford University Press, New York, USA, 1980. ISBN 978-0-195-02402-9 |
| [Alur15] | Rajeev Alur: **Principles of Cyber-Physical Systems** MIT Press, Cambridge, MA, USA, 2015. ISBN 978-0-262-02911-7. |
| [Beckers15] | Kristian Beckers: **Pattern and Security Requirements - *Engineering-Based Establishment of Security Standards*** Springer Verlag, Heidelberg, Germany, 2015. ISBN 978-3-31916663-6. |
| [Blackwell16] | Clive Blackwell, Hong Zhu (Editors): **Cyberpatterns - *Unifying Design Patterns with Security and Attack Patterns*** Springer International Publishing, Cham, Switzerland, 2016 (Softcover reprint of the original 1st edition 2014). ISBN 978-3-319-35218-3 |
| [Born1999] | Max Born, Emil Wolf: **Principles of Optics: *Electromagnetic Theory of Propagation, Interference, and Diffraction of Light*** Cambridge University Press, Cambridge, UK, 1999. ISBN 978-0-5216-4222-4 |
| [Buschmann96] | Frank Buschmann, Regine Meunier, Hans Rohnert: **Pattern-Oriented Software Architecture, Vol. 1: *A System of Patterns*** John Wiley & Sons., Inc., USA, 1996. ISBN 978-0-471-95869-7 [Also Vols. 2, 3 and 4] |

[Butler07]            Michael Butler, Cliff Jones, Alexander Romanovsky, Elena Troubytsina
                      (Editors): **Rigorous Development of Complex Fault-Tolerant
                      Systems** Springer-Verlag, Berlin (Lecture Notes in Computer Science,
                      Band 4157), 2007. ISBN 978-3-540-48265-9

[Cloutier08]          Robert Cloutier: **Applicability of Patterns to Architecting Complex
                      Systems - *Making Implicit Knowledge Explicit*** VDM Verlag,
                      Saarbrücken, Germany, 2008. ISBN 978-3-8364-8587-6

[Devons1923]          W. Stanley Devons: **The Principles of Science - *A Treatise on
                      Logic and Scientific Method*** Richard Clay & Sons, London, 1923.
                      Downloadable from: https://ia801407.us.archive.org/0/items/theprin-
                      ciplesof00jevoiala/theprinciplesof00jevoiala.pdf    [Last    accessed:
                      31.3.2018]

[Dirac1930]           P. A. M. Dirac **The Principles of Quantum Mechanics** Oxford
                      University Press, UK, 1930 [Reprint: www.Snowballpublishing.com,
                      2013. ISBN 978-1-6079-6560-2

[Fairbanks10]         George Fairbanks: **Just Enough Software Architecture - *A Risk-
                      Driven Approach*** Marshall & Brainerd Publications, Boulder, CO,
                      USA, 2010. ISBN 978-0-984-61810-1

[Fernandez-Buglioni13] Eduardo  Fernandez-Buglioni  **Security   Patterns   in   Practice:
                      *Designing Secure Architectures Using Software Patterns*** John Wiley
                      & Sons, USA, 2013. ISBN 978-1-119-99894-5

[Furrer19]            Frank J. Furrer: **Future-Proof Software-Systems - *A Sustainable
                      Evolution Strategy*** Springer Vieweg Verlag, Wiesbaden, Germany,
                      2019. ISBN 978-3-658-19937-1

[Graupe13]            Daniel   Graupe:   **Principles   of   Artificial   Neural   Networks**
                      World  Scientific  Publishing  Company,  3rd  edition,  2013.  ISBN
                      978-9-8145-2273-1

[Hanmer07]            Robert Hanmer: **Patterns for Fault-Tolerant Software** John Wiley &
                      Sons, USA, 2007. ISBN 978-0-470-31979-6

[Harms00]             A. Harms, D. R. Kingdon, K. F. Schoepf: **Principles of Fusion Energy**
                      World Scientific Publishing Company, 2000. ISBN 978-9-8123-8033-3

[Hertz19]             Heinrich  Hertz:   **Die   Prinzipien   der   Mechanik   in   Neuem
                      Zusammenhange Dargestellt** Forgotten Books Publishing, London,
                      UK, 2019 (Classic Reprint: 1891–1894). ISBN 978-0-4849-5396-2

[Hutchins18]          Greg Hutchins: **ISO 31000:2018 Enterprise Risk Management**
                      Certified Enterprise Risk Manager Academy, Portland, OR, USA, 2018.
                      ISBN 978-0-9654-6651-6

[ISO31000]            ISO 31000:2018 (ISO/TC 262 Risk Management) **Risk Management
                      —   Guidelines**   Downloadable   from:   https://www.iso.org/stand-
                      ard/65694.html [last accessed: 7.6.2020]

[Lam17]               James Lam: **Implementing Enterprise Risk Management - *From
                      Methods to Applications*** John Wiley & Sons, Inc., Hoboken, N.J.,
                      USA, 2017. ISBN 978-0-471-74519-8

[Meier03]             J.D. Meier, Alex Mackman, Srinath Vasireddy, Michael Dunner, Ray
                      Escamilla, Anandha Murukan: **Improving Web Application Security -
                      *Threats and Countermeasures (Patterns & Practices)*** Microsoft Press,
                      Seattle, OR, USA, 2003. ISBN 978-0-7356-1842-8. Downloadable
                      from:   https://www.microsoft.com/de-ch/search?q=Meier+Improving

+Web+Application+Security+-+Threats+and+Countermeasures+ [last accessed: 2.8.2020]

[Meyer16]      Thierry Meyer, Genserik Reniers: **Engineering Risk Management** Walter De Gruyter GmbH, Berlin, Germany, 2nd edition, 2016. ISBN 978-3-110-41803-3

[Morley94]     Arthur Morley, Edward Hughes, W. Bolton: **Principles of Electricity** Longman Publishing, London, UK, 5th edition (Revised by A. Morley), 1994. ISBN 978-0-582-22874-0

[Olson20]      David L. Olson, Desheng Wu: **Enterprise Risk Management Models** Springer Verlag, Berlin, Germany, 3rd edition, 2020. ISBN 978-3-662-60607-0

[Perroud13]    Thierry Perroud, Reto Inversini: **Enterprise Architecture Patterns: *Practical Solutions for Recurring IT-Architecture Problems*** Springer-Verlag, 2013. ISBN 978-3-642-37560-6

[Russell1903]  Bertrand Russell: **The Principles of Mathematics** George Allen & Unwin, London, 1903

[Schumacher05] Markus Schumacher, Eduardo Fernandez-Buglioni, Duane Hybertson, Frank Buschmann, Peter Somerlad: **Security Patterns: *Integrating Security and Systems Engineering*** John Wiley & Sons, USA, 2005. ISBN 978-0-470-85884-4

[Schumacher06] Markus Schumacher, Eduardo Fernandez-Buglioni, Duane Hybertson, Frank Buschmann, Peter Somerlad: **Security Patterns—Integrating Security and Systems** John Wiley & Sons, Chichester, UK, 2006. ISBN 978-0-470-85884-4

[Schumacher08] Markus Schumacher: **Security Engineering with Patterns - *Origins, Theoretical Models, and New Applications*** Springer Verlag, Berlin, Germany LNCS (Lecture Notes in Computer Science #2754), 2008. ISBN 978-3-540-40731-7

[Stark1910]    J. Stark: **Prinzipien der Atomdynamik** S. Hirzel Verlag, Leipzig, Germany, 1910 (3 Bände)

[Steel05]      Christopher Steel: **Core Security Patterns - *Best Practices and Strategies for J2EE(TM), Web Services, and Identity Management*** Prentice-Hall, Upper Saddle River, N.J., USA, 2005. ISBN 978-0-131-46307-3

[Varde18]      Prabhakar V. Varde, Michael G. Pecht: **Risk-Based Engineering - *An Integrated Approach to Complex Systems (Special Reference to Nuclear Plants)*** Springer Nature Singapore, Singapore, 2018. ISBN 978-9-811-30088-2

# Principles for Business and Organization

8

*Safety and security are of serious concern for all responsibilities on all levels of an organization. A holistic, organized, reproducible, and dependable execution of safety- and security-related duties enables trustworthy cyber-physical systems. Therefore, principles for safety and security must be defined and obeyed throughout the whole organization.*

## 8.1 Principle B1: Risk Culture

The management and the employees of an organization have to deal daily with *risks*. The decisions related to risk mitigation and acceptance should be guided by a well-established and accepted *risk culture* (Definition 8.1) in the organization.

▶ **Definition 8.1: Risk Culture** Risk culture denotes the combined set of corporate values, norms, attitudes, competencies, and behavior related to risk awareness (perception of risk), risk-taking (active business decisions), and risk mitigation (Controls) that determine an organization's commitment to and style of risk management.

   Adapted from: www.openriskmanual.org

The risk culture should be known and ingrained in all staff, suppliers, and partners of the organization ([Banks12], [Zhu18], [Dyer18]). This requires formalization, i.e., at least one *risk policy*. Possibly, secondary risk policies for specific areas of the organizations, such as the information system, are necessary.

F. J. Furrer, *Safety and Security of Cyber-Physical Systems*,
https://doi.org/10.1007/978-3-658-37182-1_8

> **Quote**
> *"This is, in effect, the idea of risk culture: A state where risk management processes are so intuitive and so embedded in the fabric of an institution that they exist subconsciously and are practiced as a matter of course"*
> Erik Banks, 2012

The organization's risk culture must be continuously fostered; i.e., all staff must be trained, their risk awareness must be developed, and the proper risk management procedures must be enforced.

Risk can be subdivided into *risk categories* ([Banks12]), such as financial, business, strategic, legal, and operational. This monograph deals with two specific *operational risks*: safety and security. The principle for *risk culture* is listed in Principle 8.1.

---

**Principle 8.1: Risk Culture**

1. Unambiguously define the corporate values, norms, attitudes, competencies, and behavior related to risk (= Risk Policy);
2. Carefully and timely cultivate and adapt the risk policy;
3. Communicate and enforce the risk policy throughout the organization at all times and during all activities;
4. The risk culture/risk policy must be supported, enforced, and lived by all levels of management in the organization;
5. Risk culture relies on good governance: Dedicated organizational structures, governance rules, and reporting lines must be in place;
6. The risk culture/risk policy must fully comply with the applicable regulatory/legal requirements of the specific industry and jurisdiction;
7. The risk culture should reward careful handling of risks and not facilitate unjustified risk acceptance.

---

## 8.2   Principle B2: Policies

The values, goals, constraints, long-term objectives, etc., of an organization must be *operationalized*, i.e., formulated, communicated, and enforced. All employees, stakeholders, and partners of the organization must understand and apply them. The instrument to do so is *policies* (Definition 4.8, [Williams13], [Bacik19], [Tricker19], [Dawson19], [Taylor20]). A significant number of area-specific policies exist in a larger organization, such as risk policies, safety policies, security policies, and employment policies. Policies are arranged in an—often hierarchical—scheme: The *policy architecture* (Definition 8.2).

Fortunately, helpful policy templates for various scopes are available (e.g., https://www.sans.org/information-security-policy, https://templatelab.com/security-policy-templates/).

▶ **Definition 8.2: Policy Architecture** A policy architecture defines a structured, complete, consistent, and comprehensive set of all policies issued by an organization. The policy architecture should have a comprehensive, non-overlapping hierarchy.

Various focused policy documents are required with the high-level corporate policies acting as the fundament.

The set of policies of the organization represents a vital governance instrument and a significant intellectual asset. The maintenance, i.e., the continuous and timely *adaptation* of the policies to changes in the operational environment, legal and regulatory revisions, updates in relevant standards, new risks and threats, etc., is an essential and formidable task for the successful evolution of an organization.

---

**Principle 8.2: Policies**

1.  Develop and maintain a policy architecture: Organize all necessary policies in a logical, comprehensive, non-overlapping, and complete order (= Policy architecture);
2.  Ensure that all applicable legal, regulatory, and compliance requirements are covered by the respective policy;
3.  Ensure that all applicable industry standards, certification standards, and operations standards are considered;
4.  Ensure that all chosen technology standards are evaluated and covered;
5.  Assign a policy owner to each policy, having approved management responsibility for the development, review, evaluation, evolution, implementation, and enforcement of the specific policy;
6.  Periodically review all policies and adapt them if necessary so they remain current, relevant, and effective. Closely follow and evaluate all developments in law, regulations, compliance, and technology—and their impact on the policies;
7.  Install a policy maintenance process which ensures the timely update of all policies after changes in legal, regulatory, compliance, standards, certification areas, etc., areas;
8.  Install a policy communications process that reliably instructs all stakeholders;
9.  Regularly audit the adherence to the policies of all persons and organizational units concerned. Use chartered help if necessary. Document and archive the audit results;
10. Use proven policy templates (available on the Web: e.g. from industry associations) for the composing of the individual policies.

## 8.3    Principle B3: Competence Center

The *competence center* (Definition 2.26) is necessary for any organization developing safety- or security-critical systems. The specialist knowledge is concentrated in
one, easily accessible place. Experienced experts are readily available to consult and
support the development teams. The same experts are also instrumental in project and
product reviews and advising top management in technical, mission-critical decisions.
Ensure that the competence center experts have not only excellent technical skills
but also appropriate *social skills*, such as communications skills, negotiation power,
mediation capabilities, credibility, and a certain pragmatism (e.g., [Hendricksen12],
[Hendricksen14], [Munson18], [Tiwari17], [Selinger04]). In addition, the competence center experts need a *deep understanding of the business* they are serving (e.g.,
[Candelo19], [Pugsley17], [Brunello18], [Meyer19]). The principles of a powerful competence center are presented in Principle 8.3.

---

**Principle 8.3: Competence Center**

1.   Establish and carefully evolve an in-house competence center possessing all
     knowledge for the mission-critical quality properties of processes and products,
     such as architecture, risk management, safety, security, and forensics;
2.   Imprint all relevant policies (risk, safety, security, etc.) into the competence
     center. Involve the competence center in the development or evolution of the
     policies;
3.   Staff the competence center with sufficient personnel for the qualified and timely
     support of the business units, development teams, and management;
4.   Elect only people with excellent and specialized technical skills to the compe-
     tence center;
5.   Ensure that the competence center experts also have outstanding social and per-
     sonal skills;
6.   Force the competence center experts to lifelong learning, keeping at the forefront
     of the knowledge in their field;
7.   Equip the competence center with the best possible tools facilitating their tasks;
8.   Formally integrate the competence center experts as consultants in the develop-
     ment process;
9.   Use the competence center experts for formal internal reviews of all development
     artifacts (development quality gates);
10.  Empower the competence center experts to take decisions concerning risk, safety,
     or security within their responsibility, even if they have to override a development
     team preference or a business unit demand;
11.  Provide a fair and transparent escalation path to the accountable management in
     case of disputes with the development teams or business units. Whenever possi-
     ble, support the competence center expert's decisions.

> **Quote**
> *"Indeed, the woes of Software Engineering are not due to lack of tools, or proper management, but largely due to lack of sufficient technical competence"*
> Niklaus Wirth [https://www.quotemaster.org/author/Niklaus+Wirth]

## 8.4 Principle B4: Governance

*Governance* (Definition 2.25, [Gordon20], [OECD19], [Brotby09], [Tricker19], [Ferrillo17]) is responsible for the decisions taken in an organization, also explicitly concerning *safety* and *security*. Ultimately, the decision paths and the decision-making authority determine the quality of the safety and security in a product or service—and, therefore, the responsibilities for safety accidents and security incidents.

Defensible decisions related to safety and security require:

- Complete information about the decision/trade-off, its context, and its consequences, i.e., well-informed decision-makers;
- Unambiguous assignment of the decision-making authority to specific managers (not committees) with transparent, justifiable, and auditable decision procedures.

Using a proven *framework* for safety and security governance improves the organization's governance policy (e.g., [NIST800100_06], [WEF21], [EASA19]). The adherence to the governance policy, procedures, and methods must be regularly audited ([Sigler20], [Davis21]). The main points of responsible governance are summarized in Principle 8.4.

> **Quote**
> *"Decision-makers don't understand operations in cyberspace. This lack of understanding forces leaders to speak in colorful, yet vague figures of speech"*
> Michael A. Vanputte, 2017

> **Principle 8.4: Governance**
>
> 1. Define the governance structure, processes, and responsibilities in an explicit governance policy;
> 2. Align the governance policy seamlessly to the other relevant policies, such as safety policy, security policy, and risk policy;
> 3. Document the governance in a comprehensive governance model: A governance model is the graphical representation of all managers in charge, their responsibilities, the reporting structure, and the communications paths;

4. Implement governance—i.e., all decision paths and the decision-making authority—such that risk, safety, and security concerns are prioritized, adequately addressed, consequently implemented, transparent, and auditable;
5. Assign the decision-making authority transparently and traceable to specific persons/managers (not committees). This is especially important for the reproducible acceptance of residual safety and security risks;
6. Maintain a competence center within the organization, specifically for safety concerns, security issues, and risk management;
7. Ensure that governance decisions are not based on erroneous, incomplete, or uncertain information;
8. Whenever possible, use metrics, such as safety and security management metrics and incident response metrics;
9. Periodically assess and audit the effectiveness and assertiveness of the governance policies, procedures, and accountable managers.

## 8.5    Principle B5: Record Keeping and Trustworthy Archive

Today's enterprises and organizations' actions—past and present—are under constant scrutiny by legal, regulatory, and compliance agencies. Plaintiffs, lawyers, and public authorities can exploit any perceived security incident or safety accident. An essential pillar in the defense against claims or accusations of negligence is the records and documents produced during the development and evolution of a product or service. Only when such *documentary evidence* is provided can the defense lawyer teamwork successfully.

Therefore, a complete, detailed, unbroken, and dependable documentation trail is mandatory for all activities related to critical systems' safety and security. An excellent solution to this challenge is a *trustworthy electronic archive* (Definition 8.3). A trustworthy electronic archive is much more than just a dependable storage and retrieval technology ([Borghoff05], [Millar17], [Bantin16]).

▶ **Definition 8.3: Trustworthy Archive** Electronic repository, which provides reliable, long term, unforgeable, legally permissible, auditable access to managed digital resources to its designated community, now and in the future.

This includes not only dependable storage/retrieval technology but also the archive policy and all processes to assure correct and complete information according to the mission of the organization.

Adapted from: https://www.oclc.org

> **Quote**
>
> *"Archives and records are important resources for individuals, organizations, and the wider community. They provide evidence of, and information about the actions of individuals, organizations, and communities, and the environments in which those actions occurred"*
> Laura Agnes Millar, 2017

A trustworthy archive is one of the foundations for safety and security engineering. First, the organization needs an *archival policy* (see, e.g., [Moore16]) defining the goals, processes, access rights, disaster-proofing, and retention/deletion periods.

The processes must be so reliable that the complete information required is stored now and in the future. The archive management system should conform to an international standard (e.g., [ISO14641]).

For the archive information's legal permissibility, it is mandatory that the records cannot be changed (i.e., forged) and that the original rendering remains preserved. Therefore, the records must be protected against alterations during their stay in the archive (see, e.g., [ISO16363]). Examples of *document protection* are digital signatures or embedding them in a blockchain ([Lemieux18]).

---

**Principle 8.5: Record Keeping and Trustworthy Archive**

1. Define and enforce an archive policy;
2. Define and implement all processes required to assure the complete acquisition and storage of all relevant information, meeting all business, legal, and regulatory requirements for all nation-states in which the organization is active;
3. Implement the archive such that it is auditable and legally permissible;
4. Make the archive future-proof, i.e., use formats that are standardized and designed for long-term use. Define and maintain a proactive migration strategy for a timely conversion if standards or technology changes;
5. Protect the records in the archive, both against forgery, deletion, and unauthorized access;
6. Use an adequate, user-friendly storage/retrieval management system based on international standards.

---

# 8.6   Principle B6: Product Liability

*Product liability* (Definition 8.4, [Krauss19], [Owen15]) is a looming menace for all safety- and security-critical systems. Product liability incidents can have substantial consequences following a *safety accident* or a *security incident*.

> **Quote**
> *"Product liability lawsuits have played a crucial role in ensuring public safety, encouraging – and sometimes compelling – manufacturers to put safety first"*
>     Michael I. Krauss, 2019

▶ **Definition 8.4: Product Liability** The legal liability a manufacturer or trader incurs for producing or selling a faulty product or service.

    https://www.lexico.com/definition/product_liability

No technical system can be made 100% safe or secure (Figs. 4.10, 4.24). All technical systems have residual risks for safety accidents or security incidents—and therefore, the potential for a *product liability court case*. To avoid liability court cases, first of all, responsible and competent engineering processes must be used. Such processes, especially proper *risk management*, will minimize the potential for safety accidents and security incidents. If a product liability court case has to be defended, all the related documentation) must be presented to the court, e.g., to advocate against negligence accusations or state-of-the-art violations (Note: Principle 8.1 presents an engineering viewpoint, not legal advice).

---

**Principle 8.6: Product Liability**

1. Explicitly formulate, document, and provably implement all safety and security requirements for system specification, construction, evolution, and operation;
2. Use responsible and competent engineering processes for all products and services. Place safety and security requirements higher than functional requirements;
3. Justify, fully document, and archive all architecture, design, and implementation decisions;
4. Fully document and archive all risk assessments and risk mitigation measures for these decisions, especially the resulting, acceptable residual risks;
5. Assure and demonstrate that the product meets or exceeds the *safety* state of the art. Note: Legal state of the art is often judicially defined in relevant, domain-specific standards, such as ISO/IEC 26,262;
6. Assure and demonstrate that the product meets or exceeds the *security* state of the art. Note: Legal state of the art is often judicially defined in relevant, domain-specific standards, such as [TeleTrusT21];
7. Provide sufficient documentation and training to potential users of the CPS.

## 8.7   Principle B7: Code of Ethics

Following accepted *ethical axioms* while developing and operating cyber-physical systems—primarily autonomous cyber-physical systems—is an expectation of modern society. Therefore, any organization is well-advised to formulate a *code of ethics*, explicitly targeting its products' behavior and use (Principle 8.7).

---

**Principle 8.7: Code of Ethics**

1. Develop a code of ethics for the organization concerning building and using cyber-physical systems, especially autonomous CPS's;
2. Follow the emerging ethical guidelines in fields related to the products of the organization;
3. Carefully monitor the transition of ethical guidelines into binding national law with respect to the products of the organization and their use;
4. Communicate, enforce, and document the process to assure the behavior of the CPC's being in accordance with the code of ethics, especially with the parts governed by laws or regulations;

---

## 8.8   Principle B8: People's Work Environment

The *quality of the workplace* has a strong, hidden impact on the quality of the results of the individual worker. High-quality workplaces enable high-quality work, whereas low-quality workplaces promote discontent, errors, blunders, and despair. Therefore, providing a quality workplace (Definition 8.5) is a prime responsibility of the organization and a precondition for reliable safety and security outcomes.

▶ **Definition 8.5: Quality Workplace** A quality workplace is essential to keep your employees on task and to work efficiently. An excellent work environment is marked by such attributes as competitive wages, trust between the employees and management, fairness for everyone, and a sensible workload with challenging yet achievable goals.
   https://smallbusiness.chron.com/definition-quality-workplace-13260.html

The work environment for safe and secure cyber-physical systems development includes a number of elements (Fig. 5.19). Half of them must be provided by the organization, and the other half must be brought to the workplace by the individual workers. Many theories covering quality workplace provisions exist (e.g., [Lahlou10], [Arizaleta19], [Catlin21], [Timms20], [Pfeffer18], [Joseph19], [Paine19], [McKee17]). The organization's responsibility is to organize their workplaces to fit the work and the type of employees well.

As a management task, workplace quality must regularly be analyzed and evaluated using adequate metrics, anonymous employee questionnaires, or quantitative metrics, such as the turnover rate (e.g., [Wiley10]). Perform fair, beneficial, and educational yearly personal appraisals ([Rudman20]). Note that an enjoyable work environment also reduces the risk of *insider crime*.

> **Quote**
> *"Developing talent is also recognized as one of the most significant drivers of employee engagement, which in turn is the key to the business outcomes you seek: Revenue, profitability, innovation, productivity, customer loyalty, quality, cycle time reduction, and more – everything organizations need to survive and thrive"*
>     Beverly Kaye, 2019

Workplace quality design is suggested in Principle 8.8.

---

**Principle 8.8: Workplace Quality**

1. Regularly analyze and assess the quality of the organization's workplaces, based on quantitative metrics and formal, anonymous employee questionnaires;
   **Physical Comfort**
2. Build the physical workplace to optimal comfort, i.e., related to space, light, temperature, furniture, etc.;
3. Offer good and fair pay;
4. Offer flexible work conditions suited to the individual employee (work hours, home office, etc.);
   **Intellectual Support**
5. Provide complete, consistent, and comprehensive policies, standards, and guidelines to facilitate correct design decisions;
6. Leave as much autonomy as possible to teams and individuals;
7. Encourage (if necessary: force) and facilitate lifelong learning to keep the employees current with modern developments in their fields;
8. Install the best possible tools with the highest degree of automation for their respective work;
   **Mental Health**
9. Establish a fair, transparent, and accessible management on all levels;
10. Establish effective cooperation of management and engineering in the organization to eliminate friction and prevent the loss of information;
11. Institute a productive integration of project management and systems engineering to increase the probability of appropriate decisions;
12. Set personal, well-defined, challenging, yet achievable goals;

13. Reward innovation and excellence of work in a congruous and understandable way;
14. Generate and periodically discuss a personal development plan (PDP, career plan) for the employees (e.g., [Kaye19]);
15. Perform fair, beneficial, and educational personal appraisals, preferably yearly (in critical cases in shorter intervals);
16. Conduct decent, open interviews when an employee leaves the company. Learn from the results. Use it for metrics;
17. Limit the stress level on employees;
18. Have organizational capabilities to prevent, recognize, and deal with burnouts.

## References

| | |
|---|---|
| [Arizaleta19] | Joselo Arizaleta: **In Slippers at Work - *The Good, the Bad, and the Ugly about Home Office*** Independently Published, 2019. ISBN 979-8-6626-1272-3 |
| [Bacik19] | Sandy Bacik: **Building an Effective Information Security Policy Architecture** CRC Press, Taylor & Francis Ltd., Boca Raton, FL, USA, 2008 (Paperback: 2019). ISBN 978-0-367-38730-3 |
| [Banks12] | Erik Banks: **Risk Culture - *A Practical Guide to Building and Strengthening the Fabric of Risk Management*** Palgrave Macmillan, London, UK, 2012. ISBN 978-1-137-26371-1 |
| [Bantin16] | Philip C. Bantin: **Building Trustworthy Digital Repositories -*Theory and Implementation*** Rowman & Littlefield Publishers Lanham USA 978-1-4422-6378-9 |
| [Borghoff05] | Uwe M. Borghoff, Peter Rödig, Jan Scheffczyk, Lothar Schmitz: **Long-Term Preservation of Digital Documents - *Principles and Practices*** Springer-Verlag, Berlin, Germany, 2005. ISBN 978-3-540-33639-6 |
| [Brotby09] | Krag Brotby: **Information Security Governance - *A Practical Development and Implementation Approach*** John Wiley & Sons Inc., Hoboken, NJ, USA, 2009. ISBN 978-0-470-13118-3 |
| [Brunello18] | Lara Rita Brunello: **High-Speed Rail and Access Transit Networks** Springer International Publishing AG, Cham, Switzerland, 2018. ISBN 978-3-319-61414-4 |
| [Candelo19] | Elena Candelo: **Marketing Innovations in the Automotive Industry - *Meeting the Challenges of the Digital Age*** Springer Nature Switzerland AG, Cham, Switzerland, 2019. ISBN 978-3-030-15998-6 |
| [Catlin21] | Karen Catlin: **Better Allies - *Everyday Actions to Create Inclusive 2 Engaging Workplaces*** Better Allies Press, San Mateo, CA, USA, 2nd edition 2021. ISBN ⌈978-1-7327-2335-1 |
| [Davis21] | Robert E. Davis: **Auditing Information and Cyber Security Governance - *A Controls-based Approach*** CRC Press (Taylor & Francis), Boca Raton, FL, USA, 2021. ISBN⌈978-0-367-56850-4 |
| [Dawson19] | Linda J. Dawson, Randy Quinn: **The Art of Governing Coherently - *Mastering the Implementation of Coherent Governance and Policy Governance*** Rowman & Littlefield, Lanham, MA, USA, 2019. ISBN 978-1-4758-4623-2 |

[Dyer18]        Chris Dyer: **The Power of Company Culture - *How any Business can build a Culture that improves Productivity, Performance, and Profits*** Kogan Page, New Delhi, India, 2018. ISBN 978-0-7494-8195-7

[EASA19]        EASA: **The European Plan for Aviation Safety (EPAS) 2020–2024.** European Union Aviation Safety Agency (EASA), Cologne, Germany, 2019. Downloadable from: https://www.easa.europa.eu/sites/default/files/dfu/EPAS_2020-2024.pdf [Last accessed: 31.7.2021]

[Ferrillo17]    Paul A Ferrillo, Christophe Veltsos: **Take Back Control of Your Cybersecurity Now - *Game-Changing Concepts on AI and Cyber Governance Solutions for Executives*** Advisen Ltd., New York, NY, USA, 2017. ISBN 978-1-5206-5872-8

[Gordon20]      Jeffrey N. Gordon, Wolf-Georg Ringe (Editors): **The Oxford Handbook of Corporate Law and Governance** Oxford University Press, Oxford, UK, 2020. ISBN 978-0-19874369-9

[Hendricksen12] Dave Hendricksen: **12 Essential Skills for Software Architects** Addison-Wesley Professional (Pearson Education), Upper Saddle River, NJ, USA, 2012. ISBN 978-0-321-71729-0

[Hendricksen14] Dave Hendricksen: **12 More Essential Skills for Software Architects** Addison-Wesley Professional (Pearson Education), Upper Saddle River, NJ, USA, 2014. ISBN 978-0-321-90947-3

[ISO14641]      ISO 14641:2018: *Electronic document management — Design and operation of an information system for the preservation of electronic documents* ISO Standards Organization, Geneva, Switzerland, 2018. https://www.iso.org/obp/ui/#iso:std:iso:14641:ed-1:v1:en

[ISO16363]      ISO 16363:2012 *Space data and information transfer systems — Audit and certification of trustworthy digital repositories* ISO Standards Organization, Geneva, Switzerland, 2012. https://www.iso.org/standard/56510.html

[Joseph19]      Immanual Joseph: **The Fifth Revolution - *Reinventing Workplace Happiness, Health, and Engagement through Compassion*** Independently Publisher, 2019. ISBN 978-0-5785-7789-0

[Joseph19]      Anthony D. Joseph, Blaine Nelson, Benjamin I. P. Rubinstein, J. D. Tygar: **Adversarial Machine Learning** Cambridge University Press, Cambridge, UK, 2019. ISBN 978-1-107-04346-6

[Kaye19]        Beverly Kaye, Julie Winkle Giulioni: **Help Them Grow or Watch Them Go - *Career Conversations Organizations Need and Employees Want*** Berrett-Koehler Publishers Inc., Oakland, CA, USA, 2019. ISBN 978-1-5230-9750-0

[Krauss19]      Michael I. Krauss: **Principles of Products Liability** West Academic Publishing, St. Paul, MN, USA, 3rd edition, 2019. ISBN 978-1-6402-0128-6

[Lahlou10]      Saadi Lahlou (Editor): **Designing User-Friendly Augmented Work Environments** Springer Verlag, London, UK, 2010. ISBN 978-1-447-12515-0

[Lemieux18]     Victoria L. Lemieux: **Blockchain Technology for Recordkeeping** Report - The University of British Columbia, Vancouver, BC, Canada, 2018. Downloadable from: https://www.researchgate.net [Last accessed: 28.6.2020]

[McKee17]       Annie McKee: **How to Be Happy at Work - *The Power of Purpose, Hope, and Friendship*** Harvard Business Review Press, Boston, MA, USA, 2017. ISBN 978-1-633-69225-1

[Meyer19]       Gereon Meyer, Sven Beiker (Editors): **Road Vehicle Automation 6** (Lecture Notes in Mobility) Springer Nature Switzerland, Cham, Switzerland, 2019. ISBN 978-3-030-22935-1

| | |
|---|---|
| [Millar17] | Laura Agnes Millar: **Archives (Principles and Practice in Records Management and Archives)** Facet Publishing, London, UK, 2nd edition, 2017. ISBN 978-1-7833-0206-2 |
| [Moore16] | Reagan W. Moore, Hao Xu, Mike Conway, Arcot Rajasekar, John Crabtree, Helen Tibbo: **Trustworthy Policies for Distributed Repositories** Morgan & Claypool Publishers, San Rafael, CA, USA, 2016. ISBN 978-1-6270-5885-8 |
| [Munson18] | Tony Munson: **People Skills for Engineers** Independently published, 2018. ISBN 978-1-7239-9678-8 |
| [NIST800100_06] | Pauline Bowen, Joan Hash, Mark Wilson: **Information Security Handbook - *A Guide for Managers*** Information Security, NIST Special Publication 800-100, National Institute of Standards and Technology, Gaithersburg, MD, USA, 2006. Downloadable from: https://nvlpubs.nist.gov/nistpubs/Legacy/SP/nist-specialpublication800-100.pdf [Last accessed: 31.7.2021] |
| [OECD19] | OECD: Roles and Responsibilities of Actors for Digital Security OECD (Organisation for Economic Co-operation and Development), Digital Economy Papers, No. 286, July 2019, Paris, France. Downloadable from: https://www.oecd-ilibrary.org/science-and-technology/roles-and-responsibilities-of-actors-for-digital-security_3206c421-en [Last accessed: 6.6.2021] |
| [Owen15] | David Owen, Mary Davis: **Products Liability and Safety** West Academic Publishing, St. Paul, MN, USA, New edition 2015. ISBN 978-1-6093-0228-3 |
| [Paine19] | Nigel Paine: **Workplace Learning - *How to Build a Culture of Continuous Employee Development*** Kogan Page, London, UK, 2019. ISBN 978-0-749-48224-4 |
| [Pfeffer18] | Jeffrey Pfeffer: **Dying for a Paycheck - *How Modern Management Harms Employee Health and Company Performance and What We Can Do About It*** Harper Collins Business, New York, MY, USA, 2018. ISBN 978-0-062-80092-3 |
| [Pugsley17] | Matthew Pugsley, Andrew Taylor: **Flight Deck Automation - *Development and Impact on the Role of Flight Crew*** LAP LAMBERT Academic Publishing, Riga, Latvia, 2017. ISBN 978-3-3300-3584-3 |
| [Rudman20] | Richard Rudman: **Performance Planning and Review - *Making Employee Appraisals Work*** Routledge (Taylor & Francis), Milton Park, UK, 2020. ISBN 978-0-367-71891-6 |
| [Selinger04] | Carl Selinger 2004: **Stuff You Don't Learn in Engineering School - *Skills for Success in the Real World*** Wiley-IEEE Press, New York, NY, USA, 2004. ISBN 978-0-471-65576-3 |
| [Sigler20] | Ken E. Sigler, James L. Rainey: **Securing an It Organization Through Governance, Risk Management, and Audit** CRC Press (Taylor & Francis), Boca Raton, FL, USA, 2020. ISBN 978-0-367-65865-6 |
| [Taylor20] | Andy Taylor, David Alexander, Amanda Finch, David Sutton: **Information Security Management Principles** British Computer Society (BCS), Swindon, UK, 3rd edition, 2020. ISBN 978-1-78017-518-8 |
| [TeleTrusT21] | TeleTrusT: **Guideline "State-of-the-Art" - *Technical and Organisational Measures*** IT Security Association Germany, Berlin, Germany, 2021. Downloadable from: https://www.teletrust.de/fileadmin/user_upload/2021-02_TeleTrusT-Guideline_State_of_the_art_in_IT_security_EN.pdf [Last accessed: 01.08.2021] |
| [Timms20] | Perry Timms: **The Energized Workplace -*Designing Organizations Where People Flourish*** Kogan Page, London, UK, 2020. ISBN 978-0-749-49866-5 |

[Tiwari17]        Anoop Kumar Tiwari, Prakash Y. Dhekne, Shashikanta Tarai: **Corporate Skills for Engineers** LAP LAMBERT Academic Publishing, Riga, Latvia, 2017. ISBN 978-6-2020-8018-7

[Tricker19]       Bob Tricker: **Corporate Governance -** *Principles 4 Policies and Practices* Oxford University Press, Oxford, UK, 4th edition, 2019. ISBN 978-0-198-80986-9

[WEF21]           WEF: **Safe Drive Initiative. The Autonomous Vehicle Governance Ecosystem -** *A Guide for Decision-Makers* Community Paper, World Economic Forum, Cologny/Geneva, Switzerland, April 2021. Downloadable from: http://www3.weforum.org/docs/WEF_CP_The_Autonomous_Vehicle_ Governance_Ecosystem_2021.pdf [Last accessed: 31.7.2021]

[Wiley10]         Jack Wiley: **Strategic Employee Surveys -** *Evidence-based Guidelines for Driving Organizational Success* Jossey Bass (Wiley imprint), San Francisco, CA, USA, 2010. ISBN 978-0-470-88970-1

[Williams13]      Barry L. Williams: **Information Security Policy Development for Compliance:** *ISO/IEC 27001, NIST SP 800-53, HIPAA Standard, PCI DSS V2.0, and AUP V5.0* CRC Press (Taylor & Francis), Boca Raton, FL, USA, 2013. ISBN 978-1-466-58058-9

[Zhu18]           Andy Yunlong Zhu, Max von Zedtwitz, Dimitris G. Assimakopoulos: **Responsible Product Innovation -** *Putting Safety First* Springer International Publishing, Cham, Switzerland, 2018. ISBN 978-3-319-68450-5

# General Principles

**9**

General principles for safety and security apply to the whole system; i.e., they cover many quality properties of the CPSs. Therefore, they are presented separatedly in this chapter.

## 9.1 Principle G1: Precise Safety and Security Requirements

The CPS evolution process starts with *requirements*. The new functionality or data needs a detailed definition—provided by a requirements management process ([Lamsweerde09], [Dick17]). An essential element of trustworthy CPS is the precise definition of safety and security requirements *before* any development activity starts (Fig. 2.19). The same, or even more, diligence must be applied to describing the safety and security requirements then to functional requirements. Generally, as much *formality* as possible is desirable (e.g., [Fowler10b]). This recommendation is formulated in Principle 9.1.

---

**Principle 9.1: Precise Safety and Security Requirements**

1. Explicitly formulate, justify, document, and implement all safety and security requirements. Note that safety and security requirements have a higher priority than functional requirements;
2. Explicitly include safety- and security-enhancement techniques in the requirements, such as runtime monitoring, logging, forensics, etc.;
3. In mixed-criticality systems (Definition 4.4) clearly separate the safety- and security-critical parts from the non-critical parts;

---

F. J. Furrer, *Safety and Security of Cyber-Physical Systems*, https://doi.org/10.1007/978-3-658-37182-1_9

4. Ensure that the safety and security requirements are:
   a. *Complete*: Covering all the essential safety and security aspects of the planned functionality,
   b. *Unambiguous*: Leaving no uncertainty in interpretation,
   c. *Correct*: Accurate protection of all the necessary safety and security aspects of the intended functionality,
   d. *Consistent*: No contradictions between individual requirements,
   e. *Integrated*: Seamlessly and redundancy-free fitting into the existing system,
   f. *Prioritized*: In case of compromise, safety and security requirements have precedence over functional requirements,
   g. *Conformant*: Obeying all the applicable principles, standards, laws, regulations, and best practices,
   h. *Verifiable*: Fully covered by appropriate test cases (100%),
   i. *Traceable*: Auditable link between requirement and implementation.
5. Use formal requirements description instruments for all safety and security requirements (e.g., model-based or formal requirements language);
6. Verify and validate all safety and security requirements before any development activity starts;
7. Correctly integrate all safety and security requirements into the existing system, i.e., fully align the new requirements with the already implemented safety and security measures. Avoid any redundancy, inconsistency, or unnecessary technology.

## 9.2    Principle G2: Adequate System Architecture

A well-designed, soundly maintained, adequate *system architecture* is the indispensable foundation of any trustworthy, future-proof software-system (Definition 5.1, Fig. 5.5). This is especially true for safety-critical cyber-physical systems, where any violation of proven *architecture principles*—such as disregarding the separation of concerns or introducing unmanaged redundancy—will have severe negative consequences.

Therefore, safe and secure systems must be based on adequate architectures, designed according to time-proven *architecture principles*. This monograph focuses on specific principles for the *safety* and *security* of cyber-physical systems. Excellent references explaining functional software architecture principles exist (e.g., [Hoffman01], [Furrer19], [Bass13], [Cervantes16], [Ford17], [Gorton11], [Hohmann03], [Rozanski12], [Jayaratchagan18], [Koenig19a], [Ford22] Thus, Principle 9.2 is just a strong reminder of first-rate functional software architecture's crucial importance.

**Quote**
*"When someone builds a bridge, he uses engineers who have been certified as knowing what they are doing. Yet when someone builds you a software program,*

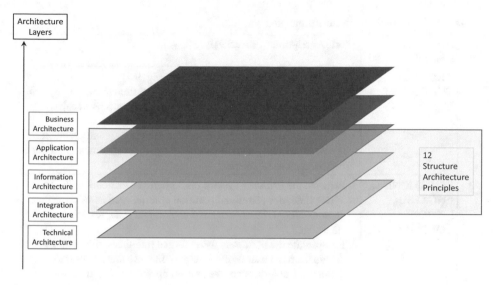

**Fig. 9.1**  Functional architecture layers

*he has no similar certification, even though your safety may be just as dependent upon that software working as it is upon the bridge supporting your weight"*
  David Parnas (https://www.brainyquote.com/quotes/dave_parnas_523901)

---

**Example 9.1: Functional Architecture Principles for Future-Proof Software-Systems[1]**

The horizontal and vertical architecture layers have been introduced in Fig. 5.5. They allow a proven partitioning of functionality into horizontal and vertical *architecture layers*. The author's experience has shown that 12 horizontal architecture principles (Fig. 9.1) constitute the foundation for optimal quality properties of the software-system's integration, information, and applications layer ([Furrer19]). These 12 architecture principles are summarized in Table 9.1.

---

**Quote**
*"It is generally agreed that increasing complexity is at the heart of most difficult problems facing today's systems architecting and engineering".*
  Mark W. Maier, Eberhardt Rechtin, 2002

---

[1] This example has been reused from the authors previous book "Future-Proof Software Systems", Springer Vieweg Verlag, Wiesbaden, Germany, 2019. ISBN 978-3-658-19937-1 ([Furrer19]).

**Table 9.1**  Functional architecture principles

| # | Architecture Principle | Short Description ([Furrer19]) |
|---|---|---|
| #1 | ARCHITEC- TURE LAYER ISOLATION | 1. Define architectural layers (e.g., as in Fig. 9.1)<br>2. ssign the functionality and data in the IT system to the appropriate architecture layer. The architecture layer then acts as a sealed container for coherent functionality and data<br>3. Isolate the data and control flow between the different architecture layers via industry-standard, technology-independent, and product-independent mechanisms |
| #2 | PARTITIONING, ENCAPSU- LATION, AND COUPLING | 1. Partition the software-system into coherent functional units. When appropriate, use additional partition criteria, such as safety/critical ⇔ non-safety-critical, real time ⇔ non-real time, fast-changing ⇔ stable, etc<br>2. Encapsulate the partitions using formal interfaces hiding all internal details from the user. Whenever possible, use formal contracts. Version the contracts and keep them upwards compatible<br>3. Build the communications and control exchange coupling between the partitions as loose as possible<br>4. Minimize the number and types of dependencies between the partitions<br>5. Use an architecture framework for the organization of functions (Principle 9.2) |
| #3 | CONCEPTUAL INTEGRITY | 1. Ensure conceptual integrity throughout the complete system by unambiguously defining—preferably formal—all the concepts, their relationships, the terminology, and the models<br>2. Consistently enforce adherence in all processes and systems to conceptual integrity<br>3. Eliminate all hidden assumptions, i.e., make them explicit and documented<br>4. Whenever possible, use domain software engineering methods |
| #4 | REDUNDANCY | 1. Carefully control the appearance of redundancy in concepts, terminology, functionality, and data during all development and evolution steps<br>2. Strictly avoid/eliminate unmanaged redundancy<br>3. Allow managed redundancy only if explicitly justified, such as fault tolerance, performance, backup, resilience, etc |
| #5 | INTEROPERA- BILITY | 1. Ensure the complete, precise definition of interoperability on all levels (technical, syntactic, semantic, and applications interoperability)<br>2. Formalize the interoperability by using (formal) interface or service contracts between the partners. Use modern constructs with pre- and post-conditions<br>3. Base interoperability whenever possible on accepted industry standards<br>4. Whenever possible, monitor the operation of the interfaces to detect and mitigate anomalies, violations, and exceptions during runtime |

(continued)

**Table 9.1**  (continued)

| # | Architecture Principle | Short Description ([Furrer19]) |
|---|---|---|
| #6 | COMMON FUNCTIONS | 1. Implement functionality and data used by multiple applications in the system as common functions, thus avoiding unmanaged redundancy<br>2. Strictly maintain a "single source of truth" for all common functionality and data<br>3. Provide standardized access mechanisms to common functionality and data |
| #7 | REFERENCE ARCHITEC-TURES, FRAME-WORKS, AND PATTERNS | 1. Make use of proven, industry-standard reference architectures, architecture frameworks, and specific architecture/design patterns<br>2. Maintain conceptual integrity and redundancy avoidance when using reference architectures, architecture frameworks, and specific architecture/design patterns<br>3. Maintain a well-organized, comprehensive repository of recommended/mandatory patterns for all levels of the software architecture |
| #8 | REUSE AND PARAMETRI-ZATION | 1. Develop and enforce a reuse strategy for the organization (covering all development and operations artifacts)<br>2. Carefully decide which applications, components, or modules shall be developed as reusable artifacts ($\Rightarrow$ more effort required)<br>3. Whenever possible, implement reusable software as configurable modules via parameters or using external business rules<br>4. Whenever possible, plan, define, and implement product families |
| #9 | INDUSTRY STANDARDS | 1. Evaluate, choose, and enforce a suitable set of proven and accepted industry standards for all areas of the CPS<br>2. Where no suitable industry standard is available, justify, generate and enforce an organization-specific standard<br>3. Maintain a complete, easily accessible, and comprehensive repository of the available/mandatory standards<br>4. Never allow the use of vendor-specific standards or standard extensions ("lock-in")<br>5. Keep the number of standards to a minimum |
| #10 | INFORMATION ARCHITECTURE | 1. Classify all your data in (a) enterprise data/information, (b) analytic data/information, (c) operational information, and (d) real-time information<br>2. Design the information architecture such that the different data/information categories are stored, processed, protected, and backed up accordingly<br>3. Provide mechanisms to maintain data/information integrity (both in time and in content) in case of failed operations |
| #11 | FORMAL MODELING | 1. Model all artifacts during requirements, specification, development, integration, and operation phases<br>2. Whenever possible, use semi-formal or formal languages and tools<br>3. Whenever possible, use model checking early in the processes<br>4. Use the specific domain knowledge from domain experts during modeling |

(continued)

**Table 9.1**  (continued)

| # | Architecture Principle | Short Description ([Furrer19]) |
|---|---|---|
| #12 | COMPLEXITY AND SIMPLI-FICATION | 1. Consciously detect and minimize accidental complexity<br>2. Actively manage and reduce essential complexity<br>3. Execute deliberate complexity reduction steps during all phases of development and evolution to simplify the system, i.e., to systematically reduce complexity (Example 4.7) |

---

**Principle 9.2: Adequate System Architecture**

1. Define, maintain, and evolve a suitable architecture for the cyber-physical system, especially the software-system (target architecture);
2. Invest sufficient effort into the definition/evolution/improvement of the system- and software architecture. Do it up-front, i.e., before any design starts;
3. Carefully and correctly integrate any new part (functionality and data) into the existing or expanded architecture;
4. Base the architecture on a set of proven, comprehensive, consistent, and enforceable architecture principles;
5. Install the necessary bodies, responsibilities, and procedures to enforce the target architecture during the evolution of the system;
6. Continuously evolve the architecture when new paradigms, concepts, or technologies become available;
7. Execute deliberate complexity reduction steps during all phases of development and evolution to simplify the architecture, i.e., to systematically reduce complexity.

---

## 9.3    Principle G3: Technical Debt

### 9.3.1  Technical Debt

*Technical debt* (Definition 9.1, [Ernst21], [Kruchten19], [Hubert02], [Beine21], [Lilienthal19], [Suryanarayana14], [Tornhill18]) is an unfortunate, avoidable, potential root cause for vulnerabilities in a CPS. If not carefully managed, technical debt accumulates slowly but mercilessly in small steps in the software-system and ultimately renders it highly endangered and unmanageable.

▶ **Definition 9.1: Technical Debt**
Technical debt in an IT system is the result of all those necessary things that you choose not to do now, but will impede future evolution or operation if left undone.
   Ward Cunningham, 2007

*Internally* generated technical debt is most often the result of uncoordinated, unprincipled, lazy, incompetent, or careless people's behavior. *External* causes for the introduction of technical debt are new architecture paradigms, technology progress, laws, and regulations, or aging IT systems.

> **Quote**
> *"The force of entropy means that disorder is the only thing that happens automatically and by itself. If you want to create a completely ad-hoc IT-architecture, you do not have to lift a finger ... it will happen automatically as a result of day-to-day IT activity"*
>     Richard Hubert, 2002

*Externally induced technical debt* is caused by sources outside of the sphere of influence of the organization, such as architecture paradigm changes (Monolith $\Rightarrow$ Components $\Rightarrow$ Service-oriented architecture $\Rightarrow$ Web-Services $\Rightarrow$ Microservices), new programming techniques/languages, emerging security threats, new legal & compliance requirements, etc.

Technical debt can be produced during all steps of development/operation (e.g., [Ernst21]), resulting in:

- *Business process debt*: Overlapping, redundant, diverging, or contradictory business processes;
- *Conceptual integrity debt*: The concepts, definitions, and semantics within the organization diverge or are not precisely defined;
- *Requirements debt*: The requirements are incomplete, unclear, unspecific, partly redundant, untraceable, or not assigned to the correct application domain (domain model);
- *Specification debt*: Ambiguity, lack of clarity, lack of integrity or consistency, incomplete or assigned to the wrong organizational unit;
- *Architecture and design debt*: Violation of architecture principles, patterns, and design best practices;
- *Defect debt*: Known defects that are not fixed;
- *Review and quality assurance debt:* Omit, shorten, or compromise necessary reviews, such as safety & security reviews, architecture reviews, compliance reviews, lack of documentation;
- *Implementation debt:* Deviation from the reviewed, accepted blueprint, shortcuts, use of non-approved software (e.g., open source, downloads from the Internet), or incomplete error and exception handling;
- *Testing debt:* Insufficient test coverage, unspecific tests, omitting regression testing, failing to use modern test techniques;

- *Deployment debt:* Deficient execution platform, inadequate business continuity planning, lack of disaster recovery mechanisms, weak runtime management;
- *Explementation debt:* Failing to remove all unused code and data from a system;
- *Documentation debt:* Missing, incomplete, unorganized, unfindable, or not current documentation of all forms (models, source code, handbooks, design decisions, etc.);
- *Tool debt:* Working with ineffectual, unsuitable, outdated, or incompatible tool-chains;
- *Infrastructure debt*: Delayed upgrade/modernization decisions;
- *Paradigm debt*: Applications that are not yet transformed to a new architecture or design paradigm.

Avoiding and eliminating technical debt requires adequate, strict processes and strong management backing (e.g., the Managed Evolution Strategy ([Murer11], [Furrer19]), Principle 9.3). Technical debt must be managed *proactively* (i.e., avoiding the generation of technical debt) and retroactively (i.e., eliminating technical debt in the system).

> **Quote**
> *"The best approach to managing technical debt is to manage it proactively. This means making deliberate and justifiable decisions to incur debt, having a reasonable strategy for measuring and monitoring it, and paying it off in the future"*
>    Neil Ernst, 2021

### 9.3.1.1  Technical Debt Management
A generic *technical debt management process* is shown in Fig. 9.2. The four steps are:

1. *Detect and identify technical debt*: The mechanisms are, e.g., code analysis (manual and automated), reviews, honest developers (who admit to having produced technical debt), system malfunctions and crashes;
2. *List and organize technical debt*: All known technical debt issues are listed and governed by the technical debt (TD) administrator;
3. *Assess and quantify technical debt*: The possible impact of technical debt issues varies broadly. Also, the cost to remedy them is considerably different. Therefore, each TD issue must be evaluated and prioritized. This is often done using templates;
4. *Remove or fix technical debt*: Once TD issues are known, evaluated, impact-analyzed, and prized, the removal/fixing of the TD issues can be decided and executed.

Note that this process is not able to manage *unknown technical debt*, i.e., technical debt items hidden in the system.

### 9.3.1.2  Technical Debt Metric
In order to properly manage technical debt, the individual technical debt issues must be quantified—a *technical debt metric* is required. Many metrics are proposed in the

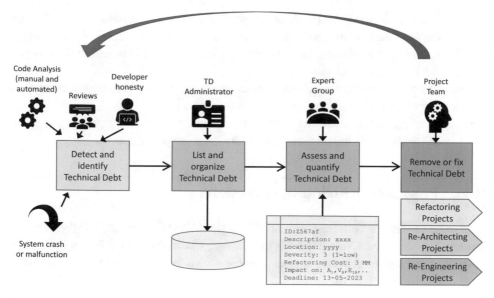

**Fig. 9.2**  Technical debt management

literature (e.g., [Nord12], [Seaman13], [Avgeriou21], [Kruchten20]). For many practical cases, a template is sufficient ([Ernst21], [Seaman13]). Technical debt management guidelines are listed in Principle 9.3.

> **Quote**
> *"«We know – but we will fix it later» – is the sure path to software hell"*
>    Anonymous

**Principle 9.3: Technical Debt**

1. Define and implement a sustainable system evolution strategy that actively manages technical debt;
2. List, organize, assess, and manage all technical debt issues in a technical debt management system (TD database);
3. Manage technical debt using a suitable metric;
4. In all phases of system creation and evolution, actively identify internal emerging technical debt and prevent its implementation;
5. Create a work environment that enables and rewards the avoidance or removal of technical debt;
6. Always allow sufficient resources (time, money, tools, material, etc.) during all phases of development, evolution, and operation. Lack of resources invariably forces the generation of internal technical debt;

7. Strictly, completely, and without exception eliminate all obsolete functionality and data from the system immediately after new functionality or data is operational;
8. For third-party products immediately, and without exceptions, install the vendor's updates or new versions;
9. Periodically execute refactoring, re-architecting, and re-engineering projects to identify and eliminate technical debt in the system;
10. After significant architecture paradigm changes or technology progresses, evaluate the impact on the system. If necessary, take corrective actions;

## 9.4    Principle G4: Architecture Erosion

Like technical debt, *architecture erosion* (Definition 9.2, [Silva10], [Li21], [Whiting20], [Knieke21]) allows new vulnerabilities to manifest, causing new risks in the CPS. If not carefully managed, architecture erosion develops slowly but mercilessly in small dissolutions in the software-system.

▶ **Definition 9.2: Architecture Erosion**
The term architecture erosion refers to the process of continuous divergence between the intended software architecture of a system and its refinement or implementation.
    Christoph Knieke, 2021

Architecture erosion is difficult to recognize: e.g., a shortcut in a development project may generate unmanaged, dangerous functional redundancy, which is subsequently not maintained because it is unknown to the configuration management system.

> **Quote**
> *"Architecture erosion is a big challenge in modern architectures leading to a deterioration of the quality properties of these systems"*
>     Christoph Knieke, 2021

There are three causes of architectural erosion:

a. *Internal causes*: These causes are under the control of the organization. They include, e.g., the violation of architecture principles, deferred refactoring or miss out on technological progress;
b. *External causes*: These are not under the control of the organization owning the software-system. They are imposed by the environment, such as evolving architecture paradigms (Example 9.2), changes in laws and regulations, or new processes (such as Agile development methods);

c. *Third-party software*: An often overlooked source of massive architecture erosion is integrating third-party software into the organization's system. The third-party software may have a very different architecture, a redundant database, duplicated functionality, or incompatible concepts and representations.

Architecture erosion can either be avoided or repaired. Managing architecture erosion is summarized in Principle 9.4.

---

**Example 9.2: Architecture Erosion external Cause—Architecture Paradigm Changes**

A strong impact on architecture erosion is *architecture paradigm changes* ([Wang04], [Chowhan18]). Fundamental architecture paradigms changes occurred many times in software engineering, e.g.,

Cobol/Fortran Monoliths (centralized) $\Rightarrow$ Component-based architecture $\Rightarrow$ Client/Server architecture (1 ... n tier) $\Rightarrow$ Service-oriented architecture $\Rightarrow$ Web-service architecture $\Rightarrow$ Cloud-based architecture $\Rightarrow$ Microservice architecture

Each paradigm change improved the productivity of development and the quality of the resulting software substantially. Software architectures based on the old paradigm became obsolete "overnight" and had to be gradually re-architected.

---

**Principle 9.4: Architecture Erosion**

1. Provide sufficient, comprehensive, complete architectural guidance for all stages of development, maintenance, and evolution to all stakeholders, such as architecture principles, models (domain model, business object model), architecture patterns, reference architecture, architecture frameworks;
2. Train awareness of all development teams for architecture erosion. Educate them to consistently use the available architecture guidance documents;

Architecture Erosion Avoidance

3. Internal architecture erosion is generated during system development and evolution. To avoid this internal architecture erosion, install and execute reviews of all development artifacts (before they are transferred to the next development step). Internal architecture erosion avoidance is a process issue;
4. External architecture erosion is unavoidable. Whenever a sufficiently strong change in the paradigms or technology materializes, the complete IT system must be analyzed, evaluating the effects of this forced architecture erosion. The consequences of the external architecture erosion must be absorbed by a planned, concentrated effort;
5. External architecture erosion may severely impact safety and security, such as new, insecure technology. Carefully evaluate the effect of the architecture erosion on security and safety. React cautiously;

Third-party software

6. Before introducing any third-party software (including open source), carefully consider the impact on the organization's IT system and its quality attributes. If the third-party software degrades the existing system architecture, especially the security and safety architectures of the system, be very cautious;

Architecture Erosion Repair

7. In the long run, architecture erosion is unavoidable. The IT system must, therefore, periodically be analyzed to identify architectural weaknesses or deviations from the intended architecture. Identified architectural erosion must then be remedied.

## 9.5    Principle G5: Separation of Concerns

Two *implementation* characteristics bedevil safety and security in a CPS:

a. The complexity of the system;
b. The amalgamation of the desired functionality and the safety/security concerns.

The strict *separation of concerns* minimizes both (Definition 5.2, [Panunzio14], [Mili04], [Kandé05], [Kandé03], [Carreira20]): The functionality of the system and the functionality required explicitly for safety/security should not be mixed in any implementation module. The separation of concerns applies to all artifacts of the development process, starting with the requirements (Principle 9.5).

> **Quote**
> *"A major cause of many complications in the field of software architectures is the lack of appropriate abstractions for separating, combining, and encapsulating concerns of various kinds in architectural descriptions"*
>     Mohamed Mancona Kandé, 2003

---

**Principle 9.5: Separation of Concerns**

1. Intentionally and precisely separate the functionality of the system from the functionality required explicitly for safety/security in all artifacts of the development process;
2. Base the system on a suitable architecture which facilitates or enforces the separation of concerns and favors safety and security concerns;

3. Use a proven architecture framework to govern and implement separation of concerns;

4. Separate the handling of functional errors, exceptions, and faults from the response to safety accidents and security incidents during operation. Have adequate procedures in place for both;

5. Use quality gates (reviews, etc.) during system development and evolution to specifically detect, avoid, or remedy separation of concerns violations;

6. Develop high awareness in the development teams for the issues of separation of concerns.

## 9.6   Principle G6: General Resilience Principles

*System resilience* results from sound engineering—based on a resilience management model and process ([Caralli11], [Ganguly18], [Hollnagel06], [Hollnagel13]). Therefore, both the development process and the operating procedures must be defined with resilience in focus ([Caralli11], [Das19]).

### 9.6.1   G6_1: Software Integrity

Software powers the functionality of most cyber-physical systems. Apart from mistakes in the software introduced during the development process, another source of *risk* exists: Alteration of the software, e.g., due to bit-flips in storage, malicious manipulation of the software, configuration mismatches, missing data, or unavailable parameter files. Such risks are due to the loss of *the software/firmware's integrity* and can have hazardous consequences (Example 4.1). To avoid software/firmware integrity failures, protection measures are available (e.g., [Yao20]). These are listed in Principle 9.6.

> **Quote**
> *"We need to recognize that we are nearly certain to experience failure modes in the field that were not realistically possible to anticipate in our design reviews"*
>    Kim Fowler, 2010

#### 9.6.1.1   Code Signing

*Code signing* (Definition 9.3, [Cooper18], [Booth14], [CASC13]) is a practical mechanism to ensure both the integrity and the trustworthiness of the creator. Firmware, operating software modules, drivers, application software, and configuration information should be digitally signed and checked before execution.

▶ **Definition 9.3: Code Signing**
Digitally signing code provides both data integrity to prove that the code was not modified, and source authentication to identify who was in control of the code at the time it was signed
  David Cooper, 2018

When signing code, a hash function is first applied to the code, generating a *hash-value* ([Rogaway04]). The hash-value allows the detection of any change made to the code, invalidating the hash-value. The hash-value of the original code is secured by the digital signature of the creator or issuer of the code; i.e., the hash-value is encrypted with the creator's private key (Definition 4.19. Before *executing* the code in the target system, the hash-value is recalculated and compared with the digitally signed hash-value provided with the code by its creator, using its public key. The code is often *time-stamped* by a trusted time-stamping authority (TSA, e.g., [Massias99]). The trusted time-stamp ensures the date of creation.

### 9.6.1.2  Control Flow Integrity
Inserting unwanted, malicious *functionality* into an IT system is the most widespread cyber-attack. However, a second attack opportunity exists: Manipulation of the *control flow* of the program execution, i.e., destructively changing the intended program flow (CFI, Definition 9.4, e.g., [Sayeed19]). The defense against malfeasant control flow alteration is control flow integrity techniques ([Lin21], [Abadi09]). Many of these techniques are based on the enforcement of the *control flow graph* of the software-system (CFG, [Lin21], [Davi15]).

▶ **Definition 9.4: Control Flow Integrity (CFI)**
Computer security techniques that prevent a wide variety of malware attacks from redirecting the flow of execution of a program
  Atilla Elçi, 2020

> **Quote**
> *"Current software attacks often build on exploits that subvert machine-code execution. The enforcement of a basic safety property, Control Flow Integrity (CFI), can prevent such attacks from arbitrarily controlling program behavior"*
>   Martin Abadi, 2007

### 9.6.1.3  Artifact History Integrity (Version control)
System engineering generates a large number of artifacts, some of them being continuously developed through many versions and possibly by different teams. The integrity in content and time of the artifacts is critical for trustworthy cyber-physical systems.

At the same time, the archived artifact development history may be required for forensic analysis or legal procedures. For each artifact, at least a unique identifier, a version
number, the date of creation/modification, the artifact owner, and the approval status
must be maintained. The appropriate tool for this task is a *version control system* (VCS,
Definition 9.5, e.g., [Sink11], [James19].

▶ **Definition 9.5: Artifact History Integrity (Version Control)**
A version control system (VCS) is a software tool that uniquely identifies all development and runtime artifacts, helps the developers on a development team to work together,
and archives the complete history of their work
   Adapted from: Eric Sink, 2011

### 9.6.1.4 Terminology
A significant danger for the *conceptual integrity* of the IT system is unprecise or even
conflicting *terminology*. Precise terminology—using a corporate terminology management system—is indispensable for smooth, comprehensible, and reliable communication
in the organization ([Großjean09], [Jackson21]).

> **Quote**
> *"Definitions for any type of terminology are necessary evils. While seemingly elemen-*
> *tary and potentially annoying, they provide a common ground from which to build"*
>    Tony Uceda Vélez, 2015

**Principle 9.6: Software Integrity**

1. Code signing: Protect the integrity of all software and firmware artifacts by a digitally signed hash (applied by the creator/developer of the soft-/firmware using his
   private key);
2. Check the integrity and intactness of all software and firmware artifacts during
   boot/start-up (using the public key of the developer);
3. Protect the integrity of all configuration and information artifacts by a digitally signed hash (applied by the developer of the soft-/firmware using his private
   key);
4. Check the integrity of all software and firmware artifacts during boot/start-up
   (using the public key of the developer);
5. Check the consistency (= match of the version numbers of all software, firmware,
   configuration files, etc.) of all artifacts during boot/start-up;
6. Implement correct error handling in case of detected integrity faults or configuration mismatches. In such a case, never start-up the system!

7. At all times, keep all related software artifacts, such as source code, models, configuration files, and documentation synchronized. Use round-trip engineering (RTE) development tools that support RTE;
8. If the risk analysis shows a probability for malfeasant control flow alteration, then use control flow integrity techniques;
9. Define and consistently apply an unequivocal and comprehensive artifact numbering/versioning scheme and use it consistently for all work products and for thorough archiving. Use an industry-standard version control system.

## 9.6.2   G6_2: Timing Integrity

Many CPS are *real-time* systems; i.e., their response time to stimuli must be upper-bounded. In a complex, mixed-criticality system, this requirement is a very strong constraint, often the *dominant* constraint (e.g., [Kopetz22], [Kopetz11], [Gomaa16], [Thiele04], [Buttazzo11], [Cooling19]). Some principles for assuring timing integrity are listed in Principle 9.7.

---

**Principle 9.7: Timing Integrity**

1. Precisely specify and document all timing requirements;
2. Never mix time-critical (real time) and non-critical functionality in the same modules;
3. Analyze the risks in case of timing violations during execution. Mitigate it when possible;
4. Exhaustively identify, analyze, and document all computing functions with real-time constraints;
5. Specify the worst-case timing requirements (execution times) of all real-time computing loops in the system;
6. Whenever possible model the real-time system and verify the timing integrity of the model;
7. Ensure that the underlying execution infrastructure meets the timing constraints of the application software;
8. Verify the timing constraints during design time by adequate tools (simulations, etc.);
9. Detect and absorb the impact of timing violations during runtime.

## 9.6.3   G6_3: Fault Containment Regions

A failure or a threat (attack) may hit any system part at any time. That part may react with a fault or error (resulting in erroneous or damaging behavior). If the system is not correctly designed, the fault in the system part can propagate to additional system parts. This may cause subsequent system parts also to fail, possibly leading to catastrophic system failure (*failure propagation, error propagation*, Fig. 9.3). In order to prevent catastrophic failure/error propagation, the mission-critical parts must be partitioned into *fault containment regions* ([Kopetz11], [Saridakis03], [Leveson11], [Kong17]).

Fault propagation must be stopped at the *interface* between two regions (Fig. 9.3). Each fault containment region must have:

- An **Output Guard**: Software module scanning the outgoing interface traffic, identifying harmful communication content, and inhibiting the transmission to the partner region;
- An **Input Guard**: Software module scanning the incoming interface traffic, identifying harmful communication content, and inhibiting the transmission from the partner region.

**Output Guard** and **Input Guard** may be formulated as interface contracts and use pre- and post-conditions (Definition 2.27).

To identify harmful traffic patterns, the *interface runtime monitor* needs a model of correct behavior: This model is provided by a detailed, complete, and precise *interface*

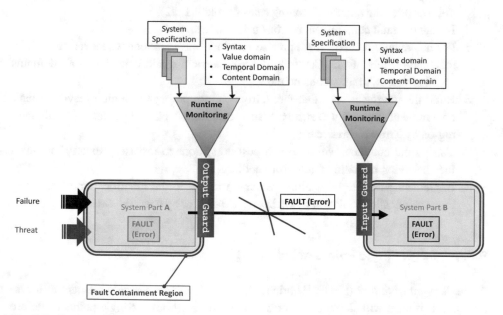

**Fig. 9.3** Fault containment regions

*specification*. The runtime monitor checks all traffic properties (syntax, values, timing, content, etc.) and blocks any non-conformant communication—thus precluding fault propagation.

---

**Quote**
*"Fault containment is an important constituent of fault tolerance. Means for fault containment allow a system to limit the impact of manifested faults to some prede-fined system boundaries"*
   Titos Saridakis, 2003

---

Whenever **Output Guard** or **Input Guard** detects any violation of the interface specification, adequate mitigation measures must be implemented to maintain system operations—possibly as degraded operation (Principle 9.8).

---

**Principle 9.8: Fault Containment Regions**

1.  Separate the mission-critical and non-mission-critical parts of the system;
2.  Partition the mission-critical parts of the system into fault containment regions;
3.  Use optimum fault containment as a partitioning rule. Fault containment partition-ing overrules the other partitioning rules (Table 9.1, #2);
4.  Isolate the fault containment reasons using interfaces;
5.  Formulate complete, detailed, precise interface specifications whenever possible in machine-readable form. Carefully describe attributes and behavior in all domains (syntax, values, timing, content, etc.);
6.  Build the interfaces such that fault/errors cannot propagate (inhibit leaving a fault containment region by **Output Guards**, inhibit entering a fault containment region by **Input Guards**);
7.  Use formal contracts with pre- and post-conditions to precisely specify the inter-face behavior, including fault/error handling;
8.  Implement adequate mitigation measures to react to the identification of an inter-face specification violation by **Output Guard** or by **Input Guard**.

---

## 9.6.4   G6_4: Single Points of Failure

A *single point of failure* (SPOF, [Ulbrich12], [Lever13], []) is a single element of a sys-tem that, if it fails, will stop mission-critical parts from working. Single points of failure can be present—often well hidden—in any layer of the architecture. Any single point of failure is a significant risk for a dependable system. Therefore, they must be avoided by careful analysis in all phases of the development cycle. Once detected, they must be eliminated by introducing adequate redundancy. The risk-based damage potential determines the extent of the necessary redundancy.

> **Quote**
> *"A single point of failure is essentially a flaw in the design, configuration, or implementation of a system, circuit, or component that poses a potential risk because it could lead to a situation in which just one malfunction or fault causes the whole system to stop working"*
>   https://avinetworks.com/glossary/single-point-of-failure/

Guidance to avoid and eliminate single points of failure is listed in Principle 9.9.

---

**Principle 9.9: Single Points of Failure**

1. Identify possible single points of failure (SPOF) during all phases of the development cycle. Locate them as early as possible;
2. Identify possible single points of failure in all levels of the architecture stack of the system (including hardware, firmware, system software, applications software, databases, etc.);
3. Investigate possible single points of failure in all third-party software
4. Identify possible single points of failure in all links to external systems (dependencies from partners, suppliers, etc.);
5. Assess the potential damage of the failure of the single point of failure on the system, especially on the mission-critical system parts (risk analysis);
6. Eliminate the identified single points of failure by the implementation of adequate redundancy (risk-based assessment);
7. Execute massive testing (fault injection, chaos engineering, overloading, stress testing, etc.) on the system to provoke SPOF failures.

## 9.6.5    G6_5: Multiple Lines of Defense

Every cyber-defense technology has specific weaknesses which malicious actors can exploit. The principle of *multiple lines of defense* (e.g., [Schmidt10]) stipulates to use a combination of different technologies to mitigate the same threats. This strategy is often called *implementation diversity* ([Hole16], [Page10], [Voges13]). The expectation is that if an actor penetrates or circumvents one technology, the second technology will not fall victim to the same strike method.

> **Quote**
> *"Defense in Depth (DiD) is an approach to cybersecurity in which a series of defensive mechanisms are layered in order to protect valuable functionality and information. If one mechanism fails, another steps up immediately to thwart an attack"*
> Adapted from: https://www.forcepoint.com/cyber-edu/defense-depth

Cascading multiple lines of defense is strongly dependent on the systems and their threats (e.g., Example 9.3). Again, an accurate risk analysis will identify the justifiable effort to implement multiple lines of defense (Principle 9.10).

---

**Example 9.3: Firewall Cascade**

*Firewalls* ([Young20], [Zwicky00]) are proven network security devices: They inspect and filter incoming and outgoing traffic on the network—allowing legitimate data to pass while blocking potentially dangerous traffic. Unfortunately, firewalls are not immune against malicious activities: Firewall vulnerabilities may allow an attacker to bypass the firewall and infiltrate unsafe network traffic into the system to be protected ([Kamara10], [Naidu02]).

A cascade of two consecutive firewalls (Fig. 9.4) based on different technologies (= Diversity, e.g., [Hole16]) throughout their complete stack significantly reduces the success of manipulating or circumventing the firewall.

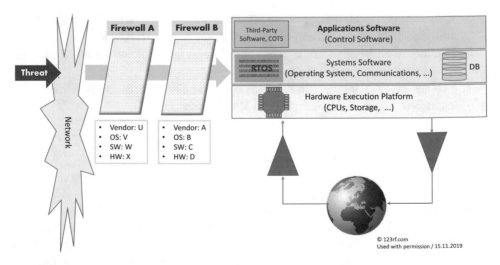

**Fig. 9.4** Double line of network defense

---

**Principle 9.10: Multiple Lines of Defense**

1. Use risk analysis and industry experience to assess the probability of a specific mitigation measure to fail in the defense of the threat;
2. Whenever necessary, implement cascaded mitigation measures to absorb a deficient operation of a mitigation measure;
3. When using cascaded mitigation measures, mandate different vendors and install thoroughly different technologies of the products on all levels of the technology stack;
4. Ensure the timely, correct, and complete maintenance of all the defense products (updates, patches, etc.).

---

## 9.7   G6_6: Fail-Safe System

The crown of safety engineering is *fail-safe systems* (Definition 9.6, e.g., [Leveson03], [Leveson16], [Fowler10b]). After one or multiple faults, the cyber-physical system will always end in a *safe state* and thus protect the users and the environment from damage.

▶ **Definition 9.6: Fail-Safe System**
A fail-safe system is one which, due to the characteristics of its equipment and components and the way in which they are integrated, is guaranteed that, in the event of any fault appearing, the system will always go to a safe status, normally affecting availability but never, and in no case, affecting safety
   https://www.leedeo.es/l/fail-safe-system

The concept of a safe state is based on the representation of the cyber-physical system as a *finite state machine* ([Wagner19], [Drusinsky06], [Kröger08], [Baranov18], [Chao20], [Börger03]). The system's operation consists of a sequence of states: Any state change is triggered by an event (Fig. 9.5). An autonomous car may be in the state `<no obstacle/constant velocity>`. The event `<forward radar detects obstacle>` transits the car into the new state `<reduce speed/maintain minimal distance>`.

Figure 9.5 emphasizes the design principle: From any state of the cyber-physical system, there should be a path to the safe state. If the cyber-physical system encounters an unresolvable problem, e.g., a failure, diverging inputs, conflicting information, unmanageable environmental conditions, etc., the CPS must immediately transit into a safe state. For the autonomous car, this state may be `<emergency braking/evade to emergency lane/stop/activate emergency indicators>`.

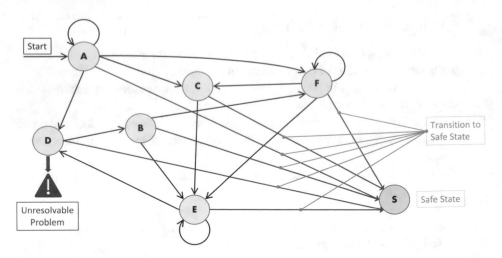

**Fig. 9.5**  Safe state

---

**Quote**
*"As engineers we sometimes find designing equipment to be well-built is much eas-
ier than designing it to fail predictably"*
   Peter Herena, 2011
   (https://www.bakerrisk.com/wp-content/uploads/Herena_Peter-resume.pdf)

---

Identifying the safe state(s) of a cyber-physical system is a very challenging task. Note that
some CPS do not have global safe states, e.g., a plane flying at 10′000 m altitude. However,
if a jet engine of this plane catches fire, the safe state is<stop  engine/launch
engine fire extinguisher/balance  thrust>. Safe states, therefore, not only
apply to the complete CPS or CPSoS but may also be very powerful for subsystems.

   The careful identification of safe states—either for the complete CPS/CPSoS—or for
critical subsystems is an essential task of safety engineering (Principle 9.11).

---

**Principle 9.11: Fail-Safe System**

1. Execute a careful hazard analysis of your full system to identify all (= goal) critical or harmful states;
2. Document all paths to the critical or harmful states in a formal way, such as state
   chart diagrams or directed graphs;
3. Model your application (or the software part of it) as a finite state machine;
4. Identify and define all fail-safe state(s), both for the complete system (if available)
   or for the mission-critical subsystems;

5. Implement reliable paths from all relevant nodes to the fail-safe state(s);
6. Extensively test (whenever possible, or otherwise simulate) the transitions to the safe state(s);
7. Use adequate tool-chains for the identification/simulation/removal of SPOF's.

## 9.7.1   G6_7: Graceful Degradation

*Graceful degradation* (Definition 9.7, e.g., [Saridakis09], [Dhall17], [Ploeg15]) is a technique to make systems user-friendly: Instead of shutting down the system in case of failures, the services of the system are gracefully reduced (Principle 9.12, Example 9.4).

▶ **Definition 9.7: Graceful Degradation**
A system that continues to run at some reduced level of functionality, performance, safety, security, or dependability after one or several of its components fail. It is a level below fault-tolerant systems, which continue running at the same rate of speed.
　　Adapted from: https://www.pcmag.com/encyclopedia/term/graceful-degradation

---

**Example 9.4: ABS Failure**

If the anti-skid braking control system (ABS) in a car fails, the software has two possibilities:

a. `<controlled braking/evade to emergency lane/stop/acti-`
   `vate emergency indicators>;`
b. `<warn the driver (indicator)/allow continuation/limit`
   `maximum speed to 80 km/h>.`

The second response constitutes a *graceful degradation*: The loss of the ABS functionality is a severe impairment of the car's safety. However, taking the car out of service would be a significant inconvenience. Therefore, allowing continuation with a warning of degraded safety and—possibly—an automatic speed limit is a defensible reaction.

**Principle 9.12: Graceful Degradation**

1. Identify the possibilities for graceful degradation in the planned system (Note: this a task of the responsible business unit);
2. Assess the risk of allowing the degraded mode of operation;
3. If the acceptable residual risk is too high, devise and implement mitigation measures to make the use of the degraded modes of operation acceptable;
4. Provide for sufficient, prominent warning to all users of the degraded mode of operation;
5. Architect and implement proven graceful degradation technologies (for specific resilience properties, such as availability, performance, safety, security);
6. Whenever possible, compensate component failures by carefully planned redundancy.

## 9.7.2   G6_8: Fault Tolerance

### 9.7.2.1  Fault Tolerance

Failures and threats in the cyber-physical system may cause *faults*. Failures of hardware or software components can occur at any level of the system stack. Threats can impact all components of the system. Therefore, faults must be expected in all elements of the CPS or CPSoS (Fig. 9.6).

Fault tolerance has two objectives:

1. Provide the requested *availability* of the CPS;
2. Compensate for *functional failures* within the CPS;

Faults can have different consequences:

1. The CPSoS, CPS, or mission-critical subsystems become inoperable (worst case);
2. The CPSoS, CPS, or mission-critical subsystems revert to degraded operation;
3. The CPSoS, CPS, or mission-critical subsystems produce faulty results and potentially endanger the user;
4. The CPSoS, CPS, or mission-critical subsystems switch to a backup system (cold or hot standby) and restart full or degraded operations;
5. The CPSoS, CPS, or mission-critical subsystems compensate the fault(s) and continue uninterrupted operation (best case).

Faults and misbehaviors can be absorbed by *fault tolerance* (Definition 9.8, [Dubrova13], [Hitt01], [Koren20], [Hole16], [Pullum01], [Butler10], [Goloubeva06]).

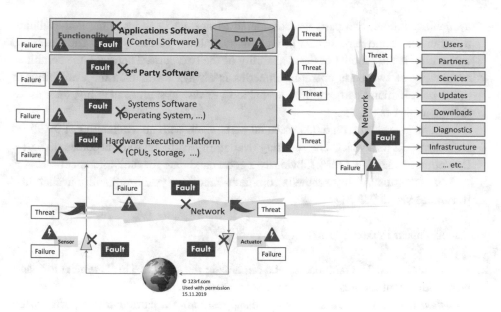

**Fig. 9.6**  Nascency of faults in a cyber-physical system

▶ **Definition 9.8: Fault Tolerance**

Fault tolerance is the ability of a system or subsystem to continue performing its intended functionality in the presence of faults.

Fault tolerance is necessary, because it is practically impossible to build a perfect system.

Elena Dubrova, 2013

> **Quote**
>
> *"As the complexity of a system grows, its reliability drastically decreases, unless compensatory measures are taken"*
>    Elena Dubrova, 2013

*System complexity*—expressed as the number of its parts and relationships—is a strong driver for low availability and thus for the introduction of fault tolerance! As the number of constituent parts raises, the system's availability decreases (Example 9.5).

**Example 9.5: Availability Calculation**

Calculating the *system availability* from the availability of its constituent parts depends on the system's topology and the availability of the individual parts. In the availability model, two parts are considered to operate in series if the non-availability of either of the parts results in the non-availability of the system. The combined availability is the product of the availability of the two parts. Therefore, the combined availability of two parts operating in series is always lower than the availability of either part. If many parts must be available to ensure system availability, their availability multiplies (Fig. 9.7).

Assume a system with 5′000 essential parts, each with an availability of 99.99 (= Downtime: 52 min/year). The resulting availability of the system becomes 60.65 (= Downtime: 143.6 days/year). Obviously a good candidate for fault tolerance!

Source:   https://www.eventhelix.com/fault-handling/system-reliability-availability/ [Last accessed: 8.9.2021]

Faults can appear in three types (Fig. 9.8):

a. *Permanent* faults: The fault occurs and remains in the system. This is typical for hardware component failures and unrecoverable software errors;
b. *Transient* faults: The fault appears and disappears after a particular time, often without intervention. This is typical for an aging sensor, for temporary unfavorable environmental conditions, or recoverable software errors;
c. *Intermittent* faults: The faults repeatedly appear for short intervals, and the system returns to regular operation. This is typical for communications outages, short power failures, or sensor behavior before they fail permanently.

Fault tolerance can be implemented in the hardware, the software, and the system level of the CPS or SPSoS (Table 9.2). In most cases, a combination will be used.

### 9.7.2.2  System-Level Implementation of Fault Tolerance
In *system-level* fault tolerance, functionally complete hardware/software subsystems are implemented using redundancy. A decision mechanism—often a voting system—then decides the correct operation (e.g., Example 9.6).

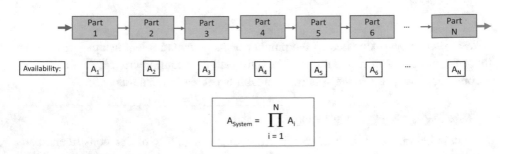

$$A_{System} = \prod_{i=1}^{N} A_i$$

**Fig. 9.7**   Availability calculation example

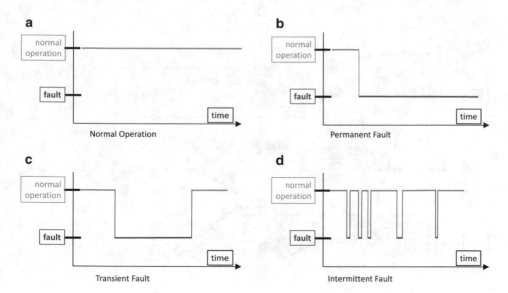

**Fig. 9.8** Fault types

**Table 9.2** Fault tolerance implementation levels

| Fault Tolerance | | Examples |
| --- | --- | --- |
| Hardware | | Use two disk drives with real-time mirrored data |
| Software | Data/Information | Multiple storage locations or error-correcting codes |
| | Functionality | Multiple implementations using diversified technologies with a voting decision device |
| System | | Use multiple subsystems with identical functionality but diversified implementation governed by voters |

**Example 9.6: System-Level Fault Tolerance (Avionics)**

Figure 9.9 shows a *triple-redundancy* system-level fault tolerance. The three subsystems have identical functionality, e.g., the autopilot of a plane ([Dubrova13], [Collinson11]). They are, however, built by different manufacturers—each one starting from the specification, with no interaction with the other manufacturers, and using a different technology (implementation diversity).

The outputs of the three parallel subsystems are fed to voters, comparing their results. The voters use a consensus protocol to agree on the correct output based on majority voting and plausibility checks. The final output from the voter $\Sigma$ is an agreed, correct computation result. Failure of one subsystem will be tolerated.

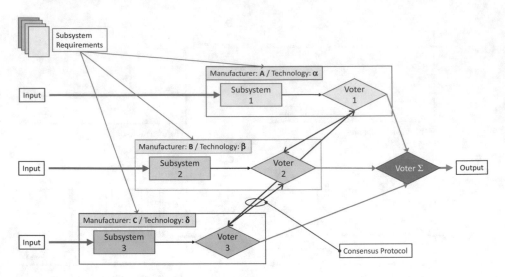

**Fig. 9.9**  Triple subsystem redundancy

### 9.7.2.3  Software-Implemented Fault Tolerance
*Software fault tolerance* has two objectives:

I.  Ensure the required *availability* of the functions;
II. Reduce the probability of *malfunctions* to an acceptable level.

Note that *availability* is a binary state: The system is either *available*, i.e., providing the intended functionality with the specified quality properties, or *unavailable*; i.e., the system has completely stopped ([Trivedi17]). A degraded operation may be offered in some cases, which would qualify as "partially available". Several metrics exist to express availability. *Malfunctions* are different: The CPS or a subsystem delivers false results, erroneous decisions, or violates a timing restriction. Malfunctions can lead to accidents or incidents, endangering life, property, or the environment.

The mechanisms to implement software fault tolerance are:

a.  *Redundancy* (Definition 9.9, Example 9.6, [Koren20]);
b.  *Diversity* (Definition 9.10, Example 9.3, [Page10], [Hole16]).

▶ **Definition 9.9: Redundancy**
The duplication of functionality, data, or information as a whole or in parts

▶ **Definition 9.10: Diversity**

A modular system has diversity when it contains differently designed, engineered, or implemented modules with (nearly) the same functional and properties specifications.

Adapted from [Hole16].

---

**Quote**

*"By simply triplicating a software module and voting on its outputs we cannot tolerate a malfunction in the module because all copies have identical faults"*
Elena Dubrova, 2014

---

Two techniques for software fault tolerance—including both redundancy and diversity—are shown in Example 9.7.

---

**Example 9.7: Software Fault Tolerance**

Figure 9.10 depicts two software fault tolerance techniques: (a) The *Recovery Block Operation* (RcB) and (b) The *N-Version Programming* (NVP). As the functional example, a sort algorithm is used.

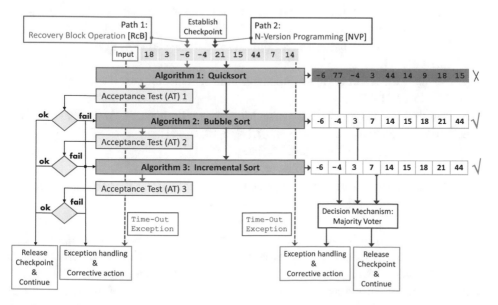

**Fig. 9.10**  Software fault tolerance

The *recovery block operation* (left, blue path in Fig. 9.10) executes the sequence:

1. Establish a checkpoint: Safe system state;
2. Calculate the cross-sum of the input $\{18,3,-6,-4,21,15,44,7,14\}=112$;
3. Calculate the result using the 1st algorithm: Quicksort (Result=$\{-6,77,$ $-4,3,44,14,9,18,15\}$);
4. Check the result with the 1st acceptance test (AT1);
5. The *acceptance test* in this case is very simple: It calculates the *cross-sum* of the 1st output $\{-6,77,-4,3,44,14,9,18,15\}=170$;
6. Compare the cross-sums of input and output: $112\neq170$: `Test failed`;
7. Alert exception handling and pass input to 2nd algorithm: Bubble Sort;
8. 2nd acceptance test: Compare cross-sum of input to cross-sum of output $\{-6,-4,$ $3,7,14,15,18,21,44\}=112$: `Test passed`;
9. Release checkpoint and continue processing;
10. If the 2nd acceptance fails, the 3rd algorithm (Incremental Sort) could be invoked;
11. The whole processing is time-bounded: A watchdog timer is set up and generates an exception in case of time overrun;

The *N-Version Programming* (right, red path in Fig. 9.10) executes the sequence:

1. Establish a checkpoint: Safe system state;
2. Calculate the result using the 1st algorithm: Quicksort (Result=$\{-6,77,-4,3,44,1$ $4,9,18,15\}$);
3. Calculate the result using the 2nd algorithm: Bubble Sort (Result=$\{-6,-4,3,7,14,$ $15,18,21,44\}$);
4. Calculate the result using the 3rd algorithm: Incremental Sort (Result=$\{-6,-4,3,$ $7,14,15,18,21,44\}$);
5. Pass the three results to the Decision Mechanism: In this case a majority voter;
6. Decide on correct result: $\{-6,-4,3,7,14,15,18,21,44\}$;
7. Alert exception handling (Result mismatch);
8. Release checkpoint and continue processing;
9. The whole processing is time-bounded: A watchdog timer is set up and generates an exception in case of time overrun.

The specification of availability and malfunction probability need a *methodology*. One possible, proven methodology is to model the system with *Reliability Block Diagrams* (RBD, Definition 9.11, [Dubrova13], [Abd-Allah97], [Denning12], [Trivedi17]). Reliability Block Diagrams allow the calculation of the availability and malfunction probability of complex systems. They also enable the calculation after introducing redundancy and diversity—and thus to verify the requirements or incrementally improve them (Example 9.8).

▶ **Definition 9.11: Reliability Block Diagram (RBD)**

Reliability Block Diagrams (RBDs) are a way of representing a system, including its subsystems and components, as a series of blocks in such a way that equipment failure rates, operating philosophies, and maintenance strategies can be quantitatively assessed in terms of the impact they are expected to have on system performance

https://www.armsreliability.com/page/resources/blog/why-do-you-need-a-reliability-block-diagram

**Example 9.8: Reliability Block Diagram**

Figure 9.11 shows part of a cyber-physical system, including *redundancy* and *diversity*. The element of the system in focus is a *function*: Individual functions constitute the full functionality of the CPS. An example of a function is a modern car's distance-keeping driver assistance function. A function has an input, executes certain computations, and generates an output. Following the function via its control flow identifies all the computational, information, and decision blocks belonging to that specific function. Each computational, information, and decision block is represented by a *reliability block* (Fig. 9.11): The function consists of a topography having serial and parallel connections between the reliability blocks.

To each reliability block, two probabilities are assigned:

1. The *availability* of the reliability block (Example 9.5);
2. The *probability of malfunction*.

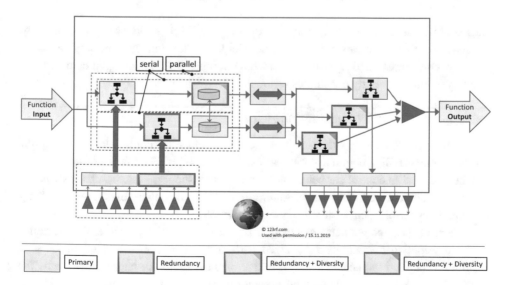

**Fig. 9.11** Reliability block diagram computation rules

Using the reliability block *composition algebra* (e.g., [Kumar10], [Denning12], [Kovacevic19], [Abd-Allah97]), i.e., successively calculating the availability and probability of malfunction values for the serial and parallel connections. When the full path is calculated, the values for the availability and the probability of malfunction for this specific function are known. If they do not match the requirements, redundancy and diversity are added, generating new reliability blocks. The calculations are repeated until the new block reliability diagram matches the requirements (Iterative approach).

**Quote**
*"Most of the software faults are design faults, activated by some non-tested or unexpected input sequence"*
  Elena Dubrova, 2014

### 9.7.2.4 Diagnosability

*Diagnostic capabilities* (Diagnosability, Definition 9.12, [Blanke15], [Isermann06]) are a crucial requirement for fault tolerance. A fault can only be correctly handled if it has been detected, identified, and classified.

▶ **Definition 9.12: Diagnosability** Diagnosability is the capability of a technical system to detect and log a permanent, intermittent or transient fault in (near-) real time, to determine its cause, and to initiate and execute corrective action.

Diagnosability is an essential quality property of the CPS for its safety and security. The required diagnostic capabilities must be adequately formulated, precisely specified, correctly engineered, and well-implemented (Example 9.9). Missing, incomplete, or untimely diagnostic ability constitutes serious risk!

**Example 9.9: Onboard Diagnostics (ODB II)**

Modern vehicles contain thousands of sensors, actuators, control loops, electronic control units (ECU), communication buses, etc. ([Ribbens17]). Many of these components are vital to control car behavior, such as driver support systems, chassis control, emission control, and many more. Automotive electronics, therefore, is an amazingly complex cyber-physical system-of-systems (CPSoS).

Each of these components can exhibit permanent, transient, or intermittent faults. The automotive community recognized early that maintenance and repair of such vehicles required online capture and storage of these faults for later evaluation by the mechanics. Additionally, upcoming regulations required an onboard diagnostic system to detect emission failures (California 1982).

The answer to these requirements was onboard diagnostics (ODB), based on an ODB-bus ([Santini10]) and a set of standardized fault codes (e.g., [DMV04]). The automotive industry fast agreed on international standards for ODB (e.g., [ISO 15031-5], [SAE J1979]). The ODB detects fault codes emitted by the components, logs them, and either directly warns the driver of a malfunction or store the fault codes for evaluation in the shop. The repair and maintenance of a modern car would be impossible without this powerful instrument.

*Fault tolerance* has become an extensive and specialized engineering domain—filling many textbooks. Different industries have varying requirements and forms of implementation of fault tolerance (e.g., [Jackson09], [Hollnagel06], [Hollnagel13], [Castano16], [Campbell17], [Beyer16], [Hole16]). Therefore, only fundamental fault tolerance principles are listed in Principle 9.13.

---

**Principle 9.13: Fault Tolerance**

1. Wholly and unambiguously specify the allowable availability and malfunction probability rates of all mission-critical functions in the CPS/CPSoS;
2. Model the functions, e.g., as a Reliability Block Diagram (RBD);
3. Add appropriate redundancy and diversity to meet the availability and malfunction probability rates and confirm them by updating/recalculating the model;
4. Carefully plan the required diagnostic capabilities of the CPS/CPSoS and include them in the requirements and specifications;
5. Repeat (1)–(4) for each change of the CPS/CPSoS;
6. Devise and implement specific test cases for the testing of redundancy and diversity;
7. Whenever the system is extended (evolution) repeat the steps (1)–(6).

---

### 9.7.3   G6_9: Dependable Foundation (Dependable Execution Infrastructure)

#### 9.7.3.1 Dependable Execution Infrastructure

The *execution infrastructure* or *execution platform* (Definition 9.13, Fig. 9.12) is the foundation of the cyber-physical system. The dependability of the foundation must match or exceed the dependability requirements (safety, security, performance, availability) of the most mission-critical software running on this execution infrastructure.

▶ **Definition 9.13: Dependable Foundation (Dependable Execution Infrastructure)**
Application execution environment that is known to be trusted from the very start and further provides system software with the means to provide a more secure system and better protect data integrity.
Adapted from: William Futral, James Greene, 2013.

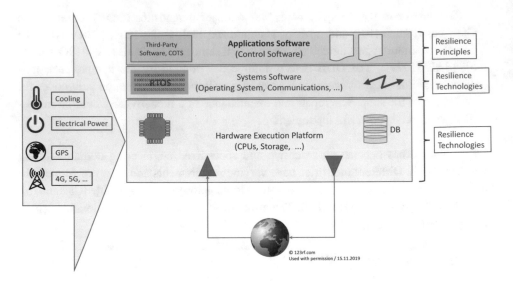

**Fig. 9.12**  Dependable execution infrastructure

The execution infrastructure also includes the external services, such as the physical elements (such as electricity and cooling), and the services (such as the Global Positioning System or the mobile phone system). The CPS/CPSoS system architecture must provide enough redundancy and diversity to match or exceed the worst-case dependability requirements (Principle 9.14).

The dependability requirements and implementation are often defined in business continuity and disaster recovery *policies* ([Snedaker13], [Schmidt10], [Kopp18]). Business continuity and disaster recovery mechanisms enable the execution platform to recover after a disaster and resume operations in the shortest possible time.

In the last decade, several specific technologies supporting the construction of dependable infrastructures were developed, such as

- *Trusted Platform Module* (TPM), Example 9.10;
- *Trusted Execution Platform* (TEP)/*Trusted Execution Environment* (TTE), Example 9.11.

**Example 9.10: Trusted Platform Module (TPM)**

The technical solution to many security challenges is *cryptography* (e.g., [Ferguson03]). Consequently, the design and implementation of powerful and high-performance cryptography have become a vital issue. For many applications, software-based solutions were either too slow or not secure enough. Therefore, the industry group "Trusted Computing Group" (https://trustedcomputinggroup.org/) was formed to develop a standard for a hardware-based trust module—the *Trusted Platform Module* TPM (Definition 9.14, [TCG19], [Arthur15]).

▶ **Definition 9.14: Trusted Platform Module (TPM)**

A Trusted Platform Module, also known as a TPM, is a cryptographic coprocessor that is present on most commercial PCs, servers, and many embedded systems.

Adapted from: Will Arthur & David Challener, 2015.

> **Quote**
> *"The rise of the Internet and the corresponding increase in security problems, particularly in the area of e-business and industrial control systems, were the main driving forces for designing TPMs. A hardware-based standardized security solution became imperative"*
>     Adapted from: Will Arthur & David Challener, 2015

The *functionality* provided by the TPM 2.0 includes ([TCG19]):

- Support for bulk (symmetric) encryption in the platform;
- High-quality random number generation;
- Cryptographic services;
- A protected persistent store for small amounts of data, sticky- bits, monotonic counters, and extendible registers;
- A protected pseudo-persistent store for unlimited amounts of keys and data;
- An extensive choice of authorization methods to access protected keys and data;
- Platform identities;
- Support for platform privacy;
- Signing and verifying digital signatures (normal, anonymous, pseudonymous);
- Certifying the properties of keys and data;
- Auditing the usage of keys and data.

Many modern implementations use *Trusted Execution Platforms* or *Trusted Execution Environments* (TEP or TEE, Definition 9.15, [Futral13], [Hussmann21], [Yao20]).

▶ **Definition 9.15: Trusted Execution Platform (TEP)/Trusted Execution Environment (TTE)**

Application execution environment that is known to be trusted from the very start and further provides system software with the means to provide a more secure system and better protect data integrity.

Adapted from: William Futral, James Greene, 2013.

The basic concept is *trusted partitioning*: IBM initially started work on partitioning in S/370™ mainframe systems in the1970s. Since then, *logical partitioning* (LPAR, e.g., [Harris02], [Irving05]) has been extended to systems software, such as operating systems. The objective of the original LPAR was:

- Protection against *inter-partition data access*: The design of partitioning-capable servers prevents any data access between partitions other than using shared networks. This design isolates the partitions against unauthorized access across partition boundaries;
- Unexpected *partition crash*: A software failure within a partition should not cause any disruption to the other partitions. Neither an application failure nor an operating system failure inside a partition interferes with the operation of other partitions;
- *Denial of service* across shared resources: The design of partitioning-capable servers prevents partitions from making extensive use of a shared resource so that other partitions using that resource become starved.

The three axioms of TEP or TEE are (Fig. 9.13):

I.   *Integrity*: The functionality and data are protected during all phases of execution against any modification; i.e., it executes exactly as specified;
II.  *Isolation*: No unintended interaction between any element in two different partitions (LPARs) is possible, including unauthorized or excessive use of resources;

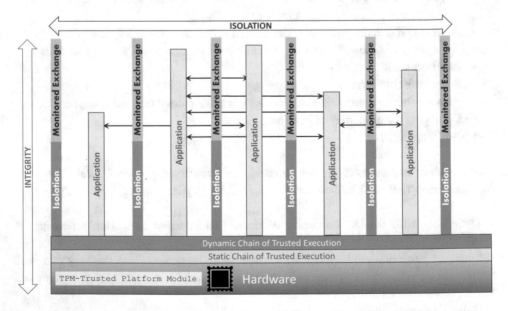

**Fig. 9.13**   Trusted execution platform

III. *Monitoring*: All exchanges (functionality, data, control flow) between elements in different partitions (LPARs) are checked and suppressed if they do not conform to the interface specification.

Different technologies have been introduced to provide TEP's or TEE's (Example 9.11).

---

**Example 9.11: Trusted Execution Technology TXT**

Figure 9.13 shows the basic architecture of a Trusted Execution Environment. This work was started by an industry consortium led by Intel ([Intel12], [Raj13]), soon followed by other specifications (e.g., [Sheperd16], [GPI18]). It relies on a Trusted Platform Module (Definition 9.14), which provides secure cryptographic functions.

When the platform powers on, the static chain of trust is executed: All firmware modules of the system are checked for their integrity (intactness) and—if all pass the integrity measurement and no unknown firmware is detected—the platform is declared as secure. The dynamic chain of trust next verifies the integrity of the operating system and then allows the start of integrity-checked applications.

---

**Quote**

*"A building is only as good as its foundation. The same is true for a computer architecture's information security"*
   Intel Corporation, 2012

---

### 9.7.3.2 Firmware

The execution platform includes "invisible" firmware. Firmware is the lowest layer of software on the platform (Fig. 9.12). The firmware is often stored in a non-volatile device that is part of the platform and is only accessible by the manufacturer. One crucial piece of firmware in ICS is the Basic Input Output System (BIOS). Research and experience have shown that also firmware can be the target of malicious activities (e.g., [Furtak10], [Yao20]).

Soon after the Shamoon virus attack (https://securityintelligence.com/the-full-shamoon-how-the-devastating-malware-was-inserted-into-networks/), a cross-industry/government collaboration started, which resulted in security guidelines for BIOS (e.g., [NIST-SP-800-147], [Yao20]).

---

**Principle 9.14: Dependable Foundation (Dependable Infrastructure)**

1. Define a business continuity/disaster recovery policy, implement and enforce it. Invest in the necessary technologies and processes;
2. Implement effective business continuity and disaster recovery mechanisms;

3. Define a Business Continuity & Disaster Management Plan and train all employees to execute it;
4. Partition the execution infrastructure into criticality regions (e.g., mission-critical, revenue-critical, compliance-critical, etc.). Identify all infrastructure elements belonging to the respective criticality level;
5. Quantitatively define and specify the dependability requirements of the execution infrastructure for each criticality level;
6. Provide sufficient redundancy, diagnosability, and diversity to match or exceed the dependability requirements;
7. Make as much use as possible of modern Trusted Execution Environment technologies (TTE & TPM);
8. Ensure that all firmware used in the execution platform is trustworthy;
9. Ensure that all third-party software used in the execution platform is trustworthy.

### 9.7.4   G6_10: Error, Exception, and Failure Management

In today's complex cyber-physical systems, errors and failures, in both hardware and software, are not the exception but a daily fact of life. For a dependable system, it is essential to provide mechanisms to detect, recognize, diagnose, and correctly handle possible errors, exceptions, and failures.

Errors, exceptions, and failures can—and will!—occur on any layer of the system stack (Fig. 5.5). Suppose errors, exceptions, and failures are not correctly handled and absorbed during design time. In that case, they will lead to malfunctions, unavailability, and outages during runtime—possibly causing safety accidents or security incidents.

> **Quote**
> *"The trouble with programmers is that you can never tell what a programmer is doing until it is too late"*
>    Seymour Cray, 1990

Therefore, architecting and designing a system with reliable error, exception, and failure management is paramount for dependable cyber-physical systems. The design teams on all levels of the CPS must invest sufficient time and effort to foresee and provide correct and timely mitigation mechanisms. This includes (Fig. 9.14):

I.  The *business process layer*: A business process consists of a series of process steps, which are executed either automatically or manually. Each process step has a desired outcome, which serves as input to the next process step. However, a process step can

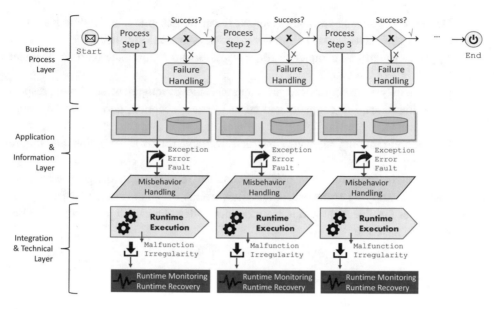

**Fig. 9.14** Levels of error, exception, and failure management

also terminate with a *failure*. Therefore, each process step must explicitly be terminated by a failure handling procedure. Note that an adequate failure handling procedure for each process step must be already fully defined in the process model (e.g., [Silver17], [Weske19]).

II. The *application and information layer*: A vast choice of causes for errors, exceptions, and failures during the implementation/programming of the business functionality exists (e.g., [Huang09], [Eloff18], [Pitakrat18], [Huckle19], [Rashid18]). They can either be discovered during design time, in the coding phase, by extensive testing, or—unfortunately—during runtime. For dependable CPS's, complete and correct error, exception, and failure management during the architecting, design, and implementation is paramount;

III. The *execution environment layer*: All diligence during the design and implementation of the CPS cannot completely eliminate the occurrence of errors, exceptions, and failures during runtime. The errors, exceptions, and failures during operation must also be appropriately absorbed to prevent such events from resulting in safety accidents or security incidents. Very often, runtime monitors are installed to detect and mitigate runtime misbehavior.

---

**Principle 9.15: Error, Exception, and Failure Management**

1. Assign the same importance to error, exception, and failure handling during all phases of development and operation of the CPS as to the functional development;

2. Define or adopt standards and guidelines for error, exception, and failure handling/management. Consistently communicate, review, and enforce them;
3. Apply thorough and complete error, exception, and failure handling/management to all layers of the system stack (Fig. 9.14);
4. Continuously train, convince, and supervise the development staff with respect to the importance of complete and correct error, exception, and failure handling/management;
5. Use all programming language constructs available for error, exception, and failure handling;
6. Use modern tools for code quality assurance (e.g., finding dangling execution paths or identifying errors/exception/failure events which are not handled);
7. Whenever feasible, deploy runtime monitors to detect and mitigate runtime misbehavior of the CPS.

### 9.7.5 G6_11: Monitoring

*Monitoring* the cyber-physical system has two objectives:

1. *Observe* the system during operation to detect anomalies and introduce countermeasures (offline or in near real time);
2. *During operation, supervise the system* to detect and prevent misbehavior, such as oncoming safety accidents or security incidents (real time).

The architecture of a CPS monitoring system is shown in Fig. 9.15: Monitoring the system means continuously comparing the actual behavior of the CPS to a *model* defining its *correct* behavior. The model consists basically of the functional and quality properties specifications, rules and standards, interface contracts, runtime infrastructure characterization, and machine learning models in many modern systems. If any deviation between the actual system behavior and the expected (= correct, model-conformant) behavior is detected, corrective action or intervention is initialized, either preventive intervention or accident/incident prevention activities (Principle 9.16).

Note that *all* layers of the cyber-physical systems must be monitored ([Julian17], [Bastos19], [Sabharwal20], [Josephsen13]).

---

**Principle 9.16: Monitoring**

1. Define the objectives of monitoring, both for technical monitoring and the business monitoring. Specify them comprehensively and completely;
2. Monitor all layers of the cyber-physical system (Fig. 9.14);
3. Carefully specify the metrics, analytics, results, and alerts to be extracted from monitoring;

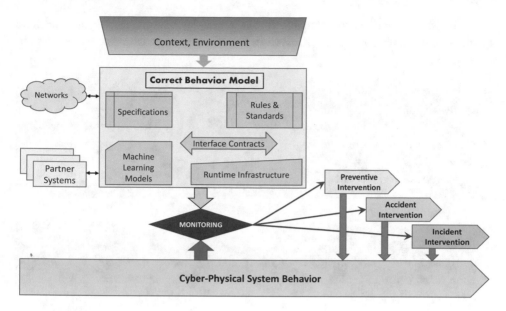

**Fig. 9.15**  Monitoring

4. Define the processes for data analysis, including incident/emergency response;
5. Specify the actions following alerts—whenever possible implement fully automated responses;
6. Implement particular mechanisms for specific properties of the system, such as general monitoring, performance, liveliness, safety monitoring (Principle 10.8), and security monitoring (Principle 11.14);
7. Recommendation: Use commercial monitoring tools whenever possible.

## 9.8   G7: Code Quality

Superb *code quality* is a de rigueur precondition for a trustworthy cyber-physical system's software. Code quality is a multifaceted concept, including, e.g., *functional code quality* meaning correct fulfillment of the requirements or *technical code quality* representing the robustness of the code implementation constructs.

Ensuring technical code quality is done via:

a. Tried and tested *coding standards* (Definition 5.5) and enforcement through automated, static code analysis;
b. Thorough testing: Functional testing and specific testing for the quality properties of the CPS (safety, security);
c. Runtime monitoring.

Tools for the support of code quality checks based on artificial intelligence/machine learning are becoming available ([Kinsbruner20]). Code quality assurance is described in Principle 9.17.

> **Quote**
> *"Everything on the SANS Institute* [https://www.sans.org] *Top 20 Internet security vulnerability list is a result of poor coding, negligent testing, and sloppy software engineering"*
>   Alan Paller in [Seacord13]

> **Quote**
> *"Test what you deploy, deploy what you test. You might have included diagnostic code for integration testing, and pull it out before deployment. WRONG! Now the code you are going to use is untested"*
>   Patrick Stakem, 2018

---

**Principle 9.17: Code Quality**

Development:

1. Establish a binding coding standard for all software development activities to support good code quality;
2. Carefully chose either an accepted industry-standard coding standard (e.g., CERT, MISRA) or assemble an own coding standard;
3. Ensure that the chosen code standard supports the required quality standards of the respective industry (e.g., safety, security, or other quality standards);
4. Regularly train the engineering staff concerning the coding standard;
5. Strictly enforce the coding standard (no exceptions);
6. Run automatic conformance checks (static code checking) of all code before a review and before all unit tests;
7. Keep the coding standard updated, i.e., aligned with the progress of industry-standard coding standards;
8. Make use of artificial intelligence/machine learning tools as soon as they become available;

Testing:

9. Specifically devise test cases not only for the functionality, but with the same (or more) care for the quality properties (performance, safety, security);

10.   Specifically devise test cases for errors, exceptions, and faults;
11.   Test third-party products with the same depth and diligence;

Runtime:

12.   Install runtime monitoring on all layers of the system.

## 9.9   G8: Modeling

Building, maintaining, and extensively using *meaningful models* for the cyber-physical system is unquestionably one of the most decisive activities for building trustworthy CPS's ([Mobus22], [Haimes19], [Olivé10], [Mittal18], [Rainey18], [Carreira20], [Jorgensen19], [Wagner19], [Selic13], [Mo21], [Taha21], [Mitra21], [Tekinerdogan20], [Mittal20], [Matulevičius17]).

> **Quote**
> *"Mathematical models are the imperative mechanisms with which to perform quantitative systems engineering"*
>    Yakov Y. Haimes, 2019

Developing, using, and maintaining appropriate models must be a central activity of the systems engineering process (Principle 9.18). An important issue is the *quality of the models* (Definition 9.16): The sufficient quality of all models must be ensured at all times. Model quality is strongly supported by enforcing modeling and diagramming guidelines ([Koenig19b]).

▶ **Definition 9.16: Model Quality**
*Expression Quality*: The model adequately represents and abstracts the targeted part of reality;

*Syntactic Quality*: The model does not violate any syntactic rules of the modeling language;

*Semantic Quality*: All the elements in the model have an unambiguously specified and agreed meaning;

*Pragmatic Quality*: The interpretation by the human stakeholders is correct with respect to what is meant to be expressed by the model. The interpretation by the tool(s) is correct with respect to the intended functionality;

*Social Quality*: The model has sufficient agreement by all stakeholders;

*Completeness Quality*: The model contains sufficient information to fulfill its role "clarity, commitment, communication, control" for the intended goal;

*Timeliness Quality*: The models are updated before any system change, i.e., in the requirements engineering phase.

---

**Principle 9.18: Modeling**

1. Develop and instill a modeling culture in the organization: Encourage and reward modeling activities wherever possible and reasonable;
2. Esteem the models as valuable, long-lived, and significant artifacts of systems development;
3. Define a model architecture, structuring all models and their relationships into a manageable, consistent model landscape;
4. Keep all models up-to-date at all times (this requires round-trip engineering);
5. Unambiguously define and document all terms and concepts before starting to model using, e.g., ontologies, taxonomies, vocabularies, etc. (= Ensure the conceptual integrity of the system);
6. Define the purpose and audience for each model before modeling starts;
7. Build and use the models explicitly for clarity, communication, commitment, and control (4 "Cs");
8. Use automated model checking for model validation and model verification (Definition 5.7);
9. Make use of well-defined views (Definition 5.11) for the organization of the models;
10. Version each model change (model history in the repository);
11. Add sufficient metadata to each model;
12. Use partitioning and a hierarchical model structure to keep the size of the models manageable;
13. Keep all models consistent at all times (Definition 5.10, Use automated consistency checking);
14. Whenever possible, use a modeling language and notation which has a mathematical logics foundation (= Use formal methods);
15. Support model quality by enforcing modeling and diagramming guidelines. Whenever possible, check the conformance to these guidelines tool-based;
16. Use industry-standard tools for modeling and model exchange;
17. Continuously assess and improve the model quality (Definition 9.16).

---

## 9.10  G9: Cloud-Based Cyber-Physical Systems

Nearly all cyber-physical systems are of mixed-criticality; i.e., they contain both mission-critical functionality/data and non-mission-critical functionality/data (Definition 4.4). When architecting a cloud-based CPS, the critical decision is the correct split between *CPS-implemented* functionality/data and the *cloud-implemented* functionality/data and the correct identification of their interrelationships. Guidance for these decisions is the result of the risk assessment.

For cloud-based cyber-physical systems, implement all applicable safety and security principles from the previous chapters. In addition, implement Principle 9.19.

The cloud landscape and technology are fast changing. This fact constitutes a danger for safety and danger. Periodical cloud security audits are required to maintain safety and security ([Majumdar20]).

---

**Principle 9.19: Cloud-Based Cyber-Physical Systems**

Overall Principles:

1. Nearly all cloud-based CPSs are mixed-criticality systems. Unambiguously identify, delineate, and specify the mission-critical parts and the non-mission-critical parts;
2. Discover *all* relationships between mission-critical and mission-uncritical parts of the system and document them accurately;
3. Faultlessly categorize *all* functions, *all* data/information (including operational, initialization, configuration data/information), and *all* relationships into mission-critical and non-mission-critical;
4. Assign all functions and all data/information to either the implementation in the CPS (= local) or the implementation in the cloud (= remote);
5. Thoroughly justify and document these assignments;
6. Execute a cloud-specific risk assessment. Re-assess the CPS to cloud assignments according to the results of the risk assessment (Note: the required, specific safety and security risk assessments are nevertheless executed with due diligence, no shortcuts allowed);
7. Repeat the verification of this principle before any change/update to the system;
8. Strictly use secure communication channels (= bilaterally authenticated, encrypted channels) between mission-critical CPS-parts and the cloud parts;

Cloud-Specific Principles:

9. Elaborate a cloud policy for the organization. Include the exit strategy for the case it should become necessary to move the cloud services to in-house or to a different cloud service provider;
10. Develop, document, and enforce a shared cloud security responsibility model and make it a binding part of the cloud contract;
11. Ensure that sufficient logs are captured and preserved for a sufficient time by the cloud provider. The logs must be stored in an unforgeable method;
12. Regularly review the logs. Audit them with respect to irregularities, possibly by a specialized third party;

13. Maintain an up-to-date data backup of the cloud data in the own organization. Use a cloud provider independent format;
14. Strictly assure least privileges. Identify, document, and approve the privileges of the cloud provider's system administrators;
15. Carefully draft the contracts with the cloud provider. Explicitly clarify all issues related to non-functional aspects, such as responsibilities, crisis management, update procedures, liability, legal and compliance;
16. Regularly audit all the processes, procedures, and contracts related to the cloud service provider (and, if applicable, its third-party subsuppliers).

## 9.11  G10: Supply Chain Confidence

Supply chains can be a significant source of severe *risk infiltration* (Definition 2.20, e.g., [Boyens21a], [Boyens21b], [Boyens12], [Boyens15], [Manners-Bell17]). The three attack vectors are (Fig. 9.16):

1. Acquisition of a *third-party product*: The product supplier may unknowingly distribute malware to its customers (Example 2.16 and 9.14). The malicious code can be introduced by an external malicious actor during the development or distribution process or by a malfeasant third-party *insider*;

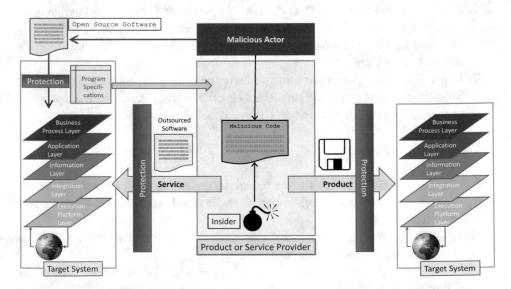

**Fig. 9.16**  Supply chain attack vectors

2. *Outsourcing* software/system development: Again, code with unwanted or destructive functionality can be introduced undetected into the software delivered to the customer. Here also, external malicious actors or malfeasant insiders in the provider company may be the source;

3. Use of *open-source software*: Unfortunately, freely downloading open-source software, e.g., from the Internet, has become a risky habit of many programmers. Such open-source program fragments may contain unwanted or dangerous functionality.

> **Quote**
> *"Identifying, assessing, and mitigating cyber supply chain risks is a critical capability to ensure business resilience"*
>    John Boyens, 2021

The key tool for managing supply chain risks is a *Cyber-Supply Chain Risk Management System* (CSCRM, Definition 9.17, [Boyson14], [NISTSCM15]).

▶ **Definition 9.17: Cyber-Supply Chain Risk Management (CSCRM)**
CSCRM is a management construct resulting from the fusion of approaches, methods, and practices from the fields of cybersecurity, enterprise risk management, and supply chain management.

It is an attempt to gain strategic management control over the rapidly globalizing cyber supply chain and to help compensate for deficiencies in purely technical approaches to security and assurance.

Sandor Boyson, 2014.

The decisive activity for defending against supply chain risks is a *specific risk assessment and mitigation*. Contrary to in-house developed systems, for third-party software, often little information is available; e.g., no source code is delivered. Therefore, proven risk management processes and methodologies (e.g., Fig. 4.26) are difficult to apply.

Two procedures are constructive for effectual management of risks potentially introduced by third parties:

I. Applying "black-box" risk analysis: Because no or insufficient information about the internals of the third-party product or service is available, vulnerabilities and potential failures are very difficult to identify. Consequently, the list of agreed *functionality* (taken, e.g., from the service level agreement, SLA) is used as the starting point for risk management.

II. Review/Audit of the third-party's risk management function (e.g., [Tschanz18], [IIA18], [Brines20]). This is an external view of the supplier's risk management policies and procedures, possibly by chartered specialists. If such is not satisfactory, they must be improved, or the supplier may not be suitable.

### 9.11.1 Black-Box Risk Analysis

The third-party product/software or service is viewed as a *black box* (Fig. 9.17). The risk analysis is applied to the functional specification and their quality attributes.

For each function, the following questions must be answered:

1. Which damage potential does a failure, degradation, or unavailability of the function have?
2. Which quality of service properties are tied to the functionality? How are they guaranteed, assured, and measured?
3. Are the residual risks of the functionality acceptable, and are the quality of service properties adequate?
4. Which additional mitigation measures do the product/software or service require for each functionality?

The results are documented in the risk analysis and mitigation matrix. This matrix is updated after each release of a new version. Based on this matrix, the decision to use the product/software or service, request or add additional protection measures (e.g., a protective shell, runtime monitoring), or avoid this product/software or service to be used.

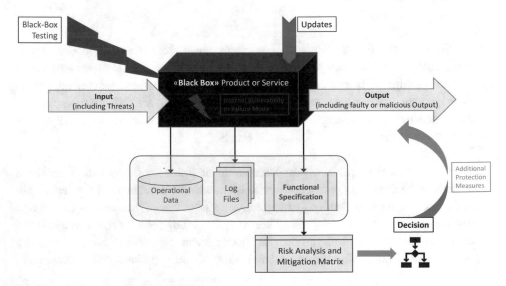

**Fig. 9.17**   Black-box risk management

## 9.11.2 Protective Shell

A proven, additional protection measure in several industries is the *protective shell* (Definition 9.18). To build a protective shell, additional hardware, software, information, and functionality are implemented. The operation of the protective shell should be made as independent from the operational system as possible. The difference to runtime monitoring (Definition 4.13 and 4.30) is the high degree of autonomy and independence of the elements of the protective shell.

▶ **Definition 9.18: Protective Shell**
Independent hardware/software layer overlaid on a system using additional information, such as supplementary sensor data or log files, to detect, analyze, interpret, and mitigate potentially dangerous or unsafe system behavior, either in real time, near-real time, or batch.

Protective shells have myriads of implementation possibilities and very various application fields. Two typical instances are shown in Example 9.12 and 9.13.

---

**Example 9.12: Asymmetric Thrust Safety**

The thrust of its engines propels a plane. The jet engines deliver the right thrust for the respective flight condition. In most flight conditions, the engines' thrust is controlled by the *autothrottle system* of the plane under the control of the autopilot.

The thrusts of multi-engine plane engines must be matched; i.e., they must deliver approximately the same thrust. If the thrusts of the engines substantially differ, the condition is called *asymmetric thrust*. Asymmetric thrust is a dangerous condition for a plane in flight and has been the cause of several severe aviation accidents (e.g., [Akbari21], [Todorova21]). If an asymmetric thrust develops, e.g., because of a performance problem of the engine, the autopilot will try to compensate for the thrust difference as long as possible. When the autopilot reaches the limits of the plane's ability, it disengages and defers to pilot control. Unfortunately, the flight condition of the plane may be precarious at this time, and the pilots may be unable to cope with the sudden, unexpected situation.

**Quote**
*"Asymmetric thrust is recognized as a contributing factor in several propulsion system malfunction plus inappropriate crew response aviation accidents"*
Amy Chicatelli, 2015

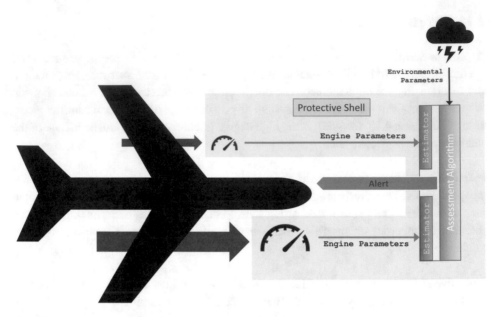

**Fig. 9.18**   Autothrottle protective shell

Flight aviation safety engineers have tried to cope with this situation for years: One solution is implementing a thrust asymmetry detection mechanism (e.g., [Chicatelli15]). This *protective shell* is wrapped around the autothrottle/autopilot system (Fig. 9.18): This additional, overlaid hardware/software-system accepts the parameters of the engines and the environment. The engine thrusts are estimated, e.g., using a Kalman filter ([Bozic18]). Estimated thrust values of all engines and the environmental parameters are evaluated by an assessment algorithm that provides early detection and warning of impending thrust asymmetry.

---

**Example 9.13: Access Control Analytics**

Figure 9.19 shows a *protective shell* for an industrial control system (ICS). Two protective safety mechanisms are emphasized (out of many possibilities):

1. The monitoring of the sensor inputs and the actuator outputs: Additional hardware and software are implemented which supervises the interface to the physical part of the system. On the input side (sensor values), checks such as range, rate of change, and additional rule-based plausibility checks are executed. The actuator signals are checked for anomalies on the output side, such as contradictoriness or out-of-bounds. This technique is imperative, e.g., in *railway infrastructure safety* ([Theeg20]);

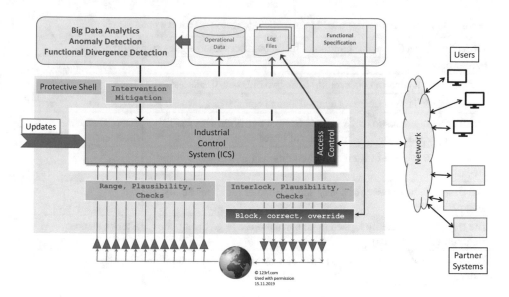

**Fig. 9.19**   ICS protective shell

2. The detection of *unauthorized accesses*: The ICS exchanges data via networks with
   users and partners. Such accesses must be strictly limited to authorized, legitimate
   users. Therefore, a dependable access control system must be implemented. Here,
   the detection of *malicious accesses* using legitimate credentials shall be presented.

The detection and inhibition of malicious accesses to the ICS using legitimate credentials, i.e., *stolen or copied access rights*, is very challenging. It is of great concern
because the longer black accesses last, the higher the damage may be.

> **Quote**
> *"While deploying real-time intrusion detection and prevention defenses is essen-*
> *tial, it is not enough. Analysts need to use creative efforts to uncover new attacks*
> *that successfully circumvent existing defenses"*
>    Mark Ryan M. Talabis, 2015

An approach suggested by [Talabis15] is *big data analysis of the access control logs* ([Talabis15], [Ghavami19]). The logs delivered by the access control system are parsed and analyzed in real time or near-real time. Anomalies are detected

**Table 9.3**  Examples of rules used in the malicious access discovery process

**Objective**: Detection and inhibition of malicious accesses using legitimate credentials

| Perception | Decision Criteria |
|---|---|
| The `UserID` logged in concurrently from two different IP addresses | Time |
| The `UserID` logged in from an out-of-bound country (e.g., Russia) | Black-List |
| The `UserID` was used twice in an hour from two different locations—separated by more than an hour of travel | Time, Haversine distance |
| The `UserID` logged in from a hundred different IP addresses during the last week | # |
| The `UserID` was active for a very short period or for an unusually long period | $\Delta t$ |
| The `UserID` repeatedly logged in at unusual times (e.g., during the night, during weekends) | Clock/calendar |
| The `UserID` does not conform to a stored user location/time/duration pattern | User profile |
| The `UserID` originates from an unusual or suspect IP address | List |
| The `UserID` accesses or tries to access an unusual or prohibited resource within the system | Authorization DB |
| The `UserID` was denied access to internal systems (1 … n times) | Authorization DB |
| The `UserID` shows anomalies in the access patterns, either in time, duration, location, or IP addresses | Time, intervals, duration, … |
| The `UserID` shows anomalies in the access patterns, such as many very short duration accesses | Frequency, time |
| The `UserID` access times out repeatedly | # |
| etc | |

automatically and trigger the immediate revocation of the respective credentials or refer back to a security officer. Some of the rules used in this anomaly discovery process are listed in Table 9.3.

Big data analysis can be done by commercial packages, or the organization can program the analysis rules, e.g., utilizing the programming languages **R** ([Walkowiak16]) or Python ([CSA20]).

### 9.11.3  Black-Box testing

Extensive *black-box testing* is a valuable instrument to detect vulnerabilities in a product/software or service ([Beizer95]). The black box is exposed to all sorts of legitimate and illegitimate, malicious, and random inputs. Typical methods are *penetration testing* (Definition 4.22), fuzzing, or *chaos engineering* (Definition 4.25).

### 9.11.4 Supplier Risk Management Review/Audit

The use of third-party products or services generates crucial dependencies: Vital parts of safety and security may be the supplier's responsibility. The organization, therefore, has to rely on the suppliers' risk management ability.

Building and maintaining confidence in the supplier's capability to identify, recognize, assess, and correctly mitigate risks requires understanding the supplier's policies and processes. The means to achieve this are *reviews* and *audits*.

First, the capabilities of the third-party *provider* must be assessed. Once the provider has been chosen, the *products* or *services* to be acquired must be individually assessed. Finally, a comprehensive *exit strategy* must be developed for the case that the cooperation with the provider needs to be terminated (e.g., [Tschanz18], [IIA18], [Brines20]).

---

## 9.12   G10a: Supply Chain Risk Management

Supplier chains introduce *risks*. Many of these risks are hidden in the chain, i.e., in the products and processes of the third-party providers and their subproviders.

> **Quote**
> *"Organizations leverage and rely on third-party providers, as well as subservice or 'fourth-party' providers, to conduct business activities. These relationships continue to expand and evolve, introducing numerous risks that must be continuously assessed and appropriately managed by the organization to achieve desired business outcomes"*
> The Institute of Internal Auditors (IIA), 2018

Supply chain risk management has been intensively studied for logistics chains. The massive damage potential of *software supply chain attacks* (Definition 9.19) has been recognized in the last years (e.g., [NCSC21a], [NCSC21b]).

▶ **Definition 9.19: Software Supply Chain Attack**
Compromising software through cyber-attacks, insider threats, or other malicious activities at any stage throughout its entire lifecycle
   UK National Cyber Security Centre, 2021

Many processes, methodologies, and frameworks have been developed to address these dangers ([Boyens12], [Boyens15], [Boyens21a], [Boyens21b], [Boyson14], [Brines20], [Christopher16]). The key findings are listed in Principle 9.20. The critical element of a provider relationship are *adequate contracts* ([Sammons17], [Tollen21], [Burden19],

[Burden21], [Klinger13], [Desai09], [Desai10]). Stipulating comprehensive, enforceable clauses for liability is essential ([Fairgrieve20], [Hunter18], [Krauss19]).

---

**Principle 9.20: Supply Chain Risk Management**

<u>"Make or Buy" Decision</u>

1. Carefully evaluate, document, and assess all "make" (in-house) or "buy" third-party provider) options. Note that third-party providers may have hidden risk or undeclared costs that are not obvious during the evaluation;

<u>Supplier Assessment</u>

1. Install and consistently use a Cyber Supply Chain Risk Management System (CSCRM) in the organization;
2. Execute regular review/audits to understand and evaluate a supplier's capability to identify, recognize, assess, and correctly mitigate risks, including an understanding of the supplier's policies and processes. Use a competent chartered company if necessary. Require improvements if necessary;
3. Apply a specific risk identification, assessment, evaluation, mitigation process to the provider's capabilities;
4. Maintain a current inventory of all third-party products, including their relationships, data inventory, and revision history;
5. Establish an undoubtful, comprehensive, and complete commitment as to which risks are mitigated by the organization and are which mitigated by the provider (Note: Objective is an acceptable residual risk in all cases);
6. Understand, assess, document, and explicitly accept the accident handling and incident response procedures of the provider;
7. Understand, assess, document, and explicitly accept the provider's update procedures (corrective, adaptive, and predictive). Especially consider the security of the update channels (authorization, code integrity, etc.);
8. Explicitly list, document, and accept (by the appropriate management level) the residual provider risks;
9. Ensure that all legal, compliance and applicable standards are adhered to by the provider and its products or services;
10. Adequate contracts are crucial. Negotiate the contracts with providers carefully and with proper legal assistance;
11. The contracts shall stipulate comprehensive, enforceable clauses for liability, e.g., in case of safety accidents and security incidents;
12. Request certified compliance to appropriate standards whenever possible;
13. Limit external and internal connections and URL's to such on an approved list for each third-party software deployment;

14. Coordinate and link the suppliers contingency handling policies & procedures with the organization's own.

Exit Strategy

1. Prepare, document, and periodically revise the exit strategy in case of a necessary replacement of the provider;
2. The contracts shall declare all steps and the timing to terminate the relationship, both in cases of an agreed separation and of a disputed termination;
3. Ensure the legal proprietary rights to all information related to the organization and (possibly) stored in the provider's IT system.

## 9.13   G10b: Supply Chain Confidence: Products

The first cyber-supply chain attack vector is integrating *third-party products* into the organization's IT system (Fig. 9.16, Example 2.16 and 9.14). Such products may introduce unwanted, malicious, or destructive functionality into any layer of the IT system (Fig. 5.5).

---

**Example 9.14: Pulse Connect Secure**

The exploitation of Pulse Connect Secure Vulnerabilities (Quoted from: https://us-cert.cisa.gov/ncas/alerts/aa21-110a)

*"The Cybersecurity and Infrastructure Security Agency (CISA, https://us-cert.cisa. gov) is aware of compromises affecting a number of U.S. government agencies, critical infrastructure entities, and other private sector organizations by a cyber threat actor—or actors—beginning in June 2020 or earlier related to vulnerabilities in certain Ivanti Pulse Connect Secure products (https://www.ivanti.com).*

*Since March 31, 2021, CISA and Ivanti have assisted multiple entities whose vulnerable Pulse Connect Secure products have been exploited by a cyber threat actor. These entities confirmed the malicious activity after running the Pulse Secure Connect Integrity Tool. To gain initial access, the threat actor is leveraging multiple vulnerabilities, including CVE-2019-11510, CVE-2020-8260, CVE-2020-8243, and the newly disclosed CVE-2021-22893 (https://cve.mitre.org). The threat actor is using this access to place webshells (https://www.rsa.com/content/dam/en/solution-brief/asoc-threat-solution-series-webshells.pdf) on the Pulse Connect Secure appliance for further access and persistence. The known webshells allow for a variety of functions, including authentication bypass, multi-factor authentication bypass, password logging, and persistence through patching".*

DHS/CISA (US Cybersecurity and Infrastructure Agency), Washington, DC, USA. Alert AA21-110A, Original release date: April 20, 2021 | Last revised: May 28, 2021)

> **Quote**
> *"Customers often accept third-party software defaults without investigating further, allowing additional accessibility vectors"*
>    CISA, 2021

Handling of the product supply chain is suggested in Principle 9.21.

### Principle 9.21: Supply Chain Confidence: Products

1. Write and enforce a third-party product acquisition and integration policy, including the specific risk identification/assessment/evaluation/mitigation/acceptance procedures;

a. <u>Supplier Assessment</u>
   See Principle 9.20

b. <u>Product Assessment</u>

2. Evaluate the quality properties of the envisaged product with the same due diligence as the functionality. Do not accept any uncommitted assurances (contract);

3. Execute a thorough risk analysis for the integration and usage of the envisaged product. Use information available in the market or public networks. Ask for references and query them;

4. Carefully investigate the access privileges of the third-party products. Be especially distrustful if privileged accesses are claimed. Restrict the access rights to the absolute minimum necessary for correct operation. Ensure this restriction immediately for every new version of the third-party product;

5. Repeat and document (2)–(4) for each major release of the product;

<u>Trustworthy Product Integration</u>

6. Strictly disable all unnecessary or unused functionality of the product;

7. Execute extensive black-box testing

<u>Operational Monitoring</u>

8. Comprehensively monitor and supervise the third-party product during operation. Have intervention mechanisms implemented for the case of misbehavior;

<u>Exit Strategy</u>

9. Prepare an exit strategy for the case of grave disputes with the vendor, deficits of the product (or its updates), or technological obsoleteness;

10. Cover such eventualities in the acquisition contract.

## 9.14   G10c: Supply Chain Confidence: Services (Outsourcing)

*Outsourcing* software/systems development to a third party can be a significant source of severe *risk infiltration* (Definition 9.20 and 2.20, Example 2.15, e.g., [Mezak06], [Ebert11], [Dumitrascu21]).

▶ **Definition 9.20: Outsourcing**
A business's lasting and result-oriented relationship with a supplier who executes business activities for an enterprise which were traditionally executed inside the enterprise, such as software development and IT-related services.
   Adapted from: Christof Ebert, 2012.

The principal risks of outsourcing software development are:

1. *Accidental insertion* of vulnerabilities (opening attack vectors or introduction of safety weaknesses);
2. *Intentional insertion* of vulnerabilities, such as back doors, logic bombs, and side channels. This is a cyber-criminal activity;
3. *Improper integration* into the existing IT system: The outsourced software cannot dependably be integrated into the existing software-system or IT infrastructure, e.g., bypassing existing authentication or authorization procedures. The outsourced software is unfit for protective mechanisms, such as a protective shell or runtime monitoring.

> **Quote**
> *"When everything goes right, you end up with high-quality software in half the time for a fraction of the cost. But over 50% of offshore outsourcing projects do not achieve their cost-saving goals or timelines... just fail completely or introduce inacceptable risks"*
>    Steve Mezak, 2006

For all outsourcing activities, a decisive shift from products to processes is mandatory: The processes for all activities, such as requirements management, acceptance procedures, and change management, must be completely, seamlessly, and effectively defined and implemented. In many such cases, accepted process quality models, such as CMMI (e.g., [Chrissis11], [Hofmann07]) are mandatory.
   One critical risk in the distributed, outsourcing mode of work is the *disintegration of architecture*: Due to the independent, loosely coordinated, and differently governed work, the all-important architecture (Fig. 5.5) cannot be maintained and leads to *architecture erosion* (Definition 9.2) and *technical debt accumulation* (Definition 9.1). The

project organization must be such that a very strong, central architecture governance is assured—with an influential and well-supported chief architect.

The pivot point between the enterprise and the supplier is the service level agreement (SLA, [Burnett09], [White17]). The quality of the SLA determines both the success of the cooperation and the desired properties of the outsourced software. Some guidance for outsourcing is listed in Principle 9.22.

---

**Quote**

*"The two most critical roles in any development project are the project manager and the chief architect. In a global software and IT project, these roles are pivotal and should be both well-trained and well-supported"*
    Christof Ebert, 2012

---

**Principle 9.22: Supply Chain Confidence: Services (Outsourcing)**

1. Write and enforce an outsourcing policy, including the specific risk identification/ assessment/evaluation/mitigation/acceptance procedures;

Supplier Assessment

2. See Principle 9.20;

Processes

3. Ensure that the processes between the enterprise and the commissioned software developer are seamless, complete, compatible, and auditable;
4. Use an industry-accepted process quality assessment, e.g., CMMI, for the alignment of the processes;
5. Whenever use process and product metrics to assess outcomes;
6. Ascertain that the requirements specifications are complete and unambiguous both for the functional requirements and the quality attributes (safety, security). Include the specification for satisfactory test cases;
7. Reach an agreement with which industry standards the outsourced work must comply, especially safety, security, and certification standards:
8. Install processes to guarantee the quality of the architecture to avoid its disintegration, architecture erosion, and the accumulation of technical debt. Charge an experienced, assertive, knowledgeable chief architect with a small team with this task. Provide central oversight of the architecture so that it will not deteriorate as lower-level designs are decided by remote teams;

9. Build sufficient quality gates into the workflow to ensure the software quality properties, especially safety and security considerations. Never allow to skip or shorten the quality gate tasks;

Integration

10. Use the entire in-house integration test chain, especially the test cases related to the quality attributes (safety, security);

Operational Monitoring

11. Comprehensively monitor and supervise the third-party software during operation. Have intervention mechanisms implemented for the case of misbehavior;

Exit Strategy

12. Prepare an exit strategy for the case of severance from the supplier. Continuously and timely ensure that all artifacts, such as source code and configuration information are fully documented and archived in the enterprise;
13. Cover such eventualities in the service level contract (SLA), including the intellectual property rights (IPR).

## 9.15   G10d: Supply Chain Confidence: Open-Source Software

Indiscriminately using *open-source software*—especially downloading free code from Internet sources—is a very severe risk. Threat actors can insert malicious code into this publicly available resource (Example 9.15, [Ahlawat21], [CISA21]). Recommendations for using open-source resources are listed in Principle 9.23. Essential is a comprehensive, well-communicated, and enforced *open-source policy* (Example 9.16, e.g., [Gatto17], [OpenLogic20]).

> **Example 9.15: Malicious Python Open-Source Libraries**

Thomas Claburn, March 2021:

"*Python Package Index nukes 3,653 malicious libraries uploaded soon after security shortcoming highlighted.*

*Unauthorized versions of CuPy and other projects flood PyPI.*

*The Python Package Index, also known as PyPI, has removed 3,653 malicious packages uploaded days after a security weakness in the use of private and public registries was highlighted.*"

Quote from: https://www.theregister.com/2021/03/02/python_pypi_purges/

> **Quote**
> *"Open-source code compromises can also affect privately owned software because developers of proprietary code routinely leverage blocks of open source in their products"*
>   CISA, 2021

---

**Example 9.16: Open-Source Core Policy**

The core of the engineering part of a good open-source usage policy includes six simple rules:

I.   Engineers must receive approval from the open-source review board (OSRB) before integrating any open-source code into a product;
II.  The software received from third parties must be audited to identify any open-source code included, which ensures license obligations can be fulfilled before product ships;
III. All software must be audited and reviewed, including all proprietary software components;
IV.  Products must fulfill open-source licensing obligations prior to customer receipt. The legal exposure must be understood and accepted. The Intellectual Property Rights (IPR) situation must be checked and respected;
V.   Approval for using a given open-source component in one product is not approved for another deployment, even if the open-source component is the same;
VI.  All changed components must go through the approval process.

Source: Adapted from https://www.linuxfoundation.org/resources/open-source-guides/using-open-source-code

Principle 9.23 shows the information with respect to open-source software.

---

**Principle 9.23: Supply Chain Confidence: Open-Source Software**

1. Rate the trustworthiness of the open-source provider. Maintain a white list of acceptable open-source providers;
2. Develop, maintain, and strictly enforce an open-source usage policy;
3. Register all open-source software in the Cyber Supply Chain Risk Management System (CSCRM) of the organization;
4. Make all open-source modules subjects in the configuration management system of the organization. Maintain versioning;

5. Scan each new version of an open-source module (code quality, vulnerabilities, etc.);
6. Archive all versions of open-source software;
7. Install an open-source review board (OSRB) which must review/assess and approve any open-source usage and deployment, also for small modules obtained via the Internet or supplier websites;
8. Provably ensure that all open-source modules used are legally acquired (license agreements? License fees? Rogue/illegally copied?);
9. Ensure that intended use of the open-source software complies with the licensing agreement, especially with the legal usage rights, the use of intellectual property, and distribution rights;
10. For potentially safety- or security-relevant open-source software, execute a specific risk identification/assessment/evaluation/mitigation/acceptance process;
11. Define under which conditions contributions to open-source platforms by employees of the organization are allowed.

> **Quote**
> *"The exponential growth of open-source projects has increased the potential attack surface and made auditing code a greater challenge"*
> UK National Cyber Security Centre (NCSC), London, UK, 2021

## 9.16 Principle G11: Trustworthy Development Process

Trustworthy cyber-physical systems are the result of a *trustworthy development process* (Definition 5.3). The trustworthy development process is safety- and security-aware, highly formalized, and avoids/eliminates code defects (Figs. 5.7 and 5.9).

Proven resilience management, models, and development processes are available (e.g., [Caralli11], [Möller16]). The organization's development process should adopt as many methods as possible from such proven reference processes.

The development process methods and technologies are not static: New, more efficient, or faster processes are introduced every year. While some are just transient, some are come to stay. Recent examples include *Agile methods* and *DevOps*. Many of these newcomers were devised for shortening development time or reducing development costs without a strong focus on safety and security. Fortunately, the safety and security concerns were added later, e.g., SafeScrum™ ([Hanssen18]) and DevSecOps ([Wilson20]).

Artificial intelligence and machine learning are ready to support the trustworthy development process, e.g., for testing, automated code reviews, and DevOps (e.g., [Kinsbruner20]),

The key characteristics of a trustworthy development process are summarized in Principle 9.24.

---

**Principle 9.24: Trustworthy Development Process**

General:

1. Use a proven, well-documented, and enforceable development process model;
2. Define and apply suitable, long-term process metrics to manage the key deliverables of the development process;
3. Install, execute, and enforce sufficient quality control gate (reviews, audits, etc.);
4. Make the development process exhaustively conformant to all legal, compliance, and certification requirements, including the necessary documentation;
5. Regularly revise, audit, and adapt the development process to new methods, tools, technologies, standards, and legal & compliance requirements;
6. Whenever a new process methodology is adopted, such as Agile or DevOps, extend it to guarantee and demonstrate the required safety and security properties (e.g., SafeScrum™, DevSecOps);

Preparation:

7. Carefully identify all safety and security requirements and specify them completely, comprehensively, and unambiguously as traceable, verifiable requirements;
8. Model the systems/subsystems, relationships, and data with an adequate degree of coverage, detail, and formalization;
9. Align the organization's management/decision structure to the development process. Define, implement, and check the performance of the deciders;
10. Insist on an adequate architecture, including the separation of concerns;
11. For all architectural and design decisions, weigh safety and security requirements higher than functional requirements. If trade-offs are necessary, carefully risk-assess them;

Execution:

12. Use the best available tools for all tasks. Ensure the conceptual integrity of all artifacts at all times (may require reverse-engineering);
13. Enforce state-of-the-art tracing and verification of all quality attributes (code checkers, extensive and specialized testing, etc.);
14. Make resilience in the development process a focus. Integrate as many methods as possible from resilience engineering (e.g., [Cavalli11]);
15. Use and maintain a metric-based resilience maturity management model;

16. Termination (application end-of-life): Carefully retire all applications, utilities, databases, and support information as fast as possible after they are no longer required;

17. Before explementing, identify and document all code, data, relationships, etc., belonging to the application to be retired;

18. Before explementing, transfer the current version of all artifacts belonging to the retiring application to the secure archive;

19. Thoroughly make sure that no code, data, or support information of the application to be retired is used by any user, partner, or other application;

20. Completely remove all code, data, and support information of the retired application from the IT system;

21. Refactor (simplifying) all databases affected by explemented data fields or schemas;

22. Update all documentation which refers to the retired application;

23. Periodically execute system scans and reviews to identify unnecessary code and data and explement them.

24. Make use of artificial intelligence/machine learning tools as soon as they become available.

## 9.17   G12: IoT Systems

The *Internet of Things* (IoT, Definition 2.9) and the *industrial control systems* (ICS, Definition 2.6) are two evident and impactful representations of cyber-physical systems. Both types of systems will take over and control more and more of our technological world (e.g., [Höller14], [Raj17]).

> **Quote**
> *"The power of false data injection attacks is high, when considering that sensors are typically in the field, they can get under the control of attackers, and their interface with the physical plant can be manipulated"*
> Marilyn Wolf, 2020

In addition to the previously considered attack paths, IoT and ICS provide new, vulnerable targets: the *sensors* and *actuators*—or, more generally, the *IoT devices* (Fig. 9.20). An attacker can:

1. Physically falsify a measurement, e.g., by placing a heat source near a temperature sensor;

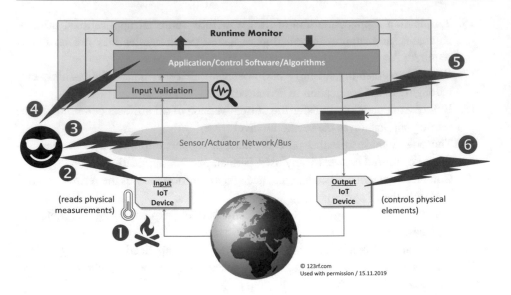

**Fig. 9.20** IoT system

2. Attack and manipulate the sensor or input IoT device (many modern sensors, actuators, and IoT devices have microcontrollers and software);
3. Attack and forge the data transmitted via the communication channel (e.g., the sensor/actuator bus or the IoT network) to the control system;
4. Attack the computer/software-controlled control system, introducing, e.g., malware or modifying the sensor input data;
5. Attack the computer/software-controlled control system, introducing, e.g., malware or modifying the actuator output data;
6. Attack and manipulate the actuator or output IoT device.

In all six cases, either fake input data or incorrect output data will lead to an erroneous—potentially dangerous—behavior of the IoT system or the ICS.

A very instructive instance for such a situation is the *Stuxnet* malware (Example 2.5, e.g., [Baezner17]). By issuing harmful actuator signals to the uranium-enrichment centrifuges, the commands destroyed a large number of the centrifuges—while showing regular operation on the ICS screens.

The protection of the IoT and ICS systems needs *additional safeguards* (Fig. 9.20, e.g., [Ashok17], [Wolf20]):

1. *Runtime monitoring* (Definition 4.13);
2. *Securing* the communications channels between sensor/actuators/IoT device and the control system (including cryptography and bilateral endpoint authentication);

3. *Validation* of the input values: A data monitor checks the incoming data and decides whether the data is acceptable or not. Methods for false data injection identification range from simple statistical outlier detection to sophisticated time series analysis, e.g., using Kalman filters ([Wolf20], [Ma20], [Bozic18]);
4. *Interception* of unplausible output values: Detect anomalies or perilous output commands or values: Runtime monitoring must detect and block potentially harmful outputs, again either by simple statistical outlier detection or time series analysis;

Unfortunately, many sensors, actuators, or IoT devices have very limited resources (energy, processing power, memory size, etc., Table 11.5). Therefore, the use of cryptographic algorithms for protection and authentication may be difficult or even impossible ([Gupta19], [Chantzis21]). In addition, the current attitude of many IoT device vendors massively lacks security awareness.

---

**Principle 9.25: IoT Systems**

1. Protect the control system and its software by all known means;
2. Execute a risk analysis to determine the danger potential of false data injection for all inputs and outputs. Define criticality classes and stipulate their minimum protection requirements;
3. Validate all input data by using an adequate validation algorithm;
4. Use a runtime monitor to detect anomalies during operation. If possible, detect and handle any anomalous situation automatically;
5. Validate all output data by using an adequate validation algorithm, including anomaly detection. Intercept non-credible data;
6. Secure all channels between the sensor/actuators/IoT devices by cryptography and authentication of both endpoints. Use at least the NIST lightweight cryptographic algorithms;
7. Sufficiently control access to the IoT data or functionality (authentication & authorization);
8. Disable all channels/ports of the IoT device which are not required for the operation (also, if possible, maintenance and configuration channels);
9. Extensively test the IoT devices (especially third-party devices). Use all modern, preferably highly automated penetration and false data injection methods;
10. Carefully and timely update/install any software change suggested by the manufacturer of the IoT device;
11. Ensure the integrity of the software during power-up, operation, and downloads;
12. Assess the security awareness of the vendor and the security mechanisms implemented by the vendor. If dubious, eliminate the vendor;
13. Implement as much logging information about the operation of the IoT devices. Get it from the IoT device if possible. Analyze the data periodically to detect anomalies.

## 9.18    G13: Impact of Artificial Intelligence

*Artificial Intelligence* (AI) will have—and already has—a tremendous impact on the safety and security of cyber-physical systems (e.g., [Jajodia20]).

AI has both positive and negative effects on *safety*. Positive effects on safety include intelligent monitoring to detect failures or malfunctions early, pre-emptive maintenance to avoid failures before they occur, smart functionality to prevent accidents, and many more. Adverse effects on safety include new types of AI failure modes, confusing complexity, incomprehensible decision algorithms, and more.

AI also has positive and negative effects on *security*—both on the attacker and the defender side. The attacker uses AI to find new, complex vulnerabilities (or aggregated chains of attack), speed up capitalizing on zero-day exploits, identify high-value targets, and many more. On the other hand, the defender deploys AI to detect unknown attacks, hardening the IT systems efficiently (e.g., intelligent penetration testing), adaptive defense strategies, and many more.

AI also can deliver valuable support of the *development process for cyber-physical systems*, i.e., in CPS software engineering (e.g., [Meziane10], [Pawar16], [Barenkamp20]). AI will contribute to the *requirements engineering phase* by disambiguating natural language requirements, providing knowledge-based systems and automated ontologies, detecting inconsistencies and contradictions, mapping to the correct domain, and tracing requirements. Valuable progress will be made in AI-supported *code generation* and code checking. AI-based tools are emerging for automated software testing, such as generating test data and test cases and estimating test coverage. AI is also currently used in estimating planning and project effort estimation, including optimized assignment of resources.

Some AI techniques have recently been introduced to facilitate the *risk management process* (e.g., [FERMA20], [D'Hoinne20], [Shackleford21]). They assist to, e.g.,

- Uncover *unknown risks* by using big data analysis on network traffic, user behavior, interface data exchanges, etc.;
- Mitigate *emergent risks* by detecting anomalies in logs, user conduct, log-in patterns, geographic analysis, etc.;
- Support the *risk management process* by introducing formal methods, precisely stating requirements, detecting inconsistencies and contradictions, etc.;

**Quote**
*"Sometimes You're the Cat and sometimes You're the Mouse: Attackers Like AI too"*
    Gartner Inc., 2020

The introduction of ML applications into an existing system needs a specific process. A suitable process for the operationalization of ML applications is *MLOps* (Machine Learning Operations, e.g., [Raj21], [Lauchande21], [Gift21]).

"Classical" programs are stable; i.e., they do not change their behavior during operation: This may not be the case with ML-based systems. Because ML systems may continue to learn and adapt during operation, their behavior may gradually change over time. Such gradual change may lead to undesired, unfavorable reactions—and thus to operational risks. Especially, the models may drift or become biased. Therefore, *operational monitoring* of the ML system is required ([Raj21], [Bartocci18], [Goodloe10], [ModelOp21]).

Last but not least, note that the introduction of ML applications can generate considerable *technical debt* ([Breck17], [Sculley18], [Ernst21]). Adequate steps must be taken during all phases of the development process to avoid and eliminate technical debt (Principle 9.3, [Kruchten19], [Lilienthal19], [Suryanarayana14], [Tornhill18]).

The life cycle of machine learning-based applications is different from classical applications and needs an adapted process—MLOps (related to DevOps). The MLOps process ([Raj21]) is highly automated and includes building, testing, and managing ML-based software.

Some guidance for the use of artificial intelligence is listed in Principle 9.26.

---

**Principle 9.26: Impact of Artificial Intelligence**

1. Carefully research the possible use of artificial intelligence in the organization for:
   a. The safety-critical products (Principle 10.10);
   b. The security-critical products (Principle 11.18);
   c. The cyber-physical systems development process;
   d. The risk management;
   e. The CPS operation.
2. Thoroughly evaluate, assess, and mitigate the risk introduced by the use of AI in products and processes;
3. If beneficial, introduce AI-based technologies and methods in products and processes;
4. Devise, implement, and maintain an adequate system architecture—both functional and with respect to safety and security;
5. Reduce the number of interconnections in the CPS to the minimum. Eliminate all communication which is not strictly needed;
6. Use an adequate process for the operationalization of ML applications, such as MLOps;
7. Use operational model monitoring to detect unfavorable model behavior, such as model drift, model bias, or model compliance;
8. Avoid the generation of technical debt during AI projects (especially ML projects);
9. Use AI-based big data analytics (online or offline) to detect anomalies or malicious patterns ([Talabis15]).

# References

| [Abadi09] | ]Martin Abadi, Mihai Budui, Ulfar Erlingsson, Jay Ligatti: *Control-Flow Integrity – Principles, Implementations, and Applications* ACM Transactions on Information System Security, Vol. 13, 2009, pp. 1–40. Downloadable from: https://www.cse.usf.edu/~ligatti/papers/cfi-tissec.pdf [Last accessed: 23.7.2021] |
|---|---|
| [Abd-Allah97] | Ahmed Abd-Allah: *Extending Reliability Block Diagrams to Software Architectures* Technical Report: USC-CSE-97-501, University of Southern California, Los Angeles, CA, USA, 1997. Downloadable from: http://citeseerx.ist.psu.edu/viewdoc/download?doi=10.1.1.365.9394&rep=rep1&type=pdf [Last accessed: 3.10.2021] |
| [Ahlawat21] | Pranay Ahlawat, Johannes Boyne, Dominik Herz, Florian Schmieg, Michael Stephan: *Why You Need an Open Source Software Strategy* BCG (Boston Consulting Group), Boston, MA, USA, April 16, 2021. Accessible at: https://www.bcg.com/publications/2021/open-source-software-strategy-benefits [Last accessed: 22.6.2021] |
| [Akbari21] | Ilyas Akbari: *The Fate of Sriwijaya Air Flight 182: A Reminder of Aviation Safety* Jurist – Legal News & Commentary, University of Pittsburgh, Pittsburgh, PA, USA, February 24, 2021. Downloadable from: https://www.jurist.org/commentary/2021/02/ilyas-akbari-sriwijaya-182-aviation-safety/ [last accessed 15.7.2021] |
| [Arthur15] | Will Arthur, David Challener: **A Practical Guide to TPM 2.0** – *Using the Trusted Platform Module in the New Age of Security* Apress Media, LLC, New York, NY, USA, 2015. ISBN 978-1-4302-6583-2. Downloadable from (Apress Open): https://www.springer.com/de/book/9781430265832 [last accessed: 25.10.2021] |
| [Ashok17] | Aditya Ashok, Manimaan Govindaras, Jianhui Wang: **Cyber-Physical Attack-Resilient Wide-Area Monitoring, Protection and Control for the Power Grid** Invited Paper, Proceedings of the IEEE, Vol. 105, No. 7, July 2017. New York, NY, USA, 2017 |
| [Avgeriou21] | Paris Avgeriou et al.: *An Overview and Comparison of Technical Debt Measurement Tools* Preprint, IEEE Software, May 2021, IEEE New York, NY, USA. Downloadable from: https://www.researchgate.net/publication/344151709_An_Overview_and_Comparison_of_Technical_Debt_Measurement_Tools/link/5f57a322458515e96d39d489/download [Last accessed: 14.11.2021] |
| [Baezner17] | Marie Baezner, Patrice Robin: *CSS CYBER DEFENSE PROJECT – Hotspot Analysis: Stuxnet* Risk and Resilience Team Center for Security Studies (CSS), ETH Zürich, Zurich, Switzerland, October 2017. Downloadable from: https://css.ethz.ch/content/dam/ethz/special-interest/gess/cis/center-for-securities-studies/pdfs/Cyber-Reports-2017-04.pdf [Last accessed: 23.01.2022] |
| [Baranov18] | Samary Baranov: **Finite State Machines and Algorithmic State Machines** – *Fast and Simple Design of Complex Finite State Machines* Independently published, 2018. ISBN 978-1-7750-9172-1 |
| [Barenkamp20] | Marco Barenkamp, Jonas Rebstadt, Oliver Thomas: **Applications of AI in classical Software Engineering** Research Article, Springer Open, Heidelberg, Germany, 2020. Downloadable from: https://aiperspectives. |

springeropen.com/track/pdf/10.1186/s42467-020-00005-4.pdf     [Last accessed: 29.10.2021]

[Bartocci18]     Ezi Bartocci, Jyotirmoy Deshmukh, Alexandre Donzé, Georgios Fainekos, Oded Maler, Dejan Ničković: **Specification-Based Monitoring of Cyber-Physical Systems** – *A Survey on Theory, Tools, and Applications* In: Bartocci E., Falcone Y. (Editors): Lectures on Runtime Verification. Lecture Notes in Computer Science, Vol 10457. Springer International Publishing, Cham, Switzerland, 2018. ISBN 978-3-319-75632-5

[Bass13]     Len Bass, Paul Clements, Rick Katzman: **Software Architecture In Practice** Pearson Education, SEI-Series, (Addison-Wesley), Upper Saddle River, NJ, USA, 3rd edition, 2013. ISBN 978-9-332-50230-7

[Bastos19]     Joel Bastos, Pedro Araujo: **Hands-On Infrastructure Monitoring with Prometheus** – *Implement and scale Queries, Dashboards, and alerting across Machines and Containers* Packt Publishing, Birmingham, UK, 2019. ISBN 978-1-7896-1234-9

[Beine21]     Gerritt Beine: **Technical Debts** – *Economizing Agile Software Architecture* De Gruyter Oldenbourg, Oldenbourg, Germany, 2021. ISBN 978-3-110-46299-9

[Beizer95]     Boris Beizer: **Black Box Testing** – *Techniques for Functional Testing of Software and Systems* John Wiley & Sons Inc., New York, NY, USA, 1995. ISBN 978-0-471-12094-0

[Beyer16]     Betsy Beyer, Chris Jones, Jennifer Petoff, Niall Richard Murphy: **Site Reliability Engineering** O'Reilly Media, Inc., Sebastopol, CA, USA, 2016. ISBN 978-1-491-92912-4

[Blanke15]     Mogens Blanke, Michel Kinnaert, Jan Lunze Marcel Staroswiecki: **Diagnosis and Fault-Tolerant Control** Springer Verlag, Berlin, Germany, 3rd edition, 2015. ISBN 978-3-662-47942-1

[Booth14]     Harold Booth, Andrew Regenscheid: **Reference Certificate Policy** Second Draft NISTIR 7924, US National Institute of Standards and Technology, Gaithersburg, MD, USA, May 2014. Downloadable from: https://csrc.nist.gov/CSRC/media/Publications/nistir/7924/draft/documents/nistir_7924_2nd_draft.pdf [Last accessed: 25.7.2021]

[Börger03]     Egon Börger, Robert Stärk: **Abstract State Machines** – *A Method for High-Level System Design and Analysis* Springer Verlag, Berlin, Germany, 2003. ISBN 978-3-540-00702-9

[Boyens12]     Jon Boyens, Celia Paulsen, Nadya Bartol, Rama Moorthy, Stephanie Shankles **Notional Supply Chain Risk Management Practices for Federal Information Systems** NISTIR 7622, NIST, Gaithersburg, MD, USA, May 2012. Downloadable from: https://nvlpubs.nist.gov/nistpubs/ir/2012/NIST.IR.7622.pdf [Last accessed: 22.6.2021]

[Boyens15]     Jon Boyens, Celia Paulsen, Rama Moorthy, Nadya Bartol: **Supply Chain Risk Management Practices for Federal Information Systems and Organizations** NIST Special Publication 800-161, NIST, Gaithersburg, MD, USA, 2015. Downloadable from: https://nvlpubs.nist.gov/nistpubs/SpecialPublications/NIST.SP.800-161.pdf [Last accessed: 15.2.2021]

[Boyens21a]     Jan Boyens: *NIST Cyber Supply Chain Risk Management – Fact Sheet* NIST, Gaithersburg, MD, USA, May 2021. Downloadable from: https://csrc.nist.gov/CSRC/media/Projects/cyber-supply-chain-risk-management/documents/C-SCRM_Fact_Sheet_Draft_May_25.pdf [Last accessed: 22.6.2021]

[Boyens21b]        Jon Boyens, Celia Paulsen, Nadya Bartol, Kris Winkler, James Gimbi: **Key Practices in Cyber Supply Chain Risk Management** – *Observations from Industry* NISTIR 8276, NIST, Gaithersburg, MD, USA, May 2021. Downloadable from: https://nvlpubs.nist.gov/nistpubs/ir/2021/NIST.IR.8276.pdf [Last accessed: 22.6.2021]

[Boyson14]         Sandor Boyson: *Cyber Supply Chain Risk Management – Revolutionizing the Strategic Control of critical IT Systems* Technovation (2014), Elsevier, Amsterdam, NL, 2014. Downloadable from: http://cjat.ir/images/PDF_English/20251.pdf [Last accessed: 22.6.2021]

[Bozic18]          S M Bozic: **Digital and Kalman Filtering** – *An Introduction to Discrete-Time Filtering and Optimum Linear Estimation* Dover Books, Mineola, New York, NY, USA, 2nd edition, 2018. ISBN 978-0-486-81735-4

[Breck17]          Eric Breck, Shanqing Cai, Eric Nielsen, Michael Salib, D. Sculley: **The ML Test Score** – *A Rubric for ML Production Readiness and Technical Debt Reduction* Proceedings of IEEE Big Data, Institute of Electrical and Electronics Engineers, New York, NY, USA, 2017. Downloadable from: https://research.google/pubs/pub46555/ [Last accessed: 4.11.2021]

[Brines20]         Miguel Brines: **Keys to Procurement Management Collection** Part I: *Introduction to Supplier Selection, Evaluation and Risk Management from suppliers in the Supply Chain and Purchasing* Independently published, 2020. ISBN 979-8-55690-445-3

[Burden19]         Kit Burden, Mark O'Conor, Duncan Pithouse: **Negotiating Technology Contracts** Global Law and Business Ltd., Horsell, UK, 2019. ISBN 978-1-7874-2322-0

[Burden21]         Kit Burden (Editor): **Outsourcing** – *A Practical Guide* Global Law and Business Ltd., Horsell, UK, 2021. ISBN 978-1-7874-243-0

[Burnett09]        Rachel Burnett: **Outsourcing IT** – *The Legal Aspects: Planning, Contracting, Managing, and the Law* Routledge (Taylor & Francis), London, UK, 2nd edition, 2009. ISBN 978-0-566-08597-0

[Butler10]         Michael Butler, Cliff B. Jones, Alexander Romanovsky, Elena Troubitsyna: **Methods, Models and Tools for Fault Tolerance** Springer-Verlag, Berlin, Germany (Lecture Notes in Computer Science, Band 5454), 2010. ISBN 978-3-642-00866-5

[Buttazzo11]       Giorgio C Buttazzo: **Hard Real-Time Computing Systems** – *Predictable Scheduling Algorithms and Applications* Springer Science + Business Media, New York, N, USA, 3rd edition, 2011. ISBN 978-1-461-40675-4

[Campbell17]       Laine Campbell, Charity Majors: **Database Reliability Engineering** – *Designing and Operating Resilient Database Systems* O'Reilly Media, Inc., Sebastopol, CA, USA, 2017. ISBN 978-1-491-92594-2

[Caralli11]        Richard A. Caralli, Julia H. Allen, David W. White: **CERT Resilience Management Model (CERT-RMM)** – *A Maturity Model for Managing Operational Resilience* SEI Series in Software Engineering, Pearson Education (Addison-Wesley Educational Publishers Inc.), Boston, MA, USA, 2011. ISBN 978-0-321-71243-1

[Carreira20]       Paulo Carreira, Vasco Amaral, Hans Vangheluwe (Editors): **Foundations of Multi-Paradigm Modelling for Cyber-Physical Systems** Springer Nature Switzerland AG, Cham, Switzerland, 2020. ISBN 978-3-030-43945-3. Downloadable from (Springer Open): https://link.springer.com/content/pdf/10.1007%2F978-3-030-43946-0.pdf [Last accessed: 15.11.2021]

| | |
|---|---|
| [CASC13] | CASC: **Code Signing** Certificate Authority Security Council, Public Key Infrastructure Consortium (PKI Consortium), San Francisco, CA, USA, 2013. Downloadable from: https://pkic.org/uploads/2013/10/CASC-Code-Signing.pdf [Last accessed: 25.7.2021] |
| [Castano16] | Victor Castano, Igor Schagaev: **Resilient Computer System Design** Springer International Publishing, Cham, Switzerland, 2016. ISBN 978-3-319-38605-8 |
| [Cavalli11] | Mariana Segovia, Jose Rubio-Hernan, Ana R. Cavalli and Joaquin Garcia-Alfaro: Cyber-Resilience Evaluation of Cyber-Physical Systems. Conference Paper, New Version in 2020 IEEE 19th International Symposium on Network Computing and Applications (NCA), Cambridge, MA, USA, Nov. 24, 2020 to Nov. 27, 2020. Downloadable from: https://www.researchgate.net/publication/344262389_Cyber-Resilience_Evaluation_of_Cyber-Physical_Systems [Last accessed: 10.6.2022] |
| [Cervantes16] | Humberto Cervantes, Rick Kazman: **Designing Software Architectures – A Practical Approach** Addison Wesley, Upper Saddle River, NJ, USA, 2016. ISBN 978-0-134-39078-9 |
| [Chantzis21] | Fotios Chantzis, Ioannis Stais, Paulino Calderon, Evangelos Deirmentzoglou, Beau Woods: **Practical IoT Hacking – The Definitive Guide to Attacking the Internet of Things** No Starch Press, San Francisco, CA, USA, 2021. ISBN 978-1-7185-0090-7 |
| [Chao20] | William S. Chao, Shuh-Ping Sun: **SBC State Machine for Model-Based Systems Engineering – Toward a Unified View of the System** Independently published, 2020. ISBN 979-8-6917-5055-7 |
| [Chicatelli15] | Amy Chicatelli, Aidan W. Rinehart, T. Shane Sowers, Donald L. Simon: **Investigation of Asymmetric Thrust Detection with Demonstration in a Real-Time Simulation Testbed** Technical Paper, American Institute of Aeronautics and Astronautics (AIAA), Reston, VA, USA, 2015. Downloadable from: https://ntrs.nasa.gov/api/citations/20150021854/downloads/20150021854.pdf [last accessed 15.7.2021] |
| [Chowhan18] | Rahul Singh Chowhan: Evolution and Paradigm Shift in Distributed System Architecture. https://www.researchgate.net/publication/330948642_Evolution_and_Paradigm_Shift_in_Distributed_System_Architecture/link/5c5d05a692851c48a9c196c4/download |
| [Chrissis11] | Mary Beth Chrissis, Mike Konrad, Sandy Shrum: **CMMI for Development – Guidelines for Process Integration and Product Improvement** Addison Wesley Publishing Inc., USA (The SEI Series in Software Engineering), 3rd revised edition, 2011. ISBN 978-0-321-71150-2 |
| [Christopher16] | Martin Christopher: **Logistics & Supply Chain Management** Pearson Education (Financial Times Press), Harlow, UK, 5th edition, 2016. ISBN 978-1-292-08379-7 |
| [CISA21] | CISA: **Defending Against Software Supply Chain Attacks** US Cybersecurity and Infrastructure Security Agency (CISA), Washington, DC, USA, 2021. Downloadable from: https://www.cisa.gov/sites/default/files/publications/defending_against_software_supply_chain_attacks_508_1.pdf [Last accessed: 23.7.2021] |
| [Collinson11] | R.P.G. Collinson: **Introduction to Avionics Systems** Springer Verlag, Dordrecht, NL, 3rd edition, 2011. ISBN 978-9-400-70707-8 |

[Cooling19]        Jim Cooling: **Software Engineering for Real-Time Systems** – *A Software Engineering Perspective toward designing real-time Systems* Packt Publishing, Birmingham, UK, 2019. ISBN 978-1-839-21658-9

[Cooper18]         David Cooper, Andrew Regenscheid, Murugiah Souppaya, Christopher Bean, Mike Boyle, Dorothy Cooley, Michael Jenkins: *Security Considerations for Code Signing* NIST, Gaithersburg, MD, USA, January 26, 2018. Downloadable from: https://nvlpubs.nist.gov/nistpubs/CSWP/NIST.CSWP.01262018.pdf [Last accessed: 23.7.2021]

[CSA20]            CSA: **Python for Data Science** – *A Crash Course for Data Science and Analysis, Python Machine Learning and Big Data* Computer Science Academy (CSA), published by GIALE LTD, London, UK, 2020. ISBN 978-1-8012-5525-7

[Das19]            Kanchan Das, Mangey Ram: **Mathematical Modelling of System Resilience** River Publishers, Gistrup, Denmark, 2019. ISBN 978-8-7702-2070-5

[Davi15]           Lucas Davi, Ahmad-Reza Sadeghi: **Building Secure Defenses Against Code-Reuse Attacks** Springer International Publishing AG Switzerland, Cham, Switzerland, 2015. ISBN 978-3-319-25544-6

[Denning12]        Richard Denning: **Reliability Block Diagrams** Applied R&M Manual for Defence Systems (GR-77 Issue 2012), Part C – Techniques, Chapter 30, UK SARS (Safety and Reliability Society), Oldham, UK, 2012. Downloadable from: https://sars.org.uk/BOK/Applied%20R&M%20Manual%20for%20Defence%20Systems%20(GR-77)/p3c30.pdf [Last accessed: 3.10.2021]

[Desai09]          Jimmy Desai: **IT Outsourcing Contracts** – *A Legal and Practical Guide* IT Governance Publishing (ITGP), Ely, UK, 2009. ISBN 978-1-8492-8029-7

[Desai10]          Jimmy Desai: **Service Level Agreements** – *A Legal and Practical Guide* IT Governance Publishing (ITGP), Ely, UK, 2010. ISBN 978-1-8492-8069-3

[Dhall17]          Rohit Dhall: *Designing Graceful Degradation in Software Systems* Proceedings of the Second International Conference on Research in Intelligent and Computing in Engineering, ACSIS 2017, March 24–26, 2017. Gopeshwar, Uttrakhand, India. Downloadable from: https://annals-csis.org/Volume_10/drp/pdf/15.pdf [Last accessed: 6.9.2021]

[D'Hoinne20]       Jeremy D'Hoinne: *The State of Artificial Intelligence in Security and Risk Management* Gartner Brief, Gartner Inc., Stamford, USA, 2020. Downloadable from: https://assets-powerstores-com.s3.amazonaws.com/data/org/20033/media/doc/the_state_of_artificial_intelligence_in_security_1599830803684001f2hn-8e45d6fd2da534e978d4decad57623e0.pdf [Last accessed: 31.10.2021]

[Dick17]           Jeremy Dick, Elizabeth Hull, Ken Jackson: **Requirements Engineering** Springer International Publishing, Cham, Switzerland, 4th edition, 2017. ISBN 978-3-319-86997-1

[DMV04]            DMV **On-Board Diagnostic Trouble Codes** Delaware Division of Motor Vehicles, Delaware City, USA, 2004. Downloadable from: https://www.dmv.de.gov/VehicleServices/inspections/pdfs/dtc_list.pdf [Last accessed: 01.10.2021]

[Drusinsky06]      Doron Drusinsky: **Modeling and Verification Using UML Statecharts** – *A Working Guide to Reactive System Design, Runtime Monitoring, and*

|  | *Execution-based Model Checking* Newnes (Elsevier), Burlington, MA, USA, 2006. ISBN 978-0-7506-7949-7 |
| [Dubrova13] | Elena Dubrova: **Fault-Tolerant Design** Springer Science & Business Media, New York, N.Y., USA, 2013. ISBN 978-1-461-42112-2 |
| [Dumitrascu21] | Sorin Dumitrascu: **Managing Software Project Outsourcing** – *A Practical Guide* Independently published, 2021. ISBN 979-8-4895-5141-0 |
| [Ebert11] | Christof Ebert: **Global Software and IT** – *A Guide to Distributed Development, Projects, and Outsourcing* Wiley-IEEE Computer Society Press, Hoboken, NJ, USA, 2011. ISBN 978-0-470-63619-0 |
| [Eloff18] | Jan Eloff, Madeleine Bihina Bella: **Software Failure Investigation** – *A Near-Miss Analysis Approach* Springer International Publishing, Cham, Switzerland, 2018. ISBN 978-3-319-87054-0 |
| [Ernst21] | Neil Ernst, Rick Kazman, Julien Delange: **Technical Debt in Practice** – *How to Find It and Fix It* The MIT Press, Cambridge, MA, USA, 2021. ISBN 978-0-262-54211-1 |
| [Fairgrieve20] | Duncan Fairgrieve, Richard S. Goldberg: **Product Liability** Oxford University Press, Oxford, UK, 3rd edition, 2020. ISBN 978-0-199-67923-2 |
| [Ferguson03] | Niels Ferguson, Bruce Schneier: **Practical Cryptography** Wiley Publishing Inc., Indianapolis, IN, USA, 2003. ISBN 978-0-471-22357-3 |
| [FERMA20] | FERMA: *Artificial Intelligence applied to Risk Management* FERMA Perspectives, Federation of European Risk Management Associations (FERMA), Brussels, Belgium, 2020. Downloadable from: https://www.eciia.eu/wp-content/uploads/2019/11/FERMA-AI-applied-to-RM-FINAL.pdf [Last accessed: 31.10.2021] |
| [Ford17] | Neal Ford, Rebecca Parsons, Patrick Kua: **Building Evolutionary Architectures** – *Support Constant Change* O'Reilly, Farnham, UK, 2017. ISBN 978-1-491-98636-3 |
| [Ford22] | Neal Ford, Mark Richards, Pramod Sadalage, Zhamak Dehghani: **Software Architecture** – *The Hard Parts: Modern Tradeoff Analysis for Distributed Architectures* O'Reilly Media, Inc., Sebastopol, CA, USA, 2022. ISBN 978-1-492-08689-5 |
| [Fowler10b] | Kim Fowler: **Mission-Critical and Safety-Critical Systems Handbook** – *Design and Development for Embedded Applications* Newnes Publishing (Elsevier), Burlington, MA, USA, 2009. ISBN 978-0-7506-8567-2 |
| [Furrer19] | Frank J. Furrer: **Future-Proof Software-Systems** – *A Sustainable Evolution Strategy* Springer Vieweg Verlag, Wiesbaden, Germany, 2019. ISBN 978-3-658-19937-1 |
| [Furtak10] | Andrew Furtak et al.: *BIOS and Secure Boot Attacks Uncovered* Ekoparty Security Conference 2010 presentation. Downloadable from: https://docplayer.net/19816640-Bios-and-secure-boot-attacks-uncovered.html [Last accessed: 14.01.2022] |
| [Futral13] | William Futral, James Greene: **Intel Trusted Execution Technology for Server Platforms** – *A Guide to More Secure Datacenters* Apress Media, LLC, New York, NY, USA, 2013. ISBN 978-1-4302-6148-3. Downloadable from (Apress Open): https://link.springer.com/book/10.1007/978-1-4302-6149-0 [last accessed: 25.10.2021] |
| [Ganguly18] | Auroop Ratan Ganguly, Udit Bathia, Stephen E. Flynn: **Critical Infrastructures Resilience** – *Policy and Engineering Principles* Routledge Publishers (Taylor & Francis), Abingdon, UK, 2018. ISBN 978-1-498-75863-5 |

[Gatto17]        James Gatto, Hean Koo: *Open Source Policies — Why You Need Them and What they Should Include* Sheppard Mullin Richter & Hampton LLP., Los Angeles, CA, USA, 2017. Downloadable from: https://www.mygamecounsel.com/wp-content/uploads/sites/32/2017/05/Open-Source-Policies-0517.pdf [Last accessed: 27.7.2021]

[Ghavami19]      Peter Ghavami: **Big Data Analytics Methods** – *Analytics Techniques in Data Mining, Deep Learning and Natural Language Processing* Walter De Gruyter Inc., Boston/Berlin, 2nd edition, 2019. ISBN 978-1-547-41795-7

[Gift21]         Noah Gift, Alfredo Deza: **Practical MLOps** – *Operationalizing Machine Learning Models* O'Reilly Media, Inc., Sebastopol, CA, USA, 2021. ISBN 978-1-098-10301-9

[Goloubeva06]    Olga Goloubeva, Maurizio Rebaudengo, Matteo Sonza Reorda, Massimo Violante: **Software-Implemented Hardware Fault Tolerance** Springer-Verlag, Berlin, Germany, 2006. ISBN 978-0-387-26060-0

[Gomaa16]        Hassan Gomaa: **Real-Time Software Design for Embedded Systems** Cambridge University Press, New York, N.Y., USA, 2016. ISBN 978-1107-04109-7

[Goodloe10]      Alwyn E. Goodloe, Lee Pike: *Monitoring Distributed Real-Time System – A Survey and Future Directions* NASA Technical Memorandum (NASA/CR–2010-216724), NASA Center for AeroSpace Information, Hanover, MD, USA, July 2010. Downloadable from: https://ntrs.nasa.gov/api/citations/20100027427/downloads/20100027427.pdf [Last accessed: 13.5.2021]

[Gorton11]       Ian Gorton: **Essential Software Architecture** Springer Verlag, Heidelberg, Germany, 2nd edition, 2011. ISBN 978-3-642-19175-6

[GPI18]          GPI: **TEE System Architecture v1.2 | GPD_SPE_009** Global Platform, Inc., Specification, Redwood City, CA, USA, 2018. Downloadable from: https://globalplatform.org/specs-library/ [Last accessed: 7.11.2021]

[Großjean09]     Ariane Großjean: **Corporate Terminology Management** – *An Approach in Theory and Practice* VDM Verlag Dr. Müller, Saarbrücken, Germany, 2009. ISBN 978-3-6391-2421-7

[Gupta19]        Aditya Gupta: **The IoT Hacker's Handbook** – *A Practical Guide to Hacking the Internet of Things* Apress Media LLC, New York, N.Y., USA, 2019. ISBN 978-1-4842-4299-5

[Haimes19]       Yacov Y. Haimes: **Modeling and Managing Interdependent Complex Systems of Systems** John Wiley & Sons, Inc. (Wiley – IEEE), Hoboken, NJ, USA, 2019. ISBN 978-1-119-17365-6

[Hanssen18]      Geir Kjetil Hanssen, Tor Stålhane, Thor Myklebust: **SafeScrum** – *Agile Development of Safety-Critical Software* Springer-Verlag, Heidelberg, Germany, 2018. ISBN 978-3-319-99333-1

[Harris02]       Nick Harris, Dale Barrick, Ian Cai, Peter G. Croes, Adriel Johndro, Bob Klingelhoets, Steve Mann, Nihal Perera, Robert Taylor: **LPAR Configuration and Management** -*Working with IBM iSeries Logical Partitions* IBM RedBook, SG24-6251-00, 2002. International Business Machines Corporation, Austin, Texas, USA, 2002. Downloadable from: http://www.redbooks.ibm.com/redbooks/pdfs/sg246251.pdf [last accessed: 26.10.2021]

[Hitt01]         Ellis F. Hitt, Dennis Mulcare: **Fault-Tolerant Avionics** CRC Press (Taylor & Francis Group), Boca Raton, FL, USA, 2001. Downloadable from: https://www.cs.unc.edu/~anderson/teach/comp790/papers/fault_tolerance_avionics.pdf [Last accessed: 31.08.2021]

[Hoffman01]     Daniel M. Hoffman, David M. Weiss (Editors): **Software Fundamentals: Collected Papers by David L. Parnas** Addison-Wesley Professional, Upper Saddle River, NJ, USA, Annotated Edition, 2001. ISBN-13: 978-0-201-70369-6

[Hofmann07]     Hubert F. Hofmann, Deborah K. Yedlin, John W. Mishler: **CMMI for Outsourcing** – *Guidelines for Software, Systems, and IT Acquisition* Addison Wesley, Upper Saddle River, NJ, USA, 2007. ISBN 978-0-321-47717-0

[Hohmann03]     Luke Hohmann: **Beyond Software Architecture** – *Creating and Sustaining Winning Solutions* Addison-Wesley Professional, Upper Saddle River, USA, 2003. ISBN 978-0-201-77594-5

[Hole16]        Kjell Jørgen Hole: **Anti-fragile ICT Systems** Simula Springer Briefs on Computing, Vol. 1, Springer International Publishing, Cham, Switzerland, 2016. ISBN 978-3-319-30068-9 Open access publication: https://link.springer.com/content/pdf/10.1007%2F978-3-319-30070-2.pdf [Last accessed: 6.9.2021]

[Höller14]      Jan Höller, Vlasios Tsiatsis, Catherine Mulligan, Stamatis Karnouskos, Stefan Avesand, David Boyle: **From Machine-to-Machine to the Internet of Things** – *Introduction to a New Age of Intelligence* Academic Press (Elsevier), Kidlington, UK, 2014. ISBN 978-0-12-407684-6

[Hollnagel06]   Erik Hollnagel, David D. Woods, Nancy Leveson (Editors): **Resilience Engineering** – *Concepts and Precepts* CRC Press (Taylor & Francis), Boca Raton, FL, USA, 2006. ISBN 978-0-754-64904-5

[Hollnagel13]   Erik Hollnagel, Jean Paries, John Wreathall (Editors): **Resilience Engineering in Practice** – *A Guidebook* CRC Press (Taylor & Francis), Boca Raton, FL, USA, 2013. ISBN 978-1-472-42074-9

[Huang09]       J. C. Huang: **Software Error Detection through Testing and Analysis** John Wiley & Sons, Inc., Hoboken, NJ, USA, 2009. ISBN 978-0-470-40444-7

[Hubert02]      Richard Hubert: **Convergent Architecture** John Wiley & Sons, New York, N.Y., USA, 2002. ISBN 978-0-471-10560-0

[Huckle19]      Thomas Huckle, Tobias Neckel: **Bits and Bugs** – *A Scientific and Historical Review of Software Failures in Computational Science* SIAM – Society for Industrial and Applied Mathematics, Philadelphia, PA, USA, 2019. ISBN 978-1-6119-7555-0

[Hunter18]      Richard J. Hunter Jr., John H. Shannon, Henry J. Amoroso: **Products Liability** – *A Managerial Perspective* Independently published, 2018. ISBN 978-1-7311-5068-4

[Hussmann21]    Sid Hussmann: **Gapfruit Trustworthy Execution Platform** White Paper, GAPFRUIT AG, Zug, Switzerland, 2021. Downloadable from: https://www.gapfruit.com/technology [last accessed: 26.10.2021]

[IIA18]         IIA: *Auditing Third-Party Risk Management* Practice Guide, The Institute of Internal Auditors, Lake Mary, FL, USA, 2018. Downloadable from: https://www.academia.edu/38307277/PG_Auditing_Third_Party_Risk_Management_pdf?auto=download [Last accessed: 29.6.2021]

[Intel12]       Intel: *INTEL© Trusted Execution Technology* Intel White Paper, Intel Corporation, Santa Clara, CA, USA, 2012. Downloadable from: https://www.intel.com/content/dam/www/public/us/en/documents/white-papers/trusted-execution-technology-security-paper.pdf [Last accessed: 6.11.2021]

[Irving05]      Nic Irving, Mathew Jenner, Arsi Kortesniemi: **Partitioning Implementations for IBM Eserver p5 Servers** IBM RedBook,

SG24-7039-02, 2005. International Business Machines Corporation, Austin, Texas, USA, 2005. Downloadable from: http://www.redbooks.ibm.com/redbooks/pdfs/sg247039.pdf [last accessed: 26.10.2021]

[Isermann06]    Rolf Isermann: **Fault-Diagnosis Systems – *An Introduction from Fault Detection to Fault Tolerance*** Springer Verlag, Berlin, Germany, 2006. ISBN 978-3-540-24112-6

[ISO 15031-5]    ISO 15031-5:2015: Road vehicles — Communication between vehicle and external equipment for emissions-related diagnostics — Part 5: Emissions-related diagnostic services. ISO Standards Organization, Geneva, Switzerland, 2015/2021. Available from: https://www.iso.org/standard/66368.html [Last accessed: 11.6.2022]

[Jackson09]    Scott Jackson: **Architecting Resilient Systems – *Accident Avoidance and Survival and Recovery from Disruptions*** John Wiley & Sons, Inc., Hoboken, NJ, USA, 2009. ISBN 978-0-470-40503-1

[Jackson21]    Daniel Jackson: **The Essence of Software – *Why Concepts Matter for Great Design*** Princeton University Press, Princeton, USA, 2021. ISBN 978-069-122538-8

[Jajodia20]    Sushil Jajodia, George Cybenko, V.S. Subrahmanian, Vipin Swarup, Cliff Wang, Michael Wellman (Editors): **Adaptive Autonomous Secure Cyber-Systems** Springer Nature Publishing, Cham, Switzerland, 2020. ISBN 978-3-030-33431-4

[James19]    Robert James: **Version Control System Intelligence – *Version Control System Explained*** Independently published, 2019. ISBN 978-1.69286466-8

[Jayaratchagan18]    Narayanan Jayaratchagan: **Elegant Software Design Principles** Independently published, 2018. ISBN 978-1-7909-4694-5

[Jorgensen19]    Paul C. Jorgensen: **Modeling Software Behavior – *A Craftsman's Approach*** CRC Press (Taylor & Francis), Boca Raton, CA, USA, 2019. ISBN 978-0-367-44604-8

[Josephsen13]    David Josephsen: **Nagios: Building Enterprise-Grade Monitoring Infrastructures for Systems and Networks** Prentice Hall Inc., USA, 2nd edition, 2013. ISBN 978-0-133-13573-2

[Julian17]    Mike Julian: **Practical Monitoring – *Effective Strategies for the Real World*** O'Reilly Media, Inc., Sebastopol, CA, USA, 2017. ISBN 978-1-4919-5735-6

[Kamara10]    Seny Kamara, Sonia Fahmy, Eugene Schultz, Florian Kerschbaum, Michael Frantzen: *Analysis of Vulnerabilities in Internet Firewalls* White paper, Center for Education and Research in Information Assurance and Security (CERIAS), Purdue University, West Lafayette, IN, USA, 2010. Downloadable from: https://www.cs.purdue.edu/homes/fahmy/papers/firewall-analysis.pdf [Last accessed: 2.8.2021]

[Kandé03]    Mohamed Mancona Kandé: **A Concern-Oriented Approach to Software Architecture** THÈSE NO 2796 (2003), ÉCOLE POLYTECHNIQUE FÉDÉRALE DE LAUSANNE, Lausanne, Switzerland, 2003. Downloadable from: https://core.ac.uk/reader/147900073 [Last accessed: 15.11.2021]

[Quote]    Mohamed Mancona Kandé, Alfred Strohmeier: *On The Role of Multi-Dimensional Separation of Concerns in Software Architecture* Position Paper for the OOPSLA'2000 Workshop on Advanced Separation of Concerns, Lausanne, September 2005. Downloadable from: https://core.ac.uk/download/pdf/147904534.pdf [Last accessed: 15.11.2021]

[Kinsbruner20]     Eran Kinsbruner: **Accelerating Software Quality** – *Machine Learning and Artificial Intelligence in the Age of DevOps* Independently published, 2020. ISBN 979-8-6711-2604-4

[Klinger13]        Paul Klinger, Rachel Burnett: **Drafting and Negotiating IT Contracts** Bloomsbury Professional Ltd., Haywards Heath, UK, 3rd edition, 2013. ISBN 978-1-8476-6712-0

[Knieke21]         Christoph Knieke, Andreas Rausch, Mirco Schindler: Tackling Software Architecture Erosion - Joint Architecture and Implementation Repairing by a Knowledge-based Approach. April 2021. Downloadable from: https://arxiv.org/abs/2104.13919 [Last accessed: 10.6.2022]

[Koenig19a]        Ian Koenig: **Principle-Based Enterprise Architecture** – *A Systematic Approach to Enterprise Architecture and Governance* Technics Publications, Basking Ridge, NJ, USA, 2019. ISBN 978-1-6346-2494-7

[Koenig19b]        Ian Koenig: **Diagramming Architecture** – *According to the Principle-Based Enterprise Architecture Method* Independently published, 2019. ISBN 978-1-7106-2010-8

[Kong17]           Shiyi Kong, Minyan Lu, Luyi Li: *Fault propagation analysis in software-intensive systems* – *A survey* Second International Conference on Reliability Systems Engineering (ICRSE), Beijing Yanqi Lake International Convention & Exhibition Center (BYCC), Huairou District, Beijing, China, July 10–12, 2017. Available from: https://ieeexplore.ieee.org/document/8030792 [Last accessed: 30.08.2021]

[Kopetz11]         Hermann Kopetz: **Real-Time Systems** – *Design Principles for Distributed Embedded Applications* Springer Science & Business Media, New York, N.Y., USA, 2nd edition, 2011. ISBN 978-1-461-42866-4

[Kopetz22]         Hermann Kopetz: **Data, Information, and Time** – *The DIT Model* Springer Briefs in Computer Science, Springer International Publishing, Cham, Switzerland, 2022. ISBN 978-3-030-96328-6

[Kopp18]           Erik Kopp: **Business Continuity Management Plain & Simple** – *How to Write A Business Continuity Plan (BCP)* Independently published, 2018. ISBN 978-1-9804-9010-4

[Koren20]          Israel Koren, C. Mani Krishna: **Fault-Tolerant Systems** Morgan Kaufmann (Elsevier), Cambridge, MA, USA, 2nd edition, 2020. ISBN 978-0-128-18105-8

[Kovacevic19]      James Kovacevic: **Understanding Reliability Block Diagrams** Eruditio, LLC, Mount Pleasant, SC, USA, 2019. Accessible at: https://hpreliability.com/understanding-reliability-block-diagrams/ [Last accessed: 14.10.2021]

[Krauss19]         Michael I. Krauss: **Principles of Products Liability** West Academic Publishing, St. Paul, MN, USA, 3rd edition, 2019. ISBN 978-1-6402-0128-6

[Kröger08]         Fred Kröger, Stephan Merz: **Temporal Logic and State Systems** Springer Verlag, Berlin, Germany, 2008. ISBN 978-3-540-67401-6

[Kruchten19]       Philippe Kruchten, Robert Nord: **Managing Technical Debt** – *Reducing Friction in Software Development* Addison-Wesley, Upper Saddle River, NJ, USA, 2019. ISBN 978-0-135-64593-2

[Kruchten20]       Philippe Kruchten: *Technical Debt – Myths and Realities* Presentation, June 2020. Downloadable from: https://pkruchten.files.wordpress.com/2020/06/kruchten-200609-technical-debt-at-xp2020.pdf [Last accessed: 14.11.2021]

[Kumar10]            Arun Kumar: **Reliability Block Diagram (RBD)** Indian Institute of
                     Technology Delhi, New Delhi, India. Lecture Note, October 25th, 2010.
                     Downloadable    from:    https://web.iitd.ac.in/~arunku/files/CEL899_Y13/
                     Reliability%20Block%20Diagram.pdf [Last accessed: 14.10.2021]

[Lamsweerde09]       Axel van Lamsweerde: **Requirements Engineering** – *From System Goals
                     to UML Models to Software Specifications* John Wiley & Sons Inc.,
                     Chichester, UK, 2009. ISBN 978-0-470-01270-3

[Lauchande21]        Natu Lauchande: **Machine Learning Engineering with MLflow** –
                     *Manage the end-to-end machine learning life cycle with MLflow* Packt
                     Publishing, Birmingham, UK, 2021. ISBN 978-1-8005-6079-6

[Lever13]            Kirsty E. Lever, Madjid Merabti, Kashif Kifayat: *Single Points of Failure
                     Within Systems-of-Systems* Conference Paper, June 2013. Downloadable
                     from: https://www.researchgate.net/publication/268684111_Single_Points_
                     of_Failure_Within_Systems-of-Systems/link/547319720cf24bc8ea19b2a0/
                     download [Last accessed: 31.8.2021]

[Leveson03]          Nancy Leveson: *White Paper on Approaches to Safety Engineering* White
                     paper, Massachusetts Institute of Technology (MIT), USA, April 23, 2003.
                     Downloadable   from:   http://sunnyday.mit.edu/caib/concepts.pdf   [Last
                     accessed: 6.9.2021]

[Leveson11]          Nancy G. Leveson: **Engineering a Safer World** – *Systems Thinking
                     applied to Safety* MIT Press, Cambridge MA, USA, 2011. ISBN
                     978-0-262-01662-9

[Leveson16]          Nancy G. Leveson: **Engineering a Safer World** – *Systems Thinking
                     Applied to Safety* MIT Press Ltd., Massachusetts, MA, USA, 2016. ISBN
                     978-0-262-53369-0

[Li21]               Ruiyin Li, Peng Liang, Mohamed Soliman, Paris Avgeriou: Understanding
                     Architecture Erosion -The Practitioners' Perceptive. Downloadable from:
                     arXiv:2103.11392v1, 21 March 2021.

[Lilienthal19]       Carola Lilienthal: **Sustainable Software Architecture** – *Analyze and
                     Reduce Technical Debt* Dpunkt Verlag GmbH, Heidelberg, Germany, 2nd
                     edition, 2019. ISBN 978-3-8649-0673-2

[Lin21]              Yan Lin: **Novel Techniques in Recovering, Embedding, and Enforcing
                     Policies for Control-Flow Integrity** Springer Nature Switzerland, Cham,
                     Switzerland, 2021. ISBN 978-3-030-73140-3

[Ma20]               Hongbin Ma, Liping Yan, Yuanqing Xia, Mengyin Fu: **Kalman Filtering
                     and Information Fusion** Springer Nature Singapore, Singapore, 2020.
                     ISBN 978-9-811-50808-0

[Majumdar20]         Suryadipta Majumdar, Taous Madi, Yushun Wang, Azadeh Tabiban,
                     Momem Oqaily, Amir Alimohammdadifar, Yosr Jarraya, Makan Pourzandi,
                     Lingyu Wang, Mourad Debbabi: **Cloud Security Auditing** Springer Nature
                     Switzerland, Cham, Switzerland, 2019. ISBN 978-3-030-23130-9

[Manners-Bell17]     John Manners-Bell: **Supply Chain Risk Management** – *Understanding
                     Emerging Threats to Global Supply Chains* Kogan Page, New Delhi,
                     India, 2nd edition, 2017. ISBN 978-0-749-48015-8

[Massias99]          H. Massias, X. Serret Avila, J.-J. Quisquater: *Timestamps* –
                     *Main   issues   on   their   use   and   implementation* Conference
                     Paper,   February   1999,   IEEE   Xplore,   New   York,   NY,   USA,
                     1999.      Downloadable      from:      https://www.researchgate.net/

publication/3824437_Timestamps_main_issues_on_their_use_and_imple-
mentation/link/02bfe51001990cf6b1000000/download    [Last    accessed:
25.7.2021]

[Matulevičius17]    Raimundas Matulevičius: **Fundamentals of Secure System Modelling**
Springer International Publishing, Cham, Switzerland, 2017. ISBN
978-3-319-87143-1

[Mezak06]    Steve Mezak: **Software without Borders** – *A Step-by-Step Guide to
Outsourcing Your Software Development* Earthrise Press, Rochester,
Michigan, USA, 2006. ISBN 978-0-9778-2680-3

[Meziane10]    Farid Meziane, Sunil Vadera: **Artificial Intelligence Applications for
Improved Software Engineering Development** – *New Prospects* Information
Science Reference, Hershey, PA, USA, 2010. Downloadable from: https://core.
ac.uk/download/pdf/81262.pdf [Last accessed: 29.10.2021]

[Mili04]    Hafedh Mili, Houari Sahraoui, Hakim Lounis, Hamid Mcheick, Amel
Elkharraz: *Concerned about Separation* Article, Université du Québec à
Montréal, Montréal (Québec), Canada, January 2004. Downloadable from:
https://www.researchgate.net/publication/244446574_Understanding_sepa-
ration_of_concerns [Last accessed: 15.11.2021]

[Mitra21]    Sayan Mitra: **Verifying Cyber-Physical Systems** – *A Path to Safe
Autonomy* The MIT Press, Cambridge, MA, USA, 2021. ISBN
978-0-262-04480-6

[Mittal18]    Saurabh Mittal, Saikou Diallo, Andreas Tolk (Editors): **Emergent
Behaviour in Complex Systems** – *A Modeling and Simulation
Approach* John Wiley & Sons, Inc., Hoboken, NJ, USA, 2018. ISBN
978-1-119-37886-0

[Mittal20]    Saurabh Mittal, Andreas Tolk (Editors): **Complexity Challenges in
Cyber-Physical Systems** – *Using Modeling and Simulation to sup-
port Intelligence, Adaptation, and Autonomy* John Wiley & Sons, Inc.,
Hoboken, NJ, USA, 2020. ISBN 978-1-119-55239-0

[Mo21]    Huadong Mo, Giovanni Sansavini, Min Xic: **Cyber-Physical Distributed
Systems** – *Modeling, Reliability Analysis and Applications* John Wiley
&Sons., Inc., Hoboken, NJ, USA, 2021. ISBN 978-1-119-68267-7

[Mobus22]    George E. Mobus: **Systems Science** – *Theory, Analysis, Modeling,
and Design* Springer Verlag, New York, NY, USA, 2022. ISBN
978-3-030-93481-1

[ModelOp21]    ModelOp: *Model Monitoring – The Path to Reliable AI* eBook, ModelOp
Corporation, Chicago, IL, USA, 2021. Downloadable from: https://www.
modelop.com/wp-content/uploads/2021/04/ebook-Model-Monitoring-The-
Path-to-Reliable-AI-1.pdf [last accessed: 27.10.2021]

[Möller16]    Dietmar P.F. Möller: **Guide to Computing Fundamentals in Cyber-
Physical Systems** – *Concepts, Design Methods, and Applications* Springer
International Publishing, Cham, Switzerland, 2016. ISBN 978-3-319-79747-2

[Murer11]    Stephan Murer, Bruno Bonati, Frank J. Furrer: **Managed Evolution** – *A
Strategy for Very Large Information Systems* Springer Verlag, Berlin,
Germany, 2011. ISBN 978-3-642-01632-5

[Naidu02]    Krishni Naidu: **Firewall Checklist** SANS Institute, Rockville Pike, North
Bethesda, MD, USA, 2002. Downloadable from: https://www.sans.org/
media/score/checklists/FirewallChecklist.pdf [Last accessed: 2.8.2021]

[NCSC21a]            NCSC: *NCSC Alert: Critical Vulnerability in Apache Log4j Library* (CVE-2021-44228, Update 2, 15. December 2021). Ireland National Cyber Security Centre, Dublin, Ireland. Downloadable from: https://www.ncsc. gov.ie/pdfs/apache-log4j-101221.pdf [Last accessed: 19.12.2021]

[NCSC21b]            NCSC: *Software Supply Chain Attacks* UK National Cyber Security Centre (NCSC), London, UK, 2021. Downloadable from: https://www.dni. gov/files/NCSC/documents/supplychain/Software_Supply_Chain_Attacks. pdf [Last accessed: 20.07.2021]

[NISTSCM15]          NIST: *Best Practices in Cyber Supply Chain Risk Management* Conference Materials, US National Institute of Standards and Technology, Workshop October 1–2, 2015, Washington, DC, USA. Downloadable from:         https://csrc.nist.gov/CSRC/media/Projects/Supply-Chain-Risk-Management/documents/briefings/Workshop-Brief-on-Cyber-Supply-Chain-Best-Practices.pdf [Last accessed: 21.2.2021]

[NIST-SP-800-147]    David Cooper, William Polk, Andrew Regenscheid, Murugiah Souppaya: **BIOS Protection Guidelines – *Recommendations of the National Institute of Standards and Technology*** NIST Special Publication 800-147, April 2011. Gaithersburg, MD, USA, 2011. Downloadable from: https:// nvlpubs.nist.gov/nistpubs/Legacy/SP/nistspecialpublication800-147.pdf [Last accessed: 14.01.2022]

[Nord12]             Robert L. Nord, Ipek Ozkaya, Philippe Kruchten, Marco Gonzalez-Rojas: *In Search of a Metric for Managing Architectural Technical Debt* 2012 Joint Working Conference on Software Architecture & 6th European Conference on Software Architecture, August 20–24, 2012, Helsinki, Finland. Downloadable from: https://resources.sei.cmu.edu/asset_files/ ConferencePaper/2012_021_001_88045.pdf [Last accessed: 14.11.2021]

[Olivé10]            Antoni Olivé: **Conceptual Modeling of Information Systems** Springer Verlag, Berlin, Germany, 2010. ISBN 978-3-642-07256-7

[OpenLogic20]        OpenLogic: *Open Source Policy Builder* OpenLogic Ltd., Minneapolis, MN, USA, 2020. Downloadable from: https://www.immagic.com/eLibrary/ARCHIVES/GENERAL/OLOGICUS/O120228P.pdf [Last accessed: 27.7.2021]

[Page10]             Scott E. Page: **Diversity and Complexity** Princeton University Press, Princeton, NJ, USA, 2010. ISBN 978-0-691-13767-4

[Panunzio14]         Marco Panunzio, Tullio Vardanega: *An architectural approach with separation of concerns to address extra-functional requirements in the development of embedded real-time software systems* Journal of Systems Architecture (Elsevier), Volume 60, Issue 9, October 2014, Pages 770–781. Access via (open access): https://www.sciencedirect.com/science/article/ pii/S1383762114000824?via%3Dihub [Last accessed: 15.11.2021]

[Pawar16]            Nahush Pawar: **Application of Artificial Intelligence in Software Engineering** IOSR Journal of Computer Engineering (IOSR-JCE), Volume 18, Issue 3, May-Jun. 2016, pp. 46–51. Downloadable from: https://www. iosrjournals.org/iosr-jce/papers/Vol18-issue3/Version-4/H1803044651.pdf [Last accessed: 29.10.2021]

[Pitakrat18]         Teerat Pitakrat: **Architecture-Aware Online Failure Prediction for Software Systems** PhD Thesis, University of Stuttgart, Stuttgart, Germany, 2018. BoD-Books on Demand, Norderstedt, Germany, ISBN 978-3-7528-7651-2. Downloadable from: https://elib.uni-stuttgart.de/bitstream/11682/9934/5/ TeeratPitakrat-Dissertation.pdf [Last accessed: 11.11.2021]

| | |
|---|---|
| [Ploeg15] | Jeroen Ploeg, Elham Semsar-Kazerooni, Guido Lijster, Nathan van de Wouw, Henk Nijmeijer: **Graceful Degradation of Cooperative Adaptive Cruise Control** IEEE Transactions on Intelligent Transportation Systems, Vol. 16, No. 1, February 2015. IEEE, New York, NY, USA, 2015. Downloadable from: http://www.dct.tue.nl/New/Wouw/IEEETITS2014_Ploeg_GracefulCACC.pdf [Last accessed: 6.9.2021] |
| [Pullum01] | Laura Pullum: **Software Fault Tolerance Techniques and Implementation** Artech House Inc., Norwood, MA, USA, 2001. ISBN 978-1-630-81234-8 |
| [Rainey18] | Larry B. Rainey, Mo Jamshidi (Editors): **Engineering Emergence** – *A Modeling and Simulation Approach* CRC Press (Taylor & Francis), Boca Raton, FL, USA, 2019. ISBN 978-1-138-04616-0 |
| [Raj17] | Pethuru Raj, Anupama C. Raman: **The Internet of Things** – *Enabling Technologies, Platforms, and Use Cases* CRC Press (Taylor & Francis Inc.), Boca Raton, FL, USA, 2017. ISBN 978-1-498-76128-4 |
| [Raj21] | Emmanuel Raj: **Engineering MLOps** – *Rapidly build, test, and manage production-ready machine learning life cycles at scale* Packt Publishing, Birmingham, UK, 2021. ISBN 978-1-8005-6288-2 |
| [Rashid18] | Ekbal Rashid: **Enhancing Software Fault Prediction With Machine Learning** – *Emerging Research and Opportunities* IGI Global Publishing, Hershey, PA, USA, 2018. ISBN 978-1-5225-3185-2 |
| [Raj13] | fTPM: H. Raj, S. Saroiu, A. Wolman, R. Aigner, J. Cox, P. England, C. Fenner, K. Kinshumann, J. Loeser, D. Mattoon, M. Nystrom, D. Robinson, R. Spiger, S. Thom, and D. Wooten: A Firmware-based TPM 2.0 Implementation. Microsoft Technical Research Note, MSR-TR-2015-84, Seattle, USA, 2015. Downloadable from: https://www.microsoft.com/en-us/research/wp-content/uploads/2016/02/msr-tr-2015-84.pdf [Last accessed: 10.6.2022] |
| [Ribbens17] | William Ribbens: **Understanding Automotive Electronics** – *An Engineering Perspective* Butterworth-Heinemann, Kidlington, OX, UK, 8th edition, 2017. ISBN 978-0-128-10434-7 |
| [Rogaway04] | Phillip Rogaway, Thomas Shrimpton: **Cryptographic Hash-Function Basics** – *Definitions, Implications, and Separations for Preimage Resistance, Second-Preimage Resistance, and Collision Resistance*. In: B. Roy and W. Meier (Editors): FSE 2004, Springer Verlag, Berlin, Germany, LNCS 3017, pp. 371–388, 2004. Downloadable from: https://link.springer.com/content/pdf/10.1007%2F978-3-540-25937-4_24.pdf [Last accessed: 25.7.2021] |
| [Rozanski12] | Nick Rozanski, Eoin Woods: **Software Systems Architecture** Addison-Wesley, Upper Saddle River, NJ, USA, 2nd edition, 2012. ISBN 978-9-332-54795-7 |
| [Sabharwal20] | Navin Sabharwal, Piyush Pandey: **Monitoring Microservices and Containerized Applications** – *Deployment, Configuration, and Best Practices for Prometheus and Alert Manager* Apress Media, LLC, New York, NY, USA, 2020. ISBN 978-1-4842-6215-3 |
| [SAE J1979] | SAE J1979/ISO 15031-5: E/E Diagnostic Test Modes J1979. SAE International, Warrendale, PA, USA, 2017. Available from: https://www.sae.org/standards/content/j1979_201702 [Last accessed: 11.6.2022] |
| [Sammons17] | Peter Sammons: **Contract Management** – *Core Business Competence* Kogan Page, New York, NY, USA, 2017. ISBN 978-0-7494-8064-6 |

[Santini10]            Al Santini: **OBD II** – *Functions, Monitors, & Diagnostic Techniques*
                       DELMAR (Cengage Learning), Boston, MA, USA, New Edition 2010.
                       ISBN 978-1-4283-9000-3

[Saridakis03]          Titos Saridakis: *Design Patterns for Fault Containment* NOKIA Design
                       Whitepaper, Espoo, Finland, 2003. Downloadable from: https://citeseerx.
                       ist.psu.edu/viewdoc/download?doi=10.1.1.415.3369&rep=rep1&type=pdf
                       [Last accessed: 30.08.2021]

[Saridakis09]          Titos Saridakis: *Design Patterns for Graceful Degradation* In: Noble J.,
                       Johnson R. (editors): Transactions on Pattern Languages of Programming
                       I. Lecture Notes in Computer Science, Vol 5770. Springer Verlag, Berlin,
                       Germany, 2009. ISBN 978-3-642-10831-0

[Sayeed19]             Sarwar Sayeed, Hector Marco-Gisbert, Ismael Ripoll, Miriam Birch:
                       *Control-Flow Integrity: Attacks and Protections* Jorunal of Applied
                       Sciences, Vol. 9, 2019, MDPI, Basel, Switzerland. Downloadable from:
                       https://www.researchgate.net/publication/336417731_Control-Flow_
                       Integrity_Attacks_and_Protections [Last accessed: 25.7.2021]

[Schmidt10]            Klaus Schmidt: **High Availability and Disaster Recovery** – *Concepts,
                       Design, Implementation* Springer Verlag, Berlin, Germany, 2010. ISBN
                       978-3-642-06379-4

[Sculley18]            David Sculley, Gary Holt, Daniel Golovin, Eugene Davydov, Todd Phillips,
                       Dietmar Ebner, Vinay Chaudhary, Michael Young: **Machine Learning** –
                       *The High-Interest Credit Card of Technical Debt* Google White Paper,
                       Google Inc., Mountain View, CA, USA, 2018. Downloadable from:
                       https://static.googleusercontent.com/media/research.google.com/de//pubs/
                       archive/43146.pdf [Last accessed: 4.11.2021]

[Seacord13]            Robert C. Seacord: **Secure Coding in C and C++** Addison-Wesley
                       Professional (SEI Series in Software Engineering), Upper Saddle River,
                       N.J., USA, 2nd edition, 2013. ISBN 978-0-321-82213-0

[Seaman13]             Carolyn Seaman: *Measuring and Monitoring Technical Debt* Presentation
                       at University 2 Maryland, Baltimore County, 7 March 2013. Downloadable
                       from:  http://www2.fct.unesp.br/grupos/lapesa/TD%20talk%20USP_0%20
                       -%20Measuring%20and%20Monitoring%20Technical%20Debt.pdf  [Last
                       accessed: 14.11.2021]

[Selic13]              Bran Selic, Sébastien Gérard: **Modeling and Analysis of Real-Time
                       and Embedded Systems with UML and MARTE** – *Developing Cyber-
                       Physical Systems* The MK/OMG Press, Morgan Kaufmann (Elsevier),
                       Waltham, MA, USA, 2013. ISBN 978-0-124-16619-6

[Shackleford21]        Dave Shackleford: *The Benefits of using AI in Risk Management* Blog,
                       TechTarget, Newton, MA, USA, 2021. Downloadable from: https://search-
                       security.techtarget.com/tip/The-benefits-of-using-AI-in-risk-management
                       [Last accessed: 31.10.2021]

[Sheperd16]            Carlton Shepherd, Ghada Arfaoui, Iakovos Gurulian, Robert P. Lee,
                       Konstantinos Markantonakis, Raja Naeem Akram, Damien Sauveron,
                       Emmanuel Conchon: Secure and Trusted Execution: Past, Present and
                       Future – A Critical Review in the Context of the Internet of Things and
                       Cyber-Physical Systems. The 15th IEEE International Conference on
                       Trust, Security and Privacy in Computing and Communications (IEEE
                       TrustCom-16), Sydney, Australia, August 2016. Downloadable from:
                       https://www.researchgate.net/publication/306039236_Secure_and_Trusted_

Execution_Past_Present_and_Future_--_A_Critical_Review_in_the_
Context_of_the_Internet_of_Things_and_Cyber-Physical_Systems/
link/5d382ed692851cd0468122fd/download [Last accessed: 10.6.2022]

[Silva10]     Victor Silva, Guilherme Horta Travassos: Technologies to Support the
Technical Debt Management in Software Projects - A Qualitative Research.
SBQS'19: Proceedings of the XVIII Brazilian Symposium on Software
Quality, Fortaleza Brazil, Brazil, October 2019, p. 314. New version availa-
ble from: https://dl.acm.org/doi/10.1145/3364641.3364679 [Last accessed:
10.6.2022]

[Silver17]    Bruce Silver: **BPMN Quick and Easy Using Method and Style –**
***Process Mapping Guidelines and Examples Using the Business Process***
***Modeling Standard*** Cody-Cassidy Press, Altadena, CA, USA, 2017. ISBN
978-0-9823-6816-9

[Sink11]      Eric Sink **Version Control by Example** Pyrenean Gold Press, Champaign,
IL, USA, 2011. ISBN 978-0-9835-0790-1. Downloadable from: https://
ericsink.com/vcbe/vcbe_usletter_lo.pdf [Last accessed: 26.12.2021]

[Snedaker13]  Susan Snedaker: **Business Continuity and Disaster Recovery Planning**
**for IT Professionals** Syngress (Elsevier), Cambridge, MA, USA, 2nd edition, 2013. ISBN 978-0-124-10526-3

[Suryanarayana14]  Girish Suryanarayana, Ganesh Samarthyam, Tushar Sharma: **Refactoring**
**for Software Design Smells** – ***Managing Technical Debt*** Morgan
Kaufmann, (Elsevier), Waltham, MA, USA 2014. ISBN 978-0-128-01397-7

[Taha21]      Walid M. Taha, Abd-Elhamid M. Taha, Johan Thunberg: **Cyber-Physical**
**Systems** – ***A Model-Based Approach*** Springer Nature Switzerland,
Cham, Switzerland, 2021. ISBN 978-3-030-36070-2. Downloadable
from    (Springer    Open):    https://link.springer.com/book/10.1007
%2F978-3-030-36071-9 [Last accessed: 17.11.2021]

[Talabis15]   Mark Ryan M. Talabis Robert McPherson, I. Miyamoto, Jason L. Martin:
**Security Analytics** – ***Finding Security Insights, Patterns, and Anomalies***
***in Big Data*** Syngress (Elsevier), Cambridge, MA, USA, 2015. ISBN
978-0-128-00207-0

[TCG19]       TCG: **TPM 2.0 – A brief Introduction** Trusted Computing Group,
Beaverton, OR, USA, 2019. Downloadable from: https://trustedcomput-
inggroup.org/wp-content/uploads/2019_TCG_TPM2_BriefOverview_
DR02web.pdf [last accessed: 25.10.2021]

[Tekinerdogan20]  Bedir Tekinerdogan, Dominique Blouin, Hans Vangheluwe, Miguel
Goulao, Paolo Carreira, Vasco Amaral (Editors): **Multi-Paradigm**
**Modelling Approaches for Cyber-Physical Systems** Academic Press,
London, UK, 2020. ISBN 978-0-128-19105-7

[Theeg20]     Gregor Theeg, Sergej Vlasenko (Editors): **Railway Signalling and**
**Interlocking** – International Compendium PMC Media House GmbH,
Leverkusen, Germany, 3rd edition, 2020. ISBN 978-3-96245-169-1. Available
at: https://www.pmcmedia.com/media/pdf/65/0b/ce/RailwaySignalling_2019_
Lesepr.pdf [last accessed 17.7.2021]

[Thiele04]    Lothar Thiele, Reinhard Wilhelm: ***Design for Timing Predictability*** Real-
Time Systems (The International Journal of Time-Critical Computing
Systems), 28. November 2004, pp. 157–177. Springer Nature Switzerland,
Cham, Switzerland

[Todorova21] Mina Todorova: *Insights into the Investigation of SRIWIJAYA AIR FLIGHT 182 Crash of Boeing 737-524 according to Safety Engineering Principles* Hauptseminar Paper, Hauptseminar SS2021, Technical University of Dresden, Dresden, Germany, 21.07.2021. Downloadable from: https://www.researchgate.net/publication/355412109_INSIGHTS_ INTO_THE_INVESTIGATION_OF_SRIWIJAYA_AIR_FLIGHT_182_ CRASH_OF_BOEING_737-524_ACCORDING_TO_SAFETY_ ENGINEERING_PRINCIPLES/link/616f3b7db148a924b8006b9d/ download [Last accessed: 20.10.2021]

[Tollen21] David W. Tollen: **The Tech Contracts Handbook** – *Software Licenses, Cloud Computing Agreements, and Other IT Contracts for Lawyers and Businesspeople* American Bar Association, Chicago, IL, USA, 3rd edition, 2021. ISBN 978-1-6410-5853-7

[Tornhill18] Adam Tornhill: **Software Design X-Rays** – *Fix Technical Debt with Behavioral Code Analysis* The Pragmatic Bookshelf, Raleigh, NC, USA, 2018. ISBN 978-1-68050-272-5

[Trivedi17] Kishor S. Trivedi, Andrea Bobbio: **Reliability and Availability Engineering** – *Modeling, Analysis, and Applications* Cambridge University Press, Cambridge, UK, 2017. ISBN 978-1-107-09950-0

[Tschanz18] Marcel Tschanz, Michael Kuss, Manuel Plattner, Patrick Akiki, Thomas Busch: *Excellence in Third-Party Risk Management (TPRM)* PricewaterhouseCoopers AG, Zurich, Switzerland, 2018. Downloadable from: https://www.pwc.ch/en/publications/2018/Excellence-in-third-party-risk-management.pdf [Last accessed: 29.6.2021]

[Ulbrich12] Peter Ulbrich, Martin Hoffmann, R. Kapitza, Daniel Lohmann, Wolfgang Schroder-Preikschat, Reiner Schmid: *Eliminating Single Points of Failure in Software-Based Redundancy* Proceedings 9th European Dependable Computing Conference, EDCC 2012, Sibiu, Romania, 2012. Downloadable from: https://www.researchgate.net/publication/254037732_Eliminating_ Single_Points_of_Failure_in_Software-Based_Redundancy [Last accessed: 31.8.2021]

[Voges13] Udo Voges: **Software Diversity in Computerized Control Systems** Springer Verlag, Wien, Austria, 2013 (Softcover reprint of the original 1st edition 1988). ISBN 978-3-709-18934-4

[Wagner19] Ferdinand Wagner, Ruedi Schmuki, Thomas Wagner,Peter Wolstenholme: **Modeling Software with Finite State Machines** – *A Practical Approach* CRC Press (Auerbach), Boca Raton, FL; USA, 2019. ISBN 978-0-367-39086-0

[Walkowiak16] Simon Walkowiak: **Big Data Analytics with R** – *Leverage R Programming to Uncover Hidden Patterns in your Big Data* Packt Publishing, Birmingham, UK, 2016. ISBN 978-1-7864-6645-7

[Wang04] Guijun Wang, C. K. Fung: Architecture paradigms and their influences and impacts on component-based software systems. 37th Annual Hawaii International Conference on System Sciences, Big Island, HI, USA, 2004. Available from: https://ieeexplore.ieee.org/document/1265643 [Last accessed: 10.6.2022]

[Weske19] Mathias Weske: **Business Process Management** – *Concepts, Languages, Architectures* Springer Verlag, Berlin, Germany, 3rd edition, 2019. ISBN 978-3-662-594315

[White17]        Robert White: **The Outsourcing Manual** Routledge (Taylor & Francis), London, UK, 2017). ISBN 978-1-138-25259-2

[Whiting20]      Erik Whiting, Sharon Andrews: Drift and Erosion in Software Architecture - Summary and Prevention Strategies. 4th International Conference on Information System and Data Mining (ICISDM), Hilo, USA, 2020. Downloadable from: https://www.researchgate.net/publication/339385701_Drift_and_Erosion_in_Software_Architecture_Summary_and_Prevention_Strategies/link/5e4e9ce3a6fdccd965b41e68/download [Last accessed: 10.6.2022]

[Wilson20]       Glenn Wilson: **DevSecOps** – *A leader's guide to producing secure software without compromising flow, feedback, and continuous improvement* Rethink Press, Gorleston, UK, 2020. ISBN 978-1-7813-3502-4

[Wolf20]         Marilyn Wolf, Dimitrios Serpanos: **Safe and Secure Cyber-Physical Systems and Internet-of-Things Systems** Springer Nature Switzerland, Cham, Switzerland, 2020. ISBN 978-3-030-25807-8

[Yao20]          Jiewen Yao, Vincent Zimmer: *Building Secure Firmware – Armoring the Foundation of the Platform* Apress Media LLC, New York, N.Y., USA, 2020. ISBN 978-1-4842-6105-7

[Young20]        Scott Young: **Designing a DMZ** SANS Institute White Paper, Bethesda, MA, USA, 2020. Downloadable from: https://www.sans.org/reading-room/whitepapers/firewalls/designing-dmz-950 [last accessed: 12.8.2020]

[Zwicky00]       Elizabeth D. Zwicky, Simon Cooper, D. Brent Chapman: **Building Internet Firewalls** O'Reilly & Associates, Sebastopol, CA, USA, 2nd edition 2000. ISBN 978-1-565-92871-8

# Principles for Safety

# 10

*Safety principles form the foundation for creating and operating trustworthy safety-critical cyber-physical systems. These principles have their roots in theoretical work and long-standing, time-tested, practical experience. Safety principles are a proven and successful way to teach, enforce, and implement safe systems.*

## 10.1 Principle S1: Safety Culture

An enforceable and robust *safety culture* is a necessary foundation for any organization involved in safety-critical cyber-physical systems. Key points of a safety culture are listed in Principle 10.1.

> **Quote**
> *"Culture is not about outcome, nor about human error. Culture is choice, framed by shared values and beliefs. Creating a strong safety culture means helping employees make good, safe choices"*
> David Marx (in [Gilbert18]), 2018

---

**Principle 10.1: Safety Culture**

1. Comprehensively formulate and document the values, attitudes, motivations, and incentives for the safety culture of the organization;
2. Involve all the stakeholders of the organization while creating the safety culture;

---

3. Follow the relevant standards (e.g., ISO 26262) while creating the safety culture;
4. Underlying and implicit assumptions enable an organization to manage and control its safety risks. Make all assumptions in the safety culture explicit and transparent;
5. Justify and document all assumptions for the safety culture;
6. Communicate and regularly check adherence to the safety culture throughout the organization. Install training programs for newcomers;
7. Define all incentives, e.g., bonuses, on all levels of the organizations to the benefit of safety—never detrimental to safety;
8. Define governance such that it strongly supports the safety culture;
9. Regularly audit, assess, and evolve the safety culture.

## 10.2   Principle S2: Safety Standards and Policies

*Safety standards* and the organization's *safety policy* are the essential documents for developing and operating trustworthy cyber-physical systems. While industry consortia predominantly supply safety standards, the safety policy is strongly geared toward the activities of the individual organization. The principle covering safety standards and the safety policy is stated in Principle 10.2.

---

**Principle 10.2: Safety Standards and Policies**

1. Develop and enforce a comprehensive safety policy covering all activities of the organization. Emphasize "Safety before Functionality";
2. Build the organigram of the organization such that the decision paths value "Safety before Cost" and "Safety before Time-to-Market";
3. Avoid technical safety debt. If technical debt is identified, remove it;
4. From the large number of available safety standards, carefully evaluate which are relevant for the organization. Respect legal, compliance, and certification requirements;
5. Build a comprehensive standards repository for easy, organized, direct access by all developers;
6. Regularly train the development staff specifically concerning the safety policy and the safety standards;
7. Insert specific safety quality gates into the development process to detect deviations from the safety policy or the mandatory safety standards (= enforcement). Apply these safety quality gates with special rigor to outsourced software development;
8. Duly reprimand/penalize infringements or disregard of the safety policy or the applicable safety standards;
9. Keep good records from the quality gate results/decisions and archive them securely for later audits or legal liability disputes;

10. Ascertain that both the safety policy and the list of applicable safety standards can stand the test of a possible, future product liability dispute;
11. Regularly revise and adjust the safety policy and the list of applicable standards (policy review board);

## 10.3   Principle S3: Safety Governance

Governance (Definition 2.25, Principle 8.4) in an organization is established to define and enforce the system of rules, practices, and processes by which a company is directed and controlled. Safety governance targets all activities related to safety, covering both the internal activities and the interaction with external entities (Principle 10.3).

**Principle 10.3: Safety Governance**

1. Exhaustively apply Principle 8.4;
   **In addition:**
2. Instill a commonly accepted safety culture in the organization (Principle 10.1);
3. Unambiguously assign clearly defined responsibilities to individual people (not committees) for all safety-related activities;
4. Install, document, and regularly audit processes and explicit responsibilities for risk management and residual risk acceptance;
5. Formally and continuously ensure the support of top management for all safety-related issues, decisions, and efforts;
6. Thoroughly and regularly train all involved persons in safety-related decision-making and the consequences of shortsighted decisions;
7. Explicitly reprimand/penalize infringements or disregard of the safety issues on all levels of the organization;
8. Participate (actively or passively) in industry consortia or standards setting organizations to gain early knowledge of safety developments;
9. Regularly audit, revise, and improve the governance processes and responsibilities, possibly using external experts;

## 10.4   Principle S4: Safety Management System

A *safety management system* (SMS, Definition 4.9) is a framework and an IT-based tool that enables an organization to manage its safety-related material, such as safety culture, policies, standards, tools, processes, and methodologies. The SMS improves the organization's ability to understand, construct, and manage trustworthy safety-critical systems.

A simple safety management system consists of a managed, organized, online repository that stores all information used during the development and operation of the cyber-physical system. It also contains the data related to activities, plans, reports, fault analysis, etc., as well as links to helpful sites and sources for safety (Principle 10.4).

---

**Principle 10.4: Safety Management System**

1. Carefully consider and evaluate the use of a safety management system (if it is not mandatory in the field of application);
2. Select an appropriate solution (Simple, managed repository or feature-rich third-party product);
3. Provide guidance and training for the use of the SMS;
4. Enforce the intended use of the SMS (Note: The value of the SMS manifests itself only if the information is complete and timely.);

---

## 10.5   Principle S5: Safety Principles

*Safety principles* are rules for the construction, evolution, and operation of safe systems. They are precisely formulated, well-justified, teachable, actionable, and enforceable (Definition 4.10). Safety principles have been distilled, applied, and proven over decades of work with trustworthy systems. In fact, safety principles are the treasure trove of safety engineering (Principle 10.5).

---

**Principle 10.5: Safety Principles**

1. From the available safety principles in books, publications, standards, field reports, case studies, accident coverage, etc., distill a set that is applicable to your field of application and the objectives of your organization (Note: The fundamental safety principles are contained in this monograph);
2. Assemble and document the mandatory, recommended, and informative safety principles for your organization in the "Safety Principles Handbook";
3. Include implementation advice in the "Safety Principles Handbook", such as patterns, guidelines, and examples;
4. Regularly revise and update the "Safety Principles Handbook";
5. For each project, clearly list which safety principles must be applied:
   - mandatory (must be unconditionally followed),
   - recommended (implement whenever possible),
   - informative (use for the guidance of design decisions).
6. In some cases, two (or more) safety principles cannot be implemented simultaneously, e.g., due to contradictions, performance, resource consumption, etc. In such cases, a justifiable, documented trade-off must be decided;

7. Any such trade-off must be risk-assessed;
8. Formulate mandatory rules for trade-offs in case of conflicting safety principles requirements;
9. Thoroughly check the adherence to the applicable safety principles during all phases of system development.

## 10.6  Principle S6: Safety Implementation

Trustworthy *safety implementation* (Fig. 4.5, Fig. 10.1) needs:

I. *Knowledge*: e.g., safety culture, policies, standards, laws, literature, education, regulations, and safety principles;
II. *Technology*: e.g., hardware, software, networks, algorithms;
III. *Processes*: e.g., governance, development, operation, maintenance;
IV. *People*: All actors involved in the value chain of the CPS;
V. *Verification*: Continuous monitoring of the safe behavior and feedback to improve I.–IV.;

Principles in this monograph have sufficiently covered the points I. to IV. However, point V. (Verification) needs additional information. Verification in the largest sense answers the question, "Are we building the system right?", i.e., Does the system correctly implement all the requirements, only the requirements, and nothing else?

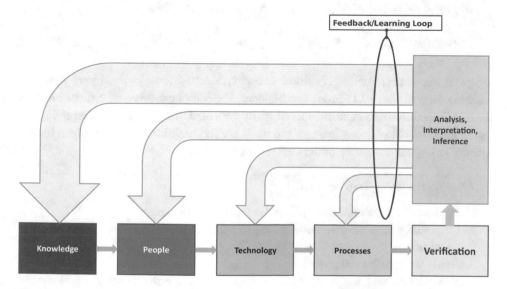

**Fig. 10.1**  Safety implementation chain

For the *verification* of a system, five techniques are available:

1. *Modeling and model checking* (e.g., [Mitra21], [Molnos17], [Logothetis04]): Build a model of the CPS with as much fineness as required for the expressiveness. Automatically check the model, if possible, with formal methods. Keep the model updated at all times;
2. *Testing* (e.g., [Bierig21], [Jose21]): Apply all the modern testing techniques, including regression and automated testing. Use specific testing procedures for the quality attributes, such as safety and security, e.g., targeted penetration, fuzzing, chaos engineering;
3. *Runtime monitoring* (Principle 10.8, e.g., [Bartocci19], [Ardagna10], [Hinrichs14], [Cheng19], [Búr20], [Logothetis04]): Supervise the running system and detect anomalies and misbehavior in real time. Whenever possible, automatically intervene in real time to prevent accidents;
4. *Offline monitoring* ([Chen98], [Sayed-Mouchaweh14], [Sampath99]): Collect and log sufficient activity and diagnosis data. Scan the log data regularly and search for anomalies or deviations from expected behavior. Take corrective action;
5. *Forensics* (e.g., [Nader20], [Brass18], [Conti21]): After an accident, near-miss, or critical incident, invoke the appropriate or compulsory forensic analysis procedures. Note that many fields of applications have their forensic procedures prescribed by laws or regulations (e.g., [ICAO01] for civil aviation, [RAIB15] for railways, [Stoop04] for maritime disasters.

> **Quote**
> *"Model Check What You Can, Runtime Verify the Rest"*
>     Timothy L. Hinrichs, 2014

Of great importance for safety-critical cyber-physical systems is the *safety feedback/ learning loop* in Fig. 10.1: From any accident, incident, near-miss, or critical incident, the maximum information must be forensically collected. Analyzing and interpreting this information leads to lessons learned, and these are fed back into the safety chain to continually strengthen the safety of the CPS (Principle 10.6).

---

**Principle 10.6: Safety Implementation**

1. Implement all the applicable principles from this monograph;
2. Continually raise the awareness for the importance of verification;
   **In addition:**
3. Apply verification techniques, i.e.,
   a. Modeling and model checking within reason for the cyber-physical system concerned,

    b. Extensive testing,

    c. Runtime monitoring,

    d. Offline monitoring,

    e. Forensics.

4. Automate verification as much as possible (e.g., in the DevOps process);
5. Carry out verification with sufficient coverage during all development and operation phases. Make adequate verification a mandatory step of the different process steps;
6. Implement the feedback/learning loop (Fig. 10.1) to track and improve the safety of the CPS continuously;
7. Know and strictly follow all the applicable forensic accident regulations and recommendations which govern the respective field of application and feedback the insights into the safety chain;
7. Follow, acquire, and use the rapid progress in verification techniques.

## 10.7   Principle S7: Safety Assessment and Audit

*Safety audits* (Definition 4.12) are executed to verify the conformance of system implementation with provisions, such as requirements, standards, and regulations. Four targets of audits exist (Fig. 10.2):

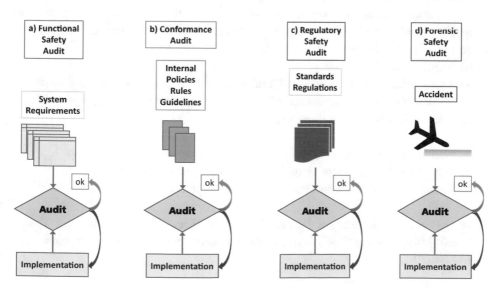

**Fig. 10.2**  Types of safety audits

1. *Functional safety audit*: The system implementation is examined concerning the requirements, especially the safety and security requirements;
2. *Conformance safety audit*: The system implementation is checked against the organization's policies, rules, guidelines, and internal standards;
3. *Regulatory safety audit*: The system implementation is gauged against industry standards, regulations, legal or certification requirements;
4. *Forensic safety audit*: After an accident, serious incident, or a near-miss, the system implementation is investigated concerning the causes, the chain of events, and the accountability of the implementation.

Each audit has two possible outcomes: a) The implementation fully conforms to the provisions, or b) the system implementation needs ameliorations. The audits can either be carried out internally (by members of the organization), externally (by chartered consultants) or by a regulatory body (such as a certification agency).

An effective audit procedure is *reviews* (e.g., [Zhu16], [Tuffley11], [Summers19], Example 10.1). Audit guidance is listed in Principle 10.7.

---

**Example 10.1: Review Types**

a) **Fagan Review**
A Fagan review ([ecci18], [Knight12], [Doolan92]) is a process to find faults and defects in human-readable documents, such as source code, formal specifications, and system files, during the software development process phases. It is named after Michael Fagan (IBM, 1976), credited as the inventor of formal software reviews. The Fagan inspection has a coherent role model, a formalized execution, and predefined result artifacts. The Fagan inspection process can well be integrated into any organization's development process.

b) **Rigorous Review**
The idea behind *rigorous reviews* ([Liu04, Chapter 17]) is to express the important properties of the specifications formally as predicate expressions (e.g., [Parnas93]). A rigorous review consists of three steps:

1. Derive the properties to be reviewed (Here: security properties);
2. Generate a graphical representation for each property to be analyzed;
3. Review all the components occurring in the graphical representation.

A rigorous review needs a formal language with suitable semantics and syntax. [Liu04] proposes the SOFL (= Structured Object-Oriented Formal Language, e.g., [Afifudi20]).

---

**Principle 10.7: Safety Assessment and Audit**

1. Define and document complete, reasonable, and comprehensive audit objectives. Revise them regularly;
2. Include and respect all laws, regulations, and standards applicable to the field of application;
3. Execute audits regularly in the organization. Execute the four types of audits:
   a. Functional safety audit,
   b. Conformance safety audit,
   c. Regulatory safety audit,
   d. Forensic safety audit.
4. Constitute specialized, well-trained audit teams (internal and external participation);
5. Carefully evaluate, install, and use defined and proven audit procedures;
6. Document and archive all audit results;
7. Immediately improve the product or process in case of weaknesses discovered by the audits.

---

## 10.8  Principle S8: Safety Runtime Monitoring

An essential element in the *safe software chain* (Fig. 10.3) is fault mitigation: Absorbing or minimizing the impact of faults and failures during *runtime*. The concept is *runtime monitoring* (Definition 4.13). The running system is continuously monitored, faults and failures are detected, diagnosed, and mitigated during runtime (Principle 10.8).

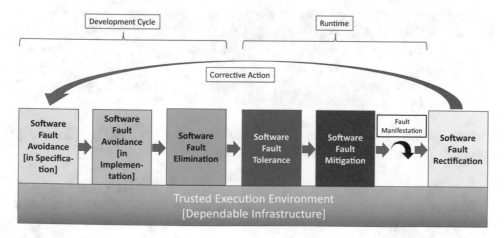

**Fig. 10.3**  Safe software chain

> **Quote**
> *"Safety is a continuous journey, not a final destination"*
>   https://inspiresafety.com/collections/safety-is-a-journey

---

**Principle 10.8: Safety Runtime Monitoring**

1. Implement as much runtime monitoring as reasonably possible in the system (detection, diagnosis, and mitigation of faults and failures during runtime);
2. Include runtime monitoring in the initial requirements and specifications;
3. Use risk assessment to justify the runtime monitoring effort;
4. Automate the response to runtime faults as much as possible (real-time response);
5. Carefully define the safe states of a system. Use runtime monitoring to transfer to a safe state whenever possible;
6. In case of irrecoverable faults, gather sufficient forensic analysis information;
7. Use operating system/language support (exceptions, faults, errors) in runtime monitoring.

---

## 10.9   Principle S9: Safe Software

*Safe software* (Definition 5.13, Fig. 5.14) is a risk-assessed element of safety-critical cyber-physical systems. The safe software prevents, absorbs, and mitigates faults and failures. In addition to the material in this monograph, the *safe software chain* (Fig. 10.3) defines a schema for reasoning about safe software. Note that safe software requires a *Trusted Execution Environment* (dependable infrastructure, Definition 9.15).

> **Quote**
> *"All of the activities that we undertake when building software might be opportunities for defects to enter the software and remain there"*
>   John Knight, 2012

The safe software chain has the following constituents (e.g., [Knight12], [Jackson09], [Gullo18], [Hollnagel14], [Leveson16], [Miller20]):

1. Software *fault avoidance* in the specifications: Precise, unambiguous, complete specifications are difficult. Unfortunately, defects in the specifications lead to the most costly consequences;

2. Software *fault avoidance* in the implementation: The unlimited malleability of software allows incredible functionality but also generates a broad spectrum of possible faults and defects;
3. Software *fault elimination*: Serious fault avoidance during development will significantly reduce—but never entirely eliminate—the number of faults and defects in the software;
4. Software *fault tolerance*: Fault tolerance is a prerequisite for the Trusted Execution Environment. However, software fault tolerance may also be needed for a trustworthy cyber-physical system;
5. Software *fault mitigation*: Unfortunately, even the best prudence during the development process cannot prevent the creation of faults and defects in the software. In some cases, these faults and defects will never become manifest. In other cases, they will surface during operation and cause malfunctions of the system;
6. Software *fault rectification*: The last and essential step in the safe software chain is rectifying any fault or defect. Not only must the defect be repaired, but all artifacts affected must be correctly updated (feedback/learning loop, reverse engineering).

The safe software chain represents a summary of the application of all safety principles in this monograph. Principle 10.9 is, therefore, a guideline.

---

**Principle 10.9: Safe Software**

1. Produce complete, consistent, unambiguous, and comprehensive requirements for the system, with particular attention to the safety requirements.
2. Transform the requirements into verifiable, testable, and traceable specifications;
3. For safe software, use the following schema:
   1. Software fault avoidance in the specifications: Use the strongest possible mechanism for the specifications, such as patterns, formal languages, formal checking, and effective reviews;
   2. Software fault avoidance in the implementation: Apply the most powerful techniques for the development process, such as "Correct-by-Construction", formal modeling and model checking, simulation, sufficient up-front effort (architecture, design expenditure), a safety-aware development process, clean code, respect all safety principles. If necessary, use design diversity;
   3. Software fault elimination: Invest sufficient effort in the fault elimination methods, such as numerous reviews of all artifacts produced, code analysis, extensive testing, specialized testing for safety. Automate the fault elimination as much as possible;
   4. Software fault tolerance: Use the risk assessment to identify necessary software fault tolerance. Employ proven fault tolerance techniques, such as redundancy and functional & data diversity;

5. Software fault mitigation: Perceive, diagnose, and mitigate faults and anomalies during runtime. Invest adequate mitigation mechanisms, such as runtime monitoring or a protective shell. Incorporate satisfactory diagnostic capabilities, both during runtime and offline;

6. Software fault rectification: Any fault or defect discovered must be repaired as soon as possible. Ensure that all components of the cyber-physical system provide sufficient, timely, and analyzable information to mend the system;

4. Strictly use a safety-aware development process;

5. Use safety-supporting tools as much as possible (DevOps tools, SMS, etc.);

6. Make extensive use of the valuable body of information from universities, government organizations, and standards (such as the extensive checklist in NASA-STD-8739.8A, [NASA20]);

## 10.10   Principle S11: Artificial Intelligence in Safety-Critical CPS

*Artificial intelligence* (AI) is invading cyber-physical systems. AI enables inspiring applications, but unfortunately, also introduces new risks. The handling of AI in trustworthy cyber-physical systems requires specific care and a deep sense of responsibility. Applicable guidance is given in Principle 10.10.

> **Quote**
> *"The only constant is change. Designing, deploying, and sustaining AI systems require engineering practices to manage inherent uncertainty in addition to constant and increased rhythm of change. Algorithms, practices, and tools to engineer AI systems are constantly evolving, and changes brought by these systems reach across the problem, technology, process, engineering, and cultural boundaries. These AI engineering practices provide a foundation for decision-makers to navigate those changes to develop viable, trusted, and extensible systems. As we build and use these systems, we will define better-codified engineering and data management practices as well as tools"*
> Angela Horneman, Andrew Mellinger, and Ipek Ozkaya, 2019

---

**Principle 10.10: AI in Safety-Critical CPS**

1. Use AI only if there is no other solution;
2. Ensure that the problem can (and should) be solved by AI;
3. Execute a thorough risk assessment concerning the AI functionality;
4. Chose AI algorithms suited to the problem;
5. Whenever possible, use XAI-algorithms (Explainable Artificial Intelligence);

6. Ensure that your training data is sufficiently complete, not corrupted, not biased, or dubious;
7. Mitigate adversarial activity during training and operation;
8. Assess and maintain sufficient model quality at all times;
9. Runtime monitor the operation of the AI algorithms and have mitigation measures in place to avoid accidents;
10. Define checkpoints to account for the potential needs of recovery, traceability, and decision justification ([Horneman19]);
11. Loosely couple the AI algorithm to the system to make it easily exchangeable;
12. Treat ethics as both a software design consideration and a policy concern (e.g., [Horneman19]).

# References

[Afifudi20]    Irfin Afifudi, Inge Martina: **Implementation of Structured Object-Oriented Formal Language for Warehouse Management System** CommIT (Communication & Information Technology) Journal, 14(1), pp. 1–8, 2020. Downloadable from: https://www.researchgate.net/publication/343234871_Implementation_of_Structured_Object-Oriented_Formal_Language_for_Warehouse_Management_System/link/5f1ecd3f299bf1720d681abe/download [Last accessed: 16.01.2022]

[Ardagna10]    Danilo Ardagna, Li Zhang (Editors): **Run-Time Models for Self-managing Systems and Applications** Birkhäuser, Springer Verlag, Basel, Switzerland, 2010. ISBN 978-3-034-60432-1

[Bartocci19]    Ezio Bartocci, Yliès Falcone (Editors): **Lectures on Runtime Verification** - *Introductory and Advanced Topics* Springer Nature, Cham, Switzerland, 2019. ISBN 978-3-319-75631-8 (LNCS 10457)

[Bierig21]    Ralf Bierig, Stephen Brown, Edgar Galvn, Joe Timoney: **Essentials of Software Testing** Cambridge University Press, Cambridge, UK, 2021. ISBN 978-1-108-83334-9

[Brass18]    Irina Brass: *Standardizing IoT Security – Implications for Digital Forensics* Digital Forensics Magazine, Issue 35, May 2018. Downloadable from: https://discovery.ucl.ac.uk/id/eprint/10050054/13/Brass_BRA001-pdf.pdf [Last accessed: 23.11.2021]

[Búr20]    Márton Búr, Gábor Szilágyi, András Vörös, Dániel Varró: *Distributed graph queries over models@run.time for runtime monitoring of cyber-physical systems* International Journal on Software Tools for Technology Transfer, Springer Verlag, Berlin, Germany, Nr. 22, pp. 79–102, 2020. Downloadable from: https://doi.org/10.1007/s10009-019-00531-5https://link.springer.com/article/10.1007/s10009-019-00531-5 [Last accessed: 23.11.2021]

[Chen98]    Jie Chen, R.J. Patton: **Robust Model-Based Fault Diagnosis For Dynamic Systems** Kluwer Academic Publisher, Dordrecht, Netherlands, 1999. ISBN 978-1-4613-7344-5

[Cheng19]     Long Cheng, Ke Tian, Danfeng (Daphne) Yao, Lui Sha, Raheem A. Beyah: *Checking is Believing - Event-Aware Program Anomaly Detection in Cyber-Physical Systems* Preprint, arXiv:1805.00074v2 [cs.CR], 25 Mar 2019. Downloadable from: https://arxiv.org/pdf/1805.00074.pdf [Last accessed: 23.11.2021]

[Conti21]     Mauro Conti, Federico Turrin: *Cyber Forensics for CPS* In: Jajodia S., Samarati P., Yung M. (editors): Encyclopedia of Cryptography, Security, and Privacy. Springer Verlag, Berlin, Germany, 2021. https://doi.org/10.1007/978-3-642-27739-9_1722-1

[Doolan92]    E.P. Doolan: *Experience with Fagan's Inspection Method* SOFTWARE—PRACTICE AND EXPERIENCE, Vol. 22, Nr. 2, pp. 173–182, February 1992. Downloadable from: https://www.ida.liu.se/~TDDC90/literature/lab-papers/doolan91.pdf [Last accessed: 27.11.2021]

[ecci18]      ecci: *Fagan Inspection* ecci group, Tutorial, Legaspi Village, Makati City, Philippines, 2018a. Downloadable from: http://eccinternational.com/iNugget/FIM.pdf [Last accessed: 27.11.2021]

[Gilbert18]   Claude Gilbert, Benoît Journé, Hervé Laroche, Corinne Bieder (Editors): **Safety Cultures, Safety Models - Taking Stock and Moving Forward** Springer Nature Switzerland AG, Cham, Switzerland, 2018b. ISBN 978-3-319-95128-7. Downloadable from: https://link.springer.com/book/10.1007%2F978-3-319-95129-4 [Last accessed: 25.5.2020]

[Gullo18]     Louis J. Gullo, Jack Dixon: **Design for Safety** John Wiley & Sons, Inc., Hoboken, NJ, USA, 2018. ISBN 978-1-118-97429-2

[Hinrichs14]  Timothy Hinrichs, A. Prasad Sistla, Lenore Zuck: *Model Check What You Can, Runtime Verify the Rest* In: Andrei Voronkov and Margarita Korovina (Editors). HOWARD-60. A Festschrift on the Occasion of Howard Barringer's 60th Birthday, Vol 42, pages 234—244. Downloadable from: https://easychair.org/publications/paper/tq7 [Last accessed: 23.11.2021]

[Hollnagel14] Erik Hollnagel: **Safety-I and Safety-II** CRC Press, Francis & Taylor, Boca Raton, FL, USA, 2014. ISBN 978-1-472-42308-5

[Horneman19]  Angela Horneman, Andrew Mellinger, Ipek Ozkaya: **AI Engineering: 11 Foundational Practices** - *Recommendations for Decision Makers from Experts in Software Engineering, Cybersecurity, and applied Artificial Intelligence* White Paper DM19-0624, 06.06.2019. CARNEGIE MELLON UNIVERSITY, Software Engineering Institute (SEI), Pittsburgh, PA, USA, 2019. Downloadable from: https://resources.sei.cmu.edu/asset_files/WhitePaper/2019_019_001_634648.pdf [last accessed: 19.8.2021]

[ICAO01]      ICAO: **Annex 13 To the Convention on International Civil Aviation Aircraft Accident and Incident Investigation** International Civil Aviation Organization, Montreal, Canada, November 2001. Downloadable from: https://www.emsa.europa.eu/retro/Docs/marine_casualties/annex_13.pdf [Last accessed: 21.11.2021]

[Jackson09]   Scott Jackson: **Architecting Resilient Systems** - *Accident Avoidance and Survival and Recovery from Disruptions* John Wiley & Sons, Inc., Hoboken, NJ, USA, 2009. ISBN 978-0-470-40503-1

[Jose21]        Boby Jose: **Test Automation - *A Manager's Guide*** BCS, The Chartered Institute for IT, Swindon, UK, 2021. ISBN 978-1-7801-7545-4

[Knight12]      John Knight: **Fundamentals of Dependable Computing for Software Engineers** CRC Press (Taylor & Francis), Boca Raton, FL, USA, 2012. ISBN 978-1-439-86255-1

[Leveson16]     Nancy G. Leveson: **Engineering a Safer World - *Systems Thinking Applied to Safety*** MIT Press Ltd., Massachusetts, MA, USA, 2016. ISBN 978-0-262-53369-0

[Liu04]         Shaoying Liu: **Formal Engineering for Industrial Software Development - *Using the SOFL Method*** Springer Verlag, Heidelberg, Germany, 2004. ISBN 978-3-540-20602-6

[Logothetis04]  Georgios Logothetis: **Specification, Modelling, Verification, and Runtime Analysis of Real-Time Systems** IOS Press, Amsterdam, Netherlands, 2004. ISBN 978-1-5860-3413-9

[Miller20]      Joseph D. Miller: **Automotive System Safety - *Critical Considerations for Engineering and Effective Management*** John Wiley & Sons, Chichester, UK, 2020. ISBN 978-1-119-57962–5

[Mitra21]       Sayan Mitra: **Verifying Cyber-Physical Systems - *A Path to Safe Autonomy*** The MIT Press, Cambridge, MA, USA, 2021. ISBN 978-0-262-04480-6

[Molnos17]      Anca Molnos, Christian Fabre (Editors): **Model-Implementation Fidelity in Cyber-Physical System Design** Springer International Publishing, Cham, Switzerland, 2017. ISDN 978-3-319-83705-5

[Nader20]       Mohamed Nader, Jameela Al-Jaroodi, Imad Jawhar: **Cyber-Physical Systems Forensics - *Today and Tomorrow*** Journal of Sensor and Actuator Networks, Basel, Switzerland, 2020, August 2020, DOI:https://doi.org/10.3390/jsan9030037. Downloadable from: https://www.mdpi.com/2224-2708/9/3/37 [Last accessed: 17.10.2020]

[NASA20]        NASA Office of Safety and Mission Assurance (OSMA): **NASA Software Assurance and Safety Standard** US National Aeronautics and Space Administration (NASA), Washington, DC, USA, 2020. Document NASA-STD-8739.8A. Downloadable from: https://standards.nasa.gov/standard/osma/nasa-std-87398 [Last accessed: 9.2.2021]

[Parnas93]      David Parnas: ***Predicate Logic for Software Engineering*** IEEE Transactions on Software Engineering, Vol. 19, No. 9, September 1993. Downloadable from: https://www.researchgate.net/publication/3187595_Predicate_Logic_for_Software_Engineering [Last accessed: 16.01.2022]

[RAIB15]        UK Rail Accident Investigation Board: ***Guidance on the Railways (Accident Investigation and Reporting) Regulations 2005*** UK Rail Accident Investigation Board, Derby, UK, Version 4.0, August 2015. Downloadable from: https://assets.publishing.service.gov.uk/government/uploads/system/uploads/attachment_data/file/456936/guidance_to_rair_regs_v4.pdf [Last accessed: 23.11.2021]

[Sampath99]     M. Sampath, R. Sengupta, S. Lafortune, K. Sinnamohideen, D. Teneketzis: **Diagnosability of discrete-event systems** IEEE Transactions on Automation, Volume 40, Issue 9, September 1999. Available from: https://ieeexplore.ieee.org/document/412626 [Last accessed: 24.11.2021]

[Sayed-Mouchaweh14]    Moamar Sayed-Mouchaweh: **Discrete Event Systems -** *Diagnosis*
                       *and Diagnosability* Springer Verlag, New York, NY, USA, 2014. ISBN
                       978-1-461-40030-1

[Stoop04]              J. A. Stoop: *Maritime Accident Investigation Methodologies* Injury
                       Control and Safety Promotion, 10. 2004, pp. 237–242. Downloadable
                       from:        https://www.researchgate.net/publication/5767263_Maritime_
                       accident_investigation_methodologies Last accessed: 13.11.2021]

[Summers19]            Boyd L. Summers: **Software Engineering Reviews and Audits**
                       CRC Press (Taylor & Francis), Boca Raton, FL, USA, 2019. ISBN
                       978-0-367-38312-1

[Tuffley11]            David Tuffley: **Software Reviews & Audits -** *A How-To Guide for*
                       *Project Staff* CreateSpace Independent Publishing Platform, 2011.
                       ISBN 978-1-4611-3046-8

[Zhu16]                Yang-Ming Zhu: **Software Reading Techniques -** *Twenty Techniques*
                       *for More Effective Software Review and Inspection* Apress Media,
                       LLC, New York, NY, USA, 2016. ISBN 978-1-484-22345-1

# Principles for Security

# 11

*Security principles form the foundation for the creation and operation of security-critical cyber-physical systems. They have their roots in theoretical work and long-standing, proven practical experience. Security principles are a proven and tested way to teach, enforce, and implement secure systems.*

> **Quote**
> *"Software weaknesses are a direct reflection of inadequacies in the state of the art and practice of software engineering, and they can affect millions of people"*
> SEI, 2021 (Architecting the Future of Software Engineering)

## 11.1 Principle E1: Security Culture

The comprehensive, enforceable, and robust *security culture* of an organization is the indispensable foundation for building trustworthy safety-critical cyber-physical systems (e.g., [Blum20]). Throughout all phases of the CPS life, people's behavior is, in many cases, decisive for the success or failure of a security incident—thus, a viable security culture is mandatory. Critical points of a safety culture are listed in Principle 11.1.

---

**Principle 11.1: Security Culture**

1. Comprehensively formulate and document the values, attitudes, motivations, and incentives for the security culture of the organization;
2. Involve all the stakeholders of the organization while creating the security culture;

3. Follow the relevant standards (e.g., ISO/IEC 27000 family of standards) while creating the security culture;
4. Underlying and implicit assumptions enable an organization to manage and control its security risks. Make all assumptions in the security culture explicit and transparent;
5. Justify all assumptions for the security culture;
6. Communicate and regularly check adherence to the security culture throughout the organization. Install training programs for newcomers;
7. Define all incentives, e.g., bonuses, on all levels of the organizations to the benefit of security—never detrimental to security;
8. Define governance such that it strongly supports the security culture;
9. Ensure that the security culture statements are consistent with other cultural values of the organization, such as performance goals. Avoid all conflicting issues in cultural values and performance metrics;
10. Because of the security culture's fundamental significance for the security of the cyber-physical systems, it must have a high priority, sufficient funding, and the necessary attention throughout the organization;
11. Nominate a responsible manager for the development and maintenance of the security culture;
12. Continuously strengthen the security culture through communications and awareness programs;

## 11.2   Principle E2: Security Standards and Policies

Security is strongly conditional on the security industry standards and the organization's internal security policy. This set of documents forms the frame of good security (Principle 11.2, [Wahe11], [Bishop18], [Arnold17]). A prerequisite to successful information security is the *information classification policy* (e.g., [UKCO18]). Each information asset in the organization must be categorized into a protection class. Once the information is *classified*, it can be correctly access-protected in rest and motion. Various information classification schemes exist (e.g., Example 11.1).

**Example 11.1: Information Classification**

One possible classification scheme consists of four classes: 1) public, 2) internal use only, 3) confidential, and 4) secret. The authorized users and protection are listed in Table 11.1.

**Principle 11.2: Security Standards and Policies**

1. Develop and enforce a comprehensive security policy covering all activities of the organization. Emphasize "security before functionality";

**Table 11.1**  Information classification example

| | Information Classification | | | |
|---|---|---|---|---|
| | Public | Internal use | Confidential | Secret |
| Definition | | | | |
| Authorized Users | Unlimited | Employees and contractors with a valid contract, authorities | Restricted group of named employees, authorities | Very small circle of "need to know" employees |
| Protection | o © (Copyright)<br>o None, public access | o Authorized access (Employee or contractor credentials)<br>o Leaving the organization permitted if required for work<br>o Mobile devices encrypted | o Authorized access - 3-way authentication<br>o Least privilege<br>o Leaving the organization needs a permit<br>o Mobile devices encrypted | o Need to know<br>o Numbered, personal copies<br>o Never in electronic form<br>o Copying strictly prohibited<br>o Leaving the organization never allowed<br>o Managed & distributed by a central authority |

2. Use industry-standard templates, such as ISO/IEC 27002, for the development of the security policy;

3. Cover the issue "security in third-party products" and "open-source software" extensively in the security policy;

4. Build the organigram of the organization such that the decision paths value "security before cost" and "security before time-to-market";

5. Avoid technical security debt;

6. If technical security debt is identified, remove it as soon as possible;

7. From the large number of available security standards, carefully evaluate which are relevant for the organization. Respect legal, compliance, and certification requirements;

8. Build a comprehensive standards repository for easy, organized, and direct access by all developers;

9. Regularly train the development staff specifically concerning the security policy and the security standards;

10. Insert specific security quality gates into the development process to detect deviations from the security policy or the mandatory security standards (= enforcement). Apply these security quality gates with special rigor to outsourced software development and third-party product acquisition;

11. Establish an approval guideline and process for all open-source software to be introduced into the organization;

12. Duly reprimand/penalize infringements or disregard of the security policy or the applicable security standards;

13. Keep good records from the quality gate results/decisions and archive them dependably for later audits or legal liability disputes;

14. Ascertain that both the security policy and the list of applicable security standards can stand the test of a possible, future liability dispute;

15. Regularly revise and adjust the security policy and the list of applicable standards (policy review board);

16. If possible, use standardized security policy languages, such as XACML (Extensible Access Control Markup Language ([Wahe11], [XACML13]);

## 11.3   Principle E3: Security Governance

Slightly modifying the general governance definition (Definition 2.25) leads to the definition of *security governance*: "Security governance is the organization, and the system of rules, practices, and processes by which a company directs and controls its security activities".

Security governance forms the stage for all security activities and the assurance for successful, effective security implementation and operation ([Ferrillo17], [Brotby09], [Blum20]). Note that the role of a competent, decisive *Chief Information Security Officer* (CISO) is of highest importance ([Badhwar21a], [Badhwar21b]).

> **Quote**
> *"Failure to implement effective information security governance will result in the continued chaotic, increasingly expensive, and marginally effective firefighting mode of operation"*
>   Krag Brotby, 2009

Therefore, establishing a complete, consistent, comprehensive, and consequently enforced security governance is a primary responsibility of senior management. Guidance for establishing security governance is listed in Principle 11.3. The first step is to establish the organization's security objectives unequivocally. For many organizations, adopting a proven *information security governance framework*  (such as [NCSP04], [ISO17799]) is highly beneficial.

**Principle 11.3: Security Governance**

Follow all principles for general governance in Principle 11.4, plus:

1. Assign the responsibility for security governance to a member (not a committee) of the top management. Request regular reporting to the board;
2. If suitable, adopt a proven, industry-standard information security governance framework;
3. Formulate the organization's security objectives in a complete, consistent, and comprehensive document. Whenever possible, reinforce the security objectives with meaningful metrics. Implement the objectives without fail;
4. Communicate, explain, and enforce the security objectives as a continuous process;
5. Maintain a list and a repository of all binding security policies, standards, and guidelines. Offer easy access to the repository in all phases of development and operation;
6. Regularly update the list of binding security policies, standards, and guidelines according to developments in law, regulations, standardization, certification, and industry;
7. Motivate, train, and support the organization's staff during all phases of development, operation, and evolution;
8. Enforce strict adherence to all security policies, standards, and guidelines in all development, operation, and evolution phases. Use adequate means, such as reviews, audits, or automated tools;
9. Screen each version of third-party products, services, or open-source software used by the organization with respect to the organization's security policies, standards, and guidelines. Risk-assess any non-compliance of third-party products or services and explicitly accept any reasonable residual risk. List any security deficiency as technical debt and risk for potential security incident;
10. Have contingency plans for cyber-incidents ready. Communicate and train their execution regularly (cyber-crisis management);

## 11.4 Principle E4: Information Security Management System

Successful *security management* in a cyber-physical system is a multifaceted task and is based on extensive information about the system (e.g., architecture, designs), the risks (e.g., methods, decisions), the standards and principles, and many more topics. This massive information must be organized, maintained, made available, and some parts of it must be enforced. In fact, much of the security of a CPS depends on the organized availability of the correct information at the right time to the concerned stakeholders. This is the task of a worthwhile *security management system* (ISMS, Definition 4.20).

The predominant perception of an ISMS is the information security *risk management handling* ([ISO2700], [BSI08], [ITG20]). The prime source for an ISMS is the ISO/IEC 27000:2018 family of standards (Example 4.15, [Gallotti19]).

However vital the information security risk management is, it is not sufficient to develop trustworthy cyber-physical systems. The ISMS must provide much more well-ordered, accessible information. Therefore, the organization must first conduct a thorough inquiry to identify all information relevant to the security. This information is then organized, tagged, catalogized, and made expediently accessible. Example 11.2 lists the primary content of an ISMS.

The security management system is *not* a single homogeneous system or product but consists of several different systems or products for specific tasks. They are often implemented as part of the *enterprise information portal* ([Firestone02]). Principle 11.4 describes basic requirements for an ISMS.

> **Quote**
> *"Practical experience has shown that optimising information security management frequently improves information security more effectively and lastingly then investing in security technology"*
>     Bundesamt für Sicherheit in der Informationstechnik, 2008

**Example 11.2: ISMS Content**

The content of an ISMS is highly conditional on the organization's needs: First, a thorough inquiry to identify all information relevant to the security tasks must be executed. Following, an ISMS can be assembled from an in-house development activity and third-party products. A selection of content of an ISMS is listed in Table 11.2.

**Principle 11.4: Information Security Management System (ISMS)**

1. Extend the security management system from risk management to a complete information resource for CPS security accessible to all stakeholders;
2. Conduct an inquiry identifying the complete information required in the organization for all security tasks. Define, assemble, and deploy the ISMS;
3. Use only proven, dependable products to assemble the security management system. Ensure that the different products never diverge in time or content;
4. Present the access to the ISMS through the enterprise information portal;
5. Strictly separate user functionality from information system management functionality. Assure that there is no common or shared functionality;

**Table 11.2**  ISMS content

**ISMS**: That part of the overall management system, based on a business-risk approach, to establish, implement, operate, monitor, review, maintain, and improve information security (ISO 17799 (https://www.iso.org/standard/39612.html))

| Topic | Content | Objective |
|---|---|---|
| Configuration | Complete, real-time information about all operational system elements (hardware, firmware, software, communications, etc.) | Exact knowledge about the elements of the operational system, including the current runtime system, the test environment, all auxiliary environments, and their connections for the purpose of analysis, audits, and targeted investigations ([Sustrico18], [Cartlidge07], [Aiello10], [ITIL19], https://www.itlibrary.org/) |
| Patch/Update Status | Precise, complete real-time information about the versions, patch, and update status of all system elements (hardware, firmware, software, communications, etc.) | Dependable overview of the situation concerning the versions, updates, functional and security patches for all system elements for the purpose of reviews and security updates for all system elements, including third-party hardware and software (routine patching, emergency patching, emergency workarounds, unpatchable assets), ([Stackpole19], [Souppaya21]) |
| Application Portfolio | Global, complete repository of all active applications, their dependencies (relationships), operative versions, parameters, owner, etc. | Support for the development of the application landscape. Providing fundamental, detailed information about all versions of the applications currently in use. Revealing complete information about context, functionality, data, relationships, technology, dependencies, and parameters of all applications |
| Mandatory Models | Current versions of all models | Domain model, architecture models, design models, application models, process models, etc. |
| Standards, Policies, Directives | Current versions of all mandatory, recommended, and informative documents | Easy access to all security-guidance documents, providing latest versions and comments Source for all stakeholders of the CPS for complying with the mandatory and recommended technical industry standards. Applicable during project work, e.g., design, reviews, and for the verification/audits of systems in operation ([Landoll16], [Scarfone10]) |
| Technical Debt | Exhaustive list of all technical debt issues in all system elements | Technical debt issue description, including risks and consequences, constraints, and commitments. Source for technical debt removal, risk analysis, security audits, and reviews ([Ernst21], [Kruchten19], Principle 9.3) |

(continued)

**Table 11.2**   (continued)

**ISMS**: That part of the overall management system, based on a business-risk approach, to establish, implement, operate, monitor, review, maintain, and improve information security (ISO 17799 (https://www.iso.org/standard/39612.html)

| Topic | Content | Objective |
|---|---|---|
| Digital Certificates | A complete register of all digital certificates used in the organization (for machines, software, organizations, and people) and exchanged with partners | Exhaustive, updated catalog of all digital certificates issued, managed and used in the organization, including their parameters (e.g., revoked or expiry date). Source for the timely update or fast revocation of the digital certificates, and for reviews and security audits ([Nash01], [Buchmann16], [NIST1800-16]) [Sectigo20]) |
| Authentication Credentials | A complete register of all authentication credentials issued or accepted in the organization (for machines, devices, IoT, software, organizations, people, and partners) | A complete register of all authentication credentials issued or accepted in the organization for the purpose of renewal, revocation, regular audits, and security assessments ([Grassi17], [Halak21]) |
| Authorization Rights | A complete account of all authorization rights to all protection assets of the organization | In-depth, complete index of all authorization rights to all protection assets of the organization, including their full parameter set. Internal and external access rights for partners, authorities, people, machines, software, and devices are included. Foundation for timely updates and restrictions, emergency withdrawal, crisis management, security audits ([Pompon16], [Rai07]) |
| Security Reviews and Audit Results | Database containing the reports/results of all security reviews and audits conducted, including their findings, required improvements, and recommendations (internal and external reviews and audits) | Input to project planning for remedying weaknesses in the system's security. The base for security and risk management ([Davis21], [Moeller10], [Sigler20]) |
| Laws and Regulations, Compliance Requirements | Well-organized list and guidance concerning all laws, regulations, and compliance requirements currently in force for the development and operation of the cyber-physical systems | Source for all stakeholders of the CPS for complying with legal provisions valid at present. Applicable during project work, e.g., reviews, and for the verification/audits of systems in operation ([Overly15], [Schreider20], [Lukings21]) |
| Certification | All applicable certification procedures, standards, guidelines, processes, and reviews | Support for all certification activities during all phases of development, evolution, and operation |

(continued)

**Table 11.2**   (continued)

**ISMS**: That part of the overall management system, based on a business-risk approach, to establish, implement, operate, monitor, review, maintain, and improve information security (ISO 17799 (https://www.iso.org/standard/39612.html))

| Topic | Content | Objective |
|---|---|---|
| Principles and Patterns | Comprehensive, annotated repository of all currently mandatory and recommended principles (Definition 4.10, Definition 4.21) and patterns (Definition 7.4) | Source for all stakeholders of the CPS for complying with the mandatory and recommended safety, security, and risk principles and patterns ([Furrer19], [Furrer22]) |
| Cyber-Crisis Handling | Procedures and guidelines for handling a cyber-crisis (Principle 11.6) | Quick, direct access to the necessary procedures for the control and mitigation of security incidents |
| Templates | Current versions of all pre-scribed templates | Templates for all software and document artifacts are a relevant means to exact completeness, precision, and uniformity between the stakeholders |

**Information Security Risk Management:**
6.  Define, document, enforce, and develop an adequate risk management process;

**Additional Functionality and Content:**
7.  Maintain a complete, timely, managed repository of all authentication credentials;
8.  Regularly execute audits to check the validity of all credentials. Force credentials protection, such as frequent password changes;
9.  Maintain a complete, timely, managed repository of all authorization rights, such as access rights;
10.  Regularly execute audits to check the validity of all authorization rights. Ensure that only authorization rights are in place that are unconditionally required for a specific task. Immediately withdraw any authorization right if a person leaves the company or changes responsibilities;
11.  Carefully manage digital certificates (for people, machines, and programs). Renew in time before expiry. Include the revocation list;

## 11.5   Principle E5: Security Principles

Like the safety principles (Principle 10.5), *security principles* (Definition 4.21) are rules for constructing, evolving, and operating secure systems. They are precisely formulated, well-justified, teachable, actionable, and enforceable. Security principles have been distilled, applied, and proven over decades of successful work with trustworthy systems. In fact, security principles form the well-distilled core of security engineering knowledge (Principle 11.5).

---

**Principle 11.5: Security Principles**

1. From the available security principles in books, publications, standards, field reports, case studies, accident coverage, etc., distill a set which applies to your field of application and the objectives of your organization (Note: the fundamental security principles are contained in this monograph);
2. Assemble and document the mandatory, recommended, and informative security principles for your organization in the "Security Principles Handbook";
3. Include implementation advice in the "Security Principles Handbook", such as patterns, guidelines, and examples;
4. Regularly revise and update the "Security Principles Handbook";
5. For each project, clearly list which security principles must be applied:
   – mandatory (must be unconditionally followed),
   – recommended (implement whenever possible),
   – informative (use for the guidance of design decisions).
6. In some cases, two (or more) security principles cannot be implemented at simultaneously, e.g., due to contradictions, performance, excessive resource consumption, usability obstruction, etc. In such cases, a justifiable, documented trade-off must be decided and implemented;
7. Any such trade-off must be risk assessed;
8. Formulate mandatory rules for trade-offs in case of conflicting safety principles requirements;
9. Record and archive any security-relevant compromise or trade-off;

---

## 11.6    Principle E6: Cyber-Crisis Management

*Cyber-crisis management* ([Kaschner20], [Ryder19]) is a specific form of *disaster recovery* ([Snedaker13]). The trigger for a cyber-crisis is a successful cyber-attack, causing either a safety accident or a security incident.

---

**Quote**
*"When your organization experiences a serious security incident (and it will), it's your level of preparedness based on the understanding of the inevitability of such an event that will guide a successful recovery"*
  Wil Allsopp, 2017

Cyber-crisis management is covered in Principle 11.6.

---

**Principle 11.6: Cyber-Crisis Management**

1. Develop, continuously test, and maintain a risk-based cyber-crisis management plan (= part of the cyber-security policy);
2. Partition your network so that infected network elements, workstations, or servers can be quickly isolated;
3. Assure the full attention and support of top management for the cyber-crisis management plan, including sufficient funding and adequate governance structures;
4. Unambiguously assign explicit and complete responsibilities for timely decisions, reporting, and communicating during all phases of a cyber-crisis (readiness, response, recovery);
5. Build, task, and train a cyber-crisis response team (incident response team);
6. Define and install a crisis management process that has to be strictly followed during crisis handling. Regularly train the process in mock situations;
7. Give special attention to the stakeholder communications activities during the crisis management;
8. Satisfy the legal and compliance reporting of the cyber-crisis and its potential damage toward the appropriate agencies (honesty and time are essential);
9. Ensure that the IT support needed during a cyber-crisis (process management tool, communications, access to logs, etc.) is available (Note: VoIP may cause a severe problem);
10. Collect sufficient information during the cyber-crisis and store it securely (e.g., for the forensic analysis);
11. Revise the crisis management plan regularly and after each security incident. Eliminate weaknesses from the plan, the execution, or the management;
12. Record and archive any execution of the crisis management plan and its results.

---

## 11.7 Principle E7: Security Implementation

The key concept in the implementation of security is the *control* (Definition 11.1, e.g., [Gallotti19], [Moyle20], [Knapp13], [Wong11]). A control is an overarching term for technical or organizational measures to counter a threat, attack, or failure. The sum of all existing controls in the CPS constitutes the *security implementation*. A complete, consistent, and effective set of controls results in a trustworthy (safe and secure) CPS. Missing, unreliable, or circumventable controls open opportunities for threats or attack vectors.

▶ **Definition 11.1: Control**

A safeguard or countermeasure prescribed for an information system or an organization designed to protect the security attributes (such as confidentiality, integrity, availability, archiving, performance) of its information and functionality and to meet a set of defined security requirements of the organization, the respective industry, the context, and the operating environment.

   Extended from: https://doi.org/10.6028/NIST.SP.800-161

A specific control often neutralizes (or significantly reduces) the impact of an identified threat, attack, or failure, such as in Example 11.3.

---

**Quote**

*"Painful experience appears to make any organization NOT practicing «best-in-class» security bordering on sheer recklessness and its management utterly failing its responsibilities"*
   Krag Brotby, 2009

---

**Example 11.3: Control for Transfer to Production Assurance**

During the evolution of a cyber-physical system, a potential, severe weakness is the transfer of unsafe or insecure software from development into the production system. Many assurance mechanisms exist to produce trustworthy CPS software (see, e.g., Principle 10.9). It is, however, of the utmost importance that all reasonable measures are entirely, consistently, and verifiably implemented. This requires a *production transfer control* (Fig. 11.2): After the safety/security assurance measures have been executed by the development/assurance team, a control gate checks the full, documented compliance. The control gate can be a manual process (review) or a fully automated procedure. No software is allowed into production if not all safety and security assurance measures have been implemented to complete satisfaction.

The primary source of reliable controls is *risk management* (Fig. 11.1): Identified threats are countered by appropriate controls. Other sources of controls are industry standards, frameworks, the policies of the organization, the dangers of the operating environment, the requirements of specific industries, and—last but not least—the applicable laws and regulations (such as in Table 11.3). A *selection* of standards and regulations is given in Table 11.4.

   Listing all possible security controls would fill 1'000s of pages ([NIST800-160B] alone has 246 pages!). Therefore, the security risk management team of the organization has the task to:

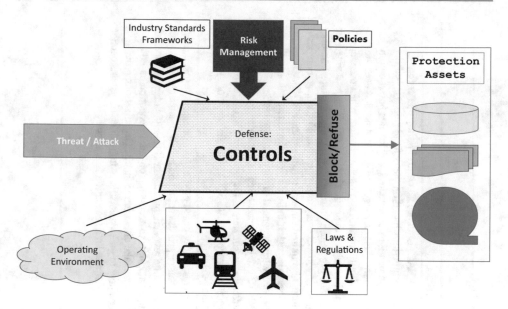

**Fig. 11.1**  Definition of controls

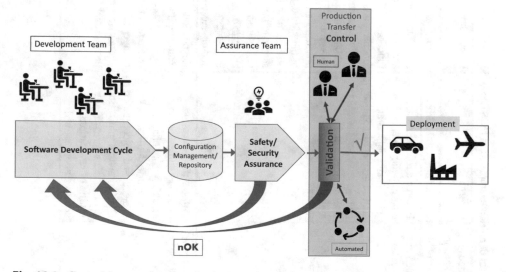

**Fig. 11.2**  Control for transfer to production assurance

1. Identify all the standards, laws, and regulations which apply to their respective field of activity;
2. Extract all the relevant *security controls*, evaluate them, and implement them;
3. Repeat this process periodically.

**Table 11.3** Security control definition example from [NIST 800–53]

| Control Identifier | Control (or Control Enhancement) Name | Control Text | Discussion |
|---|---|---|---|
| Source: https://csrc.nist.gov/publications/detail/sp/800-53/rev-5/final: Control Catalog Spreadsheet | | | |
| **AC-3(7)** | Access Enforcement I Role-based Access Control | Enforce a role-based access control policy over defined subjects and objects and control access based upon [Assignment: organization-defined roles and users authorized to assume such roles] | Role-based access control (RBAC) is an access control policy that enforces access to objects and system functions based on the defined role (i.e., job function) of the subject. Organizations can create specific roles based on job functions and the authorizations (i.e., privileges) to perform needed operations on the systems associated with the organization-defined roles. When users are assigned to specific roles, they inherit the authorizations or privileges defined for those roles. RBAC simplifies privilege administration for organizations because privileges are not assigned directly to every user (which can be a large number of individuals) but are instead acquired through role assignments. RBAC can also increase privacy and security risk if individuals assigned to a role are given access to information beyond what they need to support organizational missions or business functions. RBAC can be implemented as a mandatory or discretionary form of access control. For organizations implementing RBAC with mandatory access controls, the requirements in AC-3(3) define the scope of the subjects and objects covered by the policy |
| **AC-3(8)** | Access Enforcement I Revocation of Access Authorizations | Enforce the revocation of access authorizations resulting from changes to the security attributes of subjects and objects based on [Assignment: organization-defined rules governing the timing of revocations of access authorizations] | Revocation of access rules may differ based on the types of access revoked. For example, if a subject (i.e., user or process acting on behalf of a user) is removed from a group, access may not be revoked until the next time the object is opened or the next time the subject attempts to access the object. Revocation based on changes to security labels may take effect immediately. Organizations provide alternative approaches on how to make revocations immediate if systems cannot provide such capability and immediate revocation is necessary |

**Table 11.4**   Reference sources for security controls (examples)

| Source | Standard | Regu-lation | References |
|---|:---:|:---:|---|
| ISO 27000 Family | ✓ | | [Gallotti19], [Hintzbergen15], [ITG20], Overview: https://www.27000.org/ |
| CobiT (Control Objectives for Information Technologies) | ✓ | | [Harmer14], [ISACA19], Home: https://www.isaca.org/resources/cobit |
| CC (Common Criteria) | ✓ | | [Higaki10], [Higaki14], [ISO15408], [CC17], Home: https://www.commoncriteriaportal.org/ |
| NIST Standards & Recommendations: *A rich source of information on most areas of security* | ✓ | | [Cooper18], [Dempsey11], [Kerman20], [McCarthy21], [NIST06], [NIST14], [NIST16], [NIST800-53_20], [NISTIR8151], [Russo18], [Scarfone12], [Williams13], [SP800-53R5], [Dempsey21], Overview available at: https://www.nist.gov/publications |
| SABSA (Enterprise Security Architecture) | ✓ | | [SecDW19], Home: https://sabsa.org/ |
| The Open Group | ✓ | | [Wahe11], [Ghosh21], Home: https://www.opengroup.org/ |
| BSI (German Federal Office for Information Security) | ✓ | | [BSI12], Home: https://www.bsi.bund.de/EN/Home/home_node.html |
| PCI (Payment Card Industry) Data Security Standard | ✓ | | [Williams14], https://www.pcisccuritystandards.org |
| ISO/SAE 21,434 (Road Vehicles—Cybersecurity Engineering) | ✓ | | [Macher], [Sembera20], Home: https://www.iso.org/standard/70918.html |
| Sarbanes–Oxley Act (SOX) | | ✓ | [Anand08], [Holt06], Home: https://www.govinfo.gov/content/pkg/COMPS-1883/pdf/COMPS-1883.pdf |
| EU Privacy Directive | | ✓ | [GDPR16], [e-Directive09], [Massey20], [ITGP19], Home: https://ec.europa.eu/info/law/law-topic/data-protection/data-protection-eu_en and https://gdpr-info.eu/ |
| FISMA (Federal Information Security Modernization Act) | | ✓ | [Taylor13], Home: https://www.congress.gov/bill/113th-congress/senate-bill/2521 |
| General References | | | [Tarantino06] |

These recommendations are summarized in Principle 11.7.

In addition to the technical controls, the attacks that exploit the human element in the organization must be carefully considered (e.g., [Ferrillo17]). Careless or malicious behavior of employees opens dangerous attack paths, such as pishing, fake offerings, likejacking, fake applications, accessing an infected website, and e-mail scams.

---

**Principle 11.7: Security Implementation**

1.  Unambiguously delineate the system to be defended (system boundary, perimeter);
2.  Identify all communications paths crossing the system boundary;
3.  Apply risk management to discover the threats to the system;
4.  Define security controls to mitigate the threats;
5.  Identify all the standards, laws, and regulations that apply to the organization's respective fields of activity. This results in the catalog of applicable security controls;
6.  Extract all relevant security controls from the standards, laws, and regulations, evaluate them, and implement them completely and correctly;
7.  Use the information provided by the relevant organizations and universities, such as NIST, CERT, SEI, NSA, and CISA;
8.  Review/audit if all controls prescribed by applicable laws, regulations, and industry standards are duly considered;
9.  Document and archive the process and the controls;
10. Implement the controls;
11. Test the effectiveness of the controls, e.g., by penetration testing;
12. Repeat this control definition/implementation/testing process periodically;
13. Specifically consider and counter (as much as possible) attacks that exploit the human element;

---

## 11.8   Principle E8: Personal Data

Some of the above security principles inhibit unauthorized access to information. With cyber-physical systems, another case may arise: authorized access, but *unauthorized* or even *illegal* use of information (Example 11.4), especially personal data (Definition 11.2).

▶ **Definition 11.2: Personal Data**
Private data means any personal, personally identifiable, financial, sensitive, or regulated information (including credit or debit card information, bank account information, user names/passwords), Internet usage history, physical movement records, purchase trace, or communications activities.

https://www.lawinsider.com/dictionary/private-data

---

**Example 11.4: Personal Data in a Modern Car**

A typical example of a cyber-physical system is a modern car. Not only does it contain millions of lines of software code, connectivity to different networks, such as the Internet, the breakdown assistance center, the manufacturer support/update network—but besides several data-gathering devices, such as the GPS navigation unit or the Dashcam. Personal data can be collected (and stored), e.g., by:

- The GPS Navigation System: Route information (for all destinations reached);
- The GPS Navigation System: Guidance for finding points of interest, e.g., restaurants, routes often traveled, destinations touched;
- The Crash Recorder: Driver behavior before and during a collision. Location and severity of the collision (airbag launch and other sensor data);
- The Dashcam: Hours of videos of the road and the vehicle control by the driver, including speed limit violations, dangerous maneuvers;
- The On-board Maintenance Computer: Recording the use (and misuse) of the car;
- The Manufacturer Network: Gathering data about the car's health and anticipating problems;
- Crash or accident data (from the on-board reporting system);
- Manufacturer remote services (see, e.g., www.mercedes.me).

The law on collecting and using personal data in cyber-physical systems today (2022) is incomplete and evolving. An example is the use of dashcams (e.g., [Štitilis16]). In the European Union (EU), the preeminent regulation is the Regulation (EU) 2016/679: **G**eneral **D**ata **P**rotection **R**egulation (GDPR, https://gdpr-info.eu/, [GDPR16]). However, many countries have additional regulations (Principle 11.8).

---

**Principle 11.8: Personal Data**

**Collection:**
1. Write a comprehensive and binding policy for acquiring and using personal data in the CPS and for exporting. Write the policy in such a way that it is understandable for the "average citizen". Cover the complete life cycle of the data;
2. Prepare and document an exhaustive list of all personal data to be collected and used in the CPS. Keep this list accurate and up to date at all times;
3. Gather the currently binding laws and regulations covering the national laws in the territory intended for the CPS;

**Usage:**
4. Define and document all usages of the personal data, in the CPS and for export to connected systems;

**Deletion:**
5. Ensure that all data is definitively deleted from all storage media as soon as the data is no longer required or when a legal time limit is reached;
6. Ensure that all data is definitively deleted from all storage media when requested by the data owner;

**Conformity to Laws and Regulations:**
7. Assure that the collection and usage of the personal data are in accordance with all laws and regulations in all nations where the CPS is deployed or used;

**User consent:**
8. Whenever possible, ask for the explicit user's consent for the collection and intended usage of the personal data. Examine if such a consent overrides the respective laws and regulations;

**Traceability and Auditability:**
9.  Securely archive all agreements by the user's for the usage of their personal data;
10. Identify and securely archive the types of personal data which was used during the operation of the CPS;

## 11.9   Principle E9: Security Perimeter Protection

The "classical" protection of an IT system is the *perimeter protection*: The system to be safeguarded is explicitly delineated by its boundary, the *perimeter*. The perimeter is then secured with all the available technical means (Principle 11.9).

---

**Principle 11.9: Security Perimeter Protection**

1. Unambiguously define, document, and update the boundary (= security perimeter) of the system under consideration;
2. Determine, document, and assess all intrusion routes, i.e., all connections to the systems crossing the security perimeter;
3. Any usage of a public network (Internet, IoT, etc.) automatically and compulsory delineates the security perimeter, the public network belonging to the external, untrustworthy part;
4. Use a resilient network architecture;
5. Apply all modern defense technologies, such as firewalls, malware detection, demilitarized zones, zero trust architecture, cryptography, intrusion detection system, and extrusion detection systems;

6. Patch and upgrade all defense technology products immediately whenever updates become available (automatic updating preferred);
7. Use cryptography as much as possible to secure your perimeter (for confidentiality, privacy, authentication, etc.);
8. Seriously protect your system against targeted attacks from advanced persistent threats (APTs)—which in most cases needs specialized, competent consultants;
9. Investigate the possibility to apply security chaos engineering to your systems;
10. Never take security for granted! Securing a system is a continuous, ongoing, responsible activity that must be explicitly supported and funded by the top management.

## 11.10  Principle E10: Zero Trust Architecture

Following the gigantic US Office of Personnel Management (OPM) data breach in 2015, where more than 20 million personal records were exposed (https://www.opm.gov/cybersecurity/cybersecurity-incidents/), the US National Institute of Standards and Technology (NIST) started to research more secure network architectures. The final result—the *Zero Trust Architecture* (ZTA)—was published in 2020 as [NIST20]. ZTA focuses on the prevention of unauthorized access, and the paradigm of ZTA is making the implicit trust zones of the network as fine-granular as possible (Fig. 4.20, [Kerman20], [Gilman17]). ZTA is stated in Principle 11.10 (adapted from [NIST20]).

---

**Principle 11.10: Zero Trust Architecture (ZTA)**

1. Unambiguously define and precisely delineate the trust zones in the network;
2. View all data sources and computing resources/services as protection assets;
3. Make the implicit trust zones as fine-granular as possible;
4. Secure all communications (cryptography), regardless of the network location. Authenticate all participants when opening a connection;
5. Use the least privilege paradigm for all accesses to the protection assets, i.e., accord the lowest possible rights to execute the task, assign them only on a per-session basis, and just for the time needed ("time-boxed");
6. Limit access to the protection assets based on a dynamic access policy, i.e., include the observable state of client identity, application/service, time-of-day, requested protection asset—and possibly other behavioral or environmental attributes;
7. All admissions to protection assets are strictly enforced before access is allowed;
8. Build services designed for zero trust;
9. Create and use strong device identities;

10. Collect as much information as possible about the current state of protection assets, network infrastructure, communications, etc., and use it to improve security, especially access control;
11. Explicitly separate the trust control plane and the network plane (Fig. 4.21) in the architecture, design, and implementation;
12. Implement the trust control plane as a dependable infrastructure, not as part of the applications;
13. Focus the runtime monitoring on devices and services;
14. Maintain an up-to-date inventory of all data sources and computing resources/ services and their protection mechanisms;

## 11.11  Principle E11: Cryptography

*Cryptography* (Definition 11.3, [Hoffstein14], [Katz21a], [Paar10], [Buchanan17], [Ferguson10], and an interesting historical perspective: [Dooley18]) is a potent means to assure dependable security mechanisms. Cryptography supports and protects, e.g., privacy, confidentiality, integrity, authentication, digital signatures, etc.

▶ **Definition 11.3: Cryptography**
Cryptography is concerned with the construction of schemes that should be able to withstand any abuse. Such schemes are constructed so as to maintain a desired functionality, even under malicious attempts aimed at making them deviate from their prescribed functionality.
Oded Goldreich, 2008.

*Cryptography engineering* (e.g., [Ferguson10], [Goldreich08], [Goldreich09]) must be part of any security system engineering and also of the system's risk analysis. Protecting the security of any cyber-physical system includes the use of adequate—or even maximum—strong cryptography.

**Quote**
*"Computer security is, in many ways, a superset of cryptography"*
   *Niels Ferguson, 2010*

### 11.11.1   Lightweight Cryptography

The Internet of Things (IoT) dangerously changes the game of trade-offs between security and resources. IoT devices have severe *limitations*:

- *Processing Power Constraints*: The microcontrollers used in IoT devices have very limited processing power. Most of the processing power is used up by their primary functionality, i.e., the sensor or actuator functions. Only a fraction of the processing power remains available for security, e.g., for secure cryptography;
- *Limitations of Available Energy Supply*: Many IoT devices are battery powered and connected via wireless networks. They should have a long, maintenance-free life span. Therefore, the available energy for processing security or safety functions is very limited;
- *Cost Pressure*: The IoT will be a tremendously large mass market. IoT devices will be installed by the billions in the next decades. The fierce competition in this market generates strong cost stress—thus reducing the incentive to implement adequate safety and security measures.

> **Quote**
>
> *"In IoT, each connected device could be a potential doorway into the IoT infrastructure or personal data. The potential risks with IoT will reach new levels, and new vulnerabilities will emerge"*
>   Shancang Li, 2017

Many IoT devices' severe limitations require risky *engineering compromises* between security and resources (Table 11.5).

An IoT-based system or system-of-systems has a large number of possible *attack routes* (e.g., Fig. 4.19, [Cheruvu19]). Securing and safeguarding such a system is, therefore, a challenging engineering task ([Smith17], [Fiaschetti18], [Sabella18]). Because of some IoT devices' constraints (Table 11.5), not all desired cryptographic techniques can be implemented. The technology of *lightweight cryptography* has, therefore, been introduced (Definition 11.4, [McKay17], [Poschmann09], [Okamura17], [O'Gorman17], ([ISO29192]).

**Table 11.5**  IoT device limitations

| Device | Processor | Memory | Energy | Cryptography |
|---|---|---|---|---|
| Servers, Desktops | 32, 64 Bit | Gbytes | "unlimited" | **Conventional cryptography** |
| Tablets, Smartphones | 16, 32, 64 Bit | Gbytes | "unlimited" | |
| Embedded Systems | 8, 16, 32 Bit | Mbytes | Wh (Watt-hours) | **Lightweight cryptography** |
| RFID, Sensor Networks | 4, 8, 16 Bit, ASICs, FPGAs | kBytes | mWh (Milli-Watt hours, or less) | |

▶ **Definition 11.4: Lightweight Cryptography**

Lightweight cryptography is an encryption method that features a small footprint in terms of processing power, memory size, power consumption, and/or low computational complexity in terms of processing time.

Adapted from: https://www.nec.com/en/global/techrep/journal/g17/n01/170114.html

The security requirements of an Internet of Things system are challenging. They require much more than the traditional information security requirements of confidentiality, integrity, and availability (CIA). They must also cover, e.g., authentication, authorization, data integrity, non-repudiation, and forward and backward channel secrecy. Lightweight cryptography has gained such importance that specific standards exist (e.g., [ISO29192]). Principle 11.11 covers cryptography.

---

**Principle 11.11: Cryptography**

1. Fully integrate effective cryptography engineering into your security engineering processes;
2. Consistently use sufficiently strong cryptography for all vital functions of your system (authentication, authorization, access rights, data integrity, confidentiality, etc.). *Note*: Sufficiently = compliant to the results of the risk analysis to ensure acceptable residual risk;
3. Assure that your cryptography support processes (key generation, key distribution, PKI, certificate handling, auditing, logging, monitoring, etc.) are strong enough to assure the results of the risk analysis for acceptable residual risks;
4. Be very careful when generating and using passwords. Never use default passwords. Use strong passwords. Regularly change passwords using a secure mechanism.
5. Audit and assess your cryptography algorithms, key lengths, risk analysis, implementations, and processes at regular intervals. Correct immediately if weaknesses are prognosticated;
6. Pay special risk and engineering attention to the weakest link property of your cryptography mechanisms and processes. Improve the weakest links until the risk analysis results in acceptable residual risks;
7. In case of conflicts between available resources (processor power, energy consumption, memory usage, cost, …) and cryptographic strength, *never* violate the limits of acceptable residual risk provided by the risk assessment. Use at least recommended lightweight cryptography;
8. Use proven algorithms, industry standards, and products for the implementation of your cryptography. Develop proprietary implementations only if you have a very strong reason;
9. Update, patch, and upgrade all cryptographic products immediately when it is recommended by the vendor.

## 11.12   Principle E12: Transition to Post-Quantum Cryptography

Already in 2015, the US National Security Agency (NSA) issued warnings of the need to transition to new *quantum-resistant algorithms* ([Campagna20]). Quantum-resistant algorithms were proposed by NIST, and migration routes for the transition to the post-quantum cryptography age were proposed ([Barker21], [Housley21], [USDHS21], [TEMET21]).

> **Quote**
> *"Recent advances in quantum computing signal that we are on the cusp of our next cryptographic algorithm transition, and this transition to post-quantum cryptography will be more complicated and impact many more systems and stakeholders than any of the prior migrations"*
>    Matt Campagna, 2020

A generic, top-level process for the *transition to post-quantum cryptography* (PQS) is presented in Fig. 11.3. The process contains five steps and a feedback loop.

Guidance for the transition to post-quantum cryptography is given in Principle 11.12.

> **Quote**
> *"Quantum computing strikes at the heart of the security of the global public key infrastructure"*
>    Matt Campagna, 2020

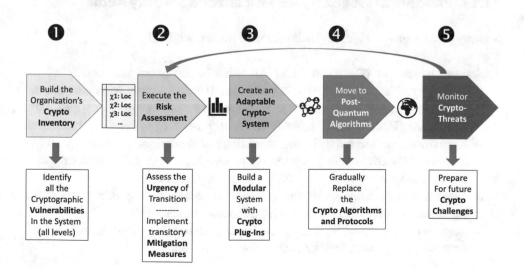

**Fig. 11.3**  Generic transition sequence to post-quantum cryptography

---

**Principle 11.12: Transition to Post-Quantum Cryptography**

1. Acknowledge the need and the timeline for the transition from classical to post-quantum cryptography on all levels of the organization;
2. Rewrite or adapt all relevant policies to completely and thoroughly cover all post-quantum aspects, implications, and consequences;
3. Carefully and exhaustively inventarize and document all cryptographic tools, products, applications, and procedures in the organization;
4. Identify automated discovery tools to assist organizations in identifying where and how public key cryptography is being used in their systems (all levels);
5. Carefully and exhaustively inventarize and document all cryptographic dependencies from external systems;
6. Apply risk analysis to future and retroactive (= information that has already been recorded by an adversary) dangers and develop mitigation measures;
7. Create a competent organizational unit within the organization for the planning, execution, and controlling of the transition to post-quantum cryptography;
8. Rely solely on accepted cryptographic standards (e.g., NIST) and proven products when transforming the organization to post-quantum;
9. Develop a security architecture to assure cyber-resilience (= ability to respond in time to future cryptographic apocalypses) and cryptographic agility (= separation of cryptographic implementations from protocols, applications, etc.);
10. Start immediately with over-encrypting (cryptographic wrapping) of the organizations' information with long-term, confidentiality or integrity requirements;

---

## 11.13   Principle E13: Security Assessment and Security Audit

*Security assessments* and *security audits* are used for (Principle 11.13):

1. *Security assessment* (Definition 11.5): All activities to assure the security of the CPS, i.e., its resilience against malicious activities. It includes penetration testing, fuzzing, chaos engineering, and additional security examination techniques (e.g., [PWC21]);
2. *Security audit* (Definition 11.6): Verify the conformance of system implementation with provisions, such as requirements, standards, and regulations (see also Fig. 10.2):
   a. *Functional Security Audit*: The system implementation is examined with respect to the requirements, especially the safety and security requirements;
   b. *Conformance Security Audit*: The system implementation is checked against the organization's policies, rules, guidelines, and internal standards;
   c. *Regulatory Security Audit*: The system implementation is gauged against industry standards, regulations, legal or certification requirements;

d. *Forensic Security Audit*: After an accident, a severe incident, or a near-miss, the system implementation is investigated concerning the causes, the chain of events, and the accountability of the implementation.

▶ **Definition 11.5: Security Assessment**

IT security assessment is a process that encompasses the discovery of vulnerabilities of an IT infrastructure and the identification of related risks at business level.
   Marco Caselli, Frank Kargl, 2016

▶ **Definition 11.6: Security Audit**

Independent review and examination of a system's records and activities to determine the compliance with established security policy, security procedures, security standards, and laws and regulations.
   Adapted from: NIST SP 800–82 Rev. 2

Effective security audits start with a goal-oriented *security audit policy* ([Moeller10]) with a strong focus on *security governance* ([Davis21]). Best results are obtained if an accepted, industry-standard *security assessment methodology* (e.g., [Toth17], [Caselli16], [Scarfone08], [OISSG06], [Schoenfield15]) and *security audit methodology* is used (e.g., [Love10], [ISO27007], [Brotby09], [Kegerreis19]). Security audits can either be carried out internally (by members of the organization), externally (by chartered consultants), or by a regulatory body (such as a certification agency).
   Very often, in-house developed software provides vulnerabilities. Therefore, the *security assessment* must be a serious part of the development process (e.g., [Dowd06]).

> **Quote**
> *"Despite the growing awareness, recent cyber and ransomware attacks are proof that many organizations are still failing to address known vulnerabilities with due care and urgency"*
>    PricewaterhouseCoopers, 2021

**Principle 11.13: Security Assessment and Security Audit**

**Security Assessment:**

1. Unambiguously define the boundary of the system under assessment (Possibly a partitioning into subsystems may be necessary);
2. Clearly identify all attack paths;
3. Assess not only the system but also all processes with respect to their impact on security;

4. Use an accepted, industry-standard security assessment methodology (possibly reasonably adapted to the organization);
5. Prioritize the system parts (applications, system software, execution infrastructure, communications channels, perimeter, etc.) according to their risk potential and exposure;
6. Make the security assessment of in-house developed software a mandatory part of the security culture;
7. Repeat the security assessment regularly;
8. Adapt the frequency of assessment to the priority of the element (weekly, monthly, yearly) and after every change to the system;
9. Automate the security assessment as much as possible;

**Security Audit:**
10. Develop, document, communicate, use, and enforce an audit policy;
11. Emphasize security governance (reporting lines, responsibilities, processes);
12. Prepare a complete, comprehensive list of all applicable laws and regulations for your application domain and designated areas of use (nation-states);
13. Prepare a complete, comprehensive list of all applicable standards for your application domain and designated areas of use (nation-states);
14. Decide which sections of the applicable standards, laws, and regulations are mandatory for your applications in the chosen areas (Internal and external provisions;
15. If applicable, include all certification requirements, standards, and regulations;
16. Use an accepted, industry-standard audit methodology (possibly reasonably adapted to the organization);
17. Repeat the security audit regularly (follow the evolution of standards, laws, regulations, and certification requirements);

## 11.14  Principle E14: Security Runtime Monitoring

*Security runtime monitoring* and *anomaly response* are the last defense against security incidents (Definition 4.30, Fig. 4.23, Table 4.8). When design-time defenses fail or are circumvented, anomalous behavior detection is the endmost operation to avert a safety accident or a security incident. Numerous runtime monitoring techniques exist, and more are researched and published every year. One promising approach for application runtime monitoring is the specification-based surveillance presented in Example 11.5. Some advice on runtime monitoring is listed in Principle 11.14.

**Quote**

*"Runtime security monitors are components of defending systems against cyber attacks and must provide fast and accurate detection of attacks"*
  Muhammad Taimoor Khan, 2016

**Example 11.5: Specification-Based Application Software Runtime Monitoring**

Specification-based application software runtime monitoring targets the detection of the anomalous behavior of the applications *during execution*. The structure of this type of runtime monitor is shown in Fig. 11.4. The basic mode of operation is the continuous comparison of the *expected behavior* (defined as an executable, formal specification of the application) and the *actual behavior,* i.e., the program execution. ([Khan16], [Shrobe79], [Drusinsky04], [Mellor04]).

   The starting point of this runtime monitor is a formal, executable specification of the program. From this, a number of program execution observation points are generated. When the program execution is started, the runtime monitor continuously compares program behavior with the program specification. Any deviation or anomaly is detected in real time, and corrective mitigation action is taken.

   Important issues for runtime monitoring are listed in Principle 11.14.

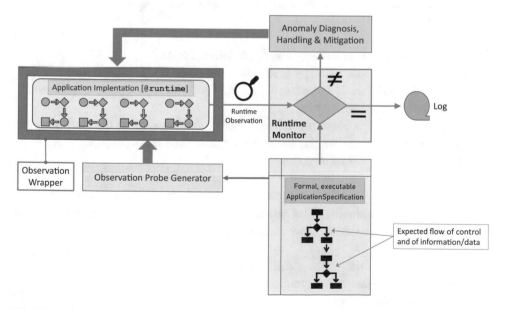

**Fig. 11.4**  Specification-based runtime monitor

---

**Principle 11.14: Security Runtime Monitoring**

1. Recognize the importance of runtime monitoring for trustworthy cyber-physical systems, especially for their safety and security;
2. Use risk management to carefully determine which risks can be reduced or eliminated by implementing an adequate runtime monitor;
3. Consider all layers of the system architecture as candidates for runtime monitoring;
4. Evaluate the available runtime monitor technologies and choose the fitting type(s);
5. Precisely formulate the runtime monitor requirements/specifications and mitigation responsibilities. Assign a high priority to its implementation;
6. Whenever possible, implement automated response to anomalies;
7. Consider and assess the additional resource consumption of the runtime monitor (CPU load, extended execution time, memory and I/O-channel load);
8. Build in and log sufficient diagnostic information for a possible safety accident reconstruction or security incident understanding, or a forensic investigation;

---

## 11.15  Principle E15: Secure Software

*Secure software* (Definition 5.19, e.g., [Ransome14], [Harrington20], [Kohnfelder21]) is a risk-assessed functional element of mission-critical cyber-physical systems. The difference between *software security* and *application security* is explained in Definition 11.7.

▶ **Definition 11.7: Software Security and Applications Security**
*Software security* is about building secure software, designing software to be secure, making sure that software is secure, and educating software developers, architects, and users about how to build security in.

*Application security* is about protecting software, and the systems that software runs after development are complete, the software is deployed and is operational.

Adapted from Gary R. McGraw, 2006

In addition to the material in this monograph, the *secure software chain* (Fig. 11.5) defines a schema for reasoning about secure software. The *secure software chain* has the same constituents as the safe software chain (shown in Fig. 10.3):

1. Software *fault avoidance* in the specifications;
2. Software *fault avoidance* in the implementation;
3. Software *fault elimination*;
4. Software *fault tolerance*;

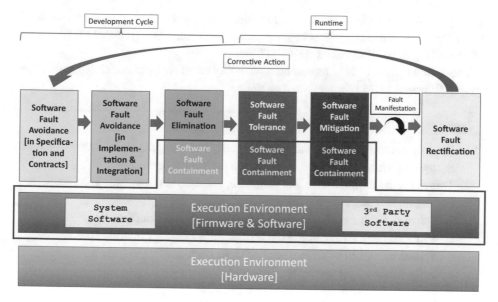

**Fig. 11.5**  Secure software chain

5. Software *fault mitigation*;
6. Software *fault rectification*.

And, in addition:

7. Software *Fault Containment*: The components of secure software are, to a large extent, third-party products, such as operating systems, browsers, database management systems, and SCADA software. Experience shows us daily that these components contain many known (but unremedied) and unknown (e.g., zero-day) vulnerabilities. These third-party products cannot be secured by the customer and constitute severe risks for the organization (e.g., Example 2.16, Example 4.20). Therefore, additional mitigation measures must be implemented to contain the negative impact of a security incident attributable to the third-party products (Fig. 11.5).

**Quote**
*"We believe software security to be the most important topic in information security for the foreseeable future"*
   James Ransome & Anmol Misra, 2014

A very useful instrument to assure software security is a *security maturity model* ([Ransome14]). Various approaches exist, such as BSIMM ([BSIMM21]), SAMM ([SAMM21]), or the ISO standard ([ISO27034]).

The secure software chain represents a summary of all security principles in this monograph. Principle 11.15 is, therefore, a guideline.

---

**Principle 11.15: Secure Software**

1. Produce complete, consistent, unambiguous, and comprehensive requirements for the system, with particular attention to the security requirements;
2. Transform the requirements into verifiable, testable, and traceable specifications;
3. For secure software, use the following schema:
   a. Software Fault Avoidance in the Specifications: see Principle 10.9;
   b. Software Fault Avoidance in the Implementation: see Principle 10.9;
   c. Software Fault Elimination: see Principle 10.9;
   d. Software Fault Tolerance: see Principle 10.9;
   e. Software Fault Mitigation: see Principle 10.9;
   f. Software Fault Rectification: see Principle 10.9;
   g. Software Fault Containment: For all third-party products, provide effective mitigation mechanisms to avoid negative impacts of a security incident as much as possible;
4. Strictly use a security-aware development process (trustworthy development process);
5. Select or adapt a security-aware development process that conforms to an industry standard;
6. Use a security maturity model to assess the security of the software;
7. Regularly review/audit the software with respect to identified, assessed, and mitigated risks and their impact on security;
8. Make extensive use of the large body of knowledge from universities and government organizations, such as SAFECode, US Department of Homeland Security Software Assurance Program, NIST, MITRE, SANS Institute, US Department of Defense Cyber Security and Information Systems Information Analysis Center (CSIAC), CERT, Bugtraq, SecurityFocus, ISO/IEC, and the German BIS([Ransome14]);

---

## 11.16  Principle E16: Insider Crime

*Insiders* are one of the very serious threats to the security of IT-based systems (Definition 2.21, e.g., [CSI20], [Ponemon20]). Not only can they generate massive damage, but their prevention and detection are rather difficult ([Cappelli12], [Gelles16], [Wong20], [Bunn16], [Thompson21]).

> **Quote**
> *"Insider threat is a high organizational risk that must be mitigated with a deliberate approach and methodology"*
>   Eleanor E. Thompson, 2021

Three sorts of insiders are dangerous (Fig. 11.6):

a. *Malicious Insiders*: employees, contractors, or suppliers with valid access credentials to the IT system of the organization. They use their credentials to inflict damage to the organization;

b. *Careless or Negligent Insiders*: employees, contractors, or suppliers with valid access credentials to the IT system of the organization. They knowingly or accidentally violate policies or execute dangerous actions, such as downloading code from the Internet, open mails of unclear origin, visiting dubious (infected) websites, or open documents from insecure sources;

c. *Infiltrators*: malfeasant, external actors that illegally obtain access credentials or circumvent the access controls and use them for cyber-sabotage, cyber-crime, or cyber-espionage.

The awkward challenge of insider threats is that they are *people problems* and cannot be solved only by technical controls: A sensible mix of technical controls and human resource management controls must be used (Principle 11.16).

## 11.16.1  Technical Controls

The appropriate technical controls are defined in Principle 11.7. In addition, an *information security tripwire* (Definition 11.8, Fig. 11.7, [Agrafiotis17]) must be explicitly

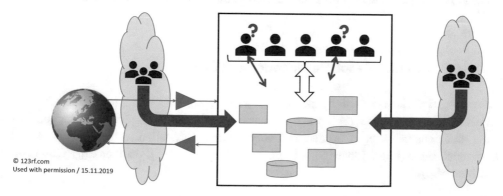

© 123rf.com
Used with permission / 15.11.2019

**Fig. 11.6**  Malicious insider threats

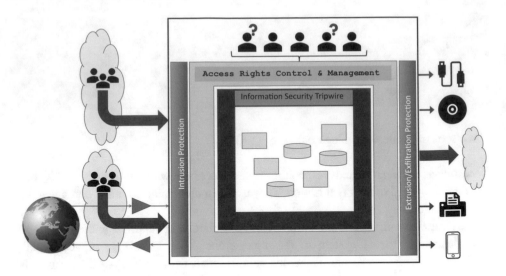

**Fig. 11.7**  Insider threat defense

incorporated for malicious insiders. A tripwire is a specific form of runtime monitor (Principle 11.14). It is designed to recognize undesired access attempts to the protected assets of the organization. The tripwire works both in real time, alerting the security officer and logging all suspicious activities for later analysis and action.

▶ **Definition 11.8: Information Security Tripwire**  Information security control for the detection of policy violations, perceive insider attack patterns, and the discovery and reporting of unauthorized or suspicious access attempts to the information assets of an organization.

Many insider attacks follow a pattern (insider attack pattern, [Agrafiotis15], [Agra-fiotis17]). These known patterns can be used to mount technical controls against insider threats.

## 11.16.2   Human Resource Management Controls

*Insider crime* is executed by people. Therefore, controls and procedures must be installed to reduce the probability of a successful insider attack. Guarding against the malicious behavior of insiders is a very difficult task, often restricted by laws and personal rights (e.g., [Shaw09], [Thompson21], [DHS14], [Ponemon20], [CERT16], [Charney14]).

> **Quote**
> *"Whether they are caused accidentally or maliciously, insider threat incidents cannot be mitigated with technology alone. Organizations need an Insider Threat Management program that combines people, processes, and technology to identify and prevent incidents within the organization"*
>    Ponemon Institute, 2020

A generic process to forfend successful insider attacks is presented in Fig. 11.8. It consists of five steps:

1. *Candidate Application*: Ask the applicant to entirely and truthfully fill in an expressive application form. Use the employment and screening policies to learn as much as possible about the applicant and his former life. Extensively utilize external information sources, such as search engines, social networks, and court and police records (Note: Local law may impose some restrictions!);
2. *Pre-employment Screening*: Carry several interviews by different people, obtain meaningful and honest references from previous employers (Example 11.6). Keep records and notes;
3. *Acclimatization*: Thoroughly introduce the successful applicant to the organization's culture, policies, best practices, code of ethics, and other relevant topics. Have him sign a memorandum of understanding;

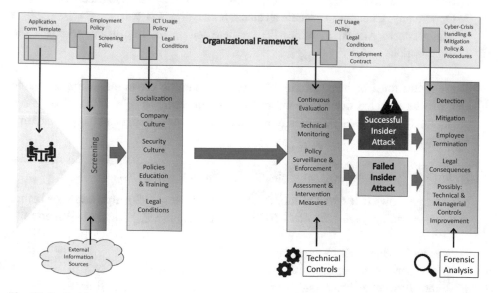

**Fig. 11.8**  Insider attack defense

4. *Continuous Observation*: Use technical monitoring, behavioral supervision, and early warning signs (Example 11.6, Example 11.7);
5. *Consequences*: For each type of insider attack (successful or unsuccessfully attempted), mitigate the impact, discipline the delinquent(s), and deduct consequences for the improvement of the controls;

---

**Example 11.6: Insider Threat Modeling**

Successfully defending against insider attacks needs the *modeling* of the potential attacker, i.e., the comprehension of his state of mind. This example shows one possible assessment scheme ([Nurse14]):

1. Precipitating Event or Catalyst: A decisive event (= the catalyst) which may push the insider over the edge, thus becoming a risk for the organization;
2. Individual's Personality Characteristics: Appraising the psychological traits and dispositions. Gather the features of an insider's personality, both of their innate self (static aspects) and their life experiences (dynamic aspects);
3. Historical Behavior: Documents the kinds of activities the insider has taken part in the past (including any criminal record or misbehavior);
4. Psychological State: Personal negative feelings, such as perceptions of organizational injustice, workplace inequality, management inequity, and being passed over;
5. Attitudes Toward Work: Degree of the insider's commitment to his work, the organization, and his environment;
6. Skill Set: Determines the insider's capability or the skills needed to execute an attack;
7. Opportunity: Chances to execute an attack on the organization;
8. Motivation to Attack: Identifies the possible reason for the insider to attack the organization.

These points must be very carefully analyzed during the pre-employment screening and also be repeated during the entire time of employment (Fig. 11.8). Note that a precipitating event (catalyst) transforming an insider into a threat can happen anytime and should be monitored.

---

**Example 11.7: Behavioral Indicators of Possibly Malicious Activity**

Experience shows that many insider delinquents exhibit questionable behavior before or while executing an insider attack ([DHS14], [Agrafiotis15], [Noonan18]) such as:

- Stop socializing with co-workers;
- Avoid friendly encounters in the cantine or coffee shop;
- Remotely accesses the network while on vacation, sick, or at odd times;

- Works odd, e.g., late hours without necessity or authorization;
- Notable enthusiasm for overtime, weekend, or unusual work schedules;
- Unnecessarily copies material (electronic or paper), especially if it is proprietary or classified;
- Exaggerated interest in matters outside of the scope of their duties;
- Curiosity in the work of others;
- Unduly interest in the security mechanisms of the organization;
- General nervousness and down slipping of work quality;
- Avoidance of eye contact;
- Not responding to a direct question (evasion or erratic response);
- Inappropriate Level of Concern: Attempts to diminish the importance of an issue. May even joke about the issue (*Why is this such a big deal? Why is everybody worried about that?*);
- Use of perception qualifiers enables an individual to enhance credibility ( *Frankly… /To be perfectly honest… / Honestly…*);
- Signs of vulnerability, such as drug or alcohol abuse, financial difficulties, gambling, illegal activities, poor mental health, or hostile behavior;
- Radicalization (political, religious, ethnical, …);
- Acquisition of unexpected wealth, unusual foreign travel, irregular work hours, or unexpected absences.

Note that trying to predict a future or detect an ongoing insider attack based on behavioral indicators is a minefield: Excessive surveillance may drive off perfectly honest employees, may violate personal rights, or generally demotivate people.

**Quote**

*"Current practice for insider threat is largely reactive—dealing with problems after they are known to have taken place. In order to move towards a proactive approach for mitigating the insider threat, organizations will have to balance technological solutions with effective monitoring of employees"*
    Christine Noonan, 2018

**Principle 11.16: Insider Crime**

1. Develop, publish, and enforce policies for:
    - Terms and conditions of employment;
    - The use of electronic media within the organization (ICT usage);
    - Especially stipulate the use of the organization's infrastructure (e-mail, web access, etc.) for private use;
    - Handling of a successful insider attack.

**Technological Defense:**

2. Classify all information assets in the organization (e.g., Table 11.1) and protect them accordingly;
3. Implement security according to Principle 11.7;
4. Use the best available technologies for intrusion and extrusion detection and keep them updated at all times;
5. Log all successful and unsuccessful accesses with sufficient information for forensic analysis after a security incident;
6. Strictly enforce the principle of least privilege. Audit and adapt the rights frequently;
7. Request three-factor authentication for all confidential accesses (User Name + Password + Digital Certificate on a smart card);
8. Implement an information security tripwire enfolding all confidential information assets. If applicable, also protect critical functionality. Incorporate known insider attack patterns;

**Human Resource Management Measures:**

9. Carefully screen the applicants (using all legally possible means and all accessible sources);
10. Obtain personal, meaningful, and honest testimonials from previous employers (Seeking personal contact);
11. Extensively and comprehensively introduce the new employee into the importance, content, and enforcement of policies, procedures, and processes;
12. Conduct continuous monitoring for signs of deviant behavior and have fair procedures in place to handle "red flags";
13. Prepare assessment and intervention measures for the cases of a successful or attempted insider crime. Cover the technical mitigation (minimization of impact) and human accountability (consequences, termination, legal ramifications;
14. After each successful or attempted insider crime, revise the technological defense mechanisms and the employment mentoring and supervision flow;
15. Develop and install a directive about acceptable gifts and gratuities;

## 11.17  Principle E17: Microservices Security

### 11.17.1  Microservices

The *granularity* of software artifacts—both in development and deployment—has long been a topic of discussion in the software community. The pressure of short time-to-market has been (partially) countered by more and more fine-granular implementations. The main milestones were the early *monolithic* architectures with a very coarse granularity (e.g., [Hasselbring18]), replaced by the *service-oriented* architectures (e.g., [Erl05], [Erl17]) with variable granularity, and culminating in the fine-granular *microservices architectures* of this decade (Definition 11.9, e.g., [Amundsen16], [Newman15], [Newman19], [Rodger17], [Stetson17]).

▶ **Definition 11.9: Microservices**
Microservices are independently releasable services that are modeled around a business domain. Microservices are an approach to distributed systems that promote the use of finely grained services that can be changed, deployed, and released independently.
   Sam Newman, 2021

**Quote**
*"The move to microservices is a seismic shift in web application development and delivery"*
   Chris Stetson, 2017

## 11.17.2   Microservices Security

Because many *microservices* are accessible over the Internet via APIs (e.g., [Higginbotham21]), sufficient security is indispensable. Unfortunately, three characteristics of microservices make the implementation of dependable *microservices security* difficult:

1. The *fine granularity* of the microservices requires an adequate fine-granular implementation of security. Managed, centralized security implementations are very difficult and inefficient to apply to microservices (Fig. 11.9, [Siriwardena20]);
2. Microservices are developed mainly by *independent, small teams* who often work in isolation and are possibly distributed over many places. The required knowledge and skills for dependable security implementations must, therefore, be available to all teams (and also be consequently enforced);
3. The very advantage of microservices—their *short delivery time* for new functionality often forces the security design and implementation to take shortcuts or skip quality assurance steps, such as security reviews or sufficient security testing.

Therefore, reliable microservices security requires different policies, architectures, designs, implementation processes, and governance ([Siriwardena20]). This is especially true if microservices are used in safety- or security-critical applications! However, all principles for security apply unrestricted.

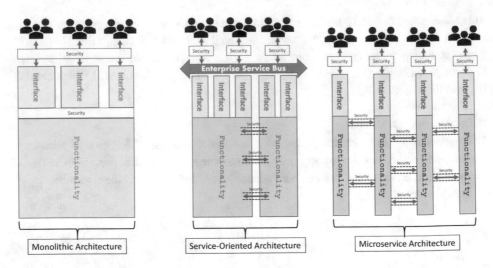

**Fig. 11.9**   Architecture granularity

### 11.17.3   Microservices Governance

Because microservices are a) quickly developed, b) grow fast in number, and are often built by individual, distributed teams, *microservices governance* ([Daya15], [Newman21]) becomes crucial for the dependable security of microservices.

> **Quote**
> *"In the world of developers, governance many times is a rather unpopular term"*
>    Shahir Daya, 2015

The most important decision is the *split of responsibility* between the central governance authority and the decentral, federated governance authorities (Example 11.8). Following this decision, the organizational structure, the responsibilities, and the reporting lines can be established. Be aware that a mismatch between microservices architecture and governance is a dangerous source of the fast accumulation of technical debt.

---

**Example 11.8: Microservices Governance**

In a microservice architecture, individual, often loosely connected teams develop and deploy software. This situation constitutes a high risk for accumulating *technical debt*, such as functional or data redundancy, technology divergence, tool proliferation, or loss of conceptual integrity. For the success of microservice architectures in an organization, two *preconditions* must be assured:

I.  An adequate *governance structure* must be in place. A proved and tested, well-defined split of responsibilities between central governance authority and decentral governance authorities must be in place;

II. Sufficient policies, guidelines, and standards must be defined and enforced to assert the desired quality properties and the conceptual integrity of the complete system.

One possible governance structure with adequate responsibilities is shown in Fig. 11.10: The activities to ensure the sound and healthy evolution of the system are assigned to the central governance authority and are unconditionally imposed on the decentral governance authorities. The decentral governance authorities receive the rights over the DevOps processes.

### 11.17.4   Migration to Microservices

A security-critical endeavor is the *migration* of an IT system based on an older architecture—such as a monolith or a service-oriented architecture—to a microservices architecture. During this migration process, the system's security may be seriously compromised. The *migration strategy* must, therefore, heavily respect the security requirements in all phases of the migration ([Siriwardena20], [Newman19], [Bucchiarone21]). Note that the migration to microservices may require a change of some of the security foundations, such as from implicit trust to the zero trust paradigm.

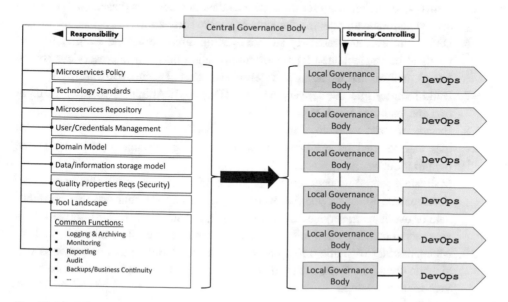

**Fig. 11.10**   Microservices governance example

**Quote**

*"You should be thinking of migrating to a microservices architecture only if you can't find any easier way to move toward your end goal with your current architecture"*

Sam Newman, 2021

**Principle 11.17: Microservices Security**

1. Carefully investigate and evaluate if microservices are suitable for the field of application and the organizational structure. If not, resign from microservices architecture or adapt the organizational structure and responsibilities (= organizational alignment);
2. Accurately, detailed, and completely define a practical split of responsibilities between central governance authority and decentral governance authorities. Install and regularly check/audit the desired impact on the performance of the development;
3. Define and enforce sufficient policies, guidelines, and standards to assert the desired quality properties and the conceptual integrity of the complete system;
4. Provide a comprehensive and complete document: "Security architecture guidance for projects" covering all issues of security implementation. Keep the document updated;
5. Define, enforce, and regularly audit a specific directive for the security mechanisms to be implemented for the access to the microservices (based on risk analysis and including all relevant elements of secure software);
6. Define, enforce, and regularly audit a specific directive for the security mechanisms to be implemented for the communication between microservices (based on risk analysis and including all relevant elements of secure software);
7. Build security into the delivery process {DevOps}: Automate security verification and testing as much as possible;
8. Execute a strict technology management, specifying precisely which technologies can be used by the DevOps teams, avoiding technology overload in the organization;
9. Maintain a central services repository for all microservices in operation and under development. Make service discovery simple. Ensure (by reviews) that no duplicate or overlapping services are developed (reuse strategy). Make informed decisions about creating a new microservice or extending an existing microservice;
10. Maintain strict versioning. Keep no more than the current and two previous versions in operation. Ensure that all applications are migrated in time before a service version is retired;

11. Precisely define the boundaries of each microservice and model the microservice in the context of the domain model;
12. Unconditionally ensure that all quality gate checks, tests, and reviews are executed to the full satisfaction on all microservices. Do not tolerate exceptions, shortcuts, or leaps;
13. Ensure that the vital data in the system is at all times consistent (in time, content, and meaning). Use a centralized repository or a distributed repository with dependable synchronization mechanisms;
14. Use the domain model to avoid unmanaged functional and data/information redundancy;
15. If migrating to a microservices architecture, the migration strategy must heavily respect the security requirements in all phases of the migration;
16. Allow sufficient time and funds to implement the security mechanisms before they go into production;

## 11.18   Principle E18: Artificial Intelligence in Security

Many security experts expect that information and computer security (ICT security) will become a *battle of artificial intelligence* (AI), i.e., a vicious fight between nearly undetectable, highly effective attack AI, and adaptive, efficient defense AI ([Chio18], [Zeadally20], [Apruzzese18], [Kim18b]).

### 11.18.1   ICT Defense Mandate

In this monograph, the interest lies in *beneficial artificial intelligence* (Definition 6.13), i.e., in AI for the *defense* of the ICT system. The *ICT defense mandate* has three missions (Fig. 11.11):

I.   *Intrusion Prevention*: Fend off all undesired traffic entering the ICT system, including unauthorized accesses, malware injection, denial-of-service attacks, external misuse of the system's functionality, etc.;
II.  *Extrusion Prevention*: Inhibit any undesired traffic leaving the ICT system, including cyber-espionage, theft of company confidential information, leakage of sensitive data, etc.;
III. *System Integrity Preservation*: Make impossible any undesired change to all ICT system elements, including programs, firmware, configuration data, information, databases, logs, documentation, etc.

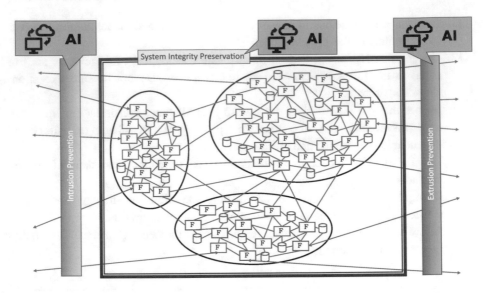

**Fig. 11.11**   ICT defense mandate

> **Quote**
> *"The use of Artificial Intelligence (AI) in cybersecurity is viewed as a race between law enforcement and cyberattackers. The leader in the race will be determined by the access to technical knowledge and the power of the supporting computing infrastructure"*
>       Sherali Zeadally, 2020

## 11.18.2   Artificial Intelligence Defense for Security

Today (2022), artificial intelligence in cyber-defense is a very active field of research and engineering (e.g., [Zhang21b]). New algorithms and methods are being discovered continuously and find their entry into products.

*Intrusion Prevention*: For intrusion prevention (Figs. 6.15, 6.16 and 6.18)—especially machine learning malware detection—much progress has been made, and a number of proven products are available. Especially their emerging ability to detect zero-day attacks is unique ([Faesen19], [Thomas20]).

*Extrusion* or *Efiltration Prevention:* Extrusion prevention focuses on preventing the illicit transfer of information assets leaving organizational boundaries or the restricted zone of authorized users (e.g., [Stalmans12]): Disappointingly, much less work has been done on the vital extrusion field than in intrusion detection.

*System Integrity Preservation*: Many elements of the cyber-physical system can be maliciously modified, thus violating the system's integrity. If the system's specified functionality is altered—e.g., by unauthorized modifications, by inadvertently using insecure software, or by added functionality in the form of malware—serious vulnerabilities can be introduced. Maliciously interfering with the system's data—such as configuration files, database information, or log files—can also open or hide dangerous attack vectors. Any intrusion or extrusion prevention failure may corrupt the system's integrity. Identifying integrity loss of large or very large systems is intricate: Often, the loss of integrity becomes only visible through a malfunction of the system. Therefore, anomaly detection is often the only method to discover integrity deficits during operation. AI methods for diagnosing behavioral anomalies are under development.

## 11.18.3   Attacks against Defense AI Machine Learning

Unfortunately, using AI (especially machine learning—ML) to defend ICT systems opens a new *attack vector*: adversarial machine learning examples (Definition 6.5, e.g., [Gardiner20], [Kloft10]).

Undoubtedly, artificial intelligence—especially machine learning—has become an indispensable field for the security of cyber-physical systems (Principle 11.18).

---

**Principle 11.18: AI in Security**

1. Periodically and carefully identify and assess the value of using artificial intelligence for cyber-defense for:
   a. Intrusion prevention,
   b. Extrusion prevention,
   c. System integrity preservation;
2. Ensure that the identified problems can (and should) be solved by AI;
3. Loosely couple the AI algorithms/products to the system to make them easily exchangeable;
4. Use an adequate system architecture with a focus on security architecture;
5. Reduce the number of interconnections (communications channels) as much as possible;
6. Generate and archive sufficient information during operation to trace, interpret, or forensically analyze the AI actions;
7. Whenever possible, use XAI-algorithms (explainable artificial intelligence);
8. Implement runtime monitoring to supervise AI decisions;
9. Ensure that the AI products are protected against malicious example attacks.

# References

[Agrafiotis15]        Ioannis Agrafiotis, Jason Nurse, OliverBuckley, Phil Legg, Sadie Creese, Michael Goldsmith: *Identifying Attack Patterns for Insider Threat Detection* Computer Fraud & Security, Volume 2015, Issue 7, July 2015, Pages 9–17. Elsevieer, Amsterdam, NL, 2015. Available from: https://www.sciencedirect.com/science/article/pii/S136137231530066X [Last accessed: 19.11.2021]

[Agrafiotis17]        Ioannis Agrafiotis, Arnau Erola, Michael Goldsmith, Sadie Creese: *Formalizing Policies for Insider-Threat Detection - A Tripwire Grammar* Journal of Wireless Mobile Networks, Ubiquitous Computing, and Dependable Applications, Innovative Information Science & Technology Research Group (ISYOU), Seoul, Republic of Korea, March 2017, pp. 26–43. Downloadable from: http://isyou.info/jowua/papers/jowua-v8n1-2.pdf [Last accessed: 18.12.2021]

[Aiello10]           Bob Aiello: Configuration Management Best Practices - *Practical Methods that Work in the Real World* Addison-Wesley Professional, Upper Saddle River, NJ, USA, 2010. ISBN 978-0-321-68586-5

[Amundsen16]         Mike Amundsen, Matt Mclarty: Microservice Architecture – *Aligning Principles, Practices, and Culture* O'Reilly Media, Inc., Sebastopol, CA, USA, 2016. ISBN 978-1-491-95625-0

[Anand08]            Sanjay Anand: The Sarbanes-Oxley Act (SOXBoK) - An Introduction. Van Haren Publishing, 's-Hertogenbosch, Nederlands, 2008. ISBN 978-9-087-53083-9

[Apruzzese18]        Giovanni Apruzzese, Michele Colajanni, Luca Ferretti, Alessandro Guido, Mirco Marchetti: *On the Effectiveness of Machine and Deep Learning for Cyber Security* 10th International Conference on Cyber Conflict, 2018, Tallinn, Estland, 29 May to 1 June 2018. Downloadable from: https://ccdcoe.org/uploads/2018/10/Art-19-On-the-Effectiveness-of-Machine-and-Deep-Learning-for-Cyber-Security.pdf [Last accessed: 21.12.2021]

[Arnold17]           Rob Arnold: Cybersecurity - *A Business Solution (An Executive Perspective on managing Cyber Risk)* Threat Sketch, LLC, Winston-Salem, NC, USA, 2017. ISBN 978-0-6929-4415-8

[Badhwar21a]         Raj Badhwar: The CISO's Next Frontier - AI, Post-Quantum Cryptography, and Advanced Security Paradigms. Springer Nature Switzerland, Cham, Switzerland, 2021. ISBN 978-3-030-75353-5

[Badhwar21b]         Raj Badhwar: The CISO's Transformation - Security Leadership in a High Threat Landscape. Springer Nature Switzerland, Cham, Switzerland, 2021. ISBN 978-3-030-81411-3

[Barker21]           William Barker, William Polk, Murugiah Souppaya: Getting Ready for Post-Quantum Cryptography - *Exploring Challenges Associated with Adopting and Using Post-Quantum Cryptographic Algorithms* NIST Cybersecurity White Paper, US National Institute of Standards and Technology, Gaithersburg, MD, USA, April 28, 2021. Downloadable from: https://nvl-pubs.nist.gov/nistpubs/CSWP/NIST.CSWP.04282021.pdf [Last accessed: 06.12.2021]

[Bishop18]           Matt Bishop: Computer Security - *Art and Science* Addison Wesley, Boston, USA, 2nd edition, 2018. ISBN 978-0-321-71233-2

[Blum20]        Dan Blum: Rational Cybersecurity for Business - *The Security Leaders' Guide to Business Alignment* Apress Media, LLC, New York, NY, USA, 2020. ISBN 978-1 484 2595 1

[Brotby09]      Krag Brotby: Information Security Governance - *A Practical Development and Implementation Approach* John Wiley & Sons Inc., Hoboken, NJ, USA, 2009. ISBN 978-0-470-13118-3

[BSI08]         BSI: BSI-Standard 101–1: *Information Security Management Systems (ISMS)* Deutsches Bundesamt für Sicherheit in der Informationstechnik, Bonn, Germany, Version 1.5, 2008. Downloadable from: https://www.bsi.bund.de/SharedDocs/Downloads/EN/BSI/Publications/BSIStandards/standard_100-1_e_pdf.pdf?__blob=publicationFile&v=1 [Last accessed: 02.12.2021]

[BSI12]         Bundesamt für Informationssicherheit BIS (German Federal Office for Information Security), Bonn, Germany. https://www.bsi.bund.de/EN/Service-Navi/Contact/contact_node.html#Start

[BSIMM21]       Sammy Migues, Eli Erlikhman, Jacob Ewers, Kevin Nassery: Building Security In Maturity Model (BSIMM) BUILDING SECURITY IN MATURITY MODEL (BSIMM) FOUNDATIONS REPORT – VERSION 12, 2021. Downloadable from: https://www.bsimm.com/download.html [Last accessed: 15.01.2022]

[Bucchiarone21] Antonio Bucchiarone et. al. (Editors): Microservices - *Science and Engineering* Springer Nature Switzerland AG, Cham, Switzerland, 2020. ISBN 978-3-030-31648-8

[Buchanan17]    William J. Buchanan: Cryptography River Publishers, Gistrup, Denmark, 2017. ISBN 978-8-7933-7910-7

[Buchmann16]    Johannes A. A. Buchmann, Evangelos Karatsiolis, Alexander Wiesmaier: Introduction to Public Key Infrastructures Springer-Verlag, Heidelberg, Germany, 2016. ISBN 978-3-662-52450-3

[Bunn16]        Matthew Bunn, Scott D. Sagan: *A Worst Practice Guide to Insier Threats* Chapter 6 (pp. 145–174) in: Matthew Bunn, Scott D. Sagan (Editors): Insider Threats Cornell University Press, Ithaca, NY, USA, 2016. ISBN 978-1501-70517-5

[Campagna20]    Matt Campagna, Brian LaMacchia, and David Ott: *Post Quantum Cryptography: Readiness Challenges and the Approaching Storm* Computing Community Consortium (CCC) Quadrennial Paper, Computing Research Association, Washington, DC, USA, 2020. Downloadable from: https://cra.org/ccc/wp-content/uploads/sites/2/2020/10/Post-Quantum-Cryptography_-Readiness-Challenges-and-the-Approaching-Storm-1.pdf [Last accessed: 06.12.2021]

[Cappelli12]    Dawn M. Cappelli, Andrew P. Moore, Randall F. Trzeciak: The CERT Guide to Insider Threats - *How to Prevent, Detect, and Respond to Information Technology Crimes (Theft, Sabotage, Fraud)* Addison-Wesley, Upper Saddle River, NJ, USA, 2012. ISBN 978-0-321-81257-5

[Cartlidge07]   Alison Cartlidge, Mark Lillycrop (Editors): An Introductory Overview of ITIL® V3 UK Chapter of the itSMF (The IT Service Management Forum), Leppington, Bracknell, UK, 2007. ISBN 0-9551245-8-1. Downloadable from: https://www.itilnews.com/uploaded_files/itSMF_ITILV3_Intro_Overview.pdf [last accessed: 18.8.2021]

[Caselli16]        Marco Caselli, Frank Kargl: *A Security Assessment Methodology for Critical Infrastructures* Conference Paper, International Conference on Critical Information Infrastructures Security, Paris, France, October 10–12, 2016. Downloadable from: https://www.researchgate.net/publication/301932855_A_Security_Assessment_Methodology_for_Critical_Infrastructures [Last accessed: 12.12.2021]

[CC17]             Common Criteria Consortium: Common Criteria for Information Technology Security Evaluation. *Part 1: Introduction and General Model* CCMB-2017-04-001, Version 3.1, Revision 5, April 2017. Downloadable from: https://www.commoncriteriaportal.org/files/ccfiles/CCPART1V3.1R5.pdf [Last accessed: 18.8.2021]

[CERT16]           CERT Insider Threat Center: Common Sense Guide to Mitigating Insider Threats The CERT Insider Threat Center Technical Note CMU/SEI-2015-TR-010, 5th edition, December 2016. Carnegie Mellon University, Pittsburg, USA, 2016. Downloadable from: https://resources.sei.cmu.edu/asset_files/technicalreport/2016_005_001_484758.pdf [Last accessed: 19.11.2021]

[Charney14]        David L. Charney: NOIR: *A White Paper on Insider Threat* US National Office for Intelligence Reconciliation (NOIR), Alexandria, VA, USA, Parts 1-3, 2014. ISBN 978-0-6922-6085-2

[Cheruvu19]        Sunil Cheruvu, Anil Kumar, Ned Smith, David M. Wheeler: Demystifying Internet of Things Security - *Successful IoT Device/Edge and Platform Security Deployment* Apress Media, LLC, New York, NY, USA, 2019. ISBN 978-1-4842-2895-1. Downloadable from (Apress Open): https://link.springer.com/content/pdf/10.1007%2F978-1-4842-2896-8.pdf [last accessed: 25.10.2021]

[Chio18]           Clarence Chio, David Freeman: Machine Learning and Security - *Protecting Systems with Data and Algorithms* O'Reilly Media Inc., Sebastopol, CA, USA, 2018. ISBN 978-1-491-97990-7

[Cooper18]         David Cooper, Andrew Regenscheid, Murugiah Souppaya, Christopher Bean, Mike Boyle, Dorothy Cooley, Michael Jenkins: *Security Considerations for Code Signing* NIST, Gaithersburg, MD, USA, January 26, 2018. Downloadable from: https://nvlpubs.nist.gov/nistpubs/CSWP/NIST.CSWP.01262018.pdf [Last accessed: 23.7.2021]

[CSI20]            Cybersecurity Insiders: *2020 Insider Threat Report* Cybersecurity Insiders, USA, 2020. Available at: https://www.cybersecurity-insiders.com/portfolio/2020-insider-threat-report/ [last accessed: 2.5.2020]

[Davis21]          Robert E. Davis: Auditing Information and Cyber Security Governance - *A Controls-based Approach* CRC Press (Taylor & Francis), Boca Raton, FL, USA, 2021. ISBN 978-0-367-56850-4

[Daya15]           Shahir Daya, et.al.: Microservices from Theory to Practice - Creating Applications in IBM Bluemix Using the Microservices Approach. Red Book SG24-8275-00, International Business Machines Corporation (IBM) Armonk, NY, USA, 2015. Downloadable from: https://www.redbooks.ibm.com/redbooks/pdfs/sg248275.pdf [Last accessed: 12.06.2022]

[Dempsey11]        Kelley Dempsey, Nirali Shah Chawla, Arnold Johnson, Ronald Johnston, Alicia Clay Jones, Angela Orebaugh, Matthew Scholl, Kevin Stine: Information Security Continuous Monitoring (ISCM) for Federal Information Systems and Organizations NIST Special Publication 800–137,

Washington, D.C., USA, 2011. Downloadable from: https://nvlpubs.nist. gov/nistpubs/Legacy/SP/nistspecialpublication800-137.pdf [Last accessed: 20.12.2020]

[Dempsey21]    Kelley Dempsey, Victoria Yan Pillitteri, Andrew Regenscheid: *Managing the Security of Information Exchanges* NIST Special Publication 800–47, Revision 1, US National Institute of Standards and Technology, Gaithersburg, MD, USA, July 2021. Downloadable from: https://nvlpubs. nist.gov/nistpubs/SpecialPublications/NIST.SP.800-47r1.pdf [Last accessed: 20.7.2021]

[DHS14]    DHS: Combating the Insider Threat US Department of Homeland Security (DHS), National Cybersecurity and Communications Integration Center, Washington, DC, USA, 2th May 2014. Downloadable from: https://www. cisa.gov/uscert/sites/default/files/publications/Combating%20the%20 Insider%20Threat_0.pdf [Last accessed: 18.12.2021]

[Dooley18]    John F. Dooley: History of Cryptography and Cryptanalysis - *Codes, Ciphers, and Their Algorithms* Springer International Publishing, Cham, Switzerland, 2018. ISBN 978-3 - 31990442-9

[Dowd06]    Mark Dowd, John McDonald, Justin Schuh: The Art of Software Security Assessment – *Identifying and Preventing Software Vulnerabilities* Addison-Wesley, Upper Saddle River, N.J., USA, 2006. ISBN 978-0-321-44442-4

[Drusinsky04]    Doron Drusinsky, J.L. Fobes: *Executable Specifications - Language and Applications* CROSSTALK, The Journal of Defense Software Engineering, September 2004, pp. 15–18. Hill Air Force Base, UT, USA, 2004. Downloadable from: https://citeseerx.ist.psu.edu/viewdoc/download?doi=10 .1.1.433.5011&rep=rep1&type=pdf [Last accessed: 13.11.2021]

[e-Directive09]    European Union: DIRECTIVE 2002/58/EC OF THE EUROPEAN PARLIAMENT AND OF THE COUNCIL, 12 July 2002, with Addendum 2009. Downloadable from: https://eur-lex.europa.eu/LexUriServ/LexUriServ. do?uri=CONSLEG:2002L0058:20091219:EN:PDF [Last accessed: 12.06.2022]

[Erl05]    Thomas Erl: Service-oriented Architecture – *Concepts, Technology, and Design* Prentice-Hall, Upper Saddle River, N.J., USA, 2005, 2nd edition, 2016. ISBN 978-0-133-85858-7

[Erl17]    Thoms Erl: Service Infrastructure- *On-Premise and in the Cloud* Pearson Education (Prentice Hall Service Technology Series), Upper Saddle River, N.J., USA, 2017. ISBN 978-0-133-85872-3

[Ernst21]    Neil Ernst, Rick Kazman, Julien Delange: Technical Debt in Practice - *How to Find It and Fix It* The MIT Press, Cambridge, MA, USA, 2021. ISBN 978-0-262-54211-1

[Faesen19]    Louk Faesen, Erik Frinking, Gabriella Gricius, Elliot Mayhew: *Understanding the Strategic and Technical Significance of Technology for Security* The Hague Centre for Strategic Studies (HCSS) and The Hague Security Delta, Den Haag, Netherlands, 2019. Downloadable from: https://securitydelta.nl/media/com_hsd/report/234/document/HSD-Rapport-AI-11619.pdf [Last accessed: 23.12.2021]

[Ferguson10]    Niels Ferguson, Bruce Schneier, Tadayoshi Kohno: Cryptography Engineering - *Design Principles and Practical Applications* Wiley Publishing Inc., Indianapolis, IN, USA, 2010. ISBN 978-0-470-47424-2

[Ferrillo17]    Paul A Ferrillo, Christophe Veltsos: Take Back Control of Your Cybersecurity Now - *Game-Changing Concepts on AI and Cyber*

*Governance Solutions for Executives* Advisen Ltd., New York, NY, USA, 2017. ISBN 978-1-5206-5872-8

[Fiaschetti18]    Andrea Fiaschetti, Josef Noll, Paolo Azzoni, Roberto Uribeetxeberria (Editors): Measurable and Composable Security, Privacy, and Dependability for Cyberphysical Systems – *The SHIELD Methodology* Taylor & Francis Ltd., USA, 2018. ISBN 978-1-138-04275-9

[Firestone02]    Joseph M. Firestone: Enterprise Information Portals and Knowledge Management Routledge (Taylor & Francis), Abingdon, UK, 2002. ISBN 978-0-750-67474-4

[Furrer19]    Frank J. Furrer: Future-Proof Software-Systems – *A Sustainable Evolution Strategy* Springer Vieweg Verlag, Wiesbaden, Germany, 2019. ISBN 978-3-658-19937-1

[Furrer22]    Frank J. Furrer: Engineering Principles for Safety and Security of Cyber-Physical Systems Springer Vieweg Verlag, Wiesbaden, Germany, 2022. ISBN 978-

[Gallotti19]    Cesare Gallotti: Information Security - *Risk Assessment, Management Systems, the ISO/IEC 27001 Standard* www.lulu.com, 2019. ISBN 978-0-244-14955-0

[Gardiner20]    Joseph Gardiner, Shishir Nagaraja: *On the Security of Machine Learning in Malware C&C Detection – A Survey* Preprint, ACM Computing Surveys, 2020. Downloadable from: http://personal.strath.ac.uk/shishir.nagaraja/ papers/secml-survey.pdf [Last accessed: 21.12.2021]

[GDPR16]    The European Parliament and Council: Regulation (EU) 2016/679 (General Data Protection Regulation - GDPR) Brussels, Belgium, 27 April 2016. Downloadable from: https://eur-lex.europa.eu/legal-content/EN/TXT/ PDF/?uri=CELEX:32016R0679 [last accessed: 29.6.2020]

[Gelles16]    Michael G. Gelles: Insider Threat - *Prevention, Detection, Mitigation, and Deterrence* Butterworth-Heinemann, Kidlington, OX, UK. 2016. ISBN 978-0-128-02410-2

[Gilman17]    Evan Gilman, Doug Barth: Zero Trust Networks - *Building Secure Systems in Untrusted Networks* O'Reilly Media Inc., Sebastopol, CA, USA, 2017. ISBN 978-1-491-96219-0

[Goldreich08]    Oded Goldreich: Foundations of Cryptography - *Volume 1: Basic Tools* Cambridge University Press, Cambridge, UK, 2008. ISBN 978-0-521-03536-1

[Goldreich09]    Oded Goldreich: Foundations of Cryptography - *Volume 2: Basic Applications* Cambridge University Press, Cambridge, UK, 2009. ISBN 978-0-521-11991-7

[Grassi17]    Paul A. Grassi et. al : NIST Special Publication 800–63B: Digital Identity Guidelines - *Authentication and Lifecycle Management* NIST, US National Institute of Standards and Technology, Gaithersburg, MD, USA June 2017. Downloadable from: https://nvlpubs.nist.gov/nistpubs/SpecialPublications/ NIST.SP.800-63b.pdf [Last accessed: 04.12.2021]

[Halak21]    Basel Halak (Editor): Authentication of Embedded Devices - *Technologies, Protocols and Emerging Applications* Springer Nature Switzerland, Cham, Switzerland, 2021. ISBN 978-3-030-60768-5

[Harmer14]    Geoff Harmer: Governance of Enterprise IT Based on COBIT 5 - *A Management Guide. IT Governance Publishing (ITGP)*, Ely, UK, 2014. ISBN 978-1-8492-8518-6

[Harrington20]     Ted Harrington: Hackable - *How to Do Application Security Right* Lioncrest Publishing, Carson City, Nevada, USA, 2020. ISBN 978-1-5445-1766-7

[Hasselbring18]    Wilhelm Hasselbring: Software Architecture - *Past, Present, Future* Chapter 10 in: Volker Gruhn, Rüdiger Striemer (Editors): The Essence of Software Engineering Springer Open, Springer International Publishing, Cham, Switzerland, 2018. ISBN 978-331-973896-3. Downloadable from: https://oceanrep.geomar.de/43455/1/Hasselbring2018_Chapter_SoftwareArchitecturePastPresent.pdf [Last accessed: 11.01.2022]

[Higaki10]         Wesley Hisao Higaki: Successful Common Criteria Evaluations - *A Practical Guide for Vendors* Createspace Independent Publishing Platform, 2010. ISBN 978-1-4528-8661-9

[Higaki14]         Wesley Hisao Higaki: Writing Common Criteria Documentation Createspace Independent Publishing Platform, 2014. ISBN 978-1-5004-1122-0

[Higginbotham21]   James Higginbotham: Principles of Web API Design - *Delivering Value With APIs and Microservices* Addison-Wesley (Pearson Education), Upper Saddle River, NJ, USA, 2021. ISBN 978-0-137-35563-1

[Hintzbergen15]    Jule Hintzbergen, Kees Hintzbergen, André Smulders, Hans Baars: Foundations Of Information Security Based on ISO27001 And ISO27002 Van Haren Publishing, Zaltbommel, Nederlands, 3rd edition, 2015. ISBN 978-94-018-0012-9

[Hoffstein14]      Jeffrey Hoffstein, Jill Pipher, Joseph H. Silverman: An Introduction to Mathematical Cryptography Springer Science & Business Media, New York, N.Y., USA, 2nd edition 2014. ISBN 978-1-493-91710-5

[Holt06]           Michael F. Holt : The Sarbanes-Oxley Act - Overview and Implementation Procedures. CIMA Professional Handbook, CIMA Publishing (Elsevier), Oxford, UK, 2005. ISBN 978-0-7506-6823-1

[Housley21]        Russ Housley: Certificates - *Transition from Traditional Algorithms to PQC Algorithms* NCCoE-Workshop Presentation, Virtual Workshop, 2021. Downloadable from: https://www.nccoe.nist.gov/sites/default/files/2021-10/10-Housley-NCCoE-Workshop-Transition-to-PQC-Certificates.pdf [Last accessed: 06.12.2021]

[ISACA19]          Ian Cooke: Enhancing the IT Audit Report using COBIT 2019. ISACA JOURNAL Vol. 4, 30 June 2020, ISACA, Schaumburg, IL, USA. Downloadable from: https://www.isaca.org/-/media/files/isacadp/project/isaca/articles/journal/2020/volume-4/enhancing-the-it-audit-report-using-cobit-2019_joa_eng_0720.pdf [Last accessed: 12.06.2022]

[ISO15408]         International Standards Organization (ISO): ISO/IEC 15408–1:2009: Information Technology - Security Techniques - Evaluation Criteria for IT Security. *Part 1: Introduction and General Model* International Standards Organization (ISO):, Geneva, Switzerland, 2015. Available at: https://www.iso.org/standard/50341.html [Last accessed: 18.8.2021]

[ISO17799]         ISO/IEC 17799–2005: Information technology - Security Techniques - Code of Practice for Information Security Management International Standards Organization (ISO), Geneva, Switzerland, 2005. Available from: https://www.iso.org/standard/39612.html [Last accessed: 10.8.2021]

[ISO27000]         ISO: ISO/IEC 27000:2018 / Information technology - *Security techniques - Information security management systems - Overview and vocabulary* International Organization for Standardization, Vernier, Switzerland, 2018.

Available from: https://www.iso.org/standard/73906.html [Last accessed: 02.12.2021]

[ISO27007]     ISO: ISO/IEC 27007:2020: Information Security, Cybersecurity, and Privacy Protection - *Guidelines for Information Security Management Systems auditing* International Standards Organization (ISO), Geneva, Switzerland, 2020. Available from: https://www.iso.org/standard/77802.html [Last accessed: 11.12.2021]

[ISO27034]     ISO/IEC 27034: Information technology - Security techniques - Application security ISO/IEC Standard (revised 2017), Geneva, Switzerland, 2017. Available at: https://www.iso.org/standard/44378.html [Last accessed: 15.01.2022]

[ISO29192]     ISO/IEC 29192-1:2012/2017: Information Technology - Security Techniques - Lightweight Cryptography - *Part 1: General* https://www.iso.org/standard/56425.html

[ITG20]        UK IT Governance: *Information Security and ISO 27001 – An Introduction* UK IT Governance, Ely, UK: Green Paper, January 2020. Downloadable from: https://www.itgovernance.co.uk/green-papers/information-security-and-iso-27001-an-introduction [Last accessed: 30.11.2020]

[ITGP19]       IT Governance Privacy Team: EU General Data Protection Regulation (GDPR) - *An Implementation and Compliance Guide* IT Governance Publishing, Ely, UK, 3rd edition, 2019. ISBN 978-1-78778-191-7

[ITIL19]       ITIL Foundation: ITIL 4th Edition The Stationery Office (STO), Norwich, UK, 2019. ISBN 978-0-1133-1607-6

[Kaschner20]   Holger Kaschner: Cyber Crisis Management - *Das Praxishandbuch zu Krisenmanagement und Krisenkommunikation* Springer Fachmedien Wiesbaden GmbH, Wiesbaden, Germany, 2020. ISBN 978-3-658-27913-4

[Katz21a]      Jonathan Katz, Yehuda Lindell: Introduction to Modern Cryptography CRC Press (Taylor & Francis), Boca Raton, FL, USA, 3rd edition, 2021. ISBN 978-0-815-35436-9

[Kegerreis19]  Mike Kegerreis, Mike Schiller, Chris Davis: IT Auditing - *Using Controls to Protect Information Assets* McGaw Hill Education, New York, NY, USA, 3rd edition, 2019. ISBN 978-1-260-45322-5

[Kerman20]     Alper Kerman: Zero Trust Cybersecurity: 'Never Trust, Always Verify' NIST Publication, October 28, 2020. Downloadable from: https://www.nist.gov/blogs/taking-measure/zero-trust-cybersecurity-never-trust-always-verify [Last accessed < >: 10.11.2020]

[Khan16]       Muhammad Taimoor Khan, Dimitrios Serpanos, Howard Shrobe: *Sound and Complete Runtime Security Monitor for Application Software* Prepring, arXiv:1601.04263v1[cs.CR],17 Jan 2016. Downloadable from: https://www.researchgate.net/publication/291229626_Sound_and_Complete_Runtime_Security_Monitor_for_Application_Software [Last accessed: 12.12.2021]

[Kim18b]       Kwangjo Kim, Muhamad Erza Aminanto, Harry Chandra Tanuwidjaja: Network Intrusion Detection using Deep Learning - *A Feature Learning Approach* Springer Nature Singapore PTE Ltd., Singapore, Singapore, 2018. ISBN 978-9-811-31443-8

[Kloft10]      Marius Kloft, Pavel Laskov: *Online Anomaly Detection under Adversarial Impact* 13th International Conference on Artificial Intelligence and Statistics (AISTATS), 2010, Chia Laguna Resort, Sardinia, Italy. Downloadable from: http://proceedings.mlr.press/v9/kloft10a/kloft10a.pdf [Last accessed: 23.12.2021]

[Knapp13]        Eric D. Knapp, Raj Samani: Applied Cyber Security and the Smart Grid - *Implementing Security Controls into the Modern Power Infrastructure* Syngress (Elsevier), Amsterdam, Netherlands, 2013. ISBN 978-1-597-49998-9

[Kohnfelder21]   Loren Kohnfelder: Designing Secure Software - *A Guide for Developers* No Starch Press, San Francisco, CA, USA, 2021. ISBN 978-1-718-50192-8

[Kruchten19]     Philippe Kruchten, Robert Nord: Managing Technical Debt - *Reducing Friction in Software Development* Addison-Wesley, Upper Saddle River, NJ, USA, 2019. ISBN 978-0-135-64593-2

[Landoll16]      Douglas J. Landoll: Information Security Policies, Procedures, and Standards - A Practitioner's Reference CRC Press, Taylor & Francis Ltd., Boca Raton, FL, USA, 2016. ISBN 978-1-482-24589-9

[Love10]         Paul Love, James Reinhard, A.J. Schwab, George Spafford: Global Technology Audit Guide (GTAG®) 15: Information Security Governance The Institute of Internal Auditors, Altamonte Springs, FL, USA, 2010. Downloadable from: https://chapters.theiia.org/montreal/ChapterDocuments/GTAG%2015%20-%20Information%20Security%20Governance.pdf  [Last accessed: 11.12.2021]

[Lukings21]      Melissa Lukings, Arash Habibi Lashkari: Understanding Cybersecurity Law and Digital Privacy - *A Common Law Perspective* Springer Nature Switzerland, Cham, Switzerland, 2021. ISBN 978-3-030-88703-2

[Macher]         Georg Macher, Christoph Schmittner, Omar Veledar, Eugen Brenner: ISO/SAE DIS 21434 Automotive Cybersecurity Standard - In a Nutshell. In: Casimiro, A., Ortmeier, F., Schoitsch, E., Bitsch, F., Ferreira, P. (editors): Computer Safety, Reliability, and Security. SAFECOMP 2020 Workshops. SAFECOMP 2020. Lecture Notes in Computer Science, Vol 12235. Springer Nature Switzerland, Cham, Switzerland. ISBN 978-3-030-55583-2_9

[Massey20]       Stephen Robert Massey: Ultimate GDPR Practitioner Guide - Demystifying Privacy & Data Protection. Fox Red Risk Publishing, London, UK, 2nd edition, 2020. ISBN 978-1-9998-2723-6

[McCarthy21]     Jim McCarthy, Don Faatz, Nik Urlaub, John Wiltberger, Tsion Yimer: Securing the Industrial Internet of Things -*Cybersecurity for Distributed Energy Resources* Volume B: Approach, Architecture, and Security Characteristic NIST Special Publication 1800–32B, Draft April 2021. Downloadable from: https://www.nccoe.nist.gov/sites/default/files/library/sp1800/energy-iiot-sp1800-32b.pdf This publication is available free of charge from https://www.nccoe.nist.gov/iiot

[McKay17]        Kerry A. McKay, Larry Bassham, Meltem Sönmez Turan, Nicky Mouha: Report on Lightweight Cryptography (US National Institute of Standards and Technology Report NISTIR 8114) CreateSpace Independent Publishing Platform, North Charleston, S.C., USA, 2017. ISBN 978-1-9811-1346-0. Downloadable from: https://nvlpubs.nist.gov/nistpubs/ir/2017/NIST.IR.8114.pdf [last accessed: 9.9.2018]

[Mellor04]       Stephen J. Mellor: *Executable and Translatable UML* CROSSTALK, The Journal of Defense Software Engineering, September 2004, pp. 19–22. Hill Air Force Base, UT, USA, 2004. Downloadable from: https://citeseerx.ist.

| | psu.edu/viewdoc/download?doi=10.1.1.433.5011&rep=rep1&type=pdf [Last accessed: 13.11.2021] |
|---|---|
| [Moeller10] | Robert R. Moeller: IT Audit, Control, and Security John Wiley & Sons, Inc., New York, N.Y., USA, 2010. ISBN 978-0-471-40676-1 |
| [Moyle20] | Ed Moyle, Diana Kelley: Practical Cybersecurity Architecture - *A Guide to creating and implementing robust Designs for Cybersecurity Architects* Packt Publishing, Birmingham, UK, 2020. ISBN 978-1-8389-8992-7 |
| [Nash01] | Andrew Nash, William Duane, Celia Joseph, Derek Brink: PKI: Implementing & Managing E-Security RSA Press, McGraw-Hill Education Ltd., New York, NY, USA, 2001. ISBN 978-0-0721-3123-9 |
| [NCSP04] | National Cyber Security Partnership (NCSP): Information Security Governance – *A Call to Action* US Corporate Governance Task Force Report, Washington, DC, USA, April 2004. Downloadable from: https://www.cyberpartnership.org/InfoSecGov4_04.pdf [Last accessed: 10.8.2021] |
| [Newman15] | Sam Newman: Building Microservices - *Designing Fine-Grained Systems* O'Reilly and Associates, Sebastopol, CA, USA, 2015. ISBN 978-1-491-95035-7 2nd edition 2021, ISBN 978-1-492-03402-5 |
| [Newman19] | Sam Newman: Monolith to Microservices - *Evolutionary Patterns to Transform Your Monolith* O'Reilly Media, Inc., Sebastopol, CA, USA, 2019. ISBN 978-1-492-04784-1 |
| [Newman21] | Sam Newman: Building Microservices - Designing Fine-Grained Systems. O'Reilly and Associates, Sebastopol, CA, USA, 2nd edition 2021. ISBN 978-1-492-03402-5 |
| [NIST06] | Karen Kent, Suzanne Chevalier, Tim Grance, Hung Dang: Guide to Integrating Forensic Techniques into Incident Response - *Recommendations of the National Institute of Standards and Technology* NIST Special Publication 800–86, 2006. Downloadable from:https://nvlpubs.nist.gov/nistpubs/Legacy/SP/nistspecialpublication800-86.pdf [last accessed: 6.7.2020] |
| [NIST14] | US National Institute of Standards and Technology: Framework for Improving Critical Infrastructure Cybersecurity NIST, Gaithersburg, MD, USA, 2014. Downloadable from: https://www.nist.gov/system/files/documents/cyberframework/cybersecurity-framework-021214.pdf [Last accessed: 24.2.2021] |
| [NIST16] | Lidong Chen, Stephen P. Jordan, Yi-Kai Liu, Dustin Moody, Rene C. Peralta, Ray A. Perlner, Daniel C. Smith-Tone: Report on Post-Quantum Cryptography US National Institute of Standards and Technology, Gaithersburg, MD, USA, 2016. Downloadable from: https://nvlpubs.nist.gov/nistpubs/ir/2016/NIST.IR.8105.pdf [last accessed: 18.4.2020] |
| [NIST1800-16] | [NIST1800-16] NIST SPECIAL PUBLICATION 1800-16: Securing Web Transactions - *TLS Server Certificate Management.* June 2020. Downloadable from: https://nvlpubs.nist.gov/nistpubs/SpecialPublications/NIST.SP.1800-16.pdf [last accessed: 17.6.2020] |
| [NIST20] | Scott Rose, Oliver Borchert, Stu Mitchell, Sean Connelly: Zero Trust Architecture NIST Special Publication 800–207 (final), August 2020. Downloadable from: https://doi.org/10.6028/NIST.SP.800-207 or https://csrc.nist.gov/publications/detail/sp/800-207/final [last accessed: 12.8.2020] |
| [NIST800-160B] | Ron Ross, Victoria Pillitteri, Richard Graubart, Deborah Bodeau, Rosalie McQuaid: Developing Cyber Resilient Systems - *A Systems Security Engineering Approach* NIST Special Publication 800–160, Volume 2, Rev. |

1, December 2021. Downloadable from: https://nvlpubs.nist.gov/nistpubs/SpecialPublications/NIST.SP.800-160v2r1.pdf [Last accessed: 09.12.2021]

[NIST800-53_20]   *Security and Privacy Controls for Information Systems and Organizations* Special Publication 800-53, Revision 5, 2020. US National Institute of Standards and Technology (NIST), Gaithersburg, MD, USA. Downloadable from: https://doi.org/10.6028/NIST.SP.800-53r5 [Last accessed: 27.9.2020]

[NISTIR8151]   Paul E. Black, Lee Badger, Barbara Guttman, Elizabeth Fong: Dramatically Reducing Software Vulnerabilities - *Report to the White House Office of Science and Technology Policy* US National Institute of Standards and Technology, NISTIR 8151, November 2016. Downloadable from: https://nvlpubs.nist.gov/nistpubs/ir/2016/NIST.IR.8151.pdf   [Last   accessed: 23.8.2020]

[Noonan18]   Christine Noonan: *Spy the Lie - Detecting Malicious Insiders* Technical Report for the U.S. Department of Energy by Pacific Northwest National Laboratory, Richland, WA, USA, March 2018. Downloadable from: https://irp.fas.org/eprint/noonan.pdf [Last accessed: 19.12.2021]

[Nurse14]   Jason R.C. Nurse, Oliver Buckley, Philip A. Legg, Michael Goldsmith, Sadie Creese, Gordon R.T. Wright, Monica Whitty: *Understanding Insider Threat - A Framework for Characterising Attacks* Conference Paper, 2014 IEEE Security and Privacy Workshops, IEEE New York, NY, USA, 2014. Downloadable from: http://www.cs.ox.ac.uk/files/6576/writ2014_nurse_et_al.PDF [Last accessed: 19.12.2021]

[O'Gorman17]   Tristan O'Gorman: *A Primer on IoT Security Risks* Security Intelligence, February 8, 2017. Downloadable from: https://securityintelligence.com/a-primer-on-iot-security-risks/ [Last accessed: 21.8.2020]

[OISSG06]   OISSG: Information Systems Security Assessment Framework (ISSAF) Open Information Systems Security Group, London, UK, Draft 0.2, 2006. Downloadable from: https://untrustednetwork.net/files/issaf0.2.1.pdf [Last accessed: 12.12.2021]

[Okamura17]   Toshihiko Okamura: *Lightweight Cryptography Applicable to Various IoT Devices* NEC Technical Journal, Vol.12, No.1, October 2017. Downloadable from: https://www.nec.com/en/global/techrep/journal/g17/n01/g1701pa.html [last accessed: 22.8.2020]

[Overly15]   Michael R. Overly: *Information Security and Legal Compliance* Kaspersky Lab, Moscow, Russian Federation, 2015. Downloadable from: http://go.kaspersky.com/rs/kaspersky1/images/Information%20Security%20and%20Legal%20Compliance%202014.pdf [Last accessed: 05.12.2021]

[Paar10]   Christof Paar, Jan Pelzl: Understanding Cryptography - *A Textbook for Students and Practitioners* Springer-Verlag, Berlin, Germany, 2010. ISBN 978-3-642-04100-6

[Pompon16]   Raymond Pompon: IT Security Risk Control Management - *An Audit Preparation Plan* Apress Media LLC, New York, N.Y., USA, 2016. ISBN 978-1-48422-139-6

[Ponemon20]   Ponemon Institute: 2020 Cost of Insider Threats Global Report Ponemon Institute, Traverse City, MI, USA, 2020. Downloadable from: https://cdw-prod.adobecqms.net/content/dam/cdw/on-domain-cdw/brands/proofpoint/ponemon-global-cost-of-insider-threats-2020-report.pdf   [Last   accessed: 17.12.2021]

[Poschmann09]          Axel Poschmann: Lightweight cryptography – *Cryptographic Engineering for a Pervasive World* Bochumer Universitätsverlag Westdeutscher Universitätsverlag, Germany, 2009. ISBN 978-3-89966-341-9

[PWC21]                PWC: *Why managing Software Vulnerabilities is Business-critical – and how to do it efficiently and effectively* PricewaterhouseCoopers AG, Zurich, Switzerland, 2021. Downloadable from: https://www.pwc.ch/en/publications/2021/ch-vulnerability-management_EN.pdf [Last accessed: 11.12.2021]

[Rai07]                Sajay Rai et. al: Identity and Access Management The Institute of Internal Auditors (The IIA), Altamonte Springs, FL, USA, 2007. Downloadable from:      https://chapters.theiia.org/montreal/ChapterDocuments/GTAG%20 9%20-%20Identity%20and%20Access%20Management.pdf [Last accessed: 05.12.2021]

[Ransome14]            James Ransome, Anmol Misra: Core Software Security - *Security at the Source* CRC Press (Taylor & Francis Group), Boca Raton, FL, USA, 2014. ISBN 978-1-032-02741-8

[Rodger17]             Richard Rodger: The Tao of Microservices Manning Publications, Shelter Island, New York, N.Y., USA, 2017. ISBN 978-1-617-29314-6

[Russo18]              Mark A. Russo: Information Technology Security Audit Guidebook - *NIST SP 800–171* Independently published, 2018. ISBN 978-1-7266-7490-4

[Ryder19]              Rodney D. Ryder, Ashwin Madhavan: Cyber Crisis Management Bloomsbury Publishing India, New Delhi, India, 2019. ISBN 978-9-3891-6550-0

[Sabella18]            Anthony Sabella, Rik Irons-Mclean, Marcelo Yannuzzi: Orchestrating and Automating Security for the Internet of Things – *Delivering Advanced Security Capabilities from Edge to Cloud for IoT* Cisco Systems Inc., USA, 2018. ISBN 978-1-5871-4503-2

[SAMM21]               Software Assurance Maturity Model: OWASP SAMM (Software Assurance Maturity Model), V2.0 Core Model Document, Open Web Application Security Project, Maryland, USA, 2021. Downloadable from: https://github.com/OWASP/samm/blob/master/Supporting%20Resources/v2.0/OWASP-SAMM-v2.0.pdf [Last accessed: 15.01.2022]

[Scarfone08]           Karen Scarfone, Murugiah Souppaya, Amanda Cody, Angela Orebaugh: Technical Guide to Information Security Testing and Assessment NIST Special Publication 800–115, Computer Security Division, Information Technology Laboratory, National Institute of Standards and Technology, Gaithersburg, MD, USA, September 2008. Downloadable from: https://nvl-pubs.nist.gov/nistpubs/Legacy/SP/nistspecialpublication800-115.pdf [Last accessed: 12.12.2021]

[Scarfone10]           Karen Scarfone, Dan Benigni, Tim Grance: Cyber Security Standards National Institute of Standards and Technology (NIST), Gaithersburg, Maryland, USA, 2010. Downloadable from: https://tsapps.nist.gov/publication/get_pdf.cfm?pub_id=152153 [Last accessed: 21.11.2020]

[Scarfone12]           Karen Scarfone, Peter Mell: Guide to Intrusion Detection and Prevention Systems (IDPS) NIST Special Publication 800–94 Revision 1 (Draft), National Institute of Standards and Technology, Washington, CD, USA, July 2012. Downloadable from: https://csrc.nist.gov/CSRC/media/Publications/sp/800-94/rev-1/draft/documents/draft_sp800-94-rev1.pdf [Last accessed: 21.5.2021]

[Schoenfield15]    Brook S. E. Schoenfield: Securing Systems – *Applied Security Architecture and Threat Models* CRC Press (Francis & Taylor), Boca Raton, FL, USA, 2015. ISBN 978-1-482-23397-1

[Schreider20a]    Tari Schreider: Cybersecurity Law, Standards and Regulations Rothstein Publishing, Brookfield, USA, 2nd edition, 2020. ISBN 978-1-9444-8056-1

[Schreider20b]    Tari Schreider: Cybersecurity Law, Standards and Regulations Rothstein Publishing, Brookfield, Connecticut, USA, 2nd edition, 2020. ISBN 978-1-9444-8056-1

[SecDW19]    SABSA: *Architecting a Secure Digital World* White Paper, SABSA Institute C.I.C (TSI), Hove, UK, 2019. Downloadable from: https://sabsa. org/architecting-a-secure-digital-world-download-request/ [Last accessed: 19.08.2021]

[Sectigo20]    Sectigo Limited: Certificate Management - *Best Practices Checklist* Sectigo Limited, Roseland, NJ, USA. Downloadable from: https://sectigostore.com/ blog/wp-content/uploads/2020/09/SectigoStore-Certificate-Management-Best-Practices-Checklist_Final.pdf [Last accessed: 13.1.2021]

[Sembera20]    Vit Sembera: ISO/SAE 21434 - Setting the Standard for Connected Cars' Cybersecurity. TrendMicro Research, Shibuya, Präfektur Tokio, Japan, 2020 . Downloadable from: https://documents.trendmicro.com/assets/white_papers/ wp-setting-the-standard-for-connected-cars-cybersecurity.pdf [Last accessed: 12.06.2022]

[Shaw09]    Eric D. Shaw, Lynn F. Fischer, Andrée E. Rose: Insider Risk Evaluation and Audit Technical Report 09–02, Defense Personnel Security Research Center, Monterey, CA, USA, August 2009. Downloadable from: https://www.dhra. mil/Portals/52/Documents/perserec/tr09-02.pdf [Last accessed: 17.12.2021]

[Shrobe79]    Howard Elliot Shrobe: *Dependency Directed Reasoning for Complex Program Understanding* Technical Report No. AITR-503, April 1979, MIT Artificial Intelligence Laboratory, Cambridge, USA, 1979. Downloadable from:    https://dspace.mit.edu/bitstream/handle/1721.1/6890/AITR-503. pdf?sequence=2&isAllowed=y [Last accessed: 13.11.2021]

[Sigler20]    Ken E. Sigler, James L.Rainey: Securing an It Organization Through Governance, Risk Management, and Audit CRC Press (Taylor & Francis), Boca Raton, FL, USA, 2020. ISBN 978-0-367-65865-6

[Siriwardena20]    Prabath Siriwardena, Nuwan Dias: Microservices Security in Action Manning Publications, Shelter Island, New York, N.Y., USA, 2020. ISBN 978-1-617-29595-9

[Smith17]    Sean Smith: The Internet of Risky Things – *Trusting the Devices That Surround Us* O'Reilly UK Ltd., 2017. ISBN 978-1-491-96362-3

[Snedaker13]    Susan Snedaker: Business Continuity and Disaster Recovery Planning for IT Professionals Syngress (Elsevier), Cambridge, MA, USA, 2nd edition, 2013. ISBN 978-0-124-10526-3

[Souppaya21]    Murugiah Souppaya, Karen Scarfone: SP 800–40 Rev. 4 (Draft) - *Guide to Enterprise Patch Management Planning: Preventive Maintenance for Technology* NIST, US National Institute of Standards and Technology, Gaithersburg, MD, USA, November 2021. Downloadable from: https://nvl-pubs.nist.gov/nistpubs/SpecialPublications/NIST.SP.800-40r4-draft.pdf [Last accessed: 04.12.2021]

[SP800-53R5]    SP 800–53 Rev. 5: Security and Privacy Controls for Information Systems and Organizations - *Control Catalog Spreadsheet (Excel Spreadsheet)* NIST,

|                     | US National Institute of Standards and Technology, Gaithersburg, MD, USA, September 2020. Downloadable from: https://csrc.nist.gov/publications/detail/sp/800-53/rev-5/final [last accessed: 19.8.2021] |
| [Stackpole19] | Bill Stackpole, Patrick Hanrion: Software Deployment, Updating, and Patching CRC Press (Taylor & Francis Group), Boca Raton, FL, USA, 2019. ISBN 978-1-138-38137-7 |
| [Stalmans12] | Etienne Stalmans, Barry Irwin: *An Exploratory Framework for Extrusion Detection* Conference paper, SATNAC 2012, Fancourt, George, South Africa, September 2012. Downloadable from: https://www.researchgate.net/publication/327622736_An_Exploratory_Framework_for_Extrusion_Detection [Last accessed: 23.12.2021] |
| [Stetson17] | Chris Stetson: MICROSERVICES Reference Architecture NGINX Ltd., Seattle, USA, 2017. Downloadable from: https://www.nginx.com [Last accessed: 11.01.2022] |
| [Štitilis16] | Darius Štitilis, Marius Laurinaitis: *Legal regulation of the use of dashboard cameras - Aspects of privacy protection* Computer Law & Security Review, Volume 32, Issue 2, April 2016, Pages 316–326. Available from: https://www.sciencedirect.com/science/article/abs/pii/S0267364916300267 [Last accessed: 06.12.2021] |
| [Sustrico18] | Paolo Sustrico: Configuration Management - *How to manage Objects of an IT Infrastructure without going crazy* Independently published, 2018. ISBN 978-1-7307-8627-3 |
| [Taylor13] | Laura P. Taylor: FISMA Compliance Handbook. Syngress (Elsevier), Amsterdam, Netherlands, 2nd edition, 2013. ISBN 978-0-124-05871-2 |
| [TEMET21] | TEMET AG: *Post-Quantum Cryptography* Presentation, TEMET AG, Zurich, Switzerland (https://www.temet.ch), 2021. Downloadable from: https://www.temet.ch/presentations/Post_Quantum_Cryptography.pdf [Last accessed: 06.12.2021] |
| [Thomas20] | Tony Thomas, Athira P. Vijayaraghavan, Sabu Emmanuel: Machine Learning Approaches in Cyber Security Analytics Springer Nature Singapore, Singapore, 2020. ISBN 978-9-811-51705-1 |
| [Thompson21] | Eleanor E. Thompson: The Insider Threat - *Assessment and Mitigation of Risks* CRC Press (Taylor & Francis), Boca Raton, FL, USA, 2021. ISBN 978-0-367-56530-5 |
| [Toth17] | Patricia Toth: NIST Handbook 162 (NIST MEP Cybersecurity) - *Self-Assessment Handbook for Assessing NIST SP 800–171 Security Requirements in Response to DFARS Cybersecurity Requirements* US National Institute of Standards and Technology, Gaithersburg, MD, USA, November 2017. Downloadable from: https://nvlpubs.nist.gov/nistpubs/hb/2017/nist.hb.162.pdf [Last accessed: 12.12.2021] |
| [UKCO18] | UK Cabinet Office: *UK Government Security Classifications* UK Policy, UK Cabinet Office, London, UK, Version 1.1, May 2018. Downloadable from: https://assets.publishing.service.gov.uk/government/uploads/system/uploads/attachment_data/file/715778/May-2018_Government-Security-Classifications-2.pdf [Last accessed: 29.11.2021] |
| [USDHS21] | USDHS: POST-QUANTUM CRYPTOGRAPHY - FREQUENTLY ASKED QUESTIONS U.S. Department of Homeland Security, November 2021. Downloadable from: https://www.dhs.gov/sites/default/files/publications/ |

post_quantum_cryptography_faq_3_seals_october_2021_508.pdf [Last accessed: 06.12.2021]

[Wahe11] Stefan Wahe: Open Enterprise Security Architecture (O-ESA) - *A Framework and Template for Policy-Driven Security* Van Haren Publishing, Zaltbommel, NL, 2011. ISBN 978-9-0875-3672-5

[Williams13] Barry L. Williams: Information Security Policy Development for Compliance: *ISO/IEC 27001, NIST SP 800–53, HIPAA Standard, PCI DSS V2.0, and AUP V5.0* CRC Press (Taylor & Francis), Boca Raton, FL, USA, 2013. ISBN 978-1-466-58058-9

[Williams14] Branden R. Williams, Anton Chuvakin: PCI Compliance - *Understand and Implement Effective PCI Data Security Standard Compliance* Syngress (Elsevier), Cambridge, MA, USA, 4th edition, 2014. ISBN 978-0-128-01579-7

[Wong11] Caroline Wong: Security Metrics - *A Beginner's Guide* McGraw-Hill (Osborne Media), New York, N.Y., USA, 2011. ISBN 978-0-071-74400-3

[Wong20] Yuk Kuen Wong: Cybersecurity: Predicting Insider Attacks - *Using Machine Learning & Artificial Intelligence Algorithms* Independently published, 2020. ISBN 979-8-6891-8127-1

[XACML13] OASIS Standard: eXtensible Access Control Markup Language (XACML) OASIS Standard, Version 3.0, 2013. Downloadable from: http://docs.oasis-open.org/xacml/3.0/xacml-3.0-core-spec-os-en.html [Last accessed: 13.01.2022]

[Zeadally20] Sherali Zeadally, Erwin Adi, Zubair Baig, Imran A. Khan: *Harnessing Artificial Intelligence Capabilities to Improve Cybersecurity* Invited Paper, IEEE Access, Volume 8, 2020, February 6, 2020. Downloadable from: https://ieeexplore.ieee.org/stamp/stamp.jsp?arnumber=8963730 [Last accessed: 21.12.2021]

[Zhang21b] Zhimin Zhang, Huansheng Ning, Feifei Shi, Fadi Farha, Yang Xu, Jiabo Xu, Fan Zhang & Kim-Kwang Raymond Choo: *Artificial intelligence in cyber security: research advances, challenges, and opportunities* Artificial Intelligence Review, 13. March 2021. Springer Verlag, Heidelberg, Germany. Available from: https://link.springer.com/article/10.1007/s10462-021-09976-0 [Last accessed: 22.12.2021]

# Principles for Risk

# 12

*In today's highly technological world, the risk is unavoidable. Therefore, handling risks correctly is paramount in all phases of developing and operating cyber-physical systems. Risk management principles, methods, and procedures must strongly support risk handling and systematic aim at known, acceptable, and documented residual risks.*

## 12.1 Risk Handling

Effective *risk management* must be executed during all phases of the life of a cyber-physical system. All risks of any specific scenario must be identified, recognized, understood, quantified, and documented. Once this is thoroughly done, the process of *derisking*, i.e., *risk management*, is initiated (Definition 12.1, [Moffat18], [Hutchins18], [ISO31000]).

▶ **Definition 12.1: Derisking**
To take steps to make (something) less risky, less likely, or less hazardous.
   Adapted from: https://www.lexico.com/definition/de-risk

> **Quote**
> *"Everything you do on a software project is risk management"*
>    Rob Moffat, 2018

The possible *derisking strategies* are listed in Table 12.1. For trustworthy cyber-physical systems, the only admissible derisking strategy consists of the four steps:

**Table 12.1**  Derisking Strategies

| De-Risking Strategy | |
|---|---|
| **Ignore** | Possible if the risk is insignificant. Otherwise a recipe for disaster |
| **Accept** | Viable if the mitigation measures have reduced the risk to an *acceptable residual risk* |
| **Avoid** | Abstain from the project or product |
| **Transfer** | Consign the risk to a 3rd party, e.g. to an insurance company or to the customer |
| **Contain** | Build the system in such a way, that the consequences of the risk are limited to a defined area |
| **Mitigate** | Add technical measures to the system to either reduce the risk, minimize the probability of materialization, or confine the consequences |

1. Build the system as resilient as possible: Apply the *resilience principles*;
2. Identify, recognize, understand, quantify, and document all risks of each possible scenario;
3. Mitigate the risks by deploying sufficient technical measures to reduce the specific risk to an acceptable, quantified *residual risk*;
4. Explicitly accept the residual risk and its possible consequences.

In the context of this monograph, only risk mitigation and residual risk acceptance are feasible!

## 12.2   Principle R1: Risk

The consequences of insufficiently mitigated risks in safety-critical and security-critical cyber-physical systems may be severe and expensive. Therefore, identifying, acknowledging, assessing, and mapping all risks are fundamental. A practical framework is the *risk landscape* (Fig. 12.1). The risk landscape maps all named risks into a coordinate system (x-axis = severity, y-axis = probability).

A great potential danger in cyber-physical systems is the *unknown* or *unrecognized* risks (red points in Fig. 12.1)! This type of risk can be partially mitigated by constructing *resilient systems*, i.e., applying general resilience principles (Table 4.1). However, no specific, targeted mitigation measures are possible. Unfortunately, often these risks are only detected after an accident or incident.

> **Quote**
> *"I see solutions for managing the risk of some very important problems that are in fact no better than astrology"*
> Douglas W. Hubbard, 2020

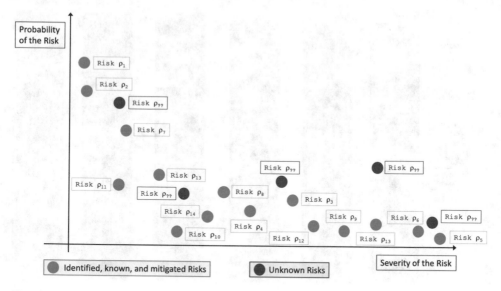

**Fig. 12.1**   Risk landscape

Therefore, the foremost responsibility in cyber-physical systems development, evolution, and operation is identifying "all"—i.e., as many as possible—risks and generating the risk landscape (Fig. 12.1 or a corresponding table).

A powerful method to ferret out risks is using specific *taxonomies*, e.g., a cyber-threat taxonomy (Example 12.1). Walk through all the possible paths of the taxonomy from left to right and perceive the risks connected to each element of the taxonomy. Execute this with different teams and domain specialists. In the end, a reasonably complete risk landscape will result, which can be maintained (improved, evolved) and used in each new project or product (Principle 12.1).

---

### Example 12.1: Cyber-Threat Taxonomy

A good starting point for the organization's (cyber-)risk taxonomy is acquiring and adapting an industry-standard *cyber-threat taxonomy*, e.g., the one extended from [Ferdinand17]. This framework is based on the *attack paths*, drawn from left to right (Fig. 12.2), i.e.,

1. The malicious attacker (man or machine) and the motivation;
2. The access/entry path;
3. The potential vulnerabilities;
4. The possible actions;
5. The imaginable targets;
6. The feared results;
7. The estimated damage.

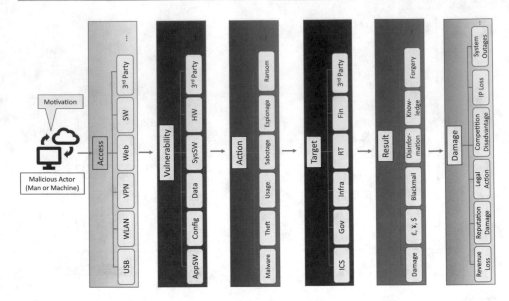

**Fig. 12.2**  Example of a cyber-threat taxonomy framework

Repeated walkthroughs from left to right through the taxonomy will provide an understanding of the risks associated with each element and attack path.

The primary metrics for the assessment of risk are as follows(Fig. 12.1):

1. The *probability* of materialization of the risk: How probable is a safety accident or a security incident caused by the respective risk?
2. The *severity* of the safety accident or the security incident: Which is the damage (usually the worst-case is assumed) caused by the respective risk?

Many *risk assessment methodologies* are available in the literature (e.g., [Freund14], [Hopkin18], [Banks12], [Flammini19], [Kohnke17], [Landoll20], [NISTSCM15], [O'Gorman17], [Vellani20], [Ruan19]). Many of these risk assessment methodologies are *qualitative*, i.e., the use of terms like "severe", "moderate", "likely", "highly unlikely", etc. More modern risk assessment methodologies apply *quantitative* metrics (Example 4.21, Table 4.14, [Hubbard20]).

Note that all assessments are heavily dependent on assumptions: Changing the assumptions, or the context, can massively impact the assessments. It is, therefore, mandatory that all assumptions presupposed are explicitly justified and documented.

---

**Principle 12.1: Risk**

1. The first step is to identify, catalogize, and value the organization's critical assets. In security, these are the (digital) protection assets. In safety, these are the elements in the mission-critical system that contribute to the residual risk;

2. Invest the maximum reasonable effort to identify and recognize as many risks as possible in the products or processes;

3. Note that—especially in security—risks may not be individual attacks, but may be composed as attack chains ([Flammini19]). Include possible attack chain scenarios in the risk analysis;

4. Define and use a precise, comprehensive terminology for all risk management activities;

5. Develop and maintain a risk landscape, either for your domain or for individual products and projects;

6. Use existing industry-proven information sources, such as cyber-threat taxonomies for the discovery and assessment of safety and security risks;

7. Use the risk landscape to identify and assess the probable risks in your creation or evolution of products. Use it as a risk inventory and risk management mechanism;

8. Evolve, i.e., complete the risk landscape continuously by analyzing accidents and incidents in the respective domain;

9. Incorporate new information from literature, standards, and principles whenever they become available (such as ISO 31000, NIST, IRM02);

10. Explicitly and completely list, justify, and record all assumptions made and the context used during risk discovery, recognition, and assessment;

---

## 12.3   Principle R2: Risk Management Process

The risk management *process* (Definition 4.34, Fig. 4.26) is at the core of reliable risk management. The risk management process is active in all CPS development and operation phases—continuously monitoring, assessing, and mitigating the risks. A necessary companion of the risk management process is the *risk management maturity model* (e.g., [Freund15], [Großjean09], [Deloitte14]). The position in the maturity model reflects an organization's risk policies, risk management capabilities, and risk governance processes—and directs its evolution. Guidance for the risk management process is listed in Principle 12.2.

**Quote**

*"The opposite of risk management is crisis management, trying to figure out what to do about the problem after it happens"*
Tom DeMarco, 2003

**Principle 12.2: Risk Management Process**

1. Based on the organization's risk policy, define, implement, enforce, and regularly audit a risk management process;
2. Use specific, tailored processes for safety risks and security risks;
3. Whenever possible, adopt industry-standard risk management standards and methodologies;
4. Apply an industry-accepted risk management maturity model. Review and assess the maturity of the risk management processes yearly and continuously improve the processes;
5. Use quantified risk management as much as applicable (i.e., use metrics);
6. Define and implement as many risk mitigation measures as are necessary to reduce all risks to acceptable, residual risks;
7. All residual risks must be understood and explicitly accepted by a higher management authority in the organization;
8. Align the organization with the risk management process, i.e., assign responsibilities and authority to the rightful units and people;
9. Explicitly assign accountability to people (not to organizational units);
10. While assigning people to the roles of the risk management process carefully match their responsibilities and their authority/decision power;
11. Periodically revise and improve the risk management process, its implementation, and effectiveness;
12. Meticulously document and archive all activities and decisions related to risk management;
13. Cultivate a risk culture in the organization: communicate, train, instruct, test, and improve the understanding and handling of risk;
14. Cultivate a responsible product innovation ("Putting Safety First") culture in the organization ([Zhu18]);

## 12.4   Principle R3: Risk Metrics

On the one hand, *metrics* are indispensable to control the evolution of properties of processes, services, and products. On the other hand, metrics are expensive and can be harmful if improperly used.

**Quote**
*"Metrics are seductive because they simplify reality and are easily communicated"*
    Rob Moffat, 2018

Quote: " "

Therefore, the metrics, their recording of the measurements, the metrics formation method, and the evaluation procedures must be carefully designed—with a clear, long-term objective in focus (Principle 12.3).

---

**Principle 12.3: Risk Metrics**

1. Carefully select a minimum set of meaningful, long-term, valuable risk metrics for the relevant areas of safety and security;
2. Strictly avoid vanity metrics—i.e., numbers that look good and seemingly provide value, but in reality, do not provide any answers;
3. Establish metrics for processes, services, and products;
4. Whenever possible, pick industry-proven, standardized metrics (from literature or standards;
5. Define a comprehensive, feasible process for acquiring the measurements. Keep this process stable for the duration of the use of the respective metric;
6. Unambiguously define what evidence is expected from each metric;
7. Restrict and specify the allowed usage of each metric;
8. Explicitly and completely list, justify, and record all assumptions made during risk metric construction and application;
9. Never (!) use risk metrics to judge or qualify people or their performance;
10. Be careful when introducing new metrics and evaluate their cost/benefit relationship;
11. Treat the metrics as a valuable asset of the organization;

---

## 12.5   Principle R4: Forensic Engineering

After a safety accident or security incident, the last activity is analyzing the cause and the chain of events (*Forensic engineering*, Definition 4.38). For mission-critical systems, forensic engineering could become an essential activity for maintaining certification, legal litigation, or countering liability claims (e.g., [Bibel18], [Porter20]). The foundation of successful forensic engineering is to gather sufficient diagnostic information during runtime to enable later analysis ([Sachowski19]). Principle 12.4 lists requirements for forensic engineering.

---

**Quote**

*"A bad day with forensic engineering is better than a good day without it"*
   Anonymous

Quote: " "

## Principle 12.4: Forensic Engineering

1. Plan the forensic engineering methods and procedures as part of the requirements gathering phase (forensic engineering requirements);
2. Explicitly formulate the forensic engineering requirements and treat them with the same importance as functional requirements;
3. Capture and store sufficient diagnostic information during runtime for subsequent analysis;
4. Securely archive the forensic analysis information;
5. After a safety accident or security incident, use specialized forensic analysis teams and tools to understand the chain of events;
6. Use all forensic results for the improvement of safety and security, both in the products/services and in the processes;
7. Carefully archive the forensic analysis results and recommendations (e.g., for a product liability claim defense);

## References

| | |
|---|---|
| [Banks12] | Erik Banks: **Risk Culture - *A Practical Guide to Building and Strengthening the Fabric of Risk Management*** Palgrave Macmillan, London, UK, 2012. ISBN 978-1-137-26371-1 |
| [Bibel18] | George Bibel: **Train Wreck - *The Forensics of Rail Disasters*** John Hopkins University Press, Baltimore, USA, 2018. ISBN 978-1-4214-2707-2 |
| [Deloitte14] | Deloitte Development LLC: *Assess the "maturity" of the risk governance process* Chapter 5 in: Risk Intelligent governance—Lessons from state-of-the-art board practices Deloitte Development LLC, London, UK, 2014. Downloadable from: https://deloitte.wsj.com/riskandcompliance/files/2014/06/riskintelligent-governance.pdf [Last accessed: 31.12.2021] |
| [Ferdinand17] | Jason Ferdinand: *The cyber security ecosystem: Defining a taxonomy of existing, emerging, and future cyber threats* SIBOS Conference Presentation, Toronto, Canada, 16–19 October 2017a. Downloadable from: https://swiftinstitute.org/wp-content/uploads/2017/11/SWI13-Cyber-Taxonomy-Vfinal.pdf [Last accessed: 23.12.2021] |
| [Flammini19] | Francesco Flammini (Editor): **Resilience of Cyber-Physical Systems - *From Risk Modelling to Threat Counteraction*** Springer Nature Switzerland, Cham, Switzerland, 2019. ISBN 978-3-319-95596-4 |
| [Freund14] | Jack Freund, Jack Jones: Measuring and Managing Information Risk - A FAIR Approach. Butterworth- Heinemann, Kidlington, UK, 2014. ISBN 978-0-124-20231-3 |
| [Freund15] | Jack Freund, Jack Jones: **Measuring and Managing Information Risk - *A FAIR Approach*** Butterworth-Heinemann (Elsevier), Kidlington, Oxford, UK, 2014. ISBN 978-0-124-20231-3 |
| [Großjean09] | Ariane Großjean: **Corporate Terminology Management - *An Approach in Theory and Practice*** VDM Verlag Dr. Müller, Saarbrücken, Germany, 2009. ISBN 978-3-6391-2421-7 |

[Hopkin18]        Paul Hopkin: **Fundamentals of Risk Management - *Understanding, Evaluating, and Implementing Effective Risk Management*** Kogan Page Ltd., New Delhi, India, 5th edition, 2018. ISBN 978-0-7494-8307-4

[Hubbard20]       Douglas W. Hubbard: **The Failure of Risk Management - *Why it's broken and how to fix it.*** John Wiley & Sons, Inc., Hoboken, New Jersey, USA, 2nd edition, 2020. ISBN 978-1-119-52203-4

[Hutchins18]      Greg Hutchins: **ISO 31000:2018 Enterprise Risk Management** Certified Enterprise Risk Manager Academy, Portland, OR, USA, 2018. ISBN 978-0-9654-6651-6

[IRM02]           Institute of Risk Management (IRM): **A Risk Management Standard** UK Institute of Risk Management (IRM), London, UK, 2002. Downloadable from: https://www.theirm.org/media/4709/arms_2002_irm.pdf [Last accessed: 10.01.2022]

[ISO31000]        ISO 31000:2018 (ISO/TC 262 Risk Management) **Risk Management — Guidelines** Downloadable from: https://www.iso.org/standard/65694.html [last accessed: 7.6.2020]

[Kohnke17]        Anne Kohnke, Ken Sigler, Dan Shoemaker: **Implementing Cybersecurity - *A Guide to the National Institute of Standards and Technology Risk Management Framework*** CRC Press, Taylor & Francis, Boca Raton, FL, USA, 2017. ISBN 978-1-498-78514-3

[Landoll20]       Douglas Landoll: **The Security Risk Assessment Handbook - *A Complete Guide for Performing Security Risk Assessments*** CRC Press (Taylor & Francis), Boca Raton, FL, USA, 2nd edition, 2020. ISBN 978-0-367-65929-5

[Moffat18]        Rob Moffat: **Risk-First Software Development - *Volume 1: The Menagerie*** Kite9 Ltd., Colchester, UK, 2018. ISBN 978-1-7174-9185-5

[NISTSCM15]       NIST: ***Best Practices in Cyber Supply Chain Risk Management*** Conference Materials, US National Institute of Standards and Technology, Workshop October 1–2, 2015, Washington, DC, USA. Downloadable from: https://csrc.nist.gov/CSRC/media/Projects/Supply-Chain-Risk-Management/documents/briefings/Workshop-Brief-on-Cyber-Supply-Chain-Best-Practices.pdf [Last accessed: 21.2.2021]

[O'Gorman17]      Tristan O'Gorman: *A Primer on IoT Security Risks* Security Intelligence, February 8, 2017. Downloadable from: https://securityintelligence.com/a-primer-on-iot-security-risks/ [Last accessed: 21.8.2020]

[Porter20]        Donald J. Porter: **Flight Failure - *Investigating the Nuts and Bolts of Air Disasters and Aviation Safety*** Prometheus Books, Buffalo, NY, USA, 2020. ISBN 978-1-6338-8622-3

[Ruan19]          Keyun Ruan: **Digital Asset Valuation and Cyber Risk Measurement - *Principles of Cybernomics*** Academic Press (Elsevier), Cambridge, MA, USA, 2019) ISBN 978-0-128-12158-0

[Sachowski19]     Jason Sachowski: **Implementing Digital Forensic Readiness - From Reactive to Proactive Process** Taylor & Francis Ltd., Boca Raton, FL, USA, 2nd edition, 2019. ISBN 978-1-138-33895-1

[Vellani20]       Karim H. Vellani: **Strategic Security Management - *A Risk Assessment Guide for Decision Makers.*** CRC-Press, Taylor & Francis Ltd., Boca Raton, FL, USA, 2nd edition, 2020. ISBN 978-1-138-58366-5

[Zhu18]           Andy Yunlong Zhu, Max von Zedtwitz, Dimitris G. Assimakopoulos: **Responsible Product Innovation - *Putting Safety First*** Springer International Publishing, Cham, Switzerland, 2018. ISBN 978-3-319-68450-5

# Final Words

**13**

## 13.1 Uncertainty

> **Quote**
> *"It ain't what you don't know that gets you into trouble—It's what you know for sure that just ain't so"*
>    Possibly Mark Twain (https://quoteinvestigator.com/)

Much of the cyber-physical systems' safety and security work deals with *uncertainty* ([Peat02]). There is uncertainty during the creation and evolution of the CPS—and even more uncertainty during its operation. Coping with uncertainty and the *risks* that uncertainty generates is the primary vocation of CPS engineering.

Uncertainty is countered by information. The more is known about the system's vulnerabilities, failure modes, and threats, the better the resulting risks can be mitigated! Therefore, gaining exhaustive, *reliable* information about the CPS and its operating environment is crucial for safety and security. However, uncertainty is the everlasting companion of cyber-physical systems engineering and must be dealt with at any time.

F. J. Furrer, *Safety and Security of Cyber-Physical Systems*, https://doi.org/10.1007/978-3-658-37182-1_13

## 13.2    Disciplined Engineering

**Quote**
*"To err is human. To really screw things up takes a computer"*
    Al Haggerty, 2017

This monograph focuses on *technical risk management*: To create, evolve, and operate cyber-physical systems in such a way, that they have identified, assessed, and acceptable residual safety and security risks.

Because most of the modern cyber-physical system's functionality is controlled by *software,* the software has become the most crucial part of the CPS. The controlling software is, in most cases, a highly complex, enormously complicated, and continuously changing constituent. The software, therefore, is a significant source of safety and security vulnerabilities. Consequently, the greatest care must be taken while cultivating the software!

The cornerstones of this monograph are *engineering principles* and *principle-based engineering*. These two practices contribute to the formalization of software engineering, hopefully making it more disciplined. Consequently and consistently applying the principles in all software life cycle phases greatly reduces the possibility of creating vulnerabilities.

As such, these *engineering principles* and *principle-based engineering* move the craft of software construction nearer to a well-founded, dependable, and highly disciplined engineering field, such as, e.g., *bridge design* (wonderfully exemplified in: [Reis19]).

## 13.3    Why?

A parting thought: "Why another book on cyber-physical systems"? "Which is the added value of this monograph?".

The objectives of the author are as follows:

I.  Present the vast *knowledge* about the safety and security of cyber-physical systems in an organized manner through definitions, principles, and examples—and glamorized by famous quotes;
II. Advance the field of *principle-based systems engineering*—thus guiding the systems development process toward integrated safety and security implementation;
III. Provide a resource for the *education* of coming generations of responsible cyber-physical systems engineers (The material is suited for a two-semester, 4 h/week, advanced lecture at a Technical University) (Fig. 13.1).

**Fig. 13.1**   Author after delivery of the manuscript in February 2022 (Fig. 13.1)

## 13.4   Final Words

This monograph closes with the *final words* in Fig. 13.2.

„In a cyber-physical system's
safety and security,
any compromise
is a planned disaster"

Frank J. Furrer

**Fig. 13.2**   Final words

# Reference

[Peat02]    F. David Peat: From Certainty to Uncertainty - The Story of Science and Ideas in the
            Twentieth Century Henry (Joseph) Press, Washingto, DC, USA, 2002) ISBN-13 :
            978-0309076418
[Reis19]    António J. Reis, José Oliveira Pedro: **Bridge Design - *Concepts and Analysis*** John
            Wiley & Sons, Chichester, UK, 2019. ISBN 978-0-470-84363-5

# References

[ACSC20a]   Australian Cyber Security Center (ACSC): Cloud Computing Security for Tenants Security guideline, July 2020. Downloadable from: https://www.cyber.gov.au/acsc/view-all-content/publications/cloud-computing-security-tenants [Last accessed: 5.5.2021]

[ACSC20b]   Australian Cyber Security Center (ACSC): Cloud Computing Security for Cloud Service Providers Security guideline, July 2020. Downloadable from: https://www.cyber.gov.au/acsc/view-all-content/publications/cloud-computing-security-cloud-service-providers [Last accessed: 5.5.2021]

[Alsmadi17]   Izzat M. Alsmadi, George Karabatis, Ahmed Aleroud (Editors): Information Fusion for Cyber-Security Analytics Springer International Publishing, Cham, Switzerland, 2017. ISBN 978–3–319–83023–0

[Aryuk21]   Aryuk Publications: Secure Coding and Application Programming: *Secure Coding - Secure Application Development* Aryuk Publications, Kindle edition, 2021. ASIN: B08SQZ4XS8

[Berg05]   Cliff Berg: High-Assurance Design - *Architecting Secure and Reliable Enterprise Applications* Addison-Wesley Longman (Pearson), Boston, USA, 2005. ISBN 978–0—321–37577–3 Gebundene Ausgabe – Illustriert, 7. Oktober 2005 Englisch Ausgabe von Cliff Berg (A

[Cabric15]   Marko Cabric: Corporate Security Management - *Challenges, Risks, and Strategies* Butterworth-Heinemann (Elsevier), Kidlington, UK, 2015. ISBN 978–0–128–02934–3

[Das21]   Ravi Das: Practical AI for Cybersecurity CRC Press (Taylor & Francis), Boca Raton, FL, USA, 2021. ISBN 978-0-367-70859-7

[DeMarco03]   Tom DeMarco, Timothy Lister: Waltzing With Bears - *Managing Risk on Software Projects* Dorset House, New York, NY, USA, 2003. ISBN 978-0-9326-3360-6

[Drucker06]   Peter F. Drucker: Managing for Results HarperCollins Business Publisher, reissue edition 2006 (Original: HarperCollins 1964). ISBN 978–0–060–87898–6

[Dubrova14]   Elena Dubrova: Software Redundancy Design of Fault-Tolerant Systems, Presentation at International Program in System-on-Chip Design, Kungl Tekniska Högskolan, Stockholm, Sweden, 2014.

© The Editor(s) (if applicable) and The Author(s), under exclusive license to Springer Fachmedien Wiesbaden GmbH, part of Springer Nature 2022
F. J. Furrer, *Safety and Security of Cyber-Physical Systems,*
https://doi.org/10.1007/978-3-658-37182-1

Downloadable from: https://people.kth.se/~dubrova/FTCcourse/ LECTURES/lecture7.pdf [Last accessed: 01.10.2021]

[Etzioni16] Amitai Etzioni, Oren Etzioni: *Designing AI Systems that Obey Our Laws and Values -Calling for Research on Automatic Oversight for Artificial Intelligence Systems* Communications of the ACM (Viewpoints), Vol. 59, No. 9, September 2016, pp. 29–31. Downloadable from: http://ai2-website.s3.amazonaws.com/publications/ai-values-etzioni.pdf [Last accessed: 28.3.2021]

[EUROCONTROL17] EUROCONTROL: Safety Audits EUROCONTROL (https://www.euro-control.int/), Brussels, Belgium, August 4, 2017. Downloadable from: https://www.skybrary.aero/index.php/Safety_Audits [Last accessed: 11.10.2020]

[Fernandez16] Eduardo B. Fernandez: Threat Modelling in Cyber-Physical Systems Conference Paper: DASC-PICom-DataCom-CyberSciTec, 2016. DOI: 10.1109. Downloadable from: https://www.researchgate.net/publication/309151336_Threat_Modeling_in_Cyber-Physical_Systems [last accessed: 2.1.2020]

[Fiskiran04] A. Murat Fiskiran, Ruby B. Lee: *Runtime Execution Monitoring (REM) to Detect and Prevent Malicious Code Execution* White Paper, Department of Electrical Engineering, Princeton University, Princeton, N.J., USA, 2004. Downloadable from: http://palms.ee.princeton.edu/PALMSopen/fiskiran04runtime.pdf [Last accessed: 23.12.2020]

[Flammini12] Francesco Flammini (Editor): Railway Safety, Reliability and Security - *Technologies and Systems Engineering* Idea Group, Hershey, PA, USA, 2012. ISBN 978–1–4666–1643–1

[Garfinkel02] Simson Garfinkel, Gene Spafford: Web Security, Privacy and Commerce - *Security for Users, Administrators, and ISPs* O'Reilly Media Inc., Sebastopol, FL, USA, 2nd edition 2002. ISBN 978–0–596–00045–5

[Godefroid01] Patrice Godefroid: *Model Checking of Software* Internal Presentation, Lucent Technologies, August 2001. Downloadable from: https://patrice-godefroid.github.io/public_psfiles/mc-of-sw-aug01.pdf [Last accessed: 23.1.2021]

[Hakulinen15] T. Hakulinen, F. Havart, P. Ninin, F. Valentini: Building an Interlock – Comparison of Technologies for constructing Safety Interlocks Proceedings of ICALEPCS2015, Melbourne, Australia, 2015. Downloadable from: https://s3.cern.ch/inspire-prod-files-5/5b8599c1832 5d0120012b33cbd994ec0 [Last accessed: 16.2.2021]

[Herdman16] Patricia Herdman: When Cars Decide to Kill - *Time for Software Safety Laws* Ethi-Teque Inc., Toronto, Canada, 2nd edition, 2016. ISBN 978–1–9884–7003–0

[Hsi11] Idris Hsi: The Conceptual Integrity of Software - *Using Ontological Excavation and Analysis Towards Ensuring Conceptual Integrity in the Design and Architecture of Computing Applications* VDM Verlag Dr. Müller, Saarbrücken, Deutschland, 2011. ISBN 978–3–8364–7507–5

[Lawless19] William Lawless, Ranjeev Mittu, Donald Sofge, Ira S. Moskowitz, Stephen Russell (Editors): Artificial Intelligence for the Internet of Everything Academic Press (Elsevier), London, UK, 2019. ISBN 978–0–12–817636–8

[Li17]       Shancang Li, Li Da Xu: Securing the Internet of Things
             Syngress Publishing (Elsevier), Cambridge, USA, 2017. ISBN
             978–0–12–804458–2

[Maleh19b]   Yassine Maleh, Mohammad Shojafar, Ashraf Darwish, Abdelkrim
             Haqiq (Editors): Cybersecurity and Privacy in Cyber-Physical Systems
             CRC Press (Taylor & Francis Ltd), Boca Raton, FL, USA, 2019. ISBN
             978-1-138-34667-3

[McGraw06]   Gary R. McGraw: Software Security - *Building Security In* Addison
             Wesley, Upper Saddle River, NJ, USA, 2006. ISBN 978–0–321–35670–3

[McIlwraith06] Angus McIlwraith: Information Security and Employee Behaviour
             - *How to Reduce Risk Through Employee Education, Training, and
             Awareness* Gower Publishing Ltd., Aldershot, UK, 2006. ISBN
             978–0–5660–8647–2

[Moe20]      Marie Elisabeth Gaup Moe: *Uncovering vulnerabilities in pacemakers -
             Results from five years of The Pacemaker Hacking Project* mnemonic as,
             Oslo, Norway, 7/1/2020. Access: https://www.mnemonic.no/blog/uncov-
             ering-vulnerabilities-in-pacemakers/ [Last accessed: 17.2.2021]

[Morillo21]  Christina Morillo: 97 Things Every Information Security Professional
             Should Know - *Collective Wisdom from the Experts* O'Reilly Media,
             Inc., Sebastopol, CA, USA, 2021. ISBN 978–1–098–10139–8

[Nahari11]   Hadi Nahari, Ronald L. Krutz: **Web Commerce Security - *Design and
             Development*** Wiley Publishing Inc., Indianapolis, IN, USA, 2011. ISBN
             978–0–470–62446–3

[Nemerov63]  Howard Nemerov: Poetry and Fiction - Essays Forgotten Books,
             London, UK, 2018 (Reprint of 1963 Original, Rutgers University Press,
             New Brunswick, N.J., USA, 1963). ISBN 978-1-334-91163-7 (Classic
             Reprint)

[Nicholson20] Mark Nicholson, Mike Parsons (Editors): Assuring Safe Autonomy
             Proceedings of the 28th Safety-Critical Systems Symposium (SSS'20),
             York, UK, 11–13 February 2020. ISBN 978–1–713305–66–8

[NISTIRIoT20] Michael Fagan, Katerina N. Megas, Karen Scarfone, Matthew Smith:
             NISTIR 8259 - Foundational Cybersecurity Activities for IoT Device
             Manufacturers May 2020. Downloadable from:: https://doi.org/10.6028/
             NIST.IR.8259 [Last accessed: 17.6.2020]

[Papernot16] Nicolas Papernot, Patrick McDaniel, Xi Wu, Somesh Jha, Ananthram
             Swami: *Distillation as a Defense to Adversarial Perturbations against
             Deep Neural Networks* 37th IEEE Symposium on Security & Privacy,
             IEEE 2016, San Jose, CA, USA, 2016. Dowloadable from: https://arxiv.
             org/pdf/1511.04508.pdf [Last accessed: 18.3.2021]

[Păsăreanu20] Corina S. Păsăreanu: Symbolic Execution and Quantitative Reasoning
             - *Applications to Software Safety and Security* Morgan & Claypool
             Publishers, San Rafael, CA, USA, 2020. ISBN 978–1–6817–3854–3

[Peltier04]  Thomas R. Peltier: Information Security Policies and Procedures - *A
             Practitioner's Reference* CRC Press, Taylor & Francis Ltd., Boca Raton,
             FL, USA, 2nd edition, 2004. ISBN 978-0-849-31958-7

[Pincus04]   J. Pincus and B. Baker: *Beyond Stack Smashing - Recent Advances in
             exploiting Buffer Overruns* IEEE Security & Privacy, Vol. 2, No. 4, pp.
             20-27, July-Aug. 2004. IEEE New York, NY, USA, doi: https://doi.
             org/10.1109/MSP.2004.36

[Reason97]        James Reason: Managing the Risks of Organizational Accidents Ashgate Publishing Limited, Aldershot, UK, new edition, 1997. ISBN 978–1–84014–105–4 (Reprinted 2008)

[Reussner19]      Ralf Reussner, Michael Goedicke, Wilhelm Hasselbring, Birgit Vogel-Heuser, Jan Keim, Lukas Märtin (Editors): Managed Software Evolution Springer Nature Switzerland, Cham, Switzerland, 2019 (Springer Open Access). ISBN 978–3–030–13498–3. Downloadable from: https://www.springer.com/de/book/9783030134983 [Last accessed: 22.3.2021]

[Seidl15]         Martina Seidl, Marion Scholz, Christian Huemer, Gerti Kappel: UML @ Classroom - *An Introduction to Object-Oriented Modeling* Springer International Publishing, Chams, Switzerland, 2015. ISBN 978–3–319–12741–5

[Shepherd16]      Carlton Shepherd, Ghada Arfaoui, Iakovos Gurulian, Robert P. Lee, Konstantinos Markantonakis, Raja Naeem Akram, Damien Sauveron, Emmanuel Conchon: *Secure and Trusted Execution: Past, Present, and Future -- A Critical Review in the Context of the Internet of Things and Cyber-Physical Systems* IEEE International Conference on Trust, Security and Privacy in Computing and Communications (TrustCom), Tianjin, China, 23–26 Aug. 2016. Downloadable from: https://www.researchgate.net/publication/306039236_Secure_and_Trusted_Execution_Past_Present_and_Future_--_A_Critical_Review_in_the_Context_of_the_Internet_of_Things_and_Cyber-Physical_Systems [Last accessed: 6.11.2021]

[Shukla10]        Sandeep Kumar Shukla, Jean-Pierre Talpins (Editors): Synthesis of Embedded Software - *Frameworks and Methodologies for Correctness by Construction* Springer Science & Business Media, New York, NY, USA, 2010. ISBN 978–1–441–96399–4

[Sikorski12]      Michael Sikorski, Andrew Honig: Practical Malware Analysis - *The Hands-On Guide to Dissecting Malicious Software* No Starch Press, San Francisco, CA, USA, 2012. ISBN 978-1-5932-7290-6

[Tate15]          Martin Tate: Off-The-Shelf IT Solutions - *A practitioner's Guide to Selection and Procurement* BCS Learning & Development (The Chartered Institute for IT), Swindon, UK, 2015. ISBN 978–1–7801–7258–3

[Torres-Pomales00] [Torres-Pomales00] Wilfredo Torres-Pomales: Software Fault Tolerance - *A Tutorial* NASA Report TM-2000–210616, NASA Langley Research Center, Hampton, Virginia, USA, 2000. Downloadable from: https://ntrs.nasa.gov/api/citations/20000120144/downloads/20000120144.pdf [Last accessed: 10.9.2021]

[US-DoD]          US Department of Defense: Systems Engineering Guide for Systems of Systems US Office of the Deputy Under Secretary of Defense for Acquisition and Technology, Version 1.0. Washington, DC, USA, Version 1.0, August 2008. Downloadable from: https://acqnotes.com/wp-content/uploads/2014/09/DoD-Systems-Engineering-Guide-for-Systems-of-Systems-Aug-2008.pdf [Last accessed: 06.01.2022]

[Vernon21]        Vaughn Vernon, Tomasz Jaskula: Strategic Monoliths and Microservices - *Driving Innovation Using Purposeful Architecture* Addison-Wesley (Pearson Education), Upper Saddle River, NJ, USA, 2021. ISBN 978–0–137–35546–4

[Wells19]   Travis Wells, Daniel Gillihan: Emotional Intelligence Mastery - *Why EQ is Important for Success and Matters More Than IQ* Pardi Publishing, 2019. ISBN 978–1–09120–443–0

[Whaiduzzaman14]   Md Whaiduzzaman, Mehdi Sookhak, Abdullah Gani, Rajkumar Buyya: *A Survey on Vehicular Cloud Computing* Journal of Network and Computer Applications, Elsevier, Amsterdam, Netherlands, Nr. 40, 2014, pp. 325–344.

[Wirth20]   Axel Wirth, Christopher Gates, Jason Smith: Medical Device Cybersecurity for Engineers and Manufacturers Artech House Publishers, Norwood, MA, USA, 2020. ISBN 978–1–630–81815–9

[Wolff18]   Josephine Wolff: You'll See This Message When It Is Too Late - *The Legal and Economic Aftermath of Cybersecurity Breaches* MIT Press, Cambridge, MA, USA, 2018. ISBN 978-0-262-03885-0

[Ziv97]   Hadar Ziv, Debra J. Richardson: *The Uncertainty Principle in Software Engineering* 1997 International Conference on Software Engineering (ICSE'97), May 17–23, 1997, Boston, MA, USA. Downloadable from: http://jeffsutherland.org/papers/zivchaos.pdf [last accessed 28.9.2020]

# Index

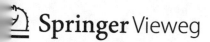

Printed in the United States
by Baker & Taylor Publisher Services